CONCRETE TECHNOLOGY

Second Edition

M L Gambhir
Thapar Institute of Engineering and Technology
Patiala

Tata McGraw-Hill Publishing Company Limited
NEW DELHI

McGraw-Hill Offices
New Delhi New York St Louis San Francisco Auckland Bogotá Guatemala
Hamburg Lisbon London Madrid Mexico Milan Montreal Panama Paris
San Juan São Paulo Singapore Sydney Tokyo Toronto

© 1995, 1986, Tata McGraw-Hill Publishing Company Limited

Second reprint 1996
RADRBRCNRXRCD

No part of this publication can be reproduced in any form or by any means without the prior written permission of the publishers

This edition can be exported from India only by the publishers,
Tata McGraw-Hill Publishing Company Limited

ISBN 0-07-462121-1

Published by Tata McGraw-Hill Publishing Company Limited,
4/12 Asaf Ali Road, New Delhi 110 002, typeset at
Urvashi Press, Vaidwara, Meerut and printed at
A P Offset, 1/11787, Panchsheel Garden, Naveen Shahdara, Delhi 110 032

*To
My Parents*

Foreword

Ever since portland cement was discovered, there has been a continuous effort on the part of engineers and scientists to replace stone as a building material. This is because cement concrete offers an alternative material, possessing more or less equal strength and capable of being moulded into any desired shape and form, whereas achieving a smooth shape in the case of stone is an expensive process.

Cement concrete has in the last 150 years been used for various purposes and has become an important building material, not only for ornamental purposes or replacing stone but also for structures which were traditionally built with steel, since reinforcing it with steel became possible. Although recent developments in plastics and other lightweight materials have made inroads into the spheres where cement concrete was used, it will be difficult to replace it at most places in view of its low cost-to-strength ratio compared with other materials—new and old. It will thus remain an important material of construction and knowledge of its economic proportioning and use will continue to be an important subject of study. In this context, the book written by Dr Gambhir is a valuable contribution, particularly because he has included in it all aspects of cement concrete. The author deserves all encouragement for his efforts in this venture.

The book covers, in detail, the properties of cement, aggregates and water as necessary for making concrete of required strength and providing it with adequate workability and durability. It also covers the properties, utility, and deleterious effects of different admixtures along with the provisions in the Indian Standards for controlling them.

A section has been devoted to the rheology of concrete describing the influence of the important factors involved. Another section has been devoted to the determination and control of engineering properties of cement concrete and bringing to the notice of readers the practical problems involved in its use including a note on the unavoidable cracks that appear in it in practice and how they can be repaired.

The author has done well in describing the variations in properties, in actual practice, with parameters and explaining what value of the real strength of concrete should be chosen when there are experimental variations. He has also included information on how to obtain the best and economical mix of aggregate and cement. Information regarding how concrete should be prepared in practice making optimum use of the inputs is useful.

The section on special concrete including information on polymer concrete and other modern developments brings the reader abreast with current knowledge in the field of concrete technology.

The book is complete in all respects and should be equally useful to students and field engineers.

JAI KRISHNA, FNA
Former Vice Chancellor
University of Roorkee
Ex-President
International Association
for Earthquake Engineering

Preface to the Second Edition

The wide acceptance of the first edition by the engineering community has motivated the author to bring out this enlarged edition. The chapters 10, 14, 15 and 17 have been rewritten and updated to bring them to the state-of-the-art status. In the chapter on the proportioning of concrete mixes, the concept of concrete as a system has been introduced. The quality management in concrete construction has been described briefly in the chapter on quality control of concrete. Ever-increasing use of nondestructive testing in the quality control has prompted the inclusion of state-of-the-art techniques of nondestructive testing. The desirability of rehabilitation of fire-damaged concrete structures during recent fires has made the author to devote more attention to this aspect of concrete technology. More information on the behaviour of concrete during fires has been added.

This edition owes much to the useful comments and suggestions from the teachers and practising engineers to whom the author expresses his grateful thanks. As suggested by the teaching community objective type questions have been added in the appendix. The study of these questions will help the readers to gain indepth understanding of the principles of concrete technology and will help the students in preparations for competitive examinations.

It is hoped that this edition will be as acceptable to the engineering community as its predecessor has been.

The author is especially grateful to his wife and children for their continued cooperation and help in bringing out this edition.

M L Gambhir

Preface to the First Edition

The rapid progress and changes in all areas of concrete technology in the recent past have motivated me to prepare this text to suit the needs of undergraduate and graduate courses of various engineering colleges and universities. Most of these institutions have introduced separate courses in concrete technology. The main purpose of this text is to provide the readers with the basic principles, and sufficient information on the state of art relating to all facets of manufacturing and production processes in the making of structural concrete. I am of the opinion that although undergraduate students can depend upon one textbook, at the same time they must realize that a lot more literature exists on the subject. A short bibliography is included at the end of the book which will provide the discerning student with the source for more detailed information and will encourage students' interest in supplementary reading. To cater to the needs of the undergraduate level course more emphasis is laid on the fundamentals and practice. A little bit of repetition has been intentionally allowed to facilitate the understanding of the subject matter.

In addition to the traditional concrete engineering topics dealing with the principles of concrete and concrete-making materials, the current state of art of special concretes and concreting techniques has been included. This will introduce the engineering faculties, practising engineers, architects, research workers, and contractors to the new developments in the production of quality concrete. Many topics of particular interest have been included because they represent areas of recent technological advances like fibre-reinforced concrete, ferrocement, fibre-reinforced ferrocement, polymer impregnated concrete, nondestructive testing, modern concepts of quality concrete, etc. The final chapter introduces the repair techniques used for concrete structures as a logical extension of the principles discussed in the preceding chapters.

As most of the information has been derived from many sources including those specially mentioned in the bibliography at the end of text, I wish to express my indebtedness to these individuals and publications contributing the information. Acknowledgements and thanks are recorded to the Bureau of Indian Standards (erstwhile Indian Standards Institution) to quote certain regulations.

I record my gratitude to Dr Jai Krishna, FNA, formerly Vice Chancellor, University of Roorkee, and Ex-president of International Association for Earthquake Engineering (IAEE) for kindly writing the foreword to the book and for his encouragement. I acknowledge with gratitude the assistance received from many quarters in the preparation of the book. Foremost thanks are due to Dr S Krishnamurthy, Professor of Civil Engineering at the Indian Institute of Technology, New Delhi, for critically reviewing the manuscript and giving valuable suggestions for the improvement of the text. Special thanks go to Dr N C Nigam, Director, Thapar Institute of Engineering and Technology, Patiala, for his encouragement and motivation.

Finally, I am grateful to my wife and children for their forbearance and patience in putting up with the inconveniences when I was engaged in the preparation of this text.

M L Gambhir

Contents

Foreword vii
Preface to the Second Edition ix
Preface to the First Edition xi

1. Concrete as Construction Material 1

   1.1 Introduction *1*
   1.2 Classification of Concrete *2*
   1.3 Properties of Concrete *3*
   1.4 Grades of Concrete *3*
   1.5 Advantages of Concrete *4*
   1.6 Disadvantages of Concrete *4*
   1.7 Concept of Quality Control *4*

2. Concrete Making Materials-I: Cement 6

   2.1 Introduction *6*
   2.2 Cement *6*

3. Concrete Making Materials-II: Aggregate 23

   3.1 Introduction *23*
   3.2 Classification of Aggregates *23*
   3.3 Characteristics of Aggregates *29*
   3.4 Deleterious Substances in Aggregates *34*
   3.5 Soundness of Aggregate *35*
   3.6 Alkali-aggregate Reaction *36*
   3.7 Thermal Properties of Aggregates *36*
   3.8 Fineness Modulus *37*
   3.9 Maximum Size of Aggregate *37*
   3.10 Grading and Surface Area of Aggregate *38*

4. Concrete Making Materials-III: Water 44

   4.1 Introduction *44*
   4.2 Quality of Mixing Water *44*
   4.3 Curing Water *47*

5. Admixtures 48

   5.1 Introduction *48*
   5.2 Functions of Admixtures *48*
   5.3 Classification of Admixtures *49*
   5.4 Indian Standard Specifications *60*

6. Properties of Fresh Concrete 63

   6.1 Introduction *63*
   6.2 Workability *64*

- 6.3 Measurement of Workability 64
- 6.4 Factors Affecting Workability 67
- 6.5 Requirements of Workability 70
- 6.6 Estimation of Errors 73

7. Rheology of Concrete   74
- 7.1 Introduction 74
- 7.2 Representation of Rheological Behaviour 76
- 7.3 Factors Affecting Rheological Properties 79
- 7.4 Mixture Adjustments 82

8. Properties of Hardened Concrete   83
- 8.1 Introduction 83
- 8.2 Strengths of Concrete 83
- 8.3 Stress and Strain Characteristics of Concrete 92
- 8.4 Shrinkage and Temperature Effects 94
- 8.5 Creep of Concrete 98
- 8.6 Permeability of Concrete 99
- 8.7 Durability of Concrete 99
- 8.8 Sulphate Attack 100
- 8.9 Acid Attack 100
- 8.10 Efflorescence 101
- 8.11 Fire Resistance 101
- 8.12 Thermal Properties of Concrete 103
- 8.13 Microcracking of Concrete 103

9. Quality Control of Concrete   105
- 9.1 Introduction 105
- 9.2 Factors Causing Variations in the Quality of Concrete 106
- 9.3 Field Control 107
- 9.4 Advantages of Quality Control 108
- 9.5 Statistical Quality Control 108
- 9.6 Measure of Variability 110
- 9.7 Applications 111
- 9.8 Quality Management in Concrete Construction 117

10. Proportioning of Concrete Mixes   122
- 10.1 Introduction 122
- 10.2 Basic Considerations for Concrete Mix Design 123
- 10.3 Factors Influencing the Choice of Mix Proportions 124
- 10.4 Methods of Concrete Mix Design for Medium Strength Concretes 133
- 10.5 Concrete Mix Design 160
- 10.6 Computer-aided Concrete Mix Design 166
- 10.7 Design of High Strength Concrete Mixes 186
- 10.8 Trial Mixes 195
- 10.9 Conversion of Mix Proportions from Mass to Volume Basis 195
- 10.10 Quantities of Materials to Make Specified Volume of Concrete 196
- 10.11 Acceptance Criteria for Concrete 197
- 10.12 Field Adjustments 198
- 10.13 Generalized Format for Concrete Mix Design 200

10.14 Design of Concrete Mix as a System *202*

11. Production of Concrete — 205

11.1 Introduction *205*
11.2 Batching of Materials *205*
11.3 Mixing of Concrete Materials *207*
11.4 Transportation of Concrete *210*
11.5 Placing of Concrete *213*
11.6 Compaction of Concrete *216*
11.7 Curing of Concrete *223*
11.8 Formwork *233*

12. Concreting under Extreme Environmental Conditions — 238

12.1 Introduction *238*
12.2 Concreting in Hot Weather *238*
12.3 Cold Weather Concreting *241*
12.4 Underwater Concreting *243*

13. Inspection and Testing — 248

13.1 Introduction *248*
13.2 Inspection Testing of Fresh Concrete *249*
13.3 Acceptance Testing of Hardened Concrete *252*

14. Special Concretes and Concreting Techniques — 272

14.1 Introduction *272*
14.2 Light-weight Concrete *274*
14.3 Ultra-light-weight Concrete *284*
14.4 Vacuum Concrete *285*
14.5 Waste Material Based Concrete *286*
14.6 Mass Concrete *291*
14.7 Shotcrete or Guniting *291*
14.8 Ferrocement *298*
14.9 Fibre-reinforced Concrete *307*
14.10 Polymer Concrete Composites (PCCs) *325*
14.11 Sulphur Concrete and Sulphur-infiltrated Concrete *333*
14.12 Jet (Ultra-rapid Hardening) Cement Concrete *339*
14.13 Gap-graded Concrete *341*
14.14 No-fines Concrete *342*

15. Deterioration of Concrete and its Prevention — 343

15.1 Introduction *343*
15.2 Corrosion of Concrete *343*
15.3 Corrosion of Reinforcement *349*

16. Repair Technology for Concrete Structures — 356

16.1 Introduction *356*
16.2 Symptoms and Diagnosis of Distress *358*
16.3 Evaluation of Cracks *365*
16.4 Selection of Repair Procedure *370*
16.5 Repair of Cracks *371*

16.6 Common Types of Repairs *389*
16.7 Typical Examples of Concrete Repair *395*
16.8 Leak Sealing *401*
16.9 Underwater Repairs *403*
16.10 Distress in Fire Damaged Structures *407*

Appendix—Objective Type Questions 409

   Cement *409*
   Aggregates *415*
   Water *417*
   Admixtures *418*
   Properties of Fresh Concrete *420*
   Rheology of Concrete *423*
   Properties Hardened Concrete *423*
   Quality Control *427*
   Proportioning of Concrete Mixes *427*
   Production of Concrete *432*
   Inspection and Testing *437*
   Special Concrete and Concreting Techniques *438*
   Deterioration of Concrete and its Prevention *443*
   Repair Technology for Concrete Structures *444*
   Answers to the Objective Type Questions *446*

Bibliography 448

Index 457

1

Concrete as Construction Material

1.1 INTRODUCTION

Concrete is the most widely used man-made construction material. It is obtained by mixing cement, water and aggregates (and sometimes admixtures) in required proportions. The mixture when placed in forms and allowed to cure becomes hard like stone. The hardening is caused by chemical action between water and the cement and it continues for a long time, and consequently the concrete grows stronger with age. The hardened concrete may also be considered as an artificial stone in which the voids of larger particles (coarse aggregate) are filled by the smaller particles (fine aggregate) and the voids of fine aggregates are filled with cement. In a concrete mix the cement and water form a paste called *cement water paste* which in addition to filling the voids of fine aggregate acts as binder on hardening, thereby cementing the particles of the aggregates together in a compact mass.

The *strength, durability* and other characteristics of concrete depend upon the properties of its ingredients, on the proportions of mix, the method of compaction and other controls during placing, compaction and curing. The popularity of the concrete is due to the fact that from the common ingredients, it is possible to tailor the properties of concrete to meet the demands of any particular situation. The advances in concrete technology have paved the way to make the best use of locally available materials by judicious mix proportioning and proper workmanship, so as to produce concrete satisfying *performance requirements*. The factors affecting the performance of concrete are shown in Fig. 1.1. The concept of treating the concrete in its entirety as a building material rather than ingredients is gaining popularity since the user is now interested in the concrete itself having desired properties without bothering about the ingredients. This concept is symbolized with the progress of *ready mixed concrete* industry where the consumer can specify the concrete of his needs, and further in the precast concrete industry where the consumer obtains finished structural components satisfying the performance requirements.

2 Concrete Technology

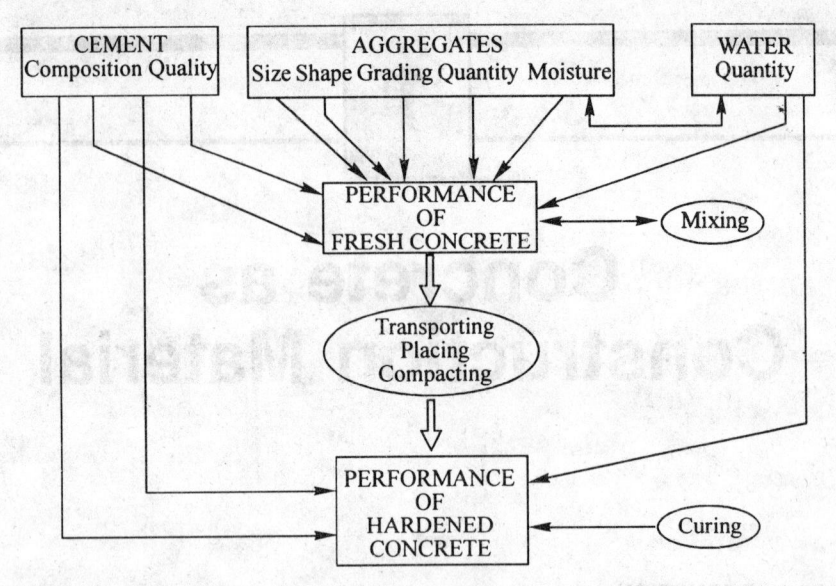

Fig. 1.1 Factors affecting performance of concrete

The various aspects covered in the following chapters are the materials, mix proportioning, elements of workmanship, e.g. placing, compaction and curing, methods of testing and relevant satistical approach to quality control. The discussions on these aspects are based on the appropriate provisions in the Indian Standard Codes.

The concrete has high compressive strength, but its tensile strength is very low. In situations where tensile stresses are developed the concrete is strengthened by steel bars forming a composite construction called *reinforced cement concrete* (RCC). The concrete without reinforcement is termed *plain cement concrete* or simply as concrete. The process of making concrete is called *concreting*. Sometimes the tensile stresses are taken care of by introducing compressive stresses in the concrete so that the initial compression neutralizes the tensile stresses. Such a construction is known as *prestressed cement concrete* construction.

1.2 CLASSIFICATION OF CONCRETE

As mentioned earlier the main ingredients of concrete are cement, fine aggregate (sand) and coarse aggregate (gravel or crushed rock). It is usual to specify a particular concrete by the proportions (by weight) of these constituents and their characteristics, e.g. a 1 : 2 : 4 concrete refers to a particular concrete manufactured by mixing cement, sand and broken stone in a 1 : 2 : 4 ratio (with a specified type of cement, water-cement ratio, maximum size of aggregate, etc.). This classification specifying the proportions of constituents and their characteristics is termed *prescripitive specifications* and is based on the hope that adherence to such prescripitive specifications will result in satisfactory performance. Alternatively, the specifications specifying the requirements of the desirable properties of concrete such as strength, workability, etc. are stipulated, and these are termed *performance oriented specifica-*

tions. Based on these considerations, the concrete can be classified either as *nominal mix concrete or designed mix concrete*. Sometimes the concrete is classified into *controlled concrete* and *ordinary concrete*, depending upon the levels of control exercised in the works and the method of proportioning concrete mixes. Accordingly, a concrete with ingredient proportions fixed by designing the concrete mixes with preliminary tests are called *controlled concrete*, whereas ordinary concrete is one where nominal mixes are adopted. In IS: 456–1978 there is nothing like uncontrolled concrete: only the *degree of control varies from very good to poor or no control*. In addition to mix proportioning, the *quality control* includes *selection of appropriate concrete materials* after proper tests, proper *workmanship in batching, mixing, transportation, placing, compaction* and *curing*, coupled with necessary checks and tests for *quality acceptance*.

1.3 PROPERTIES OF CONCRETE

Concrete making is not just a matter of mixing ingredients to produce a plastic mass, but *good concrete* has to satisfy *performance requirements* in the *plastic* or *green state* and also the *hardened state*. In the plastic state the concrete should be workable and free from *segregation* and *bleeding*. Segregation is the separation of coarse aggregate and bleeding is the separation of cement paste from the main mass. The segregation and bleeding results in a *poor quality concrete*. In its *hardened state concrete* should be *strong, durable*, and *impermeable*; and it should have *minimum dimensional changes*.

Among the various properties of concrete, its *compressive strength* is considered to be the most important and is taken as an *index of its overall quality*. Many other properties of concrete appear to be generally related to its compressive strength. These properties will be discussed in detail later in the book.

1.4 GRADES OF CONCRETE

The concrete is generally graded according to its compressive strength. The various grades of concrete as stipulated in IS: 456–1978 and IS: 1343–1980 are given in Table 1.1. In the designation of concrete mix, the letter M refers to the mix and the number to the specified characteristic strength of 150 mm work cubes at 28 days, expressed in MPa (N/mm^2). The concrete of grades M5 and M7.5 is suitable for lean concrete bases and simple foundations of masonry walls. These need not be designed. The concrete of grades lower than M15 is not suitable for reinforced concrete works and grades of concrete lower than M30 are not to be used in the prestressed concrete works.

TABLE 1.1 Grades of concrete

Grade Designation	M5	M7.5	M10	M15	M20	M25	M30	M35	M40	M45	M50	M55
Specified characteristic strength at 28 days, MPa	5	7.5	10	15	20	25	30	35	40	45	50	55

1.5 ADVANTAGES OF CONCRETE

Concrete as a construction material has the following advantages.

(i) Concrete is economical in the long run as compared to other engineering materials. Except cement, it can be made from locally available coarse and fine aggregates.

(ii) Concrete possesses a high compressive strength, and the corrosive and weathering effects are minimal. When properly prepared its strength is equal to that of a hard natural stone.

(iii) The green concrete can be easily handled and moulded into any shape or size according to specifications. The form work can be reused a number of times of similar jobs resulting in economy.

(iv) It is strong in compression and has unlimited structural applications in combination with steel reinforcement. The concrete and steel have approximately equal coefficients of thermal expansion. The concrete is extensively used in the construction of foundations, walls, roads, airfields, buildings, water retaining structures, docks and harbours, dams, bridges, bunkers and silos, etc.

(v) Concrete can even be sprayed on and filled into fine cracks for repairs by the guniting process.

(vi) The concrete can be pumped and hence it can be laid in the difficult positions also.

(vii) It is durable and fire resistant and requires very little maintenance.

1.6 DISADVANTAGES OF CONCRETE

Following are the disadvantages of concrete.

(i) Concrete has low tensile strength and hence cracks easily. Therefore, concrete is to be reinforced with steel bars or meshes.

(ii) Fresh concrete shrinks on drying and hardened concrete expands on wetting. Provision for contraction joints has to be made to avoid the development of cracks due to drying shrinkage and moisture movement.

(iii) Concrete expands and contracts with the changes in temperature. Hence expansion joints have to be provided to avoid the formation of cracks due to thermal movement.

(iv) Concrete under sustained loading undergoes creep resulting in the reduction of prestress in the prestressed concrete construction.

(v) Concrete is not entirely impervious to moisture and contains soluble salts which may cause efflorescence.

(vi) Concrete is liable to disintegrate by alkali and sulphate attack.

(vii) The lack of ductility inherent in concrete as a material is disadvantageous with respect to earthquake resistant design.

1.7 CONCEPT OF QUALITY CONTROL

Concrete, generally manufactured at the site, is likely to have variability of strength from batch to batch and also within the batch. The magnitude of this variation depends on several factors, such as the variation in the quality of constituent materials, variation in mix proportions due to batching process, variations in the quality of batching and mixing equipment available, the quality of overall workmanship and supervision at the site, and variation due to sampling and testing of concrete specimens.

The above variations are inevitable during production to varying degrees. For example, the cements from different batches or sources may exhibit different strengths. The grading and shape of aggregates even from the same source varies widely. Considerable variations occur partly due to the quality of the plant available and partly the efficiency of operation. Some of the variations in test results are due to variations in sampling, making, curing and testing the specimen even when carried out in terms of relevant specifications.

The quality control of concrete is to reduce this variation and to produce concrete of uniform quality consistent with specified minimum requirements which can be achieved by good workmanship and if the plant is maintained at peak efficiency.

2

Concrete Making Materials-I: Cement

2.1 INTRODUCTION

Cement is a well-known building material and has occupied an indispensable place in construction works. There is a variety of cements available in the market and each type is used under certain conditions due to its special properties. A mixture of cement and sand when mixed with water to form a paste is known as *cement mortar* whereas the composite product obtained by mixing cement, water, and an inert matrix of sand and gravel or crushed stone is called *cement concrete*. The distinguishing property of concrete is its ability to harden under water.

The cement commonly used is portland cement, and the fine and coarse aggregates used are those that are usually obtainable, from nearby sand, gravel or rock deposits. In order to obtain a strong, durable and economical concrete mix, it is necessary to understand the characteristics and behaviour of the ingredients. The ingredients of the concrete can be classified into two groups, namely active and inactive. The active group consists of cement and water, whereas inactive group comprises fine and coarse aggregates. The inactive group is also sometimes called the *inert matrix*. In this section the ingredients of the active group will be discussed.

Although all materials that go into a concrete mixture are essential, cement is by far the most important constituent because it is usually the delicate link in the chain. The function of cement is first, to bind the sand and coarse aggregates together, and second, to fill the voids in between sand and coarse aggregate particles to form a compact mass. Although cement constitutes only about 10 per cent of the volume of the concrete mix, it is the active portion of the binding medium and the only scientifically controlled ingredient of concrete.

2.2 CEMENT

Cement is an extremely ground material having adhesive and cohesive properties, which provide a binding medium for the discrete ingredients. It is obtained by burning together, in a definite proportion, a mixture of naturally occurring argillacious (containing alumina) and calcareous (containing calcium carbonate or lime) materials to a partial fusion at high temperature (about 1450°C). The product obtained on burning, called *clinker*, is cooled and ground to the required fineness to produce a material known as *cement*. Its inventor, Joseph Aspdin, called it portland cement because when it hardened it produced a material resembling stone from the quarries near Portland in England. During grinding of clinker gypsum or plaster of Paris ($CaSO_4$) is added to adjust the setting time. The amount of gypsum is about 3 per cent by weight of clinker. It also improves the soundness of cement.

Depending upon the location of the cement-manufacturing plant, available raw materials are pulverized and mixed in proportions such that the resulting mixture will have the desired chemical composition. The commmon calcareous materials are limestone, chalk, oyster shells and marl. The argillaceous materials are clay, shale, slate and selected blast-furnace slag. When limestone and clay are the two basic ingredients, the proportions will be approximately four parts limestone to one part of clay. Certain clays formed during volcanic eruption, known as volcanic ash or pozzolana, found near Italy, have properties similar to that of portland cement.

Since the raw materials consist mainly of lime, silica, alumina and iron oxide, these form the major constituents of cement also. Depending upon the wide variety of raw materials used in the manufacture of cements, the oxide composition of ordinary portland cement may be expressed as follows.

Oxide	Percentage	Average
Lime, CaO	60–65	63
Silica, SiO_2	17–25	20
Alumina, Al_2O_3	3–8	6
Iron oxide, Fe_2O_3	0.5–6	3
Magnesia, MgO	0.5–4	2
Sulphur trioxide, SO_3	1–2	1.5
Soda and/or potash, $Na_2O + K_2O$	0.5–1	1

These oxides interact with each other to form a series of more complex products during fusion. The compound composition will be given later.

The processes used for the manufacture of cement can be classified as dry and wet. When the basic raw material is rock, it is transported to a large gyratory, or jaw crusher for primary reduction in size (to about 150 mm). It then passes through a smaller crusher or hammer mill where further reduction takes place to a 40 mm size aggregate, and from there it goes to rock storage. From storage the crushed rock is fed to a tube mill along with clay or crushed shale. In the wet process water is added at this point to obtain a blended mixture of very finely ground raw materials and water called slurry. The slurry is stored in tanks under constant agitation and fed into huge firebrick-lined rotary kilns. In the dry process the raw materials are ground, mixed and fed to the kiln in the dry state.

The kilns are fired with crushed coal or gas from the discharge end under forced draft so that material being fed in advances against the heat blast as the kiln rotates. The kilns are mounted with the longitudinal axis inclined in such a way that the raw material or slurry is fed at the higher end. At about 425°C, the excess water is driven off, and then further along the kiln, at 875°C, the limestone breaks down into calcium oxide and carbon dioxide. Finally, at 1400°C to 1450°C, about 10 m from the discharge end, the initial melting stage of material, known as the point of incipient fusion, is reached. Sintering takes place at this point, and a substance having its own physical and chemical properties called clinker is formed. The clinker is cooled, crushed, mixed with about 3 per cent crushed gypsum, and fed into a tube mill for final grinding. The finished product known as portland cement is finally bagged.

The composition of portland cement is rather complicated but basically it consists of the following four main compounds.

Tricalcium silicate	— $3CaO \cdot SiO_2$	(C_3S)
Dicalcium silicate	— $2CaO \cdot SiO_2$	(C_2S)
Tricalcium aluminate	— $3CaO \cdot Al_2O_3$	(C_3A)
Tetracalcium alumino ferrite	— $4CaO \cdot Al_2O_3 \cdot Fe_2O_3$	(C_4AF)

The symbols in parentheses are the abbreviations generally used. To the above ingredients is added about 3 per cent gypsum ($CaSO_4$). Depending upon the wide variety of raw materials used in the manufacture of cements, typical ranges of these compounds in ordinary portland cement may be expressed as below.

Compound	Percentage by mass in cement
C_3S	30–50
C_2S	20–45
C_3A	8–12
C_4AF	6–10

The difference in the various types of ordinary portland cements are due to the variations in the relative proportions of these compounds in the cement.

2.2.1 Basic Properties of Cement Compounds

The two silicates, namely C_3S and C_2S, which together constitute about 70 to 80 per cent of the cement control the most of the strength giving properties. Upon hydration, both C_3S and C_2S give the same product called calcium silicate hydrate ($C_3S_2H_3$) and calcium hydroxide. Tricalcium silicate (C_3S) gaving a faster rate of reaction (Fig. 2.1) accompanied by greater heat evolution develops early strength. On the other hand, dicalcium silicate (C_2S) hydrates and hardens slowly and provides much of the ultimate strength. It is likely that both C_3S and C_2S phases contribute equally to the eventual strength of the cement as can be seen in Fig. 2.2. C_3S and C_2S need approximately 24 and 21 per cent water by weight, respectively, for chemical reaction but C_3S liberates nearly three times as much calcium hydroxide on hydration as C_2S. However, C_2S provides more resistance to chemical attack.

Thus a higher percentage of C_3S results in rapid hardening with an early gain in strength at a higher heat of hydration. On the other hand, a higher percentage of

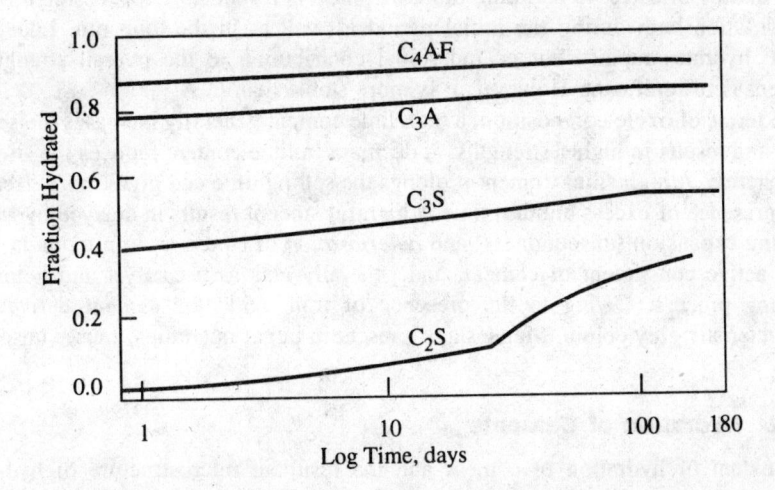

Fig. 2.1 Rate of hydration of pure cement compounds.

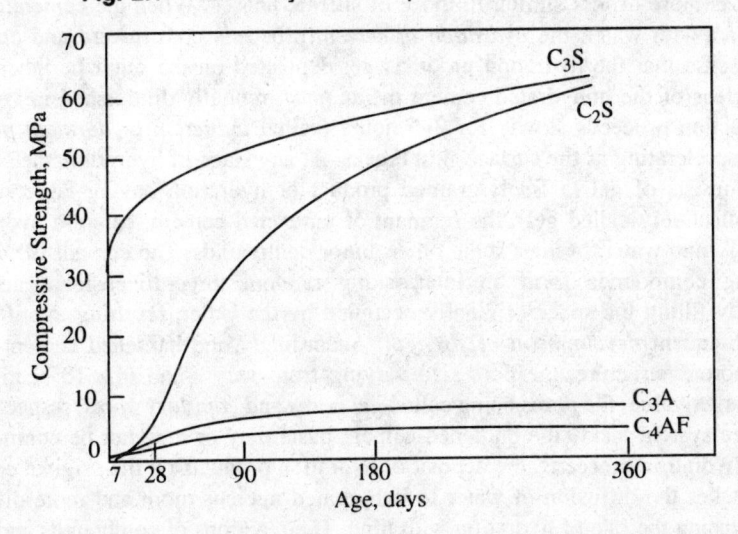

Fig. 2.2 Contribution of cement compounds to the strength of cement

C_2S results in slow hardening, less heat of hydration and greater resistance to chemical attack.

The compound tricalciumaluminate (C_3A) is characteristically fast-reacting with water and may lead to an immediate stiffening of paste, and this process is termed *flash set*. The role of gypsum added in the manufacture of cement is to prevent such a fast reaction. C_3A reacts with 40 per cent water by mass, and this is more than that required for silicates. However, since the amount of C_3A in cement is comparatively small, the net water required for the hydration of cement is not substantially affected. It provides weak resistance against sulphate attack and its contribution to the development of strength of cement is perhaps less significant than that of silicates. In addition, the C_3A phase is responsible for the highest heat of evolution both during the initial period as well as in the long run. Like C_3A, C_4AF hydrates rapidly but its individual contribution to the overall strength of cement is insignificant. However, it is more stable than C_3A.

In terms of oxide composition, a high lime content generally increases the setting time and results in higher strengths. A decrease in lime content reduces the strength of concrete. A high silica content prolongs the setting time and gives more strength. The presence of excess unburnt lime is harmful since it results in delayed hydration causing expansion (unsoundness) and *deterioration* of *concrete*. Iron oxide is not a very active constituent of cement, and generally acts as a catalyst and helps the burning process. Owing to the presence of iron oxide the cement derives the characteristic grey colour. Magnesia, if present in larger quantities, causes unsoundness.

2.2.2 Hydration of Cements

The extent of hydration of cement and the resultant microstructure of hydrated cement influences the physical properties of concrete. The microstructure of hydrated cement is more or less similar to those of silicate phases. When the cement comes in contact with water, the *hydration of cement* proceeds both inward and outward in the sense that the hydration products get deposited on the outer periphery and the nucleus of the unhydrated cement inside gets gradually diminished in volume. The reaction proceeds slowly for 2–5 hours (called *induction* or *dormant period*) before accelerating as the surface skin breaks. At any stage of hydration, the cement paste consists of gel (a finely-grained product of hydration having large surface area collectively called gel), the remnant of unreacted cement, calcium hydroxide $Ca(OH)_2$, and water, besides some other minor compounds. The crystals of various resulting compounds form an interlocking random three-dimensional network gradually filling the space originally occupied by the water, resulting in stiffening and subsequent *development of strength*. Accordingly, the hardened cement paste has a porous structure, the pore size varying from very small (4×10^{-4} μm) to a much larger value, the pores being called *gel pores* and *capillary pores*, respectively. The pore system inside the hardened cement paste may or may not be continuous. As the hydration proceeds, the deposit of hydration products on the original cement grain makes the diffusion of water to unhydrated nucleus more and more difficult thus reducing the rate of hydration with time. The reactions of compounds and their products may be symbolically represented as:

$$C_3S + H_2O \rightarrow C\text{—}S\text{—}H^* + Ca(OH)_2 \quad \text{Silicate phase}$$
$$C_2S + H_2O \rightarrow C\text{—}S\text{—}H + Ca(OH)_2$$
$$C_3A + H_2O \rightarrow C_4AH_{13} + C_2AH_8 \rightarrow C_3AH_6$$
$$C_3A + H_2O + CaSO_4 \rightarrow C_3A \cdot 3C\bar{S}H_{32} \rightarrow CA \cdot C\bar{S}H_{12}$$
$$C_4AF + H_2O \rightarrow C_3AH_6 + CFH$$

The product C—S—H represents the calcium silicate hydrate which is the *gel structure*. The above equations only refer to the processes in which the cement

---

*$H = H_2O$ and $\bar{S} = SO_3$

compounds (C_3S and C_2S, etc.) react with water to form a strong hydrated mass. The hydrated crystals are extremely small, varying from colloidal dimensions (less than 2 μm) to 10 μm or more. The $Ca(OH)_2$ liberated during the reaction of *silicate phase* crystallizes in the available free space.

Rate of Hydration

As mentioned earlier, the reaction of the compound C_3A with water is very fast in that *flash setting*, i.e., siffening without strength development, can occur because the C—A—H phase prevents the hydration of C_3S and C_2S. However, some of the $CaSO_4$ ground in the clinker dissolves immediately in water and the sulphate ions in the solution react with C_3A to form insoluble calcium sulphoaluminate which deposits on the surface of the C_3A to form a protective colloidal membrane and thus retard the direct hydration reaction. When all the sulphate is consumed, hydration can accelerate. The amount of sulphate must, therefore, be carefully controlled to leave little excess C_3A to hydrate directly. The hardening of C_3S appears to be catalyzed by C_3A so that C_3S becomes almost solely responsible for the gain of strength up to about 28 days by growth and interlocking of C—S—H gel. The later age increase in strength is due to the hydration of C_2S. The *rate of strength development* can, therefore, be modified by changes in the relative quantities of these compounds.

Mechanism of Hydration

C_3A reacts from beneath the thin membrane of calcium sulphoaluminate formed on the C_3A surface. Owing to the larger volume of calcium sulphoaluminate, pressure develops and the membrane eventually bursts, allowing the sulphate in solution to come in contact with unreacted C_3A to reform the membrane. The cyclic process continues until all the sulphate in solution is consumed, whereupon the C_3A can hydrate directly at a faster rate and the transformation of calcium sulphoaluminate into needle like monosulphate crystals leads to the loss of *workability* and to setting. This gives rise to the *induction period* which ends when the protective membrane is disrupted. Although the reaction between C_3S and water proceeds at the same time, in a properly retarded cement the end of induction period of C_3S hydration coincides with the point at which the sulphate in solution is no longer available for reaction. Setting, now, is due to the simultaneous growth of aluminate hydrate, monosulphate and silicate hydrate in the inter-particle space. The above theory is termed the *protective membrane* layer theory.

Effect of Admixtures on Hydration

Some *admixtures* may reduce the electric repulsion between the individual positively charged hydrating cement particles, so that they approach closer and stick to form agglomerates which grow and eventualy settle out. This process is termed *flocculation* and the agglomerate flocs. The anions may flocculate the *colloidal membrane*, thus making it more permeable. The rapid diffusion of water through the permeable membrane increases hydrostatic pressure beneath the membrane till it reaches a level sufficient to rupture it at an earlier stage in hydration, thus accelerating the hydration of cement.

2.2.3 Water-cement Ratio and Compressive Strength

A cement of average composition requires about 25 per cent of water by mass for chemical reaction. In addition, an amount of water is needed to fill the gel pores. The total amount of water thus needed for chemical reaction and to fill the gel pores is about 42 per cent. The general belief that a *water-cement ratio* of less than 0.40 or so should not be used in concretes, because for the process of hydration, the gel pores should be saturated, is not valid. This is because, even in the presence of excess water, the complete hydration of cement never takes place due to the decreasing porosity of the hydration products. As a matter of fact, a *water-cement ratio* of less than 0.40 is quite common in high-strength structural concretes.

In concrete, as explained earlier, the hardened cement paste is a porous ensemble. Also, the concentration of the solid products of hydration in the total space or volume available (the original water and hydrated cement) is an index of *porosity*. Like any other porous solid, the compressive strength of cement paste (or concrete) is related to the parameter *gel-space ratio* or *hydrate-space ratio*. The *water-cement ratio* which governs the compressive strength, is really an expression of the concentration of hydration products in the total volume at a particular age for the resultant degree of hydration.

2.2.4 Physical Properties of Portland Cement

The cement to be used in construction must have certain given qualities in order to play its part effectively in structure. When these properties lie within a certain range, the engineer is confident that in most of the cases the cement performance will be satisfactory. Also, based on these properties, it is possible to compare the quality of cement from different sources. Frequent tests are carried out on the cement either on dry powder or hardened cement paste, and sometimes on the concrete made from the cement, to maintain quality within specified limits. The important physical properties of a cement are as follows.

Fineness

The fineness of a cement is a measure of the size of particles of cement and is expressed in terms of *specific surface of cement*. It can be calculated from particle size distribution or one of the air permeability methods. It is an important factor in determining the *rate of gain of strength* and *uniformity* of quality. For a given weight of cement, the surface area is more for a finer cement than for a coarser cement. The finer the cement, the higher is the rate of hydration, as more surface area is available for chemical reaction. This results in the early development of strength. The effect of fineness on the compressive strength of cement is shown in Fig. 2.3. If the cement is ground beyond a certain limit its cementative properties may be adversely affected due to prehydration by atmospheric moisture. As per Indian Standard Specifications the residue of cement should not exceed 10 per cent when sieved on a 90-micron IS sieve. In addition, the amount of water required for constant slump concrete decreases with the increase in the fineness of cement.

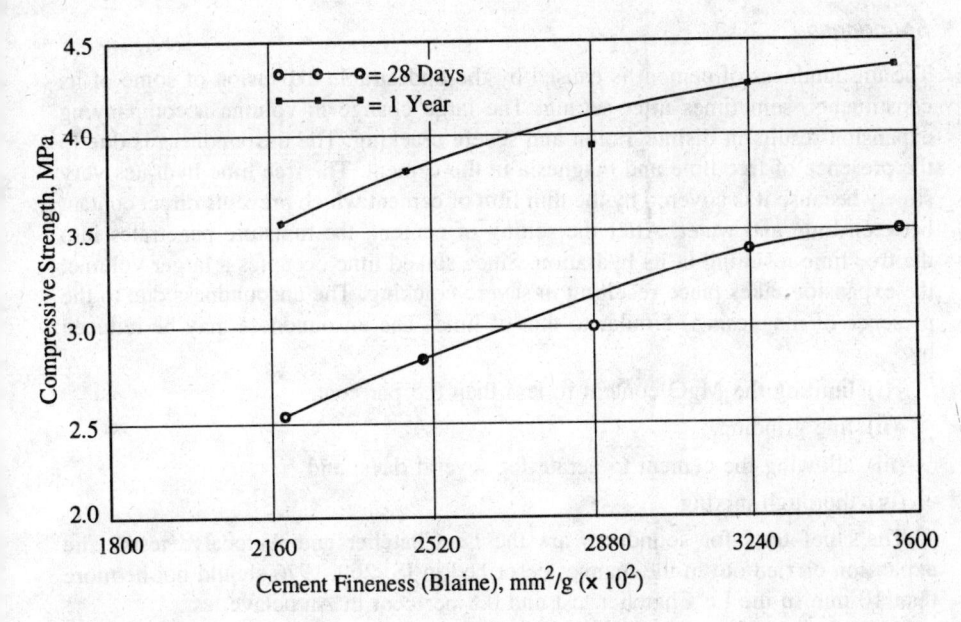

Fig. 2.3 The effect of fineness of cement on the compressive strength of concrete

Setting Time

Cement when mixed with water forms paste which gradually becomes less plastic, and finally a hard mass is obtained. In this process of setting, a stage is reached when the cement paste is sufficiently rigid to withstand a definite amount of pressure. The time to reach this stage is termed setting time. The time is reckoned from the instant when water is added to the cement. The setting time is divided into two parts, namely, the *initial* and the *final setting times*. The time at which the cement paste losses its plasticity is termed the *initial setting time*. The time taken to reach the stage when the paste becomes a hard mass is known as the *final setting time*.

It is essential for proper concreting that the initial setting time be sufficiently long for finishing operations, i.e. transporting and placing the concrete. The setting process is accompanied by temperature changes. The temperature rises rapidly from the initial setting to a peak value at the final setting. The setting time decreases with rise in temperature up to 30°C and vice versa. The setting times specified for various types of cements are given in Table 2.2. For an ordinary portland cement, the initial setting time should not be less than 30 minutes and final setting time should not be more than 600 minutes. A phenomenon of abnormal premature hardening within a few minutes of mixing the water is termed *false set*. However, not much heat is evolved and remixing the paste without water restores the plasticity and then the cement sets in the normal manner with no appreciable loss of strength.

In practice, the length of time for which a concrete mixture will remain plastic is usually more dependent on the amount of mixing water used and atmospheric temperature than on the setting time of cement.

Soundness

The unsoundness of cement is caused by the undesirable expansion of some of its constituents, sometimes after setting. The large change in volume accompanying expansion results in disintegration and severe cracking. The unsoundness is due to the presence of free lime and magnesia in the cement. The free lime hydrates very slowly because it is covered by the thin film of cement which prevents direct contact between lime and water. After the setting of cement, the moisture penetrates into the free lime resulting in its hydration. Since slaked lime occupies a larger volume, the expansion takes place resulting in severe cracking. The unsoundness due to the presence of magnesia is similar to that of lime. The unsoundness may be reduced by:

(i) limiting the MgO content to less than 0.5 per cent,
(ii) fine grinding,
(iii) allowing the cement to aerate for several days, and
(iv) thorough mixing.

The chief tests for soundness are the Le Chatelier and Autocalve tests. The expansion carried out in the manner described in IS: 269–1976 should not be more than 10 mm in the Le Chatelier test and 0.8 per cent in Autoclave test.

Compressive Strength

It is one of the important properties of cement. The strength tests, generally carried out in tension on samples of neat cement, are of doubtful value as an indication of ability of the cement to make concrete strong in compression. Therefore, these are largely being superseded *by the mortar cube* crushing tests and *concrete compression tests*. These are conducted on standardized aggregates under carefully controlled conditions and therefore give a good indication on strength qualities of cement. Cement mortar cubes (1 : 3) having an area of 5000 mm^2 are prepared and tested in compression testing machine. For ordinary portland cement the compression strength at 3 and 7 days curing should not be less than 16 MPa and 22 MPa, respectively. The graded standard sand used for preparing the cubes should conform to IS: 650–1966.

Standard Sand

A particular variety of sand obtainable at Ennore in Tamil Nadu is used as standard sand which closely resembles the Leighton Buzzard sand (the British Standard Sand) in its properties. The imported Leighton sand has been replaced by Ennore sand. The standard sand has following properties.

(i) The standard sand shall be of quartz, of light grey or whitish variety and shall be free from silt.
(ii) The sand grains shall be angular with shape approximating to spherical forms.
(iii) The sand shall pass through IS: 850-µm sieve and not more than 10 per cent shall pass through IS: 600-µm sieve.
(iv) It shall be free from organic impurities.

Heat of Hydration

The silicates and aluminates of cement react with water to form a binding medium which solidifies into a hardened mass. This reaction by virtue of which portland cement becomes a binding medium is termed *hydration*. The hydration of cement is exothermic with approximately 120 cal/g being liberated. In the interior of a large concrete mass, the hydration can result in a large rise in temperature. At the same time, the exterior of the concrete mass losses some heat so that a steep temperature gradient may be established, and during the subsequent cooling of the interior, severe cracking may occur. On the other hand, the heat of hydration may be advantageous in preventing the freezing of water in the capillaries of freshly placed concrete in cold weather.

The heat of hydration is defined as the quantity of heat, in calories per gram of hydrated cement, evolved upon complete hydration at a given temperature. It is determined by measuring the heats of evolution of unhydrated and hydrated cements in a mixture of nitric and hydrofluoric acids, the difference between the two values represents the heat of hydration. The heat of hydration for low-heat portland cement should not be more than 66 and 75 cal/g for 7 and 28 days, respectively.

The *heat of hydration* increases with temperature at which hydration takes place. For OPC it varies from 37 cal/g at 5°C to 80 cal/g at 40°C. For common types of portland cements, about 50 per cent of the total heat is liberated between 1 and 3 days; about 75 per cent in 7 days and 83 to 91 per cent in six months. The rate of heat evolution as well as total heat depends on the composition of cement. By restricting the quantities of compounds C_3A and C_3S in cement, the high rate of heat liberation in early stages can be checked. The rate of hydration and the heat evolved increases with the fineness of cement but the total amount of heat liberated is unaffected by fineness.

Specific Gravity

The specific gravity of portland cement is generally about 3.15, but that of cement manufactured from materials other than limestone and clay, the value may vary. Specific gravity is not an indication of the quality of cement. It is used in calculation of mix proportions.

2.2.5 Chemical Properties of Cements

The *loss on ignition* test is carried on portland cement to determine the loss of weight when the sample is heated to 900 to 1000°C. The loss in weight occurs as the moisture and carbon dioxide which are present in combination with free lime or magnesia evaporate. The presence of mositure causes prehydration of cement and may be absorbed from atmosphere during manufacturing or afterwards. The carbon dioxide is also taken from the atmosphere. The loss in weight is a measure of the freshness of cement. Since the hydroxides and carbonates of lime and magnesium have no cementing property, they are termed inert substances. The less the loss on ignition, the less is the quantity of these inert substances and better the cement.

The loss on ignition is determined by heating one gram of cement sample in a platinum crucible at a temperature of 900°C to 1000°C for minimum of 15 minutes.

Normally the loss will be in the neighbourhood of 2 per cent. Maximum allowable is 4 per cent.

Insoluble Residue

The insoluble material is an inactive part of cement. It is determined by stirring one gram of cement in 40 ml of water and adding 10 ml of concentrated HCl. The mix is boiled for 10 minutes maintaining constant volume. Any lump, if present, is broken and the solution filtered. The residue on filter is washed with Na_2CO_3 solution, water and HCl in the given order and, finally, again with water. The filter paper is dried, ignited, and weighed to give an insoluble residue. The minimum the residue, the better is the cement. The maximum allowable value is 0.85 per cent.

2.2.6 Various Types of Cements

By changing the chemical composition of the cement by varying the percentage of the four basic compounds it is possible to obtain several types of cements, each with some unique characteristics. Hence the cement can be manufactured for the required performance. The oxide and compound composition of some of the commonly used portland cements is given in the Table 2.1.

TABLE 2.1 Composition of portland cements

Type of cement	Oxide composition, per cent				Compound composition, per cent			
	CaO	SiO_2	Al_2O_3	FeO	C_3S	C_2S	C_3A	C_4AF
Normal or ordinary cement	63	20.6	6.3	3.6	40	30	11	11
Rapid-hardening cement	64.5	20.7	5.2	2.9	50	21	9	9
Low-heat cement	60	22.5	5.2	4.6	25	45	5	14
Sulphate-resisting cement	64	24.5	3.7	3.0	40	40	5	9

The following are the main types of portland cement.

Rapid-hardening Portland Cement

This cement is similar to ordinary portland cement (OPC) but with higher C_3S content and finer grinding. It gains strength more quickly than OPC, though the final strength is only slightly higher. The one-day strength of this cement is equal to the three-day strength of OPC with the same *water-cement ratio*. This cement is used where a rapid strength development is required. The rapid gain of strength is accompanied by a higher rate of heat development during the *hydration of cement*. This may have advantages in cold weather concreting, but a higher concrete temperature may lead to cracking due to subsequent thermal contraction, and hence should not be used in mass concreting or thick structural sections. The composition, fineness and other properties are governed by IS: 8041–1976. It is only about 10 per cent costlier than OPC.

Portland-slag Cement

This type of cement is made by intergrinding not less than 35 per cent of ordinary portland cement clinker and granulated *blast-furnace slag* (a waste product consisting of a mixture of lime, silica and alumina obtained in the manufacture of pig iron). The slag can also be used together with limestone as a raw material for the conventional manufacture of portland cement resulting in clinker which when ground gives *portland-slag cement*. This cement is less reactive than OPC and gains strength a little more slowly during the first 28 days, and adequate curing is essential. It has the advantages in generating heat less quickly than OPC. It is suitable for mass concreting but unsuitable in cold weather. Because of its fairly high sulphate resistance it is used in sea-water construction. The composition and properties are governed by IS: 455–1976.

Low-heat Portland Cement

This cement is less reactive than OPC and is obtained by increasing the proportion of C_2S and reducing C_3S and C_3A. This reduction in the content of more rapidly hydrating compounds C_3S and C_3A results in a slow development of strength but the ultimate strength is the same. In any case, to ensure a sufficient rate of development of strength, the *specific surface* of cement must not be less than 320000 mm^2/g. The *initial setting time* is greater than OPC. The properties and composition are governed by IS: 269–1976.

Portland-pozzolana Cement

Portland-pozzolana cement can be produced either by grinding together portland cement clinker and pozzolana with the addition of gypsum or calcium sulphate, or by intimately and uniformly blending portland cement and fine pozzolana. While grinding two materials together presents no difficulty, the mixing of dry powders intimately is extremely difficult. The blending should be resorted to only when the grinding rechniques prove uneconomical in a particular case and requisite machinery to ensure uniformity of production is available. If the blending is not uniform, it is reflected in the performance tests.

Portland-pozzolana cement produces less heat of hydration and offers greater resistance to the attack of impurities in water than normal portland cement. It is particularly useful in marine and hydraulic constructions, and other mass concrete structures. The portland-pozzolana cement can generally be used wherever ordinary portland cement is usable under normal conditions. However, all the pozzolanas need not necessarily contribute to the strength at early ages. IS: 1489–1976 gives the specifications for the production of portland-pozzolana cement equivalent to ordinary portland cement on the basis of seven-day compressive strength. The compressive strength of portland-pozzolana cement at 28 days also has been specified to enable the portland-pozzolana cement to be used as substitute for ordinary portland cement in plain and reinforced concrete works. The portland-pozzolana cement should conform to the requirements specified in IS: 1489–1976.

The average compressive strength of mortar cubes (area of face 50 cm^2) composed of one part of cement, three parts of standard sand (conforming to IS: 650–1966) by mass and $(p/4) + 3.0$ per cent (of combined mass of cement and sand) water obtained in the manner described in IS: 4031–1968 should be as follows:

(i) at $168 \pm 2h$ 22 MPa, minimum and
(ii) at $672 \pm 4h$ 31 MPa, minimum

where p is the percentage of water to produce a paste of standard consistency.

High-strength Portland Cement

The high-strength portland cement is produced from the same materials as in the case of ordinary portland cement. The higher strengths are achieved by increasing the tricalcium silicate (C_3S) content and also by finer grinding of the clinker. The fineness obtained by *Blaine's air permeability test* is specified to be of the order of 350000 mm^2/g. The initial and final setting times are same as that of ordinary portland cement. The *minimum compressive strengths* of the cement are 23 MPa and 33 MPa at the end of one day and seven days, respectively. The use of this cement was originally restricted to the production of railway sleepers and generally referred to as sleeper cement. The railway specifications require that the initial setting time should not be less than 90 minutes. At higher *water-cement ratios*, the high-strength concrete has about 80 per cent higher strength and at lower *water-cement ratios* 40 per cent higher strength than OPC. The cost of high-strength portland cement is only marginally higher than the OPC. The use of this cement in the usual 1 : 2 : 4 nominal mix, with a *water-cement ratio* of 0.60 to 0.65 can easily yield M25 concrete. Its composition and properties are governed by IS: 8112–1976.

Super Sulphate Cement

This cement is manufactured from well granulated slag (80 to 85 per cent) and calcium sulphate (10 to 15 per cent) together with 1 to 2 per cent of portland cement. Its specific surface is between 350000 and 500000 mm^2/g. It has an initial *setting time* between $2\frac{1}{2}$ to 4 hours and *final setting* between $4\frac{1}{2}$ to 7 hours. The total *heat of hydration* is very low, about 38 cal/g after seven days and 42 cal/g at 28 days, which make it suitable for mass concreting. However, the cement requires great care while concreting in cold weather. Its setting action is different from the other cements and the admixtures should not be used. If cured in air, the surface of concrete gets softened by atmospheric carbon dioxide, and hence water curing is preferable. The supersulphate cement concrete may expand or contract slightly on setting according to conditions and hence should be properly cured. The *rate of hardening* increases with temperature up to about 38°C but decreases above that.

For a normal mix of proportion 1 : 2 : 4 with a *water-cement ratio* of 0.55, the strengths are 35 MPa after seven days, 50 MPa after 28 days, and between 50 to 70 MPa after six months.

High-alumina Cement

This cement is basically different from OPC and concrete made with it has properties different from OPC. High-alumina cement (HAC) is very reactive and produces very high early strength. About 80 per cent of the ultimate strength is developed at the age of 24 hours and even at 6 to 8 hours. High-alumina cement has an initial *setting time* of about 4 hours and the final *setting time* about 5 hours. Generally no additives are added to alumina cement. For the same *water-cement ratio*, the alumina cement is more workable than portland cement. The strength is adversely affected by rise in tempreature. HAC is extremely resistant to chemical attack and is suitable for under sea water applications.

The approximate chemical oxide composition is as follows

Alumina (Al$_2$O$_3$)   39 per cent
Ferric Oxide (Fe$_2$O$_3$)   10 per cent
Lime (CaO)   38 per cent
Ferrous Oxide (FeO)   4 per cent
Silica (SiO$_2$)   6 per cent

The raw materials are limestone or chalk and bauxite which are crushed into lumps not exceeding 100 mm. The materials are heated to the fusion point at about 1600°C. The solidified material is fragmented and then ground to a fineness of 250000–320000 mm^2/g. The very dark grey powder is passed through magnetic separators to remove metallic iron. The alumina cement is considerably more expensive.

Waterproof Cement

Waterproof cement is manufactured by adding a waterproof substance to ordinary cement during mixing. The common admixtures are calcium stearate, aluminium stearate and the gypsum treated with tannic acid.

White Portland Cement

The process of manufacturing white cement is the same but the amount of iron oxide which is responsible for greyish colour is limited to less than 1 per cent. This is achieved by careful selection of raw materials and often by the use of oil fuel in place of pulverized coal in the kiln. The suitable raw materials are chalk and limestones having low iron contents, and white clays.

Coloured Portland Cement

These are basically portland cements to which pigments are added in quantities up to 10 per cent during the process of grinding the cement clinker. A good pigment should be permanent and chemically inert when mixed with cement. For lighter colours white cement has to be used as basis.

Hydrophobic Cement

This type of cement is obtained by adding the substances like stearic acid, boric acid and oleic acid to OPC during grinding of cement clinker. These acids form a film around the cement particles which prevent the entry of atmospheric moisture and the film breaks down when the concrete is mixed, and then the normal hydration takes place.

The physical and chemical requirements for some of the commonly used cements are summarized in the Tables 2.2 and 2.3.

2.2.7 Storage of Cement

It is often necessary to store cement for a long period, particularly when deliveries are irregular. Although cement retains its quality almost indefinitely if moisture is kept away from it, the cement exposed to air absorbs moisture slowly, and this causes its deterioration. An absorption of 1 or 2 per cent of water has no appreciable effect, but a further amount of absorption retards the hardening of cement and

20　Concrete Technology

TABLE 2.2 Physical requirements for different portland cements

Characteristic	Ordinary portland cement (IS: 269–1976)	Rapid hardening portland cement (IS: 8041–1976)	Low-heat portland cement (IS: 269–1976)	High-strength portland cement (IS: 8112–1976)	Portland-pozzolana cement (IS: 1484–1976)	Portland slag cement (IS: 455–1976)
1	2	3	4	5	6	7
Fineness						
Specific surface (mm²/g), min.	225000	325000	320000	350000	300000	225000
Setting time						
Initial setting time, minutes, min.	30	5	60	30	30	30
Final setting time, minutes, max.	600	30	600	600	600	600
Soundness						
Maximum expansion Le Chatelier method (mm)	10 (5)⁺	10 (5)	10 (5)	10 (5)	10 (5)	10 (5)
Autoclave method (per cent)	0.8	0.8	0.8	0.8	0.8	0.8
Heat of hydration, (cal/g), max.						
7 days	—	—	65	—	—	—
28 days	—	—	75	—	—	—
Compressive strength, MPa						
1 day	—	16	10	—	—	—
3 days	16	27.5	16	23	16	16
7 days	22	—	35	33	22	22
28 days	—	—	—	43	31	—
Drying shrinkage (per cent) max.	—	—	—	—	0.15	—

⁺ for aerated sample.

TABLE 2.3 Chemical requirements for different portland cements

Characteristic	Ordinary portland cement (IS: 269-1976)	Rapid hardening portland cement (IS: 8041-1976)	Low-heat portland cement (IS: 269-1976)	High-strength portland cement (IS: 8112-1976)	Portland-pozzolana cement (IS: 1489-1976)	Portland-slag cement (IS: 455-1976)
Maximum percentage of						
Magnesia	6.0	6.0	6.0	6.0	6.0	8.0
Sulphur anhydrite SO_3	2.75 (3.0)*	2.75 (3.0)	2.75 (3.0)	2.75 (3.0)	2.75 (3.0)	3.0
Insoluble residue	2.0	2.0	2.0	2.0	α**	2.5
Loss on ignition	5.0	5.0	5.0	4.0	5.0	4.0
Permitted additives (other than gypsum)	1.0	1.0	1.0	1.0	1.0	1.0
Content of slag, (per cent)	—	—	—	—	—	25-65
Content of pozzolana	—	—	—	—	10-25	—
Lime saturation factor	0.66 to 1.02	0.66 to 1.02	—	0.66 to 1.02	—	—
Ratio of percentage of alumina to that of iron oxide, min.	0.66	0.66	0.66	0.66	—	—

* When the compound C_3A is more than 7 per cent

$$\text{Lime saturation factor} = \frac{CaO - 0.7\,SO_3}{2.8\,SiO_2 + 1.2\,Al_2O_3 + 0.65\,Fe_2O_3}$$

**Insoluble residue in portland-pozzolana cement is given by: $p + \dfrac{2.0(100 - p)}{100}$

where p is the declared percentage of pozzolana in cement.

reduces its strength. The more finely a cement is ground the more reactive it is, and consequently the more rapidly does it absorb moisture from damp surroundings.

Cement, in bulk, can best be stored in bins of depth 2 m or more. Usually a crust about 5 cm thick forms, and this must be removed before cement is taken for use. The bagged cement may also be kept safely for many months if stored in waterproof shed with nonporous walls and floors, the windows being tightly shut. Once the cement has been properly stored it should not be disturbed until it is to be used. The practice of moving and restacking the bags to reduce warehouse pack only exposes fresh cement to air.

2.2.8 Rejection

Due to defective storage for long periods, cement is adversely affected. The cement remaining in bulk storage with the manufactures for more than six months or cement in jute bags in local storage in the hands of dealers for more than three months after completion of tests may be retested before use and rejected if they fail to conform to any of the requirements of IS: 269–1976.

3

Concrete Making Materials-II: Aggregate

3.1 INTRODUCTION

As explained in Chapter 2, concrete can be considered to be an artificial stone obtained by binding together the particles of relatively inert fine and coarse materials with cement paste. Aggregates are generally cheaper than cement and impart greater volume stability and durability to concrete. The aggregate is used primarily for the purpose of providing bulk to the concrete. To increase the density of the resulting mix, the aggregate is frequently used in two or more sizes. The most important function of the fine aggregate is to assist in producing workability and uniformity in mixture. The *fine aggregate* also assists the cement paste to hold the *coarse aggregate* particles in suspension. This action promotes plasticity in the mixture and prevents the possible *segregation* of paste and coarse aggregate, particularly when it is necessary to transport the concrete some distance from the mixing plant to point of placement.

The aggregates provide about 75 per cent of the body of the concrete and hence its influence is extremely important. They should therefore meet certain requirements if the concrete is to be workable, strong, durable, and economical. The aggregate must be of proper shape (either rounded or approximately cubical), clean, hard, strong and well graded. It should possess chemical stability and, in many cases, exhibit abrasion resistance and resistance to freezing and thawing.

3.2 CLASSIFICATION OF AGGREGATES

The classification of the aggregates is generally based on their geological origin, size, shape, unit weight, etc.

3.2.1 Classification According to the Geological Origin

The aggregates are usually derived from natural sources and may have been naturally reduced to size (e.g. gravel or shingle) or may have to be reduced by crushing. The suitability of the locally available aggregate depends upon the geological history of the region. The aggregate may be divided into two categories, namely the natural aggregate and artificial aggregates.

Natural Aggregate

These aggregates are generally obtained from natural deposits of sand and gravel, or from quarries by cutting rocks. Cheapest among them are the natural sand and gravel which have been reduced to their present size by natural agents, such as water, wind and snow, etc. The river deposits are the most common and have good quality. The second most commonly used source of aggregates is the quarried rock which is reduced to size by crushing. Crushed aggregates are made by breaking rocks into requisite graded particles by blasting, crushing and screening, etc. From the petrological standpoint, the natural aggregates, whether crushed or naturally reduced in size, can be divided into several groups of rocks having common characteristics. Natural rocks can be classified according to their geological mode of formation, i.e. igneous, sedimentary or metamorphic origin, and each group may be further divided into categories having certain petrological characteristics in common. Such a classification has been adopted in IS: 383–1970.

Artificial Aggregate

The most widely used artificial aggregates are clean broken bricks and air-cooled fresh blast-furnace-slag. The broken bricks of good quality provide a satisfactory aggregate for the mass concrete and are not suitable for reinforced concrete work if the crushing strength of brick is less than 30 to 35 MPa. The bricks should be free from lime mortar and lime sulphate plaster. The brick aggregate is not suitable for waterproof construction. It has poor resistance to wear and hence is not used in concrete for the road work.

The blast-furnace-slag is the byproduct obtained simultaneously with pig iron in the blast furnace, which is cooled slowly in air. Carefully selected slag produces concrete having properties comparable to that produced by using gravel aggregate. However, the corrosion of steel is more due to sulphur content of slag, but the concrete made with blast-furnace-slag aggregate has good fire resisting qualities. The other examples of the artificial slag are the expanded shale, expanded slag, cinder, etc.

3.2.2 Classification According to Size

The size of aggregates used in concrete range from few centimetres or more, down to a few microns. The maximum size of the aggregate may vary, but in each case it is to be so graded that the particles of different size fractions are incorporated in the mix in appropriate proportions. The particle size distribution is called the grading of the aggregate. According to size the aggregate is classified as: fine aggregate, coarse aggregate and all-in-aggregate.

Fine Aggregate

It is the aggregate most of which passes through a 4.75 mm IS sieve and contains only so much coarser material as is permitted by the specifications. Sand is generally considered to have a lower size limit of about 0.07 mm. Material between 0.06 mm and 0.002 mm is classified as silt, and still smaller particles are called clay. The soft deposit consisting of sand, silt and clay in about equal proportions is termed loam. The fine aggregate may be one of the following types:

(i) natural sand, i.e. the fine aggregate resulting from natural disintegration of rock and/or that which has been deposited by stream and glacial agencies,

(ii) crushed stone sand, i.e. the fine aggregate produced by crushing hard stone, or

(iii) crushed gravel sand, i.e. the fine aggregate produced by crushing natural gravel.

According to size, the fine aggregate may be described as *coarse, medium and fine sands*. Depending upon the particle size distribution, IS: 383–1970 has divided the fine aggregate into four grading zones. The grading zones become progressively finer from grading zone I to grading zone IV.

Coarse Aggregate

The aggregates most of which are retained on the 4.75 mm IS sieve and contain only so much of fine material as is permitted by the specifications are termed coarse aggregates. The coarse aggregate may be one of the following types:

(i) crushed gravel or stone obtained by the crushing of gravel or hard stone,

(ii) uncrushed gravel or stone resulting from the natural disintegration of rock, or

(iii) partially crushed gravel or stone obtained as a product of the blending of the above two types.

The graded coarse aggregate is described by its *nominal size*, i.e. 40 mm, 20 mm, 16 mm, and 12.5 mm, etc. For example, a graded aggregate of nominal size 12.5 mm means an aggregate most of which passes the 12.5 mm IS sieve. Since the aggregates are formed due to natural disintegration of rocks or by the artificial crushing of rock or gravel, they derive many of their properties from the parent rocks. These properties are chemical and mineral composition, petrographic description, specific gravity, hardness, strength, physical and chemical stability, pore structure, and colour. Some other properties of the aggregates not possessed by the parent rocks are particle shape and size, surface texture, absorption, etc. All these properties may have a considerable effect on the quality of concrete in fresh and hardened states.

All-in-aggregate

Sometimes combined aggregates are available in nature comprising different fractions of fine and coarse aggregates, which are known as *all-in-aggregate*. In such cases adjustments often become necessary to supplement the grading by addition of respective size fraction which may be deficient in the aggregate. Like coarse aggregate, the all-in-aggregate is also described by its nominal size. The all-in-aggregates are not generally used for making high quality concrete.

Single-size-aggregate

Aggregates comprising particles falling essentially within a narrow limit of size fractions are called *single-size-aggregates*. For example, a 20 mm single-size-aggregate means an aggregate most of which passes through a 20 mm IS sieve and the major portion of which is retained in a 10 mm IS sieve.

3.2.3 Classification According to Shape

The particle shapes of aggregates influence the properties of fresh concrete more than those of hardened concrete. Depending upon the particle shape, the aggregate may be classified as rounded, irregular or partly rounded, angular or flaky.

Rounded Aggregate

The aggregate with rounded particles (river or seashore gravel) has minimum voids ranging from 32 to 33 per cent. It gives minimum ratio of surface area to the volume, thus requiring minimum cement paste to make good concrete. The only disadvantage is that the interlocking between its particles is less and hence the development of the bond is poor, making it unsuitable for high strength concrete and pavements.

Irregular Aggregate

The aggregate having partly rounded particles (pitsand and gravel) has higher percentage of voids ranging from 35 to 38. It requires more cement paste for a given workability. The interlocking between particles, though better than that obtained with the rounded aggregate, is inadequate for high strength concrete.

Angular Aggregate

The aggregate with sharp, angular and rough particles (crushed rock) has a maximum percentage of voids ranging from 38 to 40. The interlocking between the particles is good, thereby providing a good bond. The aggregate requires more cement paste to make workable concrete of high strength than that required by rounded particles. The angular aggregate is suitable for high strength concrete and pavements subjected to tension.

Flaky and Elongated Aggregates

An aggregate is termed flaky when its least dimension (thickness) is less than 3/5th of its mean dimension. The mean dimension of the aggregate is the average of the sieve sizes through which the particles pass and retained, respectively. The particle is said to be elongated when its greatest dimension (length) is greater than nine-fifths its mean dimension.

The angularity of aggregate affects the workability or stability of the mix which depends on the interlocking of the particles. The elongated and flaky particles also adversely affect the durability of concrete as they tend to be oriented in one plane with water and air voids forming underneath. The presence of these particles should be restricted to 10 to 15 per cent. This requirement is particularly important for crushed-fine aggregate, since the material made in this way contains more flat and elongated particles. The angularity of the aggregate can be estimated from the proportion of voids in a sample compacted as prescribed in IS: 2386 (Part-I)-1963. The higher the *angularity number*, the more angular is the aggregate. The *elongation index* of an aggregate is defined as the percentage by weight of particles present in it whose greatest dimension (length) is greater than nine-fifths their mean dimension.

Whereas, the *flakiness index* is the percentage by weight of particles having least dimension (i.e. thickness) less than three-fifths their mean dimension.

The surface texture of the aggregate depends on the hardness, grain size and pore characteristics of the parent rocks, as well as the type and magnitude of the disintegrating forces. Based on the surface characteristics, IS: 383–1963 classifies the aggregates as glassy, smooth, granular, crystalline, honeycombed, porous, etc.

The shape and surface texture of aggregate influence the workability of fresh concrete and the compressive strength of hardened concrete, particularly in high strength concrete. The strength of concrete, especially the flexural strength, depends on the bond between the aggregate and cement paste. The bond is partly due to the interlocking of the aggregate and paste. A rough surface results in a better bond. The bond is also affected by the physical and chemical propeties, mineralogical and chemical composition, and the electrostatic condition of the particle surface, e.g. a chemical bond may exist in the case of a limestone aggregate.

3.2.4 Classification Based on Unit Weight

The aggregates can also be classified according to their unit weights as normal-weight, heavyweight, and lightweight aggregates.

Normal-weight Aggregate

The commonly used aggregates, i.e. sands and gravels; crushed rocks such as granite, basalt, quartz, sandstone and limestone; and brick ballast etc. which have specific gravities between 2.5 and 2.7 produce concrete with unit weight ranging from 23 to 26 kN/m^3 and crushing strength at 28 days between 15 to 40 MPa are termed normal-weight concrete. The properties and the requirements of normal-weight aggregate will be discussed in details in the succeeding sections.

Heavyweight Aggregate

Some heavyweight aggregates having specific gravities ranging from 2.8 to 2.9 and unit weights from 28 to 29 kN/m^3 such as magnetite (Fe_3O_4), barytes ($BaSO_4$) and scarp iron are used in the manufacture of heavyweight concrete which is more effective as a radiation shield. Concretes having unit weight of about 30 kN/m^3, 36 kN/m^3 and 57 kN/m^3 can be produced by using magnetite, baryte and scrap iron, respectively. The compressive strength of these concretes is of the order of 20 to 21 MPa. The cement-aggregate ratio varies from 1 : 5 to 1 : 9 with a *water-cement ratio* between 0.5 to 0.65. They produce dense and crack-free concrete. The main drawback with these aggregates is that they are not suitably graded and hence it is difficult to have adequate *workability* without *segregation*.

Lightweight Aggregate

The lightweight aggregates having unit weight up to 12 kN/m^3 are used to manufacture the structural concrete and masonry blocks for reduction of the self-weight of the structure. These aggregates can be either natural, such as diotomite, pumice, volcanic cinder, etc. or manufactured, such as bloated clay, sintered flyash or foamed blast-furnace-slag. In addition to reduction in the weight, the concrete produced by using light-weight aggregate provides better thermal insulation and improved fire resistance.

The main requirement of the lightweight aggregate is its low density; some specifications limit the unit weight to 12 kN/m^3 for fine aggregate and aproximately 10 kN/m^3 for coarse aggregates for the use in concrete. Because of high water absorption, the workable concrete mixes become stiff within a few minutes of mixing, thus requiring the wetting of the aggregates before mixing in the mixer. In the mixing operation, the required water and aggregate are usually premixed prior to the addition of cement. Approximately, 6 litres of extra water are needed per cubic metre of lightweight aggregate concrete to enhance the workability by 25 mm. To produce satisfactory strength of concrete, the cement content may be 3.5 kN/m^3 or more. Due to the increased permeability and rapid carbonation of concrete, the cover to the reinforcement using lightweight aggregate in concrete should be increased. The other characteristics of concrete using lightweight aggregates are reduced workability due to rough surface texture, lower tensile strength, lower modulus of elasticity (50 to 75 per cent of that of normal concrete) and higher creep and shrinkage. However, the ratio of creep strain to the elastic strain is the same for both the lightweight and normal-weight concretes. They have the tendency to segregate. Some of the important lightweight aggregates are as follows.

Bloated Clay Aggregate

The particles of such an aggregate range from 5 to 20 mm size. They are approximately spherical in shape, hard, light and porous. The water absorption is about 8 to 20 per cent. They produce concrete weighing up to about 19 kN/m^3. The bloated clay aggregates are used in places where the cost of crushed stone aggregate is high, and suitable clays, especially silts from waterworks, are easily available. The *sintered flyash aggregate* having a unit weight of about 10 kN/m^3 produces structural concrete with a unit weight of 12 to 14 kN/m^3.

The *expanded shale* known as herculite or haydite is produced by passing the crushed shale through a rotary kiln at 1100°C. Gases within the shale expand forming millions of tiny air cells within the mass. The cells are surrounded by a hard vitreous waterproof membrane. The product is carefully screened into commercial sizes. This aggregate is used to a large extent, to replace the stone aggregate in the production of structural concrete because it reduces the weight by about one-third for no loss of strength for comparable cement content. It has high resistance to heat and is used for refractory lining, fireproofing of structural steel, and for the construction of other concrete surfaces exposed to high temperatures. In addition they have better sound absorption.

Vermiculite is another artifical lightweight aggregate which produces low-strength and high-shrinkage concrete. It is not used for structural concrete but is widely used for insulating concrete roof decks.

The compressive strength of lightweight concrete is comparable to that of normal-weight concrete. The concrete produced with *artificial* and *processed aggregates* needs extra cover to the reinforcement due to high absorption of the aggregate. The other factor limiting their usefulness is the relatively low modulus of elasticity. The flexural strength of light-weight concrete is of the same order as that of normal concrete at early ages but does not improve significantly with long moist curing. The shear strength on the average is as high as for normal-weight concrete. The shrinkage and creep strains are usually somewhat higher for lightweight concretes. *Air-entrainment* can be used to make the concrete to perform like good quality normal weight concrete.

IS: 9142–1979 covers the specifications for artificial lightweight aggregate for concrete masonary units, while IS: 2686–1977 covers cinder aggregate for use in lime concrete for the manufacture of precast blocks. The use of bloated clay and sintered flyash aggregate has been envisaged in IS: 546–1978.

3.3 CHARACTERISTICS OF AGGREGATES

As explained earlier, the properties and performance of concrete are dependent to a large extent on the characteristics and properties of the aggregates themselves. In general, an aggregate to be used in concrete must be clean, hard, strong, properly shaped and well graded. The aggregate must possess chemical stability, resistance to abrasion, and to freezing and thawing. They should not contain deleterious material which may cause physical or chemical changes, such as cracking, swelling, softening or leaching. The properties of aggregate for concrete are discussed under the following heads.

Strength of Aggregate

The strength of concrete cannot exceed that of the bulk of aggregate contained therein. Therefore, so long as the strength of aggregate is of an order of magnitude stronger than that of the concrete made with them, it is sufficient. However, in the cases of high strength concretes subjected to localized stress concentration leading to stresses higher than the overall strength of concrete, the strength of aggregate may become critical. Generally three tests are prescribed for the determination of strength of aggregate, namely, aggregate crushing value, aggregate impact value and ten per cent fines value. Of these, the crushing value test is more popular and the results are reproducible. However, the ten per cent fines value test which gives the load required to produce ten per cent fines from 12.5 mm to 10 mm particles, is more reliable. IS: 383–1970 prescribes a 45 per cent limit for the crushing value determined as per IS: 2386 (Part-IV)-1963 for the aggregate used for concrete other than for wearing surfaces and 30 per cent for concrete for wearing surfaces, such as runways, roads and pavements. BS: 882–1965 prescribes a minimum value of ten tonnes in the 10 per cent fines test for aggregate to be used in wearing surfaces and five tonnes when used in other concretes.

The other related *mechanical properties* of aggregate which are important especially when the aggregate is subjected to high wear are *toughness* and *hardness*.

The *toughness* of aggregate which is measured as the resistance of the aggregate to failure by impact, determined in accordance with IS: 2386 (Part-IV)-1963 may be used instead of its crushing value. The *aggregate impact value* shall not exceed 45 per cent by weight for aggregate used for concrete other than those used for wearing surfaces and 30 per cent for concrete for wearing surfaces.

The *hardness* of the aggregate defined as its resistance to wear obtained in terms of *aggregate abrasion value* is determined by using the Los Angeles machine as described in IS: 2386 (part-IV)-1963. The method combines the test for *attrition* and *abrasion*. A satisfactory aggregate should have an abrasion value of not more than 30 per cent for aggregates used for wearing surfaces and 50 per cent for aggregates used for non wearing surface.

The strength of an aggregate as measured by its *resistance to freezing* and *thawing* is an important characteristic for a concrete exposed to severe weather. The resistance to freezing and thawing is related to its *porosity, absorption,* and *pore structure.* In a fully saturated aggregate, there is not enough space available to accommodate the expansion due to freezing of water resulting in the failure of the particles.

An aggregate with higher *modulus of elasticity* generally produces a concrete with higher modulus of elasticity. The modulus of elasticity of aggregate affects also the magnitude of *creep* and *shrinkage* of concrete. The *compressibility of aggregate* would reduce distress in concrete during its volume changes while a strong and rigid aggregate might lead to the *cracking* of the surrounding cement paste. Thus the aggregate of moderate or low strength and modulus of elasticity can be valuable in preserving the *durability of concrete*.

Particle Shape and Texture

The physical characteristics such as *shape, texture and roughness* of aggregates, significantly influence the *mobility* (i.e. the workability) of fresh concrete and the *bond* between the aggregate and the mortar phase. As described earlier the aggregates are generally divided into four categories, namely, *rounded, irregular, angular* and *flaky*. The *rounded aggregates* are available as river or seashore gravel which are fully waterworn or completely shaped by attrition, whereas *irregular* or *partly rounded aggregate* (pitsands and gravels) are partly shaped by attrition and have rounded edges. The *angular aggregate* possessing well-defined edges formed at the intersection of roughly planer faces are obtained by crushing the rocks. The angular aggregates obtained from laminated rocks having thickness smaller than the width and/or length are termed *flaky*. The rounded aggregates require lesser amount of water and cement paste for a given *workability*. The water content could be reduced by 5 to 10 per cent, and the sand content by 3 to 5 per cent by the use of rounded aggregate. On the other hand, the use of crushed aggregate may result in 10 to 20 per cent higher *compressive strength* due to the development of stronger *aggregate-mortar bond*. This increase in strength may be up to 38 per cent for the concrete having a *water-cement ratio* below 0.4. The elongated and flaky particles, having a high ratio of surface area to volume reduce the workability appreciably. These particles tend to be oriented in one plane with water and air voids underneath. The *flakiness index* of coarse aggregate is generally limited to 25 per cent. The *surface texture* is a measure of the smoothness or roughness of the aggregate. Based on the visual examination of the specimen, the *surface texture* may be classified as *glassy, smooth, granular, rough, crystalline, porous* and *honeycombed*. The strength of the *bond* between aggregate and cement paste depends upon the surface texture. The bond is the development of mechanical anchorage and depends upon the surface roughness and surface porosity of the aggregate. An aggregate with a rough, porous texture is preferred to one with a smooth surface as the former can increase the *aggregate-cement bond* by 75 per cent, which may increase the compressive and flexural strength of concrete up to 20 per cent. The surface pores help in the development of good bond on account of suction of paste into these pores. This explains the fact that some aggregates which appear smooth still bond strongly than the one with rough surface texture.

The shape and surface texture of fine aggregate govern its void content and thus affect the water requirement of mix significantly.

Specific Gravity

The *specific graivty* of an aggregate is defined as the ratio of the mass of solid in a given volume of sample to the mass of an equal volume of water at the same temperature. Since the aggregate generally contains voids, there are different types of specific gravities.

The *absolute specific gravity* refers to the volume of solid material excluding the voids, and therefore, is defined as the ratio of the mass of solid to the weight of an equal void-free volume of water at a stated temperature. If the volume of aggregate includes the voids, the resulting specific gravity is called the *apparent specific gravity*. As the aggregate generally contains both impermeable and capillary voids (voids between the particles), the apparent specific gravity refers to volume including impermeable voids only. It is therefore the ratio of the mass of the aggregate dried in an oven at 100 to 110°C for 24 hours to the mass of the water occupying a volume equal to that of solids including impermeable voids or pores. The specific gravity most frequently and easily determined is based on the saturated surface dry condition of the aggregate because the water absorbed in the pores of the aggregate does not take part in the chemical reaction of the cement and can therefore be considered as a part of the aggregate. This specific gravity is required for the calculations of the yield of concrete or of the quantity of aggregate required for a given volume of concrete. The specific gravity of an aggregate gives valuable information on its quality and properties. It is seen that the higher the specific gravity of an aggregate, the harder and stronger it will be. If the specific gravity is above or below that normally assigned to a particular type of aggregate, it may indicate that the shape and grading of the aggregate has changed.

The specific gravity is determined as described in IS: 2386 (Part-III)-1963. The specific gravity is given by:

$$\text{Specific gravity} = \frac{c}{a-b}$$

and

$$\text{Apparent specific gravity} = \frac{c}{c-b}$$

where a = mass of saturated surface dry aggregate in air,
b = mass of saturated surface dry aggregate in water, and
c = mass of ovendry aggregate in air.

The specific gravity of majority of natural aggregates lie between 2.6 and 2.7.

Bulk Density

The *bulk density* of an aggregate is defined as the mass of the material in a given volume and is expressed in kilograms/litre. The bulk density of an aggregate depends on how densely the aggregate is packed in the measure. The other factors affecting the bulk density are the *particle shape, size*, the *grading of the aggregate* and the *moisture content*. The shape of the particles greatly affects the closeness of the packing that can be achieved. For a *coarse aggregate* of given specific gravity, a higher bulk density indicates that there are fewer voids to be filled by sand and cement.

The bulk density of an aggregate can be used for judging the quality of aggregate by comparison with normal density for that type of aggregate. It determines the type of concrete for which it may be used. The bulk density is also required for converting proportions by weight into the proportions by volume. The bulk density is determined as described in IS: 2386 (Part-III)-1963.

Voids

The empty space between the aggregate particles are termed *voids*. It is the difference between the gross volume of aggregate mass and the volume occupied by the particles alone. The void ratio of an aggregate can be calculated from the specific gravity and bulk density of aggregate mass as follows:

$$\text{Void ratio} = 1 - \frac{\text{bulk density}}{\text{apparent specific gravity}}$$

Porosity and Absorption of Aggregates

Due to the presence of air bubbles which are entrapped in a rock during its formation or on account of the decomposition of certain constituent minerals by atmospheric action, *minute holes* or *cavities* are formed in it which are commonly known as *pores*.

The pores in the aggregate vary in size over a wide range, the largest being large enough to be seen under a microscope or even with the naked eye. They are distributed through out the body of the material, some are wholly within the solid and the others are open to the surface of the particle. The porosity of some of the commonly used rocks varies from 0 to 20 per cent. Since the aggregate constitute about 75 per cent of the concrete, the *porosity of aggregate* contributes to the overall *porosity of concrete*. The *permeability* and *absorption* affect the *bond* between the aggregate and the cement paste, the *resistance* of concrete to *freezing and thawing, chemical stability, resistance to abrasion*, and the *specific gravity* of the aggregate.

The pores may become reservoirs of free moisture inside the aggregate. The percentage of water absorbed by an aggregate when immersed in water is termed the *absorption of aggregate*. The aggregate which is saturated with water but contains no *surface free moisture* is termed the *saturated surface dry aggregate*. The method for determining the water absorption of an aggregate is described in IS: 2386 (Part-III)-1963. If the aggregate is previously dried in an oven at 105°C to a constant weight before being immersed in water for 24 hours, the absorption is referred to as on ovendry basis. On the other hand, the percentage of water absorbed by an air dried aggregate when immersed in water for 24 hours is termed absorption of aggregate (air dry basis). The knowledge of the absorption of an aggregate is important for concrete mix design calculations.

Moisture Content of Aggregate

The *surface moisture* expressed as a percentage of the weight of the *saturated surface dry aggregate* is termed as *moisture content*. Since the absorption represents the water contained in the aggregate in the saturated-surface dry condition and the moisture content is the water in excess of that, the total water content of a moist aggregate is equal to the sum of absorption and moisture content. IS: 2386 (Part-III)-1963 describes the method to determine the moisture content of concrete aggregate.

The determination of moisture content of an aggregate is necessary in order to determine the net *water-cement ratio* for a batch of concrete. A high moisture content will increase the effective water-cement ratio to an appreciable extent and may make the concrete weak unless a suitable allowance is made. IS: 2386 (Part-III)-1963 gives two methods for its determination. The first method, namely, the *displacement method*, gives the moisture content as a percentage by mass of the saturated surface dry sample whereas the second method namely the *drying method*, gives the moisture content as a percentage by mass of the dried sample. The moisture content obtained by these two methods are quite different. The moisture content given by the drying method will normally be the total moisture content due to free plus absorbed water. The accuracy of the displacement method depends upon the accurate information of the *specific gravity* of the material in a saturated-surface dry condition.

Bulking of Fine Aggregate

The increase in the volume of a given mass of fine aggregate caused by the presence of water is known as *bulking*. The bulking of fine aggregate is caused by the films of water which push the particles apart. The extent of bulking depends upon the *percentage of moisture* present in the sand and its *fineness*. It is seen that the bulking increases gradually with moisture content up to a certain point and then begins to decrease with further addition of water due to the merging of films, until when the sand is inundated. At this stage the bulking is practically nil. With ordinary sands the bulking usually varies between 15 and 30 per cent. A typical graph shown in Fig. 3.1 gives the variation of per cent bulking with moisture content. Finer sand

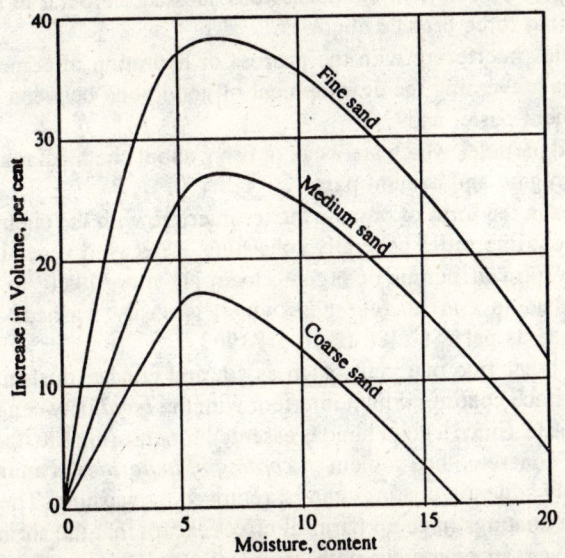

Fig. 3.1 Effect of moisture content on the bulking of sand

bulks considerably more and the maximum bulking is obtained at a higher water content than the coarse sand. In extremely *fine sand* the bulking may be of the order of 40 per cent at a moisture content of 10 per cent but such a sand is unsuitable for concrete. In the case of *coarse aggregate,* the increase in volume is negligible due to the presence of free water as the thickness of the moisture film is very small compared with particle size. The precentage bulking is obtained in accordance with IS: 2386 (Part-III)-1963.

If the sand is measured by volume and no allowance is made for bulking, the mix will be richer than that specified because for given mass, moist sand occupies a considerably larger volume than the same mass of the dry sand. This results in a mix deficient in sand increasing the chances of the *segregation* and *honey-combing* of concrete. The *yield of concrete* will also be reduced. It is necessary, in such a case, to increase the measured volume of the sand by the percentage bulking, in order that the amount of sand put into concrete to be the amount intended for the *nominal mix* used (based on dry sand). If no allowance is made for the bulking of sand a nominal concrete mix 1 : 2 : 4, for example, will correspond to 1 : 1.74 : 4 for a bulking of 15 per cent. An increase in bulking from 15 to 30 per cent will result in an increase in the *concrete strength* by as much as 14 per cent. If no allowance is made for bulking the concrete strength may vary by as much as 25 per cent.

3.4 DELETERIOUS SUBSTANCES IN AGGREGATES

The *materials* whose presence may adversely affect the *strength, workability* and long-term *performance of concrete* are termed *deleterious materials*. These are considered undesirable as constituent because of their intrinsic weakness, softness, fineness or other physical or chemical characteristics harmful to the concrete behaviour.

Depending upon their action, the deleterious substances found in the aggregates can be divided into three broad categories:
  (i) impurities interfering with the process of hydration of cements,
  (ii) coatings preventing the development of good bond between aggregate and the cement paste, and
  (iii) unsound particles which are weak or bring about chemical reaction between the aggregate and cement paste.

The impurities in the form of organic matter interfere with the chemical reactions of *hydration*. These impurities generally consisting of decayed vegetable matter and appearing in the form of humus or organic loam are more likely to be present in *fine aggregate* than in *coarse aggregate* which is easily washed. The effect of impurities is tested as per IS: 2386 (Part- II)-1963.

The clay and other fine materials, such as silt and crusher dust may be present in the form of surface coatings which interfere with the *bond* between the aggregate and the cement paste. Since a good bond is essential for ensuring satisfactory *strength* and *durability* of concrete, the problem of *coating of impurities* is an important one. The soft or loosely adherent coatings can be removed by washing. The well-bonded chemically stable coatings have no harmful effect except that the shrinkage may be increased. However, an aggregate with chemically reactive coatings can lead to serious trouble. The silt and the fine dust, if present in excessive amounts, increase the *specific surface of the aggregate* and hence the amount of water required to wet all particles in the mix, thereby reducing the strength and durability of concrete.

The total amount of deleterious material should not exceed 5 per cent as per IS: 383–1970. The limits of deleterious materials are given in Table 3.1.

TABLE 3.1 Limits of deleterious materials (maximum percentage by mass)

Deleterious substances	Fine aggregate Uncrushed	Fine aggregate Crushed	Coarse aggregate Uncrushed	Coarse aggregate Crushed
Coal and lignite	1.0	1.0	1.0	1.0
Clay lumps	1.0	1.0	1.0	1.0
Soft fragments	—	—	3.0	—
Material passing 75 μm IS Sieve	3.0	3.0	3.0	3.0
Shale	1.0	—	—	—
Total of all deleterious materials	5.0	2.0	5.0	5.0

The sand obtained from a seashore or a river estuary contains salt and sometimes its percentage may be as high as 6 per cent of mass of sand. The salt can be removed from the sand by washing it with fresh water before use. If salt is not removed, it absorbs moisture from air and may cause *efflorescence;* and slight *corrosion of reinforcement* may also occur.

Unsound particles are broadly grouped as (i) the particles failing to maintain their integrity, and (ii) particles leading to disruptive expansion on freezing or exposure to water.

The *shale* and other particles of low density, such as *clay lumps, wood, coal,* etc. are regarded as unsound as they lead to *pitting and scaling.* If the percentage of these particles exceeds 2 to 5 per cent of the mass of aggregate they may adversely affect the strength of concrete. The presence of mica in fine aggregate has also been found to considerably reduce the compressive strength of concrete. Hence, if mica is present in fine aggregate, a suitable allowance for the possible reduction in strength of concrete should be made. Likewise, gypsum and other sulphates must not be present in the aggregates.

Iron pyrites and marcasite are the most common *expansive inclusions* in the aggregate. These sulphides react with water and oxygen in the air resulting in the *surface staining of concrete* and *pop-outs.* The effect is more under warm and humid conditions.

The majority of these impurities are found in natural aggregate deposits, rather than crushed aggregate.

3.5 SOUNDNESS OF AGGREGATE

The soundness indicates the ability of the aggregate to resist excessive changes in volume due to changes in environmental conditions, e.g. freezing and thawing, thermal changes, and alternating wetting and drying. The aggregate is said to be unsound when volume changes result in the *deterioration of concrete.* This may appear in the form of local scaling to extensive surface cracking or to disintegration over a considerable depth, and thus vary from an impaired appearance to a structurally dangerous situation.

IS: 2386 (Part-V)-1963 describes a method to determine the resistance to disintegration of aggregates by saturated solution of sodium sulphate (Na_2SO_4) or magnesium sulphate ($MgSO_4$). According to IS: 383–1963 the average loss of weight after ten cycles should not exceed 12 and 18 per cent when tested with sodium sulphate and magnesium sulphate, respectively.

3.6 ALKALI-AGGREGATE REACTION

The *alkali-aggregate reaction* is the reaction between the active silica constituents of the aggregate, and the alkalies in cement. The reactive forms of silica occur in opaline or chalcedonic cherts, siliceous limestones, rhyolites and rhyolitic tuffs, andesite and andesite tuffs, phyllites, etc. The reaction starts with the attack on the siliceous minerals in the aggregate by the alkaline hydroxides derived from the alkalies in cement. As a result of this reaction, an *alkali silicate gel* is formed altering the borders of the aggregate. The gel is confined by the surrounding cement paste and an internal pressure is developed leading to expansion resulting in *cracking* and *disruption of cement paste*. This expansion appears to be due to hydraulic pressure generated through osmosis, but can also be due to the swelling pressure of the still solid products of the alkali-silica reaction. The *reactivity of aggregate* depends upon its *particle size* and *porosity* as these influence the area over which the reaction can take place. The *expansion of cement* also depends upon the *alkali content* and *fineness of cement*.

The rate of alkali-aggregate reaction is also affected by the availability of non-evaporable water in the paste and the reaction is accelerated under the condition of alternating wetting and drying. The temperature in the range of 10 to 40°C accelerates the reaction.

The expansion due to *alkali-aggregate reaction* can be reduced by adding the reactive silica in a finely powdered form to the concrete mix. This is because the addition of *reactive silica* increases the *surface area,* increasing the calcium hydroxide/alkali ratio of the solution at the boundaries of the aggregate. Under such circumstances, a nonexpanding calcium alkali silicate product is formed. It is generally recommended that 20 g of reactive silica be added for each gram of alkali in excess of 0.5 per cent of the mass of the cement IS: 2386 (part-VII)-1963 describes two methods for the determination of the potential reactivity of the aggregate.

3.7 THERMAL PROPERTIES OF AGGREGATES

The *thermal properties* of the aggregates affect the durability and the other qualities of concrete. The investigations reported to date do not present a clear-cut picture of the effects that might be expected.

The principal thermal properties of the aggregate are: (i) coefficient of thermal expansion, (ii) specific heat, and (iii) thermal conductivity.

The *coefficient of thermal expansion* of the concrete increases with the coefficient of thermal expansion of aggregate. If the coefficient of expansion of coarse aggregate and of cement paste differs too much, a large change in temperature may introduce differential movement which may break the *bond* between the aggregate and the paste. If the coefficients of the two materials differ by more than 5.4×10^{-6} per °C, the *durability* of concrete subjected to freezing and thawing may be affected.

The coefficient of expansion of the aggregate depends on the parent rock. For majority of aggregates, the coefficient of thermal expansion lies between approximately 5.4×10^{-6} and 12.6×10^{-6} per °C. For hydrated portland cement the coefficient varies between 10.8×10^{-6} and 16.2×10^{-6} per °C. The coefficient of thermal expansion varies with the degree of saturation and can be determined by Verbeck's dilatometer.

The *specific heat* of the aggregate is a measure of its heat capacity, whereas the *thermal conductivity* is the ability of the aggregate to conduct the heat. These properties of the aggregate influence the specific heat and thermal conductivity of the concrete, and are important in the case of mass concrete and where insulation is required.

3.8 FINENESS MODULUS

The *fineness modulus* is a numerical index of fineness, giving some idea of the *mean size of the particles* present in the entire body of the aggregate. The determination of the fineness modulus consists in dividing a sample of aggregate into fractions of different sizes by sieving through a set of standard test sieves taken in order. Each fraction contains particles between definite limits. The limits being the opening sizes of standard test sieves. The material retained on each sieve after sieving represent the fraction of aggregate coarser than the sieve in question but finer than the sieve above. The sum of the *cumulative percentages* retained on the sieves divided by 100 give the fineness modulus. The sieves that are to be used for the sieve analysis of the aggregate (coarse, fine or all-in-aggregate) for concrete as per IS: 2386 (Part-I)-1963, are 80 mm, 40 mm, 20 mm, 10 mm, 4.75 mm, 2.36 mm, 1.18 mm, 600 μm, 300 μm and 150 μm.

The fineness modulus can be regarded as a *weighted average size* of a sieve on which material is retained, and the sieves being counted from the finest. For example, a fineness modulus of 6.0 can be interpreted to mean that the sixth sieve, i.e., 4.75 mm is the average size. The value of fineness modulus is higher for coarser aggregate. For the aggregates commonly used, the fineness modulus *of fine aggregate* varies between 2.0 and 3.5, for *coarse aggregate* between 5.5 and 8.0, and from 3.5 to 6.5 for *all-in-aggregate*.

The object of finding fineness modulus is to grade the given aggregate for the most economical mix for the required *strength* and *workability* with minimum quantity of cement. If the test aggregate gives higher fineness modulus, the mix will be harsh and if, on the other hand, gives a lower fineness modulus it gives an uneconomical mix. For *workability*, a coarser aggregate requires less *water-cement ratio*. The fineness modulus is also important for measuring the slight variations in the aggregate from the same source.

3.9 MAXIMUM SIZE OF AGGREGATE

In general, the larger the *maximum size of the aggregate*, the smaller is the cement requirement for a particular water-cement ratio. This is due to the fact that the workability of concrete increases with the increase in the maximum size of the aggregate. In a mass concrete work, the use of a larger size aggregate is beneficial due to the lesser consumption of cement. This will also reduce the *heat of hydration* and corresponding *thermal stresses* and *shrinkage cracks*. Moreover, due to the smaller surface area of the larger size aggregate, the *water-cement ratio* can be decreased which increases the strength. However, in practice, the size of aggregate is limited depending upon the size of mixing, handling, and placing equipment. The maximum size of aggregates also influences the compressive strength of concrete in that, for a particular volume of aggregate, the compressive strength tends to increase with the decrease in the size of the *coarse aggregate*. This is due to the fact that smaller size aggregates provide larger *surface area* for bonding with the mortar matrix. In addition, the *stress concentration* in the *mortar-aggregate interface* increases with the maximum size of the aggregate. Thus, for concrete with a higher water-cement ratio, the nominal size of the coarse aggregate must be as large as possible, whereas for high strength concretes a 10 or 20 mm size of aggregate is preferable. In general, for strengths up to 20 MPa aggregates up to 40 mm may be used, and for strengths above 30 MPa, aggregates up to 20 mm may be used.

According to IS: 456-1978, the maximum nominal size of coarse aggregate should not be greater than one-fourth the minimum thickness of the member, and should be restricted to 5 mm less than the minimum clear distance between the main bars or 5 mm less than the minimum-cover-to-reinforcement distance and 5 mm less than the spacing between the cables, strands or sheathing in case of prestressed concrete. Within these limits, the nominal maximum size of coarse aggregates may be as large as possible.

3.10 GRADING AND SURFACE AREA OF AGGREGATE

The *particle size distribution* of an aggregate as determined by sieve analysis is termed *grading of the aggregate*. If all the particles of an aggregate are of uniform size, the compacted mass will contain more *voids*, whereas an aggregate comprising particles of various sizes will give a mass containing lesser voids. The particle size distribution of a mass of aggregate should be such that the smaller particles fill the voids between the larger particles. The proper grading of an aggregate produces *dense concrete* and needs less quantity of *fine aggregate* and *cement paste*. It is, therefore, essential that the coarse and fine aggregates be well graded to produce *quality concrete*.

The grading of an aggregate is expressed in terms of percentages by weight retained on or passing through a series of sieves taken in order, 80 mm, 40 mm, 20 mm, 10 mm, 4.75 mm for coarse aggregate, and 10 mm, 4.75 mm, 2.36 mm, 1.18 mm, 600 microns, 300 microns and 150 microns for fine aggregate. The sieves are arranged in such an order that the square openings are half for each succeeding smaller size.

The curve showing the cumulative percentages of the material passing the sieves represented on the ordinate with the sieve openings to the logarithmic scale represented on the abscissa is termed the *grading curve*. The grading curve indicates whether the grading of a given sample conforms to that specified, or is too coarse or too fine, or deficient in a particular size.

(i) In case the actual grading curve is lower than the specified grading curve, the aggregate is coarser and *segregation* of mix might take place.

(ii) In case the actual grading curve lies well above the specified curve, the aggregate is finer and more water will be required, thus increasing the quantity of cement for a constant *water-cement ratio*. Therefore, this is uneconomical.

(iii) If the actual grading curve is steeper than the specified, it indicates an excess of middle-size particles and leads to harsh mix.

(iv) If the actual grading curve is flatter than the specified grading curve, the aggregate will be deficient in middle-size particles.

The *grading of the aggregate* affects the *workability* which, in turn, controls the water and cement requirements, *segregation,* and influences the *placing* and *finishing* of concrete. These factors represent the important characteristics of fresh concrete and affect its properties in the hardened state. The main factors governing the desired aggregate grading are: the *surface area of aggregate,* the *relative volume occupied by the aggregate,* the *workability of the mix,* and the *tendency to segregate.*

A number of methods have been proposed for arriving at an ideal grading that would be applicable for all aggregates. None of these has been universally successful because of economic considerations, effect of *particle shape* and *texture* of the aggregate, and differences in cement from different sources. Grading specifications have been developed, however, which on the average will produce a concrete of satisfactory properties from materials available in a particular area. The *surface area* is affected by the maximum size of aggregates. If a sphere of diameter d is taken as representative of the shape of aggregate, the ratio of surface area to the volume is $6/d$. This ratio of surface of the particles to their volume is called *specific surface.* The surface area will vary with the shape but is inversely proportional to the particle size. The smaller the size of aggregate, the greater is the surface area per unit mass or unit volume. The aim must, therefore, be to have as large a maximum aggregate size as possible and to grade it down in such a way that the voids in the *coarse aggregate* are filled with the minimum amount of *fine aggregate.* This arrangement, however, cannot be carried too far as an aggregate graded in this way would be too harsh and a slight excess of fines is necessary to prevent this. The greatest contribution to this total surface area is made by the smaller size aggregate and, therefore particular attention should be paid to the proportion and grading of fine aggregate. The mortar consisting of fine aggregate and cement should be slightly in excess of that just required to fill the voids in the coarse aggregate. Too coarse a fine aggregate results in *harshness, bleeding* and *segregation* and too fine an aggregate requires too large a *water-cement ratio* for adequate *workability.* The *surface area* of aggregate also influence the amount of mixing water and cement required. Generally, the water-cement ratio is fixed from strength consideration. However, the amount of cement paste should be sufficient to cover the surface of all the particles for proper workability and bond. The *drying shrinkage* is less with a smaller amount of mixing water, and the temperature rise due to *hydration* and, therefore, *cracking* on subsequent cooling is less with the smaller proportion of cement in the mix.

Economical and uniform concrete cannot be produced with pit-run or crusher-run aggregate, and it is necessary that the aggregate be separated into its component sizes so that it can be combined in the concrete mix within the limits of variation permitted by the specifications.

The *grading* of fine aggregate has a much greater effect on workability of conerete than does the grading of the coarse aggregate. Experience has shown that usually very coarse sand or very fine sand is unsatisfactory for concrete. Fine grading conforming to the specifications laid by IS: 383–1963 shall be satisfactory for most concretes.

In the case of *graded aggregate,* the grading and the overall *specific surface area* are related to one another, although there can be many grading curves corresponding to same specific surface. If the grading extends to a larger maximum particle size, the overall specific surface is reduced decreasing the water requirements, but the relation is not linear. As discussed earlier the maximum size of the aggregate that can be used for a certain job depends upon the size of the member and of the reinforcement used. For reinforced concrete work, the aggregate having a maximum size of 20 mm is generally considered satisfactory.

3.10.1 Gap-graded Aggregate

Gap-grading is defined as a grading in which one or more intermediate-size fractions are absent. The term 'continuously graded' is used to distinguish the conventional grading from gap-grading. On a grading curve, gap-grading is represented by a horizontal line over the range of size omitted. Some of the important features of gap-graded aggregate are as follows.

(i) For the given *aggregate-cement* and *water-cement ratios* the highest workability is obtained with lower sand content in the case of gap-graded aggregate rather than when continuously graded aggregate is used.
(ii) In the more workable range of mixes, gap-graded aggregates show a greater tendency to segregation. Hence gap-grading is recommended mainly for mixes of relatively low workability that are to be compacted by vibration.
(iii) The gap-graded aggregate does not affect compressive or tensile strengths.

3.10.2 Grading Limits

There is no universal ideal grading curve. The concrete for satisfactory performance can be obtained with various gradings of aggregate. However, IS: 383–1970 has recommended certain limits within which the grading must lie to produce satisfactory concrete, subject to the fulfilment of certain desirable properties of aggregate, such as *shape, surface texture, type of aggregate* and amount of *flaky* and *elongated materials.* The grading of coarse aggregate may be varied through wider limits than that for fine aggregate since it does not affect much the workability, uniformity and finishing qualities. The grading limits for coarse aggregate are given in Table 3.3.

It is difficult to control the grading of fine aggregate. If it is done on large scale, it can be effected by combining two or more different kinds of sand from different sources.

The sands are generally divided into different zones according to the percentage passing the IS: 600 micron sieve. IS: 383-1970 classifies the sand into four zones, I, II, III and IV so that the range of percentage passing the 600 micron sieve in each zone does not overlap. The *grading limits* of four zones are given in Table 3.4. From grading zone I to IV, the fine aggregate becomes progressively finer, and the ratio of fine to coarse aggregate should be progressively reduced as suggested in Table 3.6.

An aggregate is considered as belonging to the zone in which its percentage passing the 600 micron sieve falls and it is allowed to fall outside the limits fixed for other sieves by not more than a total of 5 per cent. For crushed stone sands the permissible limit on IS: 150 micron sieve is increased by 20 per cent. However, this does not affect the 5 per cent allowance permitted above as applied to other sieve sizes.

It has been noticed that if the *specific surface of aggregate* is kept constant, wide difference in grading does not affect the *workability* appreciably. However, the variation of fine to coarse aggregate ratio to keep total surface constant cannot be pushed too far and if a very fine sand is being used this process may result in the mix being *under-sanded* with a serious risk of *segregation*, specially in cases of lean mixes.

In the case of *all-in-aggregate*, the necessary adjustments may be made in the grading by the addition of a single-size aggregate without separating into fine and coarse aggregates.

The grading limits for various types of aggregates are given in Tables 3.2 to 3.5.

TABLE 3.2 Grading limits for single size coarse aggregate of nominal size

IS sieve designation	Percentage passing for single sized aggregate of nominal size, mm				
	40	20	16	12.5	10
80 mm	—	—	—	—	—
40 mm	85–100	100	—	—	—
20 mm	0–20	85–100	100	—	—
16 mm	—	—	85–100	100	—
12.5 mm	—	—	—	85–100	100
10 mm	0–5	0–20	0–30	0–45	85–100
4.75 mm	—	0–5	0–5	0–10	0–20
2.36 mm	—	—	—	—	0–5

TABLE 3.3 Grading limits for coarse aggregates

IS sieve designation	Percentage passing for graded aggregate of nominial size, mm			
	40	20	16	12.5
80	100	—	—	—
40	95–100	100	—	—
20	30–70	95–100	100	100
16	—	—	95–100	—
12.5	—	—	—	90–100
10.0	10–35	25–55	30–70	40–85
4.75	0–5	0–10	0–10	0–10
2.36	—	—	—	—

TABLE 3.4 Grading limits for fine aggregates

IS sieve designation	Percentage passing by weight			
	Grading (Zone I)	Grading (Zone II)	Grading (Zone III)	Grading (Zone IV)
10 mm	100	100	100	100
4.75 mm	90–100	90–100	90–100	95–100
2.36 mm	60–95	75–100	85–100	95–100
1.18 mm	30–70	55–90	75–100	90–100
600 µm	15–34	35–59	60–79	80–100
300 µm	5–20	8–30	12–40	15–50
150 µm	0–10	0–10	0–10	0–15
Fineness modulus	4.0–2.71	3.37–2.10	2.78–1.71	2.25–1.35

TABLE 3.5 Grading limits for all-in-aggregate

IS sieve designation	Percentage by weight passing for all-in-aggregate of	
	40 mm nominal size	20 mm nominal size
80 mm	100	—
40 mm	95–100	100
20 mm	45–75	95–100
4.75 mm	25–45	30–50
600 µm	8–30	10–35
150 µm	0–5	0–6

TABLE 3.6 Suggested proportion of fine to coarse aggregate for different size of aggregates

Nominal size of graded coarse aggregate, mm	Fine aggregate: Coarse aggregate, for the sand of zone			
	Zone I	Zone II	Zone III	Zone IV
10	1 : 1	1 : 1.5	1 : 2	1 : 3
20	1 : 1.5	1 : 2	1 : 3	1 : 3.5
40	1 : 2	1 : 4	1 : 3.5	—

The properties of aggregates are summarized in the Table 3.7

TABLE 3.7 Summary of aggregate properties

Characteristics	Significance	Test	Specifications
1. Particle shape and texture	Affects workability of fresh concrete	Visual inspection, flakiness and elongation test IS: 2386 (Part I)-1963	Limits on flaky or enlogated particles, flakiness index not greater than 30 to 40 per cent is desirable
2. Resistance to crushing	In high strength concrete, aggregate low in crushing value will not give high strength even though cement strength is higher	Aggregate impact value test IS: 2386 (Part IV)-1963	30% impact value for payment; 45% for other applications
3. Specific gravity	Required in mix design calculation; unit weight of concrete; yield of concrete	Specific gravity determination IS: 2386 (Part III)-1963)	—
4. Bulk density	Rodded bulk density is useful as a check on the uniformity of aggregate grading; loose bulk density is useful to convert masses into bulk volumes on site or vice versa	Test for bulk density IS: 2386 (Part III)-1963	—
5. Absorption and surface moisture	Affects the mix proportions; to control water content to maintain water-cement ratio constant	Test for absorption and surface moisture IS: 2386 (Part III)-1963	
6. Deleterious substance	Organic impurities and coatings interfere with hydration of cement	Test of impurities IS: 2386 (Part III)-1963	Limits on impurities have been prescribed in IS: 383–1970
7. Grading	Economizes cement content and improves workability	IS: 2386 (Part I)-1963	Grading limits for coarse, fine and all-in-aggregate are laid down in IS: 383–1970
8. Chemical stability	Significant for strength and durability of all types of structures specially subjected to chemical attack	IS: 2386 (Part VII)-1963	—
9. Resistance to freezing and thawing	Significant to cold countries where frost action deteriorates concrete due to alternate freezing and thawing	Test for soundness IS: 2386 (Part VII)-1963	—

4

Concrete Making Materials-III: Water

4.1 INTRODUCTION

Water is the most important and least expensive ingredient of concrete. A part of mixing water is utilized in the hydration of cement to form the binding matrix in which the inert aggregates are held in suspension until the matrix has hardened. The remaining water serves as a lubricant between the fine and coarse aggregates and makes concrete workable, i.e. readily placeable in forms.

Generally, cement requires about 3/10 of its weight of water for hydration. Hence the minimum *water-cement ratio* required is 0.35. But the concrete containing water in this proportion will be very harsh and difficult to place. Additional water is required to lubricate the mix, which makes the concrete workable. This additional water must be kept to the minimum, since too much water reduces the strength of concrete. The water-cement ratio is influenced by the grade of concrete, nature and type of aggregates, the workability and durability, etc.

If too much water is added to concrete, the excess water along with cement comes to the surface by *capillary action* and this cement-water mixture forms a scum or thin layer of chalky material known as *laitance*. This laitance prevents bond formation between the successive layers of concrete and forms a plane of weakness. The excess water may also leak through the joints of the formwork and make the concrete *honeycombed*. As a rule, the smaller the percentage of water, the stronger is the concrete subject to the condition that the required workability be allowed for.

4.2 QUALITY OF MIXING WATER

The water used for the mixing and curing of concrete should be free from injurious amounts of deleterious materials. The unwanted situations, leading to the distress of concrete, have been found to be a result of, among others, the mixing and curing water being not of appropriate quality. Potable water is generally considered satisfactory for mixing concrete. In the case of doubt about the suitability of water, particularly in remote areas or where water is derived from sources not normally utilized for domestic purposes, water should be tested.

4.2.1 Effect of Impurities in Water on Properties of Concrete

The strength and durability of concrete is reduced due to the presence of impurities in the mixing water. The effects are expressed mainly in terms of difference in the setting times of portland cement mixes containing impure mixing water as compared to clean fresh water, and concrete strengths compared with those of control specimens prepared with good water. A difference in 28 days compressive strength up to 10 per cent of control test is generally considered to be a satisfactory measure of the quality of mixing water. IS: 456–1978 prescribes a difference in initial setting time of ±30 minutes with initial setting not less than 30 minutes. The effluents from sewerage works, gas works, and from paint, textile, sugar and fertilizer industry are harmful to concrete. The tests show that water containing excessive amounts of dissolved salts reduces compressive strength by 10 to 30 per cent of that obtained using fresh water. In addition, water containing large quantities of chlorides, e.g. sea water, tends to cause persistent dampness and surface *efflorescence*. Sea water increases the corrosion of the reinforcing steel. The danger is more in tropical regions, particularly with lean mixes.

The adverse effects on compressive strength of concrete due to various dissolved salts are given in Table 4.1.

TABLE 4.1 Effects of dissolved salts in water on compressive strength

Percentage of salt in water	*Percentage reduction in compressive strength*
0.5 SO_4	4
1.0 SO_4	10
5.0 NaCl	30
CO_2	20

The effect of various impurities on the properties of concrete are summarized below.

(i) Suspended Particles

The presence of suspended particles of clay and silt in the mixing water up to 0.02 per cent by weight of water does not affect the properties of concrete. Even higher percentage can be tolerated so far as strength is concerned, but other properties of concrete are affected. IS: 456–1978 allows 2000 mg/litre of suspended matter. The muddy water should, however, remain in settling basins before use.

(ii) Miscellaneous Inorganic Salts

The presence of salts of manganese, tin, zinc, copper and lead in water causes reduction in the strength of concrete. The zinc chlorides retard the set of concrete to such an extent that no strength tests are possible at 2 and 3 days. The effect of lead nitrate is completely destructive.

Some salts like sodium iodate, sodium phosphate, sodium arsenate and sodium borate reduce the initial strength of concrete to a very low degree.

The carbonates of sodium and potassium may cause extremely rapid setting and, in large concentrations, reduce the concrete strength. On the other hand, the presence of calcium chloride accelerates setting and hardening. The quantity of calcium chloride is restricted to $1\frac{1}{2}$ per cent by weight of cement.

(iii) Salts in Sea Water

The sea water generally contains 3.5 per cent of dissolved salts, about 78 per cent of which is sodium chloride, and 15 per cent chloride and sulphate of magnesium. The salts present in sea water reduce the ultimate strength of concrete. The reduction in strength of concrete may be of the order of 10 to 20 per cent. However, the major concern is the risk of corrosion of reinforcing steel due to chlorides. In general, the risk of corrosion of steel is more when the reinforced concrete member is exposed to air than when continuously submerged under water, including sea water. The presence of chlorides in water is responsible for *efflorescence*. It should not therefore be used where surface finish is of importance. The sea water should not be used in reinforced concrete and prestressed concrete constructions. Under unavoidable circumstances, it may be used for plain concrete when it is constantly submerged in water.

(iv) Acids and Alkalies

The industrial waste water containing acids or alkalies is usually unsuitable for concrete construction. Regarding acidity the water having pH value higher than 6 can be used. However, the pH value may not be a satisfactory measure of the amount of acid. The effect of acidity in water is best gauged on the basis of total acidity, the extent of which should satisfy the following requirement.

The amount of 0.1 normal NaOH required to neutralize 100 ml sample of water using phenolphthalein as indicator should not be more than 1 ml. This acidity is equivalent to 49 ppm of H_2SO_4 or 36 ppm of HCl.

The limit of alkalinity is guided by the requirement that *the amount of 0.1 normal HCl required to neutralize 100 ml of sample should be less than 5 ml.* This alkalinity is equivalent to 265, 420 and 685 ppm of carbonates (as Na_2CO_3), bicarbonates (as $NaHCO_3$) and the sum of the two, respectively.

(v) Algae

Algae may be present in mixing water or on the surface of aggregate particles. It combines with the cement and reduces the bond between aggregates and cement paste. The water containing algae has the effect of entraining large quantities of air in the concrete and thus lowering the strength of concrete.

(vi) Sugar

If the amount of sugar present in the mixing water is less than 0.05 per cent by weight of water there is no adverse effect on the strengh of concrete. Small amounts of sugar up to 0.15 per cent by weight of cement retard the setting of cement and the early strengths may be reduced whereas the 28-day strength may be improved. When the quantity of sugar is increased to 0.20 per cent by weight of cement, setting is accelerated. When quantity is further increased, rapid setting may result and 28-day strength is reduced.

(vii) Oil Contamination

Mineral oils not mixed with animal or vegetable oils have no adverse effect on the strength of concrete. If the concentration of mineral oil is up to 2 per cent by weight of cement, a significant increase in strength has been noticed. For a percentage of mineral oil more than 8 per cent, the strength is slightly reduced. The vegetable oils have detrimental effect on the strength of concrete, particularly at later ages.

Limit of impurities in water. The impurities in water should be restricted as per the Table 4.2. The pH value of water suitable for concrete construction shall generally be between 6 and 8. The water which is fit for drinking purposes will be fit for concrete construction.

TABLE 4.2 Limits of permissible impurities

Type of impurities	Permissible percentage of solids by weight of water
Organic	0.02
Inorganic	0.30
Sulphates	0.05
Alkali Chlorides	
(a) Plain concrete	0.20
(b) Reinforced concrete	0.10

4.3 CURING WATER

The water which is satisfactory for mixing concrete can also be used for curing it but should not produce any objectionable stain or unsightly deposit on the surface. Iron and organic matter in the water are chiefly responsible for staining or discoloration and especially when concrete is subjected to prolonged wetting, even a very low concentration of these can cause staining. According to IS: 456–1978, the presence of tannic acid or iron compounds in the curing water is objectionable.

5

Admixtures

5.1 INTRODUCTION

Admixtures are the materials other than the basic ingredients of concrete—cement, water, and aggregates— added to the concrete mix immediately before or during mixing to modify one or more of the specific properties of concrete in the fresh or hardened state. The use of admixture should offer an improvement not economically attainable by adjusting the proportions of cement and aggregates, and should not adversely affect any property of the concrete. Admixtures are no substitute for good concreting practice. An admixture should be employed only after an appropriate evaluation of its effects on the particular concrete under the conditions in which the concrete is intended to be used. It is often necessary to conduct tests on the representative samples of the materials for a particular job under simulated job conditions in order to obtain reliable information on the properties of concrete containing admixtures.

The admixtures ranging from addition of chemicals to waste materials have been used to modify certain properties of concrete. The properties commonly modified are that *rate of hydration* or *setting time, workability, dispersion* and *air-entrainment*. The admixture is generally added in a relatively small quantity. A *degree of control* must be exercised to ensure proper quantity of the admixture, as an excess quantity may be detrimental to the properties of concerte. In using any admixture, careful attention should be given to the instructions provided by the manufacturer of the product.

5.2 FUNCTIONS OF ADMIXTURES

Some of the important purposes for which the *admixtures* could be used are:
  (i) to accelerate the initial set of concrete, i.e. to speed up the rate of development of strength at early ages,
 (ii) to retard the initial set,
(iii) to increase the strength of concrete,
 (iv) to improve the workability,
  (v) to reduce the heat of evolution,

(vi) to increase the durability of concrete, i.e. its resistance to special conditions of exposure, like repeated freezing and thawing cycles,
(vii) to control the alkali-aggregate expansion,
(viii) to decrease the capillary flow of water through concrete and to increase its impermeability to liquids,
(ix) to improve the penetration and pumpability of concrete,
(x) to reduce the segregation in grout mixtures,
(xi) to increase the bond between old and new concrete surfaces,
(xii) to increase the bond of concrete to the steel reinforcement,
(xiii) to inhibit the corrosion of concrete,
(xiv) to increase the resistance to chemical attack,
(xv) to produce cellular concrete,
(xvi) to produce coloured concrete or mortar for coloured surfaces,
(xvii) to produce concrete of fungicidal, germicidal and insecticidal properties,
(xviii) to produce nonskid surfaces, and
(xix) to decrease the weight of concrete per cubic metre.

5.3 CLASSIFICATION OF ADMIXTURES

According to the characteristic effects produced by them, the admixtures may be broadly classifed as:
(i) accelerating admixtures,
(ii) retarding and water reducing admixtures,
(iii) grouting admixtures,
(iv) air-entraining admixtures,
(v) air-detraining admixtures,
(vi) gas forming admixtures,
(vii) expansion-producing admixtures,
(viii) waterproofing and permeability reducing admixtures,
(ix) corrosion inhibiting admixtures,
(x) fungicidal, germicidal and insecticidal admixtures,
(xi) bonding admixtures,
(xii) pozzolanic admixtures,
(xiii) colouring admixtures or pigments,
(xiv) concrete hardening admixtures, and
(xv) super plasticizers.

5.3.1 Accelerating Admixture

An admixture used to speed up the initial set of concrete is called an accelerator. These are added to concrete either: (a) to increase the *rate of hydration of cement*, and hence to increase the rate of *development of strength*, or (b) to shorten the *setting time*. An increase in the rate of early strength development may help in: (a) earlier removal of forms, (b) reduction of required period of curing, and (c) earlier placement of structure in service. Accelerating admixtures are also used when the concrete is to be placed at *low temperatures*. The benefits of reduced time of setting may include: (a) early finishing of surface, (b) reduction of pressure on

50 Concrete Technology

forms or of period of time during which the forms are subjected to hydraulic pressure; and (c) more effective plugging of leaks against hydraulic pressure. With proper protection these admixtures will partly compensate for the retardation of strength-development due to low temperatures.

The most widely used accelerator is *calcium chloride* ($CaCl_2$). It is available as flakes (77 per cent $CaCl_2$) or in the fused form (92 per cent $CaCl_2$). Calcium chloride can generally be used in amounts up to 2 per cent by mass of cement, but IS: 7861 (Part II)-1981 recommends a maximum of 1.5 per cent of $CaCl_2$ for plain and reinforced concrete works in cold weather conditions. However, $CaCl_2$ or admixtures containing soluble chlorides are not permitted to be used in prestressed concrete due to the possibility of stress corrosion. The benefits of the use of calcium chloride are usually more pronounced when it is employed for concreting at temperatures below 25°C. Calcium chloride should not be used in concrete which will be subjected to alkali-aggregate reaction or exposed to soils or water containing sulphates, in order to avoid lowering of the resistance of concrete to sulphate attack.

The use of 2 per cent calcium chloride by mass of cement can reduce the *setting time* by one-third and raise the one to seven day *compressive strength* by 3 to 8 MPa. An increase in *flexural strength* of 40 to 80 per cent of one day and up to 12 per cent at 28 days is obtained. The selection of the optimum amount should be based on the type of cement, temperature of concrete and the ambient temperature. Large doses of $CaCl_2$ result in *flash set* of concrete and also in increased shrinkage. The effect of $CaCl_2$ on the compressive strength of concrete is shown in Fig. 5.1.

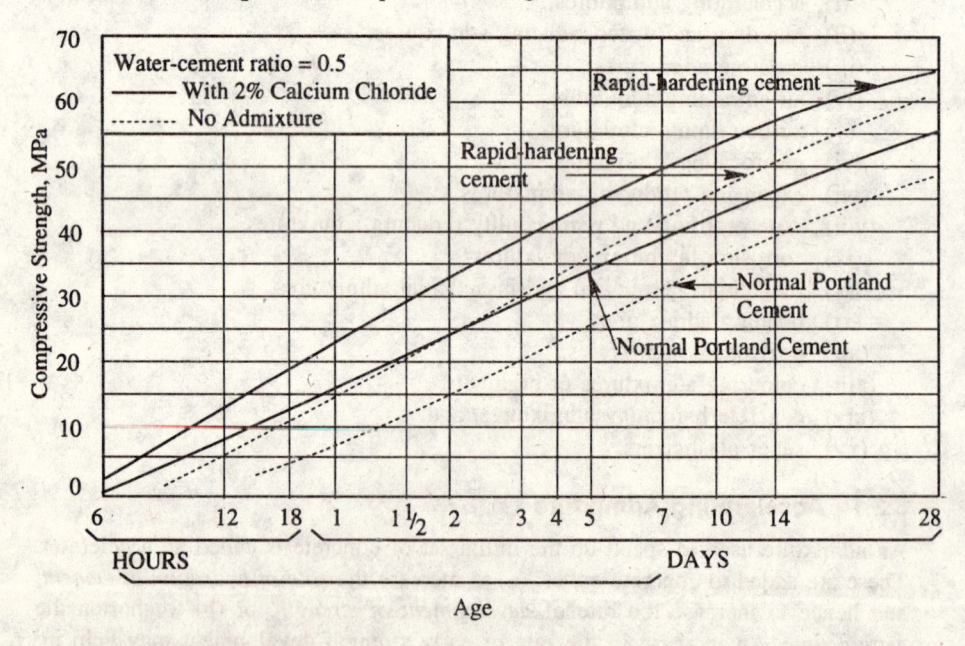

Fig. 5.1 Effect of admixtures on the compressive strength of concrete

Some of the accelerators containing fluosilicates and trietholamine are capable of reducing the period during which concrete remains plastic to less than 10 minutes.

Some set accelerators are produced under the trade name 'Quickset' which when added to neat cement produce setting in a matter of seconds. This makes this type of accelerator valuable for making cement plugs to stop pressure leaks.

The other less commonly used accelerators consist of NaCl, Na_2SO_4, NaOH, Na_2CO_3, K_2SO_4 and KOH. In contrast to $CaSO_4$, the effect of Na_2SO_4 and K_2SO_4 is the acceleration of hydration of cement. Rapid hydration can be achieved in the first two hours by the addition of NaOH or KOH. It has been noticed that two per cent of calcium chloride has the same effect on the acceleration of hydration as a rise in temperature of about 11°C.

5.3.2 Retarding and Water-reducing Admixtures

The *retarding admixtures* delay or prolong the setting of the cement paste in concrete. They are used primarily to offset the accelerating and damaging effect of high temperature and to keep concrete workable during the entire placing period which should be sufficiently long so that the succeeding lifts can be placed without the development of cold joints or discontinuities in the structural unit. The speeding up of hydration means that some of the water usually available to provide workability is used by the cement. Therefore, more water is required to maintain the slump at an appropriate level, which in turn, means lower concrete strength. High temperatures, low humidity, and wind cause rapid evaporation of water from the mix during summer. This drying of concrete leads to the *cracking* and *crazing* of the surface.

A retarding admixture holds back the hydration process, leaving more water for *workability* and allowing concrete to be finished and protected before drying out. Some of the retarding admixtures also reduce the water requirement of the mixture making further reductions possible in the *water-cement ratio*. These admixtures increase the *compressive strength* and also *durability* under freezing and thawing. The retarders are also helpful for the concrete that has to be hauled long distances in transit mix trucks, to ensure that it remains in plastic and placeable condition. The materials generally used as *water-reducing* and *set controlling admixtures* belong to the following groups:

   (i) lignosulphonic acids and their salts,
   (ii) modifications and derivatives of lignosulphonic acids and their salts (e.g. Ca, Na, NH_4 salts),
   (iii) hydroxylated carboxylic acids and their salts, and
   (iv) modifications and derivatives of hydroxylated carboxylic acids and their salts.

The lignosulphonates and carboxylic acid derivatives are water reducing and set retarding admixtures, and they are known to reduce *setting times* by two to four hours and water requirement by 5 to 10 per cent. The *compressive strength* at 2 or 3 days are usually equal to or higher than those of corresponding concrete without the admixture and the strength at 28 days or latter may be 10 to 20 per cent higher. The modifications and derivatives of lignosulphonic acids and hydroxylated carboxylic acids are the *water-reducing admixtures*. They may be used with accelerating or retarding admixtures. Calcium sulphate (gypsum), sugar and carbohydrates also retard the set. The carbohydrate derivatives and calcium lignosulphonate are used in fractions of a per cent by mass of the cement.

The *water-reducing admixtures* are mainly used to improve the quality of concrete, obtain specified strength at a lower cement content, or increase the slump of a given concrete mix without increasing the water content. The specific effect of water-reducing and set-controlling admixtures vary with the type of cement, water-cement ratio, mixing temperature, ambient temperature and other job conditions, and therefore, it is generally recommended that the retarder used be adjusted to meet the job conditions.

5.3.3 Grouting Admixtures

Some retarders are especially useful in cement grout slurries, particularly where the grouting is prolonged, or in the cases where the grout must be pumped for a considerable distance, or where hot water is encountered underground. Cement grouts containing pozzolanic materials are often used in cementing oil wells. Admixtures are also used to prevent the rapid loss of water from cement paste to the surrounding formation. Some of the grouting admixtures are gels, clays, pregelatinized starch and methyl cellulose.

5.3.4 Air-entraining Admixtures

These admixtures help incorporate a controlled amount of air in the form of minute bubbles in concrete during mixing, without significantly altering the setting or the rate of hardening characteristics of concrete. It is generally recognized that a proper amount of entrained air results in improved workability, easier placing, increased durability, better resistance to frost action and reduction in bleeding and water gain. The entrained air bubbles (approximately 0.01 to 0.25 mm diameter) reduce the capillary forces in concrete. The capillaries are interrupted by relatively large air voids in *air-entrained concrete*. The air voids are not interconnected. Air-entrainment, while improving durability and plasticity, may have an adverse effect on the strength of concrete. The decrease in strength is usually proportional to the amount of entrained air. Within the normal range of air content, the maximum reduction in compressive strength and flexural strength rarely exceeds 15 and 10 per cent, respectively. The beneficial amount of entrained air depends upon the maximum size of coarse aggregate used in the mix. The optimum percentage of air giving a balance between compressive strength and durability is given in Table 5.1. Thus air-entrainment alters properties of both the freshly mixed and hardened concrete. Air-entrained concrete is considerably more plastic and workable than non-air-entrained concrete. The durability of hardened concrete is improved by increased uniformity, decreased absorption and permeability, and by elimination of planes of weaknesses at the top of lifts. The *air-entraining agents* also find very useful application in making *cellular concrete* and *lightweight aggregate concrete*.

As mentioned above, air-entrained concrete contains microscopic bubbles of air formed with the aid of chemicals called surface active agents. These materials have the property of reducing surface tension of water enabling the water to hold air when agitated, resulting in a foam. The structure of air-entrained concrete is shown in Fig. 5.2. A satisfactory air-entraining agent must not react chemically with cement. It must be able to produce air bubbles of a definite size which must not break too rapidly. These entrained air bubbles constitute a definite part of fine aggregate and

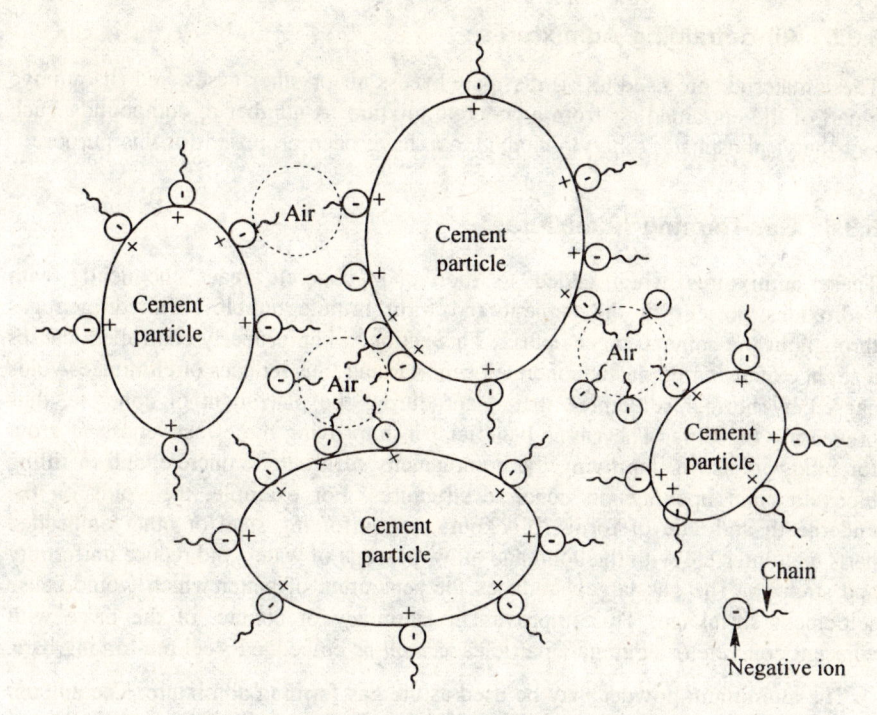

Fig. 5.2 Structure of air-entrained of concrete

lubricate the concrete. The bubbles act like ball-bearings to help increase the *mobility* of concrete by reducing friction between the particles.

The compounds used for air-entrainment are a number of natural wood resins, various sulphonated compounds, and some fats and oils. The air-entraining agents made from modified salts of a sulphonated hydro-carbon tend to plasticize a concrete mix. These are particularly useful where aggregates which tend to produce harsh concrete or natural sand deficient in fines are used in producing concrete. Another type of air-entraining admixture made from neutralized vinsol resin is used in mass concrete and concrete used in highway pavements.

TABLE 5.1 Optimum air content of concrete

Maximum size of aggregate, mm	Air-content, per cent by volume	
	Naturally entrained	Optimal total
Sand-cement mortar	4.0	14 ± 2
10	3.0	8 ± 1.5
12.5	2.5	7.5 ± 1.5
20	2.0	7.0 ± 1.5
25	1.5	6 ± 1.5
40	1.0	4.5 ± 1.5
50	0.5	4.0 ± 1.0
70	0.3	3.5 ± 1.0
150	0.2	3.0 ± 0.5

5.3.5 Air-detraining Admixtures

These materials are used to: (i) dissipate excess air or other gases, and (ii) remove a part of the entrained air from a concrete mixture. A number of compounds, such as tributyl phosphate, dibutylphthalate, etc. have been proposed for this purpose.

5.3.6 Gas-forming Admixtures

These admixtures when added to mortar or concrete react chemically with hydroxides present in the cement and form minute bubbles of hydrogen gas throughout the cement-water matrix. This action, when properly controlled, causes a slight expansion in plastic concrete or mortar and thus reduces or eliminates voids caused by normal settlement that occur during the placement of concrete, thus preventing bleeding. The gas is beneficial in improving the effectiveness of grout for filling joints, in improving the homogeneity of grouted concrete, and in filling blockouts and openings in concrete structures. For example, the voids on the underneath and side of forms, blockouts on reinforcing steel or other embedded parts may interfere with the bond and allow passage of water, and reduce uniformity and strength. The gas largely reduces the separation of water which would cause settlement shrinkage. This improves the intimacy of contact of the paste with adjacent concrete or aggregate particles as well as embedded steel reinforcing bars.

The aluminium powder may be used as the gas forming admixture. The amount of powder added usually varies from 0.005 to 0.02 per cent by mass of cement. Zinc and magnesium powders are also used for this purpose while hydrogen peroxide and bleaching powder can be used in combination to produce oxygen instead of hydrogen bubbles in concrete.

The effect on the strength of the concrete depends to a large extent on the restraint offered to expansion. With complete restraint imposed, the strength is not affected appreciably with very small amounts of aluminium powder. Larger amounts of powder increase the expansion appreciably resulting in a lightweight, low-strength concrete.

5.3.7 Expansion-producing Admixtures

The *expansion-producing admixtures* either expand themselves or react with other constituents of concrete with resulting expansion. This expansion may be of about the same magnitude as the *drying shrinkage* at later ages or may be little greater. This concept has been used in the development of *non-shrinking cement* wherein the expansion-producing compound is mixed with cement in appropriate proportion to get the desired expansion or shrinkage compensation.

A number of expansion agents have been reported, such as granulated iron and chemicals, and anhydrous sulphoaluminate, etc. Granulated iron and chemicals promote oxidation of iron resulting in the formation of iron oxide which occupies an increased solid volume. These admixtures are employed in laying heavy machine foundations, patching, production of *shrinkage-compensating concrete* which is free from shrinkage cracks, and production of self-stressing and pre-stressed concrete.

5.3.8 Waterproofing and Permeability Reducing Admixtures

Water under pressure and in contact with one surface of concrete, can be forced through channels between the two surfaces. The water passing in this manner is a measure of the *permeability of concrete*. Water can also pass through concrete by the action of capillary forces. The materials used to reduce the water flow by the first method are termed *permeability reducer*, whereas the materials used to reduce second type of flow are more properly called *dampproofers*.

A concrete having proper mix design, low *water-cement ratio* and sound aggregate will be impervious and need no additives. However, the resistance of concrete to the penetration of moisture can be improved by adding chemically active water-repelling agents like soda and potash soaps to which are sometimes added lime or calcium chloride. These admixtures prevent the water penetration of dry concrete, or stop the passage of water through unsaturated concrete.

As explained earlier, the *air-entraining agents* increase the plasticity of concrete and therefore help place concrete more uniformly. They also reduce *bleeding* by holding the water in films around the air bubbles, thus reducing the permeability. The small disconnected voids produced by air-entrainment also break up the capillaries in concrete and, therefore, offer a barrier to the passage of water by capillary action. For these reasons air-entraining agents may also be considered as permeability-reducing and dampproofing agents.

Another type of concrete waterproofer consists of a film applied to the surface, preferably the one adjacent to the water source. The asphaltic products, thick viscous liquids, form an impervious coating over the surface. Sodium silicate compounds enter the surface pores and form a gel which prevent water from entering the concrete.

5.3.9 Corrosion Inhibiting Admixtures

Compounds, such as sodium benzoate, sodium nitrate, etc. can be used to prevent the corrosion of reinforcement. A 2 per cent benzoate solution in mixing water may be used to prevent corrosion of reinforcement. Sodium nitrate has been found to be effective in preventing corrosion of steel in concrete containing calcium chloride.

5.3.10 Fungicidal, Germicidal and Insecticidal Admixtures

Certain materials like polyhalogenated phenols, dieledren emulsions and copper compounds when added as admixtures impart fungicidal, germicidal or insecticidal properties to the hardened cement pastes, mortars or concretes.

5.3.11 Bonding Admixtures

When fresh concrete is placed over a concrete surface already set and at least partially cured, the fresh concrete shrinks while setting which makes the new concrete pull away from the old surface. Due to this reason the old surfaces are usually prepared so that the aggregates are exposed and clean which makes the cement paste in the freshly placed concrete bond the aggregate in the same way as it bonds the aggregates in the new mix. A cement paste slurry is often applied to the prepared old surface immediately prior to pouring new concrete to increase the amount of paste available at the surface for bonding purposes. In the situations where such a treatment cannot be applied, the bonding admixtures can be used to join two surfaces. These admixtures increase the bond strength between the old and new concrete.

There are two tupes of bonding admixtures in common use. In the first type the bonding is accomplished by a metallic aggregate and in the other synthetic latex emulsions are used. The metallic aggregate type of admixture consists of fine cast-iron particles to which is added a chemical that causes them to oxidize rapidly when mixed with portland cement and water. The rapid oxidation of the iron particles in the cement slurry applied over the old concrete surface results in the expansion of iron particles. The tiny fingers that thrust out into both the old and the new concrete bind them together. This admixture can also be used as waterproofer by applying additional coats. Successive coats build up a thin but dense watertight film over the surface.

There are a number of types of synthetic latex bonding admixtures, which essentially consist of highly polymerized synthetic liquid resins dispersed in water. The commonly used *bonding admixtures* are made from natural rubber, synthetic rubber or any of a large number of organic polymers or copolymers. The polymers include polyvinyl chloride, polyvinyl acetate, acrylics and butadienestyrene copolymers. These admixtures are *water-emulsions* which are generally added to the mixture in proportions equivalent to 5 to 20 per cent by mass of cement depending upon the actual bonding requirements. Since these admixtures are emulsions, the bonding agent must lose water for its adhesive ingredients to set. When a bonding agent is sprayed on a concrete surface, the pores in the concrete absorb the water and allow the resin particles to coalesce and bond. When a bonding agent is mixed with cement paste or a mortar, the water is used in the hydration of cement and the resin is left to bind both the surfaces.

5.3.12 Pozzolana Admixtures

A *pozzolana* is a finely ground siliceous material which, as such, does not possess cementitious property in itself, but reacts in the presence of water with lime (calcium hydroxide) at normal temperature to form compounds of low solubility having cementitious properties. The action is termed *pozzolanic action*.

A number of natural materials, such as diatomaceous earth, opaline cherts and shales, tuffs and pumicites, and some artificial materials, such as fly ash, granulated blast furnace slag, etc. are used as pozzolanas.

The pozzolanas can be used in combination with or for the partial replacement of portland cement. The pozzolanic materials when used as replacement are generally substituted for 10 to 35 per cent of cement. This substitution produces concrete that is more permeable but much more resistant to the action of salt, sulphate or acid water. Strength gain is usually slower than for the normal concrete.

Pozzolanas when added to concrete mixes, rather than substituted for a part of the cement, improve workability, impermeability, and resistance to chemical attack. The overall effect depends on the aggregates used in the concrete. The aggregates deficient in fine material give the best results.

Some pozzolanas reduce the expansion caused by the *alkali-aggregate-reaction* in concrete. The expansion is caused by the interaction of alkali in portland cement and certain siliceous types of aggregates. Excessive expansion causes *pattern cracking* of concrete. This expansion can usually be controlled by using the amounts of pozzolana ranging from 2 to 35 per cent by mass of cement depending upon the type of aggregate and alkali content of cement.

Thus some of the advantages obtained through their use are:
(i) improved workability with lesser amount of water,
(ii) reduction in heat of hydration,
(iii) increased resistance to attack from salts and sulphates, and
(iv) prevention of calcium hydroxide leaching.

The undersirable side effects of pozzolana are the reduction in the rate of development of strength, an increase in the *drying shrinkage* and reduction in *durability*. In the construction of massive structures, such as dams, the pozzolana can be used advantageously resulting in a saving of cement, and the reduction in the *heat of hydration*.

IS: 456–1978 permits the use of pozzolana like fly ash conforming to IS: 3812 (Part II)-1981 or burnt clay conforming to IS: 1344–1982 as admixtures for concrete.

Fly Ash

The *fly ash* or *pulverized fuel ash* (PFA) is the residue from the combustion of pulverized coal collected by the mechanical or electrostatic separtaors from the fuel gases of thermal power plants. Its composition varies with the type of fuel burnt, load on boiler and type of separator, etc. Fly ash consists mainly of spherical glassy particles ranging from 1 to 150 μm in diameter, of which the bulk passes through a 45 μm sieve. The fly ash obtained from electrostatic precipitators may have a *specific surface* of about 350000 to 500000 mm^2/g, i.e., it is finer than portland cement. The fly ash obtained from cyclone separators is comparatively coarser and may contain larger amounts of unburnt fuel.

Like portland cement, fly ash contains oxides of calcium, aluminium and silicon, but the amount of calcium oxide is considerably less. The principal constituents of fly ash are.

(i) Silicon dioxide—SiO_2 — 30 to 60 per cent
(ii) Aluminium oxide—Al_2O_3 — 15 to 30 per cent
(iii) Unburnt fuel (Carbon) — up to 30 per cent
(iv) Calcium oxide—CaO — 1 to 7 per cent
(v) Magnesium oxide—MgO — small amounts
(vi) Sulphur trioxide—SO_3 — small amounts

The carbon content in fly ash should be as low as possible, whereas the silica content should be as high as possible.

The *fly ash* may be used in concrete either as an *admixture* or in-part replacement of cement. The pozzolanic activity is due to the presence of finely divided glassy silica and lime which produce calcium silicate hydrate as is produced in portland cement. The growth and interlocking of this hydrate gives mechanical strength. The lime produced in portland cement *hydration* provides the right enviroment for *pozzolanic action* to proceed. Since similar hydrates are produced, the combination of two reactions in mixed cement-fly ash concrete results in improved mechanical strength. The fly ash is generally used in the following three ways.

(i) *As a part replacement of cement* This simple replacement for up to 15 to 30 per cent of cement reduces the strength at ages up to 3 months, but once sufficient lime has been liberated to start the pozzolanic action, the rate of development of strength increases rapidly and equality can be attained after 1–3 months. After this stage, the pozzolanic reaction continues at a rate higher than the cement hydration, and higher strength can be obtained.

The optimum amount of pozzolana as a replacement may normally range between 10 and 30 per cent, but is usually nearer the lower limit and may be as low as 4 to 6 per cent for natural pozzolana. A fine grinding of silica and *high temperature curing* increase the reactivity of pozzolana. The part replacement results in an increased *workability* which can be used to reduce water content and, in turn, increase strength. Thus, the water content can be sufficiently reduced to limit the loss at early ages to 25 per cent.

(ii) *As a part replacement of fine aggregate* This substitution of fly ash for sand has a beneficial effect on the strength even at early ages, but is rather uneconomical. The procedure is little used.

(iii) *As a simultaneous replacemnt of cement and fine aggregate* This replacement enables the strength at a specified age to be equalled depending on the water-content.

The practice of adjusting the water-content and the sum of cement and fly ash contents to produce the concrete of same workability and 28 day compressive strength but increased strength at later ages, is very useful. The factors suggested for this adjustment are given in Table 5.2.

TABLE 5.2 Multiplying factors to equal the 28 day compressive strength

Cement replacement by fly ash $(15 + 5n)$*, per cent	Multiplying factor for water-content $(0.980 - 0.022 n)$, kg/m^3	Multiplying factor for total amount of cement and fly ash, $(1.035 + 0.015 n)$, kg/m^3
15	0.980	1.035
20	0.958	1.050
25	0.936	1.065
30	0.914	1.080
35	0.892	1.095
40	0.870	1.110
45	0.848	1.125

*n is integer varying from 1 to 6

Due to different densities of cement and fly ash 3100–3200 kg/m^3 and 2200–2400 kg/m^3, respectively, a part replacement by equal mass increases the volume of cementitious material, whereas replacement by equal volume reduces the mass. In practice the replacement is usually on a mass basis. The use of fly ash influences the volume yield of concrete. It has little effect on the drying shrinkage of concrete.

Granulated Blast Furnace Slag

It is a byproduct obtained during the production of iron. The oxide composition of blast furnace slag is similar to portland cement so far as the oxides of calcium, aluminium and silicon are concerned, but it contains less calcium oxide. If the slag is cooled slowly the resulting material has no cementing properties. However, when cooled rapidly it solidifies in a glassy form which is reactive with water having alkaline medium. As explained earlier the granulated slag can be used for the production of *blast furnace cement*. To produce this cement the slag is mixed with portland cement clinker and gypsum, and ground together. The grinding in this procedure can result in different particle sizes of two constituents due to difference in their hardness. However, the blast furnace slag ground alone to cement fineness can be used as *admixture* and mixed with portland cement at the point of use. The alkaline medium required to initiate the hydration can be provided by lime, sodium hydroxide or gypsum. The slags with high sodium or potassium oxide content are self activating as the alkaline medium is provided by these oxides. Like portland cement the growth and interlocking of calcium silicate hydrate gives mechanical strength. The exact hydration products formed depend upon the type of activ. r used. The portland cement component always hydrates first generating an alkaline medium for the slag to hydrate. The main advantage in using blast furnace slag is in the lower rate of heat evolution and its improved resistance to sulphate attack.

The effects of blast furnace slag on the workability are much less than those of fly ash due to lesser specific surface of 325000 ± 20000 mm^2/g. The improvement in *workability* is probably equivalent to about 5 per cent increase in water content which is insignificant.

5.3.13 Concrete Colouring Admixtures or Pigments

Pigments are the admixtures added to produce coloured cements. One of the methods of producing coloured concrete surfaces in modern construction is to use concrete paint applied after the concrete surface has been neutralized, either through exposure or by using a neutralizing agent like zinc sulphate. The other most commonly used method involves integrating colour into the surface of concrete while it is still fresh. This can be accomplished by mixing natural metallic oxides of cobalt, chromium, iron, etc. called pigments into a topping mix. This is the best way of distributing the colour evenly throughout the concrete.

The colouring admixture made with synthetic oxides mixed with one or more additional drying ingredients are also available. But the *pigments* used must be permanent and should not react with free lime in concrete. To obtain a good colouring affect, the pigments should be ground with the cement in a ball mill. Sometimes they can be mixed with fillers, like chalk, but the excessive use of fillers may affect *strength of concrete*. The chief pigments used in concrete are as follows.

(i) *Black*: The best permanent black pigment is carbon black, but manganese black gives a brown tint whereas magnetic ferrous oxide has a purple tint.

(ii) *Blue*: The materials used are barium manganate and ultramarine. The former is adversely affected by sulphur fumes in polluted atmosphere. Ultramarine is suitable for concrete used in nonwearing surfaces.

(iii) *Brown*: Raw umber or burnt umber form satisfactory brown pigments.

(iv) *Green*: Artificially produced chromium oxide and chromium hydroxide are suitable.

(v) *Red*: The most commonly used material is the naturally occurring red oxide of iron.

(vi) *Yellow*: Hydroxides of iron give yellow colour.

5.3.14 Concrete-hardening Admixtures

The plain concrete surfaces which are subjected to heavy traffic begin to deteriorate by wearing after a period of time. The hardener commonly used to prevent the destruction of the surface can be divided into two groups, namely, the chemical hardeners and fine metallic aggregates. The liquid chemical hardeners consists of silicofluorides or fluosilicates and a wetting agent. The latter reduces the surface tension of liquid and allows it to penetrate the pores of the concrete more easily. The chemicals combine with free lime and calcium carbonate present in concrete to bind the fine particles into highly wear resistant flintlike topping.

On the other hand, the metallic hardeners consisting of specially processed grade iron particles are dry-mixed with portland cement which is spread evenly over to freshly floated concrete surface and are worked into concrete by floating. This gives highly wear resistant and less brittle concrete topping.

Sometimes abrasive materials like fine particles of flint, aluminium oxide, silicon carbide, or emery are used in the topping applied as dry shake to obtain water-resistant nonskid surfaces. The most commonly used concrete admixtures are given in Table 5.3.

5.4 INDIAN STANDARD SPECIFICATIONS

The Indian Standard Specifications for admixture for concrete, IS: 7861 (Part II)-1981 covers the chemical and *air-entraining admixtures,* solid or liquid emulsions, to be added to the concrete at the time of mixing so as to achieve the desired property in the concrete, in the plastic or hardened state. The admixtures covered in this standard are as given:
  (i)   accelerating admixtures,
  (ii) retarding admixtures,
  (iii) water-reducing admixtures, and
  (iv) air-entraining admixtures.

Admixtures 61

TABLE 5.3 Common types of concrete admixtures

Admixtures	Functions	Typical compounds	Applications	Disadvantages
Accelerators	(i) More rapid gain of strength	Calcium chloride, Sodium nitrite	(i) Normal rate of strength development at low temperature	(i) Possible cracking due to heat evolution
	(ii) More rapid setting	Sodium sulphate Sodium aluminate Sodium silicate Sodium carbonate Potassium hydroxide	(ii) Shorter stripping times (iii) Plugging of pressure leaks (iv) Sprayed concreting	(ii) Possibility of corrosion of embedded reinforcement
Retarders	Delayed setting	Hydroxylatedcarboxylic acids, sugars	(i) Maintain workability at high temperatures (ii) Reduce rate of heat evolution (iii) Extend placing times	May promote bleeding
Water-reducing accelerators	Increased workability with faster gain of strength	Mixtures of calcium chloride and lignosulphonate	Water-reducers with faster strength development	Risk of corrosion
Water-reducing retarders	Increased workability and delayed setting	Mixtures of sugars or hydroxylatedcarboxylic acids and lignosulphonate	Water-reducers, with slower loss of workability	
Air-entraining agents	Entrainment of air into concrete	Wood resins, fats, lignosulphonates	Increase durability to frost without increasing cement content, cellular concrete	Careful control of air content and mixing time necessary

(cond.)

Admixtures	Functions	Typical compounds	Applications	Disadvantages
Water-proofers	(i) Prevention of water from entering capillaries of concrete (ii) Reduced permeability of concrete	Potash soaps, butylstearate, petroleum waxes	Reduce permeability Reduce surface staining Watertightness of structures without using very low water-cement ratio	
Plasticizers (Water-reducers)	Increased workability	Calcium and sodium lignosulphonate	(i) Higher workability with strength unchanged (ii) Higher strength with workability unchanged (iii) Less cement for same strength and workability	Retardation at high dosages Tendency to segregate Premature stiffening under certain conditions
Superplasticizers (Super-water-reducers)	Greatly increased workability	Suphonated melamineformaldehyde resin, Suphonated naphthaleneformaldehyde resin, Mixtures of saccharates and acid amides	(i) Water-reducers, but over a wider range (ii) Facilitate production of flowing concrete	Tendency to segregate May increase rate of loss of workability

6

Properties of Fresh Concrete

6.1 INTRODUCTION

The performance requirements of hardened concrete are more or less well defined with respect to shape, finish, strength, durability, shrinkage and creep. To achieve these objecvies economically, the fresh concrete, in addition to having a suitable composition in terms of quality and quantity of cement, aggregate and admixtures, should satisfy a number of requirements from the mixing stage till it is transported, placed in formwork and compacted. The requirements may be summarized as follows.

(i) The mix should be able to produce a homogeneous fresh concrete from the constituent materials of the batch under the action of the mixing forces. A less mixable concrete mix requires more time to produce a homogeneous and uniform mix. The property is termed *mixability*. The power requirement in a mixing plant is a function of *time of mixing*. The power input curve rises to a maximum during mixing and then falls off to a steady value for the concrete containing too much water, indicating that the concrete is thoroughly mixed. For concretes of normal workability there is no drop off after the maximum has been reached. On the other hand, for concrete containing insufficient water, the normal maximum is not reached. These observations enable to control and improve reproducibility of batches. It is generally noticed that the power consumed in merely turning over the machinery ranges from about 25 to 50 per cent of gross power.

(ii) The mix should be *stable*, in that it should not segregate during transportation and placing when it is subjected to forces during handling operations of limited nature. Any *segregation* that is caused during the transportation operation does not correct during remaining operations that follow. The tendency of *bleeding* should be minimized.

(iii) The mix should be cohesive and sufficiently mobile to be placed in the form around the reinforcement and should be able to cast into the required shape without losing continuity or homogeneity under the available techniques of placing the concrete at a particular job. The property is termed *flowability* or *mobility of fresh concrete*.

(iv) The mix should be amenable to proper and thorough compaction into a dense, compact concrete with minimum voids under the existing facilities of compaction at the site. A best mix from the point of view of compactability should achieve a 99 per cent elimination of the original voids present. This property is termed the *compactability of concrete*.

(v) It should be possible to attain a satisfactory surface finish without honeycombing or blowing holes from formwork and on free surface by trowelling and other processes. This capability is termed finishability.

6.2 WORKABILITY

The diverse requirements of *mixability, stability, transportability, placeability mobility, compactability* and *finishability* of fresh concrete mentioned above are collectively referred to as *workability*. The workability of fresh concrete is thus a composite property. It is difficult to define precisely all the aspects of the workability in a single definition. IS: 6461 (Part VII)-1973 defines *workability* as *that property of freshly mixed concrete or mortar which determines the ease and homogeneity with which it can be mixed, placed, compacted and finished*. The *optimum workability* of fresh concrete varies from situation to situation, e.g., the concrete which can be termed as workable for pouring into large sections with minimum reinforcement may not be equally workable for pouring into heavily reinforced thin sections. A concrete may not be workable when compacted by hand but may be satisfactory when vibration is used.

Sometimes the terms *consistency and plasticity* are used to denote the workability of a concrete mix. The *consistency* of the mix really means the wetness of the mix, and a wetter mix need not have all the above desired properties. On the other hand, an extremely wet mix may cause segregation and may be difficult to place in moulds. *Plasticity* is the cohesiveness of the mix to hold the individual grains together by the cement matrix.

6.3 MEASUREMENT OF WORKABILITY

The quantitative assessment describing the concrete as being of high or low workability or semi-dry or plastic, etc. may mean different things to different people. The commonly used practice of defining this physical property by a numerical scale based on the empirical tests for its measurement has been found to be unsatisfactory in many situations, thus restricting its applications, in that many builders prefer to rely on subjective assessment rather than on empirical tests.

A number of different empirical tests are available for measuring the workability of fresh concrete, but none of them is wholly satisfactory. Each test measures only a particular aspect of it and there is really no unique method which measures the workability of concrete in its totality. However, by checking and controlling the uniformity of the workability, it is easier to ensure a uniform quality of concrete

and hence uniform strength for a particular job. The empirical tests widely used are:
  (i) the slump test,
  (ii) the compacting factor test,
  (iii) the Vee-Bee consistency test, and
  (iv) the flow test.

Out of these four, the slump test is perhaps the most widely used, primarily because of the simplicity of the apparatus required and the test procedure. The slump test indicates the behaviour of a compacted concrete cone under the action of gravitational forces. The test is carried out with a mould called the slump cone. The slump cone is placed on a horizontal and nonabsorbent surface and filled in three equal layers of fresh concrete, each layer being tamped 25 times with a satndard tamping rod. The top layer is struck off level and the mould lifted vertically without disturbing the concrete cone. The subsidence of concrete in millimeters is termed the *slump*. The concrete after the test when slumps evenly all around is called *true slump*. In the case of very lean concrete, one-half of the cone may slide down the other which is called a *shear slump;* or it may collapse in case of very wet concretes. The slump test is essentially a measure of *consistency* or the *wetness* of the mix. The test is suitable only for concretes of medium to high workabilities (i.e. having slump values of 25 mm to 125 mm). For very stiff mixes having zero slump, the slump test does not indicate any difference in concretes of different workabilities. It must be appreciated that the different concretes of the same slump may, indeed, have different workabilities under the site conditions. However, the slump test has been found to be useful in ensuring the uniformity among different batches of supposedly similar concrete under field conditions. The slump test is limited to concretes with maximum size of aggregate less than 38 mm.

The *compacting factor test* gives the behaviour of fresh concrete under the action of external forces. It measures the *compactability* of concrete which is an important aspect of workability, by measuring the amount of compaction achieved for a given amount of work. The compacting factor test has been held to be more accurate than slump test, especially for concrete mixes of *medium* and *low workabilities*, i.e.

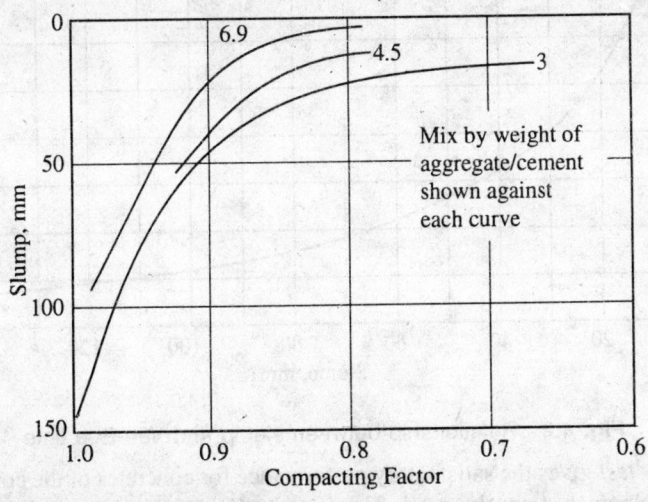

Fig. 6.1 Relation between slump and compacting factor

compacting factor of 0.9 to 0.8, because the test is more sensitive and gives more consistent results. The test has been more popular in laboratory conditions. For concrete of very low workabilities of the order of 0.70 or below, the test is not suitable, because this concrete cannot be fully compacted for comparison in the manner described in the test. The relationship between slump and compacting factor is given in Fig. 6.1.

The *Vee-Bee test* is suitable for stiff concrete mixes having *low and very low workability*. Compared to the slump test and compacting factor test, the Vee-Bee test has the advantage that the concrete in the test receives a similar treatment as it would in actual practice. The test consists in moulding a fresh concrete cone in a cylindrical container mounted on a vibrating table. The concrete cone when subjected to vibration by starting the vibrator starts to occupy the cylindrical container by the way of getting remoulded. The *remoulding* is considered complete, when the concrete surface becomes horizontal. The time required for complete remoulding in seconds is considered as a measure of workability and is expressed as the number of *Vee-Bee seconds*. Since the endpoint of the test when the concrete surface becomes horizontal is to be ascertained visually, it introduces a source of error which is more pronounced for concrete mixes of *high workability* and consequently records low Vee-Bee time. For concrete of slump in excess of 125 mm, the remoulding is so quick that time cannot be measured. The test is therefore, not suitable for concrete of higher workability, i.e., slump of 75 mm or above. An approximate relationship between slump and Vee-Bee time is given in the Fig. 6.2.

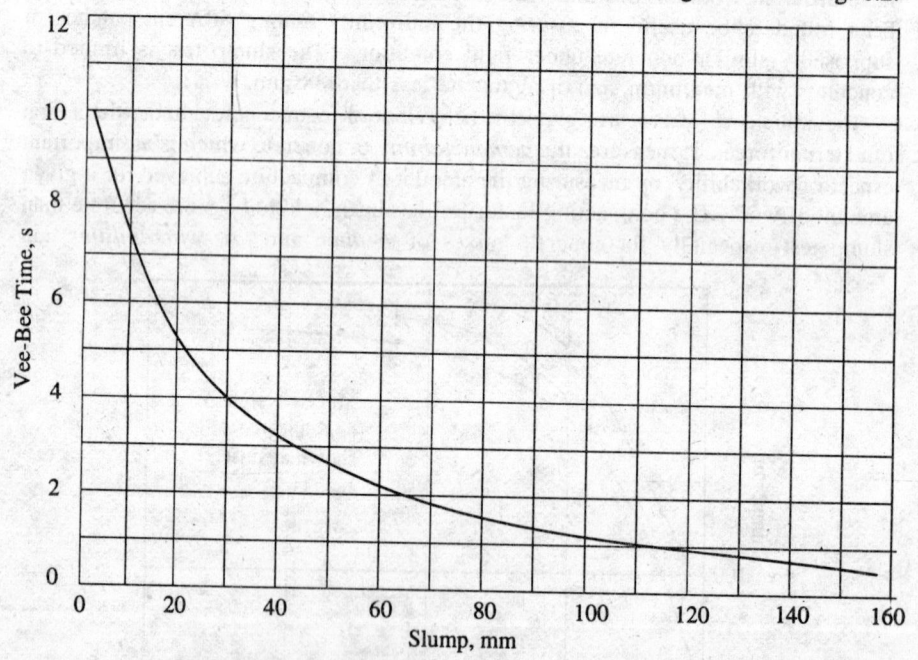

Fig. 6.2 Relationship between slump and Vee-Bee time

The *flow test* gives the satisfactory performance for concretes of the consistencies for which slump test can be used. The test consists of moulding a fresh concrete cone on the top of the platform of flow table, and in giving 15 jolts of 12.5 mm magnitudes. The spread of the concrete, measured as the increase in diameter of concrete heap and expressed as the percentage of the original base diameter of cone, is taken as a measure of the *flow* or *consistency of the concrete*. The test suffers from the drawback that the concrete may scatter on the flow table with a tendency towards *segregation*.

As each of the above tests measures only a particular aspect of workability, there is no rigid correlation between the workabilities of concrete as measured by different test methods. In the absence of definite correlations between different measures of workability under different conditions, it has been recommended that, for a given concrete, the appropriate test method be decided beforehand and workability expressed in terms of such a test only rather than be interpreted from the results of other tests. Table 6.1 gives the range of the expected values of workability measured by different test methods for comparable concretes.

In addition to the specific faults inherent to each test the major drawbacks are summarized below.

(i) The tests are quite arbitrary and empirical as far as the measurement of workability is concerned because each of these tests is a single-point test measuring a single quantity which at times may classify as identical two concretes that may behave quite differently on the job.

(ii) The results from these tests are influenced by minor variations in techniques of carrying out the test, i.e. they are operator sensitive.

(iii) None of the tests is capable of dealing with concrete of whole range of workabilities, e.g. the slump test is quite incapable of differentiating between two concretes of very low workability (zero slump) or two concretes of very high workability (collapse slump). Moreover, the test results could be used as a simple statement of qualitative behaviour of concrete under particular circumstances.

However, with all their faults, the empirical tests have made the progress in concrete mix design possible. There is a strong need for the development of a new rational test based on rheological techniques. The recent attempts to rigorously define the workability in terms of one or more physical constants by an idealized model is a distinct possibility.

6.4 FACTORS AFFECTING WORKABILITY

The workability of fresh concrete depends primarily on the materials, mix proportions, and environmental conditions.

6.4.1 Influence of Mix Proportions

In the concrete comprising a cement-aggregate-water system, the *aggregate* occupy approximately 70 to 75 per cent of the total volume of concrete and economy demands that the volume of aggregates should be as large as possible. The total *specific area* of the aggregate is to be minimized to the extent possible by proper choice of size, shape and proportion of fine and coarse aggregates. Different size fractions are so chosen as to minimize the *void content*, and such a mixture will need more water for lubricating effects to overcome the reduction in mobility due

TABLE 6.1 Suggested values of workability of fresh concrete for different placing conditions

Placing condition	Degree of workability	Compacting factor, maximum size of aggregate 10 mm	20 mm	40 mm	Vee-Bee time, slump for 20 mm aggregate
(i) Hand compaction of heavily reinforced sections	High (flowing)	0.95	0.95	0.95	125–150 mm slump
(ii) Concreting of lightly reinforced section by hand or vibration of heavily reinforced sections	Medium (plastic)	0.88	0.90	0.92	5–2 s Vee-Bee time, 25–75 mm slump
(iii) Concreting of lightly reinforced sections with vibration; road pavements and slabs with hand-operated vibrators; and vibration of mass concrete	Low (stiff plastic)	0.82	0.84	0.85	10–5 s Vee-Bee time, 5–50 mm slump
(iv) Concreting of shallow sections with vibrations	Very low (stiff)	0.75	0.78	0.80	20–10 s Vee-Bee time, 0–25 mm slump
(v) Concreting by intensive vibrations with vibropressing, centrifugation, etc.	Extremely low (very stiff)	0.65	0.68	Not used	30–20 s Vee-Bee time

to dense packing of particles. The *water-cement ratio* in itself determines the intrinsic properties of cement paste and the requirements of workability such that there is sufficient cement paste to surround the aggregate particles as well as fill the voids in the aggregate. It has been noticed that the change in the measured value of workability due to relative change in water content in concrete is independent of the composition of concrete within wide limits. An increase of *water content* results in monotonous increase in *workability* but eventually a stage is reached where *segregation* and *bleeding* occur, and use of higher water content will result in the more serious problems of shrinkage and creep of hardened concrete. However, the water content is limited to some maximum value given by the *water-cement ratio* which is dependent on the *target design strength* of hardened concrete, making it imperative to study the effect on workability of other factors.

6.4.2 Influence of Aggregate Properties

The effect of aggregate properties on the workability of fresh concrete can be summarized as follows.

(i) For the same volume of aggregate in concrete, the use of coarse aggregate of larger size and/or rounded aggregate gives higher workability because of reduction in total *specific surface area* and particle interference. The use of elongated aggregates results in low workability, primarily due to increase in particle interference.

(ii) The use of finer sand increases the specific surface area, there by increasing the water demand for the same workability. In the other words, for the same water content, the use of finer sand decreases workability.

(iii) Because of the greater contribution to the total specific surface area, the grading of fine aggregate is more critical than the grading of coarse aggregate. Nevertheless, the proportion of fine to coarse aggregates should be so chosen as neither to increase the total specific surface area (by excess of fine aggregate) nor to increase the particle interference (due to deficiency in fine aggregate). An unsuitable choice of overall grading can produce honeycombing or segregation. In

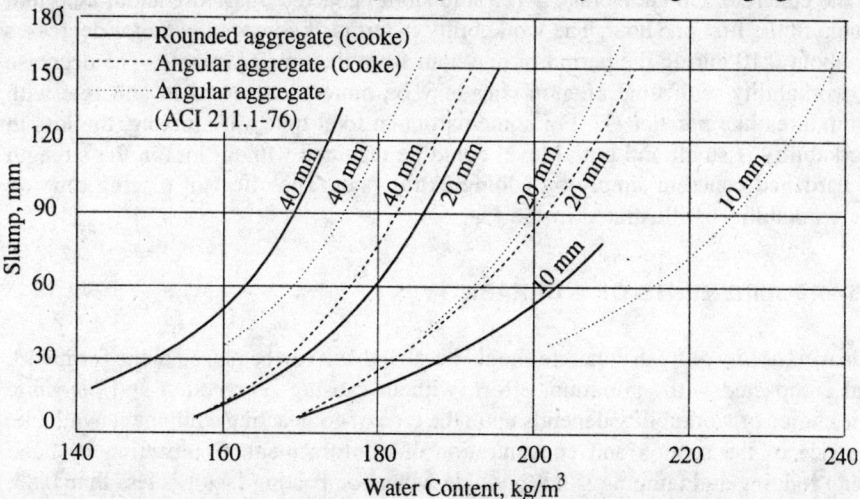

Fig. 6.3 Effect of water content and aggregate size on the workability of concrete

normal range of mixes though an increase in fines content decreases workability, but in practice there is an optimum fines content for maximum workability, such that either an increase or decrease of fines reduces workability.

(iv) Generally, the mixes with higher *water-cement ratio* would require a somewhat finer grading and for mixes with low water-cement ratio (as in case of high-strength concrete) a coarser grading is preferable. The effect of water content and aggregate size is shown in Fig. 6.3.

(v) The workability is also affected by the properties of cement both physical and chemical, but to a much lesser extent than that by the aggregate properties. The influence of cement properties may have to be taken into account especially for richer mixes. A rapid-hardening cement will reduce workability as compared to ordinary portland cement because of its higher specific surface and the fact that it hydrates more rapidly, and also the fineness of cement has an influence on bleeding.

(vi) The presence and nature of admixtures or cement replacement may affect the workability considerably.

6.4.3 Effect of Environmental Conditions

The workability of a concrete mix is also affected by the temperature of concrete and, therefore, by the ambient temperature. On a hot day it becomes necessary to increase the water content of the concrete mix in order to maintain the desired warkability. The amount of mixing water required to bring about a certain change in workability also increases with temperature.

6.4.4 Effect of Time

The fresh concrete loses workability with time mainly because of the loss of moisture due to evaporation. A part of mixing water is absorbed by aggregate or lost by evaporation in the presence of sun and wind, and part of it is utilized in the chemical reaction of hydration of cement. The loss of workability varies with the type of cement, the concrete mix proportions, the initial workability and the temperature of the concrete. On an average a 125 mm slump concrete may lose about a 50 mm slump in the first one hour. The workability in terms of *compacting factor* decreases by about 0.10 during the period of one hour from the time of mixing. The decrease in workability with time after mixing may be more pronounced in concrete with admixtures like *plasticizers*. For some particular total time after mixing, the loss in workability is small and initial level could be regained without loss in the strength of hardened concrete simply by adding extra water. The effect of placing time on the workability is illustrated in the Fig. 6.4.

6.5 REQUIREMENTS OF WORKABILITY

The *workability* of fresh concrete should be such that it can be placed in the formwork and compacted with minimum effort, without causing *segregation* and *bleeding*. The choice of workability depends upon the type of compacting equipment available, the size of the section and concentration of reinforcement. Compaction by hand using rodding and tamping is not possible when compacting factor is less than 0.85. Ordinary techniques of vibration are not applicable if the compacting factor falls

Properties of Fresh Concrete 71

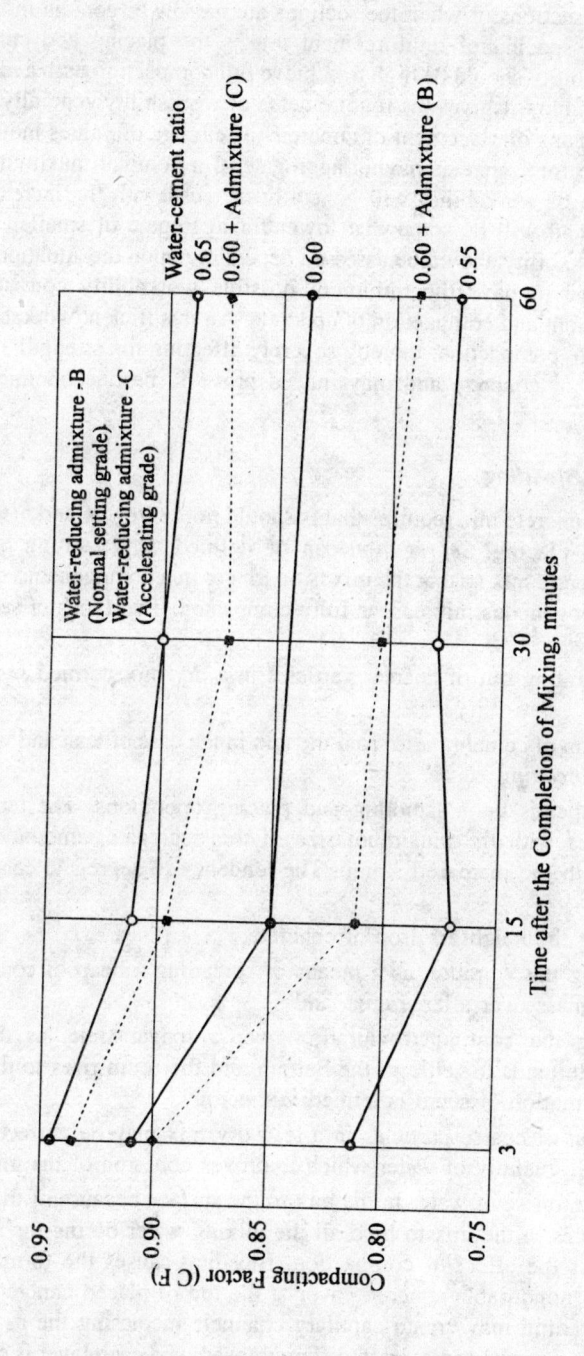

Fig. 6.4 Effect of placing time on the workability

below 0.70. In such cases techniques like vibro-pressing have to be adopted. For heavily reinforced sections or when the sections are narrow or contain inaccessible parts or when the spacing of reinforcement makes the placing and compaction difficult, the workability should be high to achieve full compaction with a reasonable amount of effort. Table 6.1 gives the requirements of workability generally required for different conditions of placement of concrete. The range of values indicated are considered suitable for concretes having aggregate of a nominal maximum size of 20 mm. The value of workability will generally increase with the increase in the size of aggregate and will be somewhat lower for aggregate of smaller size than indicated. The workability should be assessed depending upon the situation at hand. The aim should be to have the minimum possible workability consistent with satisfactory placement and compaction of concrete. An insufficient workability may result in incomplete compaction, thereby severely affecting the strength, durability and surface finish of concrete and may indeed prove to be uneconomical in the long run.

Segregation and Bleeding

The *stability* of a concrete mix requires that is should not *segregate* and *bleed* during transportation and placing. *Segregation* can be defined as separating out of the ingredients of concrete mix so that the mix is no longer in a homogeneous condition. Only a stable homogeneous mix can be fully compacted. Two types of segregation can occur:

(i) the separating out of coarser particles in a dry mix, termed *segregation*, and

(ii) separation of cement paste from the mix in the case of lean and wet mixes, termed *bleeding*.

Segregation depends upon handling and placing operations. The tendency to segregate increases with the maximum size of the aggregate, amount of coarse aggregate, and with the increased slump. The tendency to segregate can be minimized by:

(i) reducing the height of drop of concrete,

(ii) not using the vibration as a means of spreading a heap of concrete into a level mass over a large area, and

(iii) reducing the continuted vibration over a longer time, as the coarse aggregate tends to settle to the bottom and the scum rises to the surface (this formation of scum is termed *laitance*).

The *segregation* of coarse particles in a lean dry mix may be corrected by the addition of a small quantity of water which improves cohesion of the mix.

Bleeding is due to rise of water in the mix to the surface because of the inability of the solid particles in the mix to hold all the mixing water during the settling of the particles under the effect of compaction. Bleeding causes the formation of a porous, weak and nondurable concrete layer at the top of placed concrete. In case of lean mixes bleeding may create capillary channels increasing the *permeability* of concrete. When concrete is placed in different layers and each layer is compacted after allowing certain time to lapse before the next layer is laid, bleeding may also result in a plane of weakness between two layers. Any *laitance* formed should be removed by brushing and washing before a new layer is added. Overcompacting the surface should be avoided.

6.6 ESTIMATION OF ERRORS

The workability tests are operator sensitive. Due to inherent errors in experimental measurements the results of workability tests are generally stated to the following accuracies:

Slump	±	5mm
Compacting factor	±	0.01
Vee-Bee time	±	0.5 s

The experimental errors estimated by replicate measurements as reported in the literature are given in the Table 6.2.

TABLE 6.2 Statistical estimate of errors in workability tests

Type of workability test	Range of values	Standard deviation
Slump	0–100 mm	11.00
Compacting factor	0.75–1.0	0.024
Vee-Bee time	1–24 s	0.25σ

7

Rheology of Concrete

7.1 INTRODUCTION

Rheology may be defined as the science of the deformation and flow of materials, and is concerned with relationships between stress, strain, rate of strain, and time. The term rheology deals with the materials whose flow properties are more complicated than those of simple fluids (liquids or gases). The rheological principles and techniques as applied to concrete include the deformation of hardened concrete, handling and placing of freshly mixed concrete, and the behaviour of its constituent parts, namely, cement slurries and pastes. However, in this section, only the rheological properties of fresh concrete are considered. The *rheology* of fresh concrete like workability includes the parameters of *stability, mobility* and *compactability*, which are necessary to determine the suitability of any concrete mix as shown in Fig. 7.1. For the purpose of discussion of rheological properties of fresh concrete these parameters are redefined in terms of forces involved in the transmission of mechanical stresses on the concrete. The fresh concrete is subjected to normal and shearing forces during its handling and placing.

Stability is defined as a condition in which the aggregate particles are held in homogeneous dispersion by matrix, and random sampling shows the same particle size distribution during transportation, placing and compacting. The stability of concrete is measured by its *segregation* and *bleeding* characteristics. Segregation is defined as the mixture's instability caused by weak matrix that cannot hold individual aggregate particles in homogeneous dispersion. The resistance to segregation depends upon the *cohesion* between the particles of the mix. Segregation can occur in concretes of both wet and dry consistencies. Segregation in wet mix can occur when the water content is such that the paste cannot hold aggregate particles in the distributed position while the concrete is transported, placed and compacted. Dry segregation takes place when a concrete with low *water-cement ratio* results in a crumbly mix during handling. However, the crumbly mixes are often satisfactory when the concrete is vibrated, as during vibration the matrix becomes fluid momentarily and develops cohesion and shear resistance. On the other hand, bleeding occurs when mortar is unstable and releases free water. Bleeding should be controlled and reduced to a minimum.

Rheology of Concrete 75

```
VISCOSITY ──┐
COHESION ───┼──► FLOWABILITY OR MOBILITY ──┐
INTERNAL FRICTION ──┘                       │
                                            ▼
RELATIVE DENSITY ──► COMPACTABILITY ──► RHEOLOGY OF FRESH CONCRETE
                                            ▲
SEGREGATION ──┐                             │
BLEEDING ─────┴──► STABILITY ───────────────┘
```

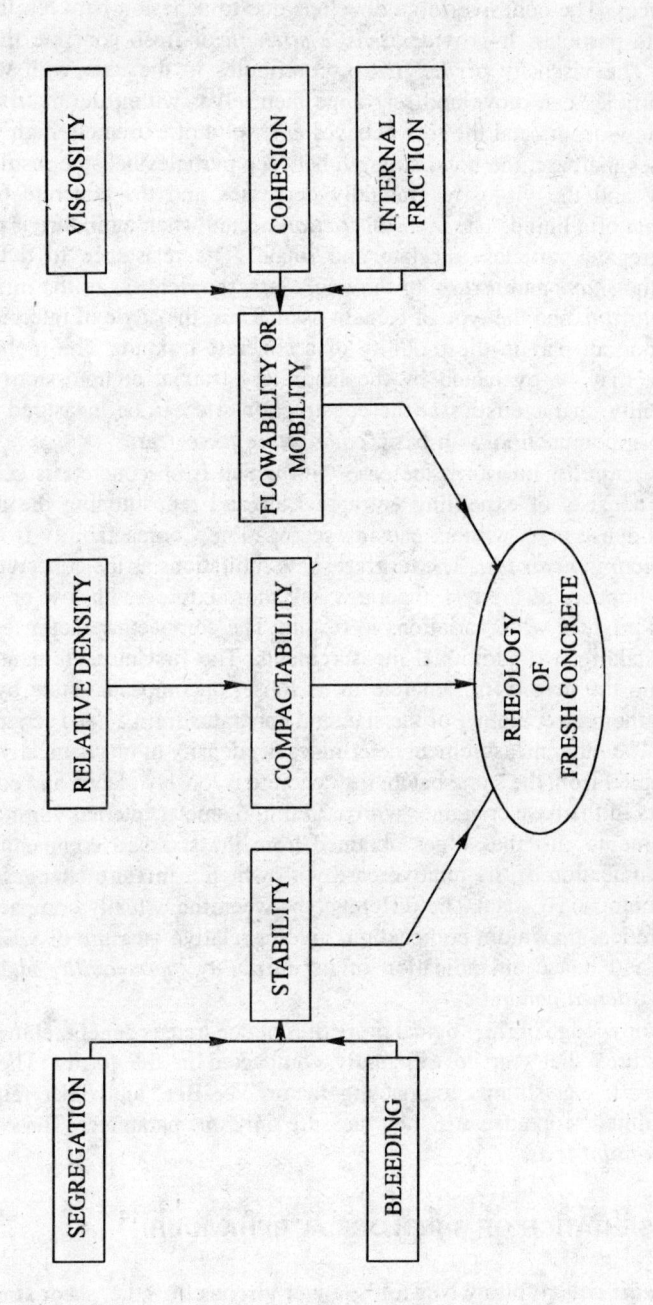

Fig. 7.1 Parameters defining the rheology of fresh concrete

The *mobility* of fresh concrete is its ability to flow under *momentum transfer*, i.e. under mechanical stresses. The flow is restricted by *cohesive, viscous* and *frictional* forces. The cohesive force develops due to adhesion between the matrix and aggregate particles. It provides *tensile strength* of fresh concrete that resists segregation. The viscosity of the matrix contributes to the ease with which the aggregate particles can move and rearrange themselves within the matrix. At low stresses no flow occurs and the mix behaves as a solid of extremely high viscosity. As the stresses increase, the bond strength between particles become insufficient to prevent flow and the viscosity gradually decreases and the concrete behaviour changes to that of a liquid. The *internal friction* occurs when a mixture is displaced and the aggregate particles translate and rotate. The resistance to deformation depends on the *shape* and *texture of the aggregate*, the richness of the mixture, the *water-cement ratio*, and the type of cement used. Thus, the angle of internal friction plays an important part in the mobility of a concrete mixture. The mobility of a concrete mix may be evaluated by the laboratory triaxial compression test. The relative mobility characteristics at the construction site can be measured by using *Vee-Bee test* in conjunction with basic *compacting factor test*.

The *compactability* measures the ease with which fresh concrete is compacted. Compacting consists of expellling entrapped air and repositioning the aggregate particles in a dense mass without causing segregation. Compactability is measured by the *compacting factor test*. The test has some limitations as the cohesive mixture sticks in the hoppers of the test apparatus and the mixtures with low or very low workabilities produce wide variations in results. The compacting factor test can be extended by taking two additional measurements. The first measurement consists in determining the density of concrete in its loose, uncompacted state by placing concrete into the base container of the standard apparatus from a hand scoop without compaction. The other measurement determines the density of mechanically vibrated concrete sampled from the same batch; the concrete is loosely placed and compacted in three layers in the base container with a 25 mm diameter internal vibrator. These two measurements plus the values obtained from the standard compacting factor test give an indication of the relative ease with which a mixture changes from its loose to the compacted state. The difference between the actually compacted state and the theoretical maximum compaction, gives a relative measure of *void content of concrete*, and hence an indication of its *durability, permeability* and relative *strength of hardened concrete*.

Thus the knowledge of rheological properties of concrete is beneficial in selecting concrete mixtures that can be efficiently compacted in the forms. The current *workability tests*, e.g. slump, compacting factor, Vee-Bee, and other remoulding tests are of limited scope because they measure only one parameter. These tests are termed *single-point tests*.

7.2 REPRESENTATION OF RHEOLOGICAL BEHAVIOUR

The *ideal liquids* which follow Newton's law of viscous flow, i.e. shear stress being proportional to the *rate of shear strain*, are termed *Newtonian liquids*. The constant of proportionality may be used as a physical constant characteristic of the materials. The flow behaviour of fresh concrete does not conform to it. The ratio of shear stress to shear rate is not constant but depends upon the shear rate at which it is measured, and may also depend on the shear history of the concrete sample being investigated. However, at low shear rates that are important in practice, the behaviour can be represented by a straight line which does not pass through the origin, i.e. which has an intercept on the stress axis. The intercept indicates the minimum stress below which no flow occurs.

The fact that concrete can stand in a pile (as in the case of the slump test) suggests that there is some minimum stress necessary for flow to occur at all. The minimum stress is called *yield stress* and designated by the symbol τ_0. Thus the simplest flow equation of concrete, Fig. 7.2, can be written as:

$$\tau = \tau_0 + \mu\dot{\gamma}$$

where τ_0 is the yield value indicating the cohesion of the material, μ is a constant having the dimensions of viscosity and termed *plastic viscosity*. This mathematical relationship is called the *Bingham model*. The constants τ_0 and μ are the parameters characterizing the flow properties of the material. Thus Bingham model distinguishes between the shear stress of the material expressed in terms of its cohesion, plastic viscosity, and the rate at which the shear load is applied. To establish a straight line, at least two points are required. Accordingly, the *workability* of concrete cannot be defined by the *single-point tests* that determine only one parameter, i.e. produce only single point, and therefore have to be used in combination with other tests to achieve a better understanding of concrete rheology. For example, the Vee-Bee test can be used with compacting factor test to measure *mobility* and *compactability*. Tattersal measured mobility characteristics by a single test method which provided two or more points (shear conditions). The procedure is based on determining the power required to mix concrete at various speeds and then calculating torque by

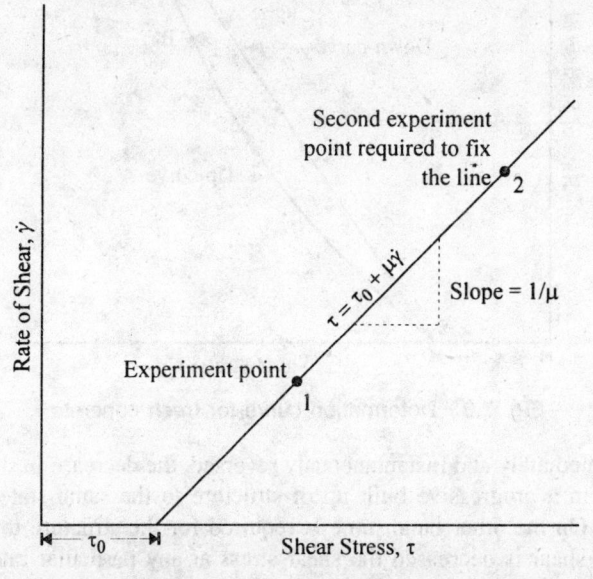

Fig. 7.2 The Bingham model

78 Concrete Technology

dividing the power by the speed. The torque is related to the mixing speed by the following relation.

$$T = g + hN$$

where T is the torque measured, at N rps, and g and h are constants proportional to the cohesion and plastic viscosity, respectively of the mixture.

It has been found that the two mixtures having identical values of g and h will also have identical values for consistency, compacting factor, and Vee-Bee time. On the other hand, when these values differ, two mixtures may show similarity in any one of the three standard tests, but will behave differently in the other two.

The behaviour of concrete departs significantly from the *Bingham model* in at least one of the following respects.
  (i) yield stress is not well defined, and
  (ii) the flow curve is not linear except over a very limited range of shear rates.

In case the flow curve is concave upward, i.e. one for which the slope increases with shear rate, it indicates that the shearing forces are destroying some structure that existed in the material when it was unstressed, the progress of destruction being greater for higher shear rates. The curve *ABC* of Fig. 7.3 is obtained as the rate of shear is gradually increased up to *C*, if the rate of shear is now decreased steadily to zero, the down curve may not be coincident with the up curve. If the point *C* represents the breakdown in the structure under shear and the structural breakdown

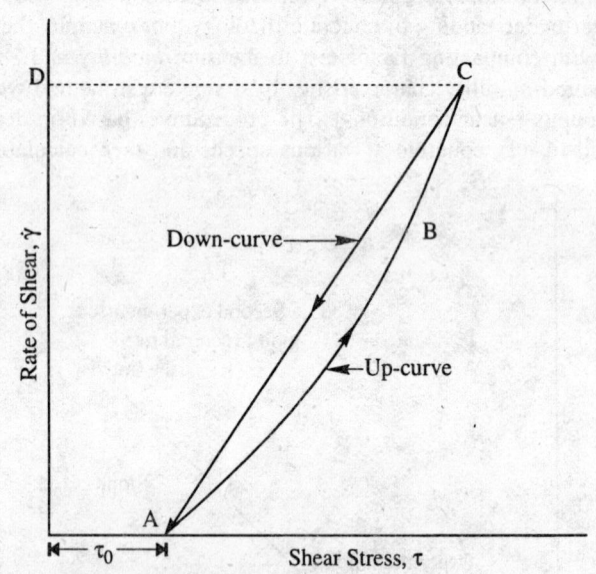

Fig. 7.3 Deformation curve for fresh concrete

process is immediately and instantaneously reversed, the decrease in shear rate from *D* will result in a progressive built up of structure to the same state as it had on the up curve. On the other hand, time is required for the structure to rebuild, thus as the rate of shear is decreased the shear stress at any particular rate of shear on the down curve will be less than the shear at the same rate on the up curve and the two curves will not be coincident and hysteresis loop will form. Provided its limits are recognized, the Bingham model can be applied to fresh concrete under practical circumstances.

7.3 FACTORS AFFECTING RHEOLOGICAL PROPERTIES

Like workability the rheological properties of fresh concrete are affected by the mix proportions, i.e. the amount of each constituent, properties of the ingredients, the presence of admixtures, the amount of mixing, and the time elapsed after mixing. Though these factors have already been explained in earlier sections, in this section the factors are discussed in relation to their effects on the flow properties of concrete.

7.3.1 Mix Proportions

The concrete mixture is proportioned to provide the workability needed during construction and to assure that the hardened concrete will have the stipulated performance characteristics.

A concrete mix having an excess amount of coarse aggregate will lack sufficient mortar to fill the void system, resulting in a loss of cohesion and mobility. Such a mix is termed harsh and requires a greater amount of effort to place and compact. The harshness may also be caused by a low air content; an increase in entrained air may alleviate the excessive use of fine aggregate. On the other hand, an excessive amount of fine aggregate or entrained air in a concrete mixture will greatly increase the cohesion and render the concrete difficult to move. A high fine aggregate content increases the surface area of particles, which increases the amount of paste required to coat these surfaces to have the same amount of *mobility*. This, in turn, can result in increased *drying shrinkage* and *cracking*.

In practice the concrete is usually proportioned with an excess of fine aggregate and with a higher cement content than would be necessary for a concrete mixture of optimum fine aggregate content.

A higher cement content makes the concrete sticky and sluggish particularly in the normal range of slumps for cast-in-situ concrete. Furthermore, the lower *water-cement ratio* and higher cement content reduces workability of rich mixes from that measured immediately after mixing.

7.3.2 Consistency

The *consistency* of concrete, as measured by the slump test, is an indicator of the relative water content in the concrete mix. An increase in the water content or slump above that required to achieve a workable mix produces greater fluidity and decreased internal friction. The reduction in the cohesion within the mixture increases the potential for segregation and excessive bleeding. Thus a water content more than that needed will not improve the rheological properties of concrete. On the other hand, too low a slump or water content will reduce the mobility and compactability which pose difficulties in placement and consolidation. As far as rheological properties of concrete are concerned an increase of one per cent in *air entrainment* is equivalent to an increase of one per cent in *fine aggregate* or an increase of the *water content* by three per cent. An excessively dry mix may also result in a loss of cohesion and increase in dry segregation.

7.3.3 Hardening and Stiffening

Elevated temperature, use of rapid-hardening cement, cement deficient in gypsum, and use of accelerating admixtures, increase the rate of hardening which reduces the mobility of concrete. The dry and porous aggregate will rapidly reduce workability by absorbing water from the mixture or increasing the surface area to be wetted.

7.3.4 Aggregate Shape and Texture

The shape of the aggregate particles and *aggregate texture* influence the rheology of concrete appreciably. The rough and highly *angular aggregate* particles will result in higher percentage of voids being filled by mortar, requiring higher fine aggregate contents and correspondingly higher *water content*. Similarly, an *angular fine aggregate* will increase internal friction in the concrete mixture and require higher water contents than well-rounded natural sands.

7.3.5 Aggregate Grading

A well-graded aggregate gives good *workability*. The absence of a particular size of aggregate (gap-graded) or a change in the size distribution may have appreciable effect on the void system and workability. These effects are greater in the *fine aggregate* than in *coarse aggregate*. As the fine aggregate becomes finer, the water requirement increases and concrete mixture becomes increasingly sticky. As the fine aggregate becomes coarser, cohesion is reduced, the mixture becomes harsh and tendency for bleeding increases. Adjustment of the grading of fine aggregate will be necessary to maintain workability as the above mentioned changes occur.

7.3.6 Maximum Aggregate Size

An increase in the maximum size of aggregate will reduce the fine aggregate content required to maintain a given workability, and will thereby reduce the surface area to be wetted and hence the cement content necessary for a constant *water-cement ratio*.

7.3.7 Admixtures

Out of the large number of admixtures used in concrete to obtain improved performance characteristics, the admixtures which have significant effect on the rheology of concrete are plasticizers and super-plasticizers, air-entraining agents, accelerators and retarders. These admixtures are used in three ways:
  (i) to give increased workability with the same long-term strength and durability,
  (ii) to give same workability with less water content and hence higher strength, and
  (iii) to give the same workability and strength with less cement content; however, the reduced cement content should be enough from durability considerations.

The commercially available plasticizers based on lignosulphonate salt (0.15 per cent) can reduce the water content by 10 per cent without any detrimental effect. Super-plasticizers have made it possible to obtain *flowing* or *self-levelling concrete* which needs minimum compaction effort to compact it into corners of the forms and around the congested reinforcement. Generally, a yield value of 350 Pa is proposed as the maximum value for the concrete to be self-levelling. Such a concrete can be obtained by the addition of ultrapure lignosulphate or synthetic napthalene or metamine resins. However, the concrete with very high workabilities obtainable with superplasticizers are likely to segregate.

As explained earlier the *water-cement system* consititutes a system of particles suspended in a high electrolyte concentration bringing the particles close enough to stick. This process is termed *flocculation*. The admixtures modify the condition in the cement-water suspension to prevent the formation of flocculated structure by changing the interparticle attraction/repulsion.

The *plasticizer* or *super-plasticizers* interact with cement particles, introducing a *membrane* of absorbed charged molecules around each particle, which prevent physically the particles approaching each other so closely as to stick, so that flocculation is prevented. The membrane thus forms a steric barrier to close contact. By increasing the thickness of *absorbed membrane*, the attraction between particles can be reduced to zero. The process reduces the yield value by the mechanism of *deflocculation* or dispersion of cement particles. In addition repulsive electrical forces are also generated between particles due to the absorption of ionized compounds, which increase the *plastic viscosity* through the operation of secondary electroviscous effect. Thus the *admixtures* of the group prevent close contact betwen the cement particles by a combination of electrostatic or steric repulsions, thus weakening the structure which can form at rest and reducing the yield value. Beyond an optimum percentage, the increase in the concentration of the plasticizer does not decrease the yield rate appreciably. However, reduction in yield value increases with increasing cement content of the mix. The accompanying change in *plastic viscosity* varies from an increase at low sand content (coarse overall grading) to a decrease at higher sand content because when the concrete with low sand content flows, it is flocculated cement in the mix that separates the coarse particles. If the cement is deflocculated these coarse particles come closer and touch each other generating a greater resistance to flow resulting in an increase in the plastic viscosity of concrete even though the cement paste is reduced. On the other hand, concrete with high sand content relies less on cement because there is sufficient sand to fill the voids in the coarse aggregate, so that the dispersion of cement does not bring the aggregate any closer together.

The *air-entraining agents* introduce spherical air bubbles 10 to 250 μm in diameter by modifying the surface tension of the aqueous phase in the mix. This is accomplished by compounds having negatively charged head group ($-COO^-$, SO_3^{2-}, SO_4^{2-}) which is *hydrophilic* (water attracting) and a nonpolar tail which is *hydrophobic* (water repelling). The materials most commonly used are abietic and pemeric acid salts, sodium olerate, sodium caprate, alkyle sulphate, etc. These compounds generate air bubbles with an apparent negative charge which form bridges between cement particles giving an increased yield value. The bubbles act like ball-bearings to allow larger particles to flow past each other more easily, thus decreasing the plastic viscosity. The *air-entrainment* changes the rheological properties of concrete very significantly by increasing *cohesion* and reducing tendency for *bleeding*. For example, 4 to 6 per cent air-entrainment by volume increases the cohesion of mix permitting a reduction in sand content by about 5 per cent.

The *accelerators* or *retarders* will reduce or extend the workability time for a given mixture. The time of retention of improved workability is critical for the uses where concrete is to be transported or manipulated before placement. The reduction in workability is similar to the stiffening of concrete due to slow chemical reaction taking place during the induction period. Since the admixture changes the nature of membrane around the cement particles and the composition of aqueous solution, they change the rate of stiffening and this has been especially noticed with *super-plasticizers*. A retarder plasticizer reduces workability loss and lengthens the retention time due to the slowing down of the process of setting. The superplasticizers give large improvements in the workability without retarders that would be needed if conventional plasticizers were used at such high dosages. It is also seen that the original flowing consistency could be retained by adding a second dose after up to 60 minute for metamine resin and up to 150 minute for naphthalene resin. Sometimes superplasticizers and retarders are used in combination to obtain slower stiffening than that obtained when they are used alone.

7.4 MIXTURE ADJUSTMENTS

Proper attention to the rheological properties of a mixture can effectively reduce construction and material costs. The changes in rheological properties of concrete will often be detected visually. It is essential to make adjustments as the properties of material and field conditions change.

8

Properties of Hardened Concrete

8.1 INTRODUCTION

The principal properties of concrete which are of practical importance are those concerning its strength; stress-strain characteristics; shrinkage and creep deformations; response to temperature variation; *permeability* and *durability*. Of these, the strength of concrete assumes a greater significance because the strength is related to the structure of hardened cement paste and gives an overall picture of the quality of concrete. The strength of concrete at a given age under given curing conditions is assumed to depend mainly on *water-cement ratio* and degree of compaction. Abrams *water-cement law* in this connection is well-known. Probably it is more correct to relate the strength of concrete to the concentration of the solid products of hydration of cement in the space available for these products, and the Power's gel/space ratio versus strength is more relevant in these studies. The *voids* present in concrete mass have been found to influence greatly the strength of concrete.

8.2 STRENGTHS OF CONCRETE

8.2.1 Compressive Strength

Of the various strengths of concrete the determination of compressive strength has received a large amount of attention because the concrete is primarily meant to withstand compressive stresses. Cubes, cylinders and prisms are the three types of compression test specimens used to determine the compressive strength. The cubes are usually of 100 mm or 150 mm side, the cylinders are 150 mm diameter by 300 mm height; the prisms used in France are 100 mm × 100 mm × 500 mm is size. The specimens are cast, cured and tested as per standards prescribed for such tests. When cylinders are used, they have to be suitably capped before the test, an operation not required when other types of specimens are tested.

The compressive strengths given by different specimens for the same concrete mix are different. The cylinders and prisms of a ratio of height or length to the lateral dimension of 2 may give a strength of about 75 to 85 per cent of the cube strength of normal-strength concrete. The effect of height/lateral dimension ratio of specimen on compressive strength is given in Fig. 8.1.

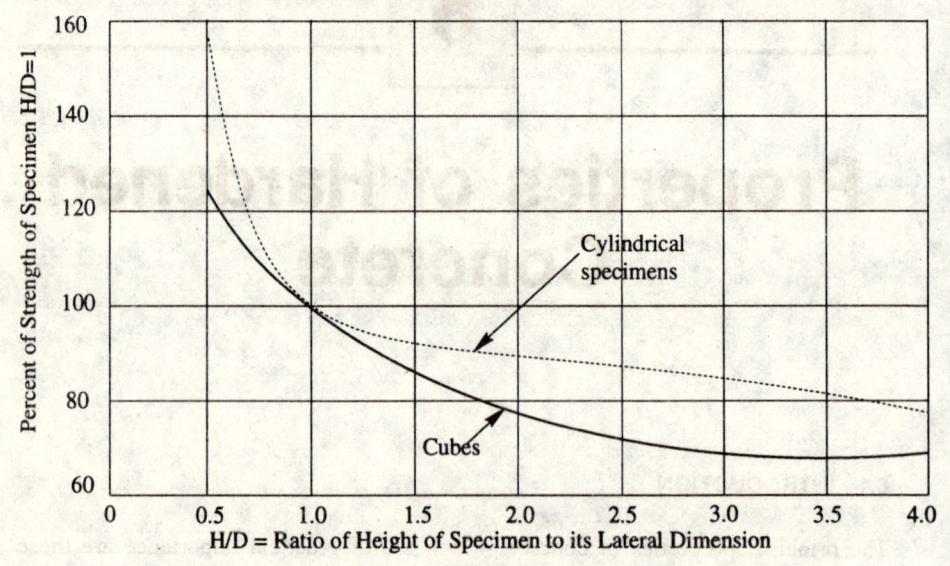

Fig. 8.1 Effect of height/lateral dimension ratio of specimen on the compressive strength

8.2.2 Flexural Strength

The determination of *flexural tensile strength* is essential to estimate the load at which the concrete members may crack. As it is difficult to determine the *tensile strength* of concrete by conducing a direct tension test, it is computed by flexure testing. The flexural tensile strength at failure or the *modulus of rupture* is thus determined and used when necessary. Its knowledge is useful in the design of pavement slabs and airfield runway as flexural tension is critical in these cases. The modulus of rupture is determined by testing standard test specimens of 150 mm × 150 mm × 700 mm over a span of 600 mm or 100 mm × 100 mm × 500 mm over a span of 400 mm, under symmetrical two-point loading. The modulus of rupture is determined from the moment at failures as $f_r = M/Z$. Thus the computation of f_r assumes a linear behaviour of the material up to failure which is only a rough estimation. The results are affected by the size of the specimens; casting, curing and moisture conditions; manner of loading (third point or central point loading); rate of loading, etc. The test is conducted and the strength determined according to the prescribed standards. The strength estimated by flexure test is higher than the tensile strength of concrete because of the assumption of the linear behaviour of material up to failure in the computation of f_r. On the other hand, the direct test gives lower apparent tensile strength. The accidental eccentricity in the direct tension test may also lower the apparent tensile strength. In the direct tension test, as the entire volume of specimen is under maximum stress, the probability of weak element occurring is high. The relationships of flexural strength with compressive and tensile strengths are given Figs. 8.2 and 8.3, respectively.

Properties of Hardened Concrete 85

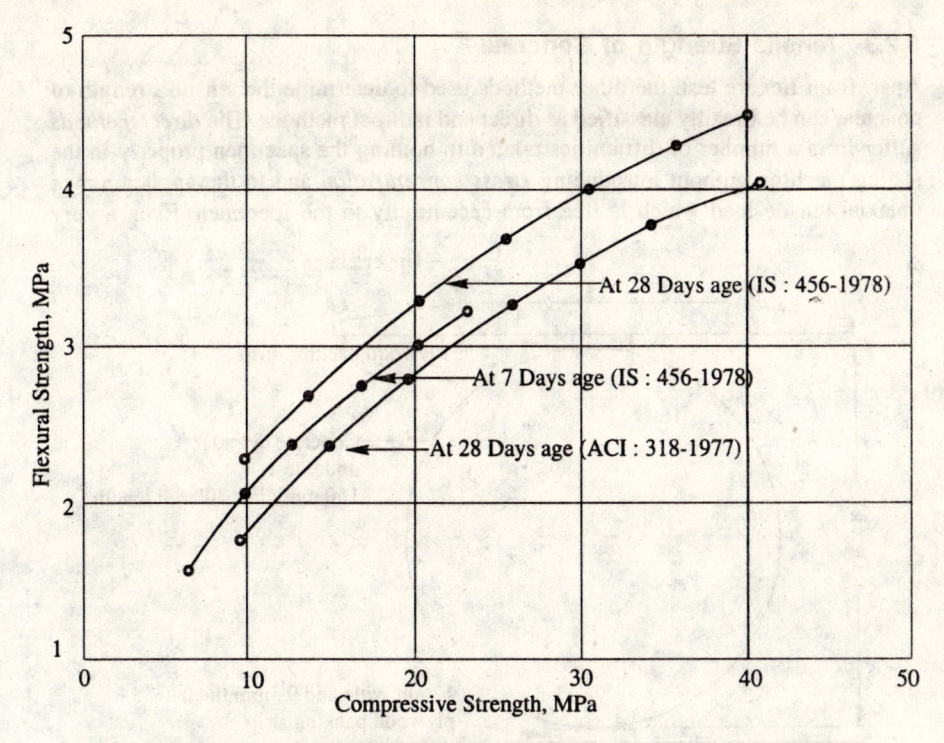

Fig. 8.2 Relationship between flexural strength and compressive strength at various ages

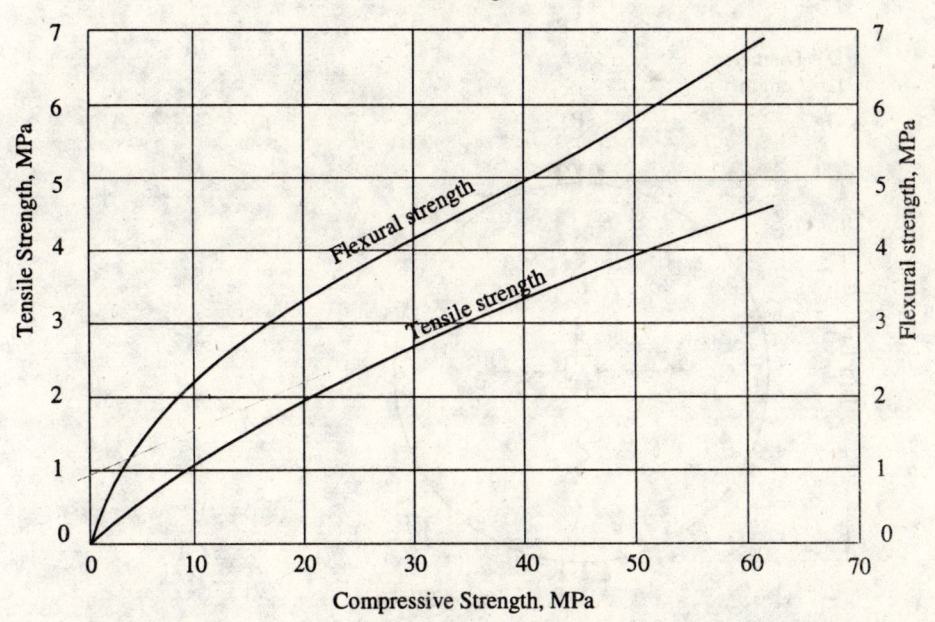

Fig. 8.3 Relationship between compressive strength, tensile strength and flexural strength

8.2.3 Tensile Strength of Concrete

Apart from flexure test, the other methods used to determine the tensile strength of concrete can be broadly classified as direct and indirect methods. The *direct methods* suffer from a number of difficulties related to holding the specimen properly in the testing machine without introducing *stress concentration* and to the application of uniaxial tensile load which is free from eccentricity to the specimen. Even a very

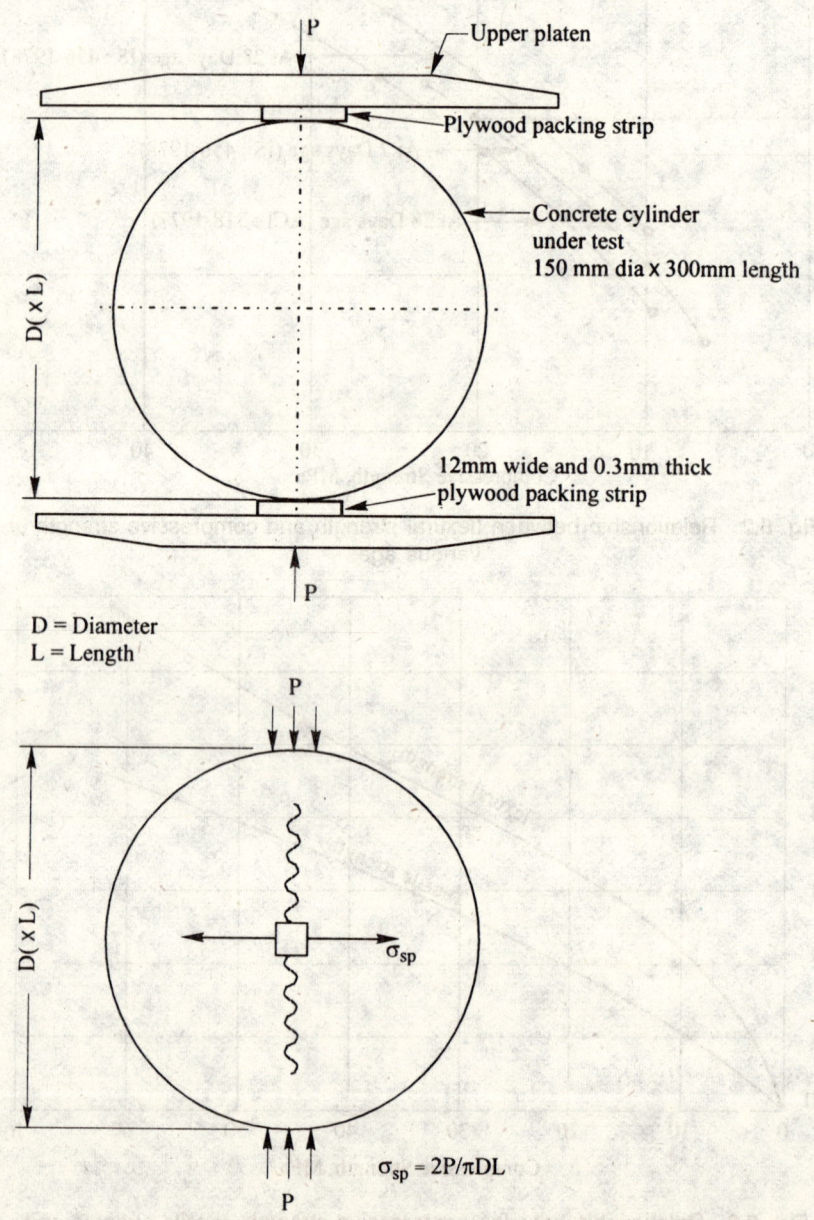

Fig. 8.4 Loading arrangement for split strength determination

small eccentricity of load will induce bending and axial force conditions and the concrete fails at apparent tensile stress other than the tensile strength.

Because of the difficulties involved in conducting the *direct tension* test, a number of *indirect methods* have been developed to determine the tensile strength. In these tests, in general a compressive force is applied to a concrete specimen in such a way

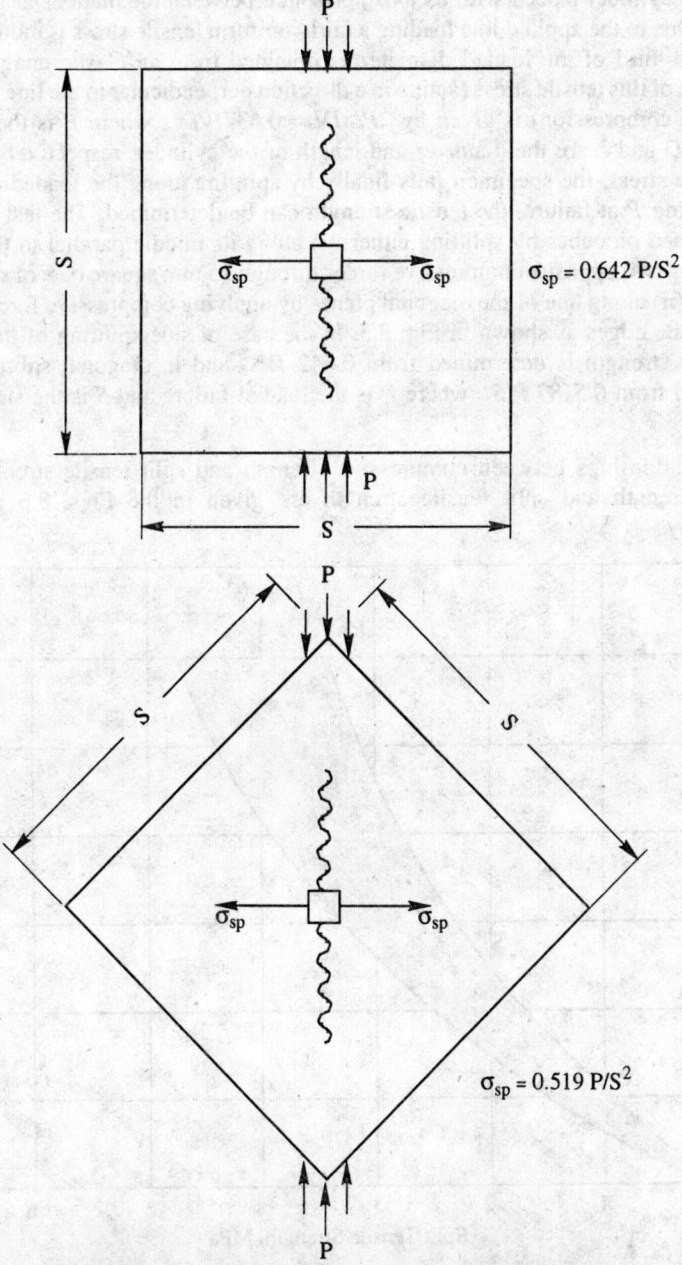

$\sigma_{sp} = 0.642\ P/S^2$

$\sigma_{sp} = 0.519\ P/S^2$

Fig. 8.5 Split strength from testing of cubes

that the specimen fails due to tensile stresses induced in the specimen. The tensile stress at which failure occurs is the *tensile strength* of concrete.

The *splitting tests* are well-known indirect tests used for determining the tensile strength of concrete, sometimes referred to as the *splitting tensile strength* of concrete. The test consists of applying compressive line loads along the opposite generators of a concrete cylinder placed with its axis horizontal between the plattens as shown in Fig. 8.4. Due to the applied line loading a fairly uniform tensile stress is induced over nearly two-third of the loaded diameter as obtained from an elastic analysis. The magnitude of this tensile stress (acting in a direction perpendicular to the line of action of applied compression) is given by $2P/\pi DL = 0.637 P/DL$, where P is the applied load, and D and L are the diameter and length of the cylinder, respectively. Due to this tensile stress, the specimen fails finally by splitting along the loaded diameter and knowing P at failure, the tensile strength can be determined. The test can also be performed on cubes by splitting either: (i) along its middle parallel to the edges by applying two opposite compressive forces through 15 mm square bars of sufficient length, or (ii) along one of the diagonal planes by applying compressive forces along two opposite edges as shown in Fig. 8.5. In the case of side-splitting of the cubes, the tensile strength is determined from $0.642\ P/S^2$ and in diagonal splitting it is determined from $0.5187\ P/S^2$, where P is the load at failure and S is the side of the cube.

The relationships between compressive strength and split tensile strength; and flexural strength and split tensile strength are given in the Figs. 8.6 and 8.7, respectively.

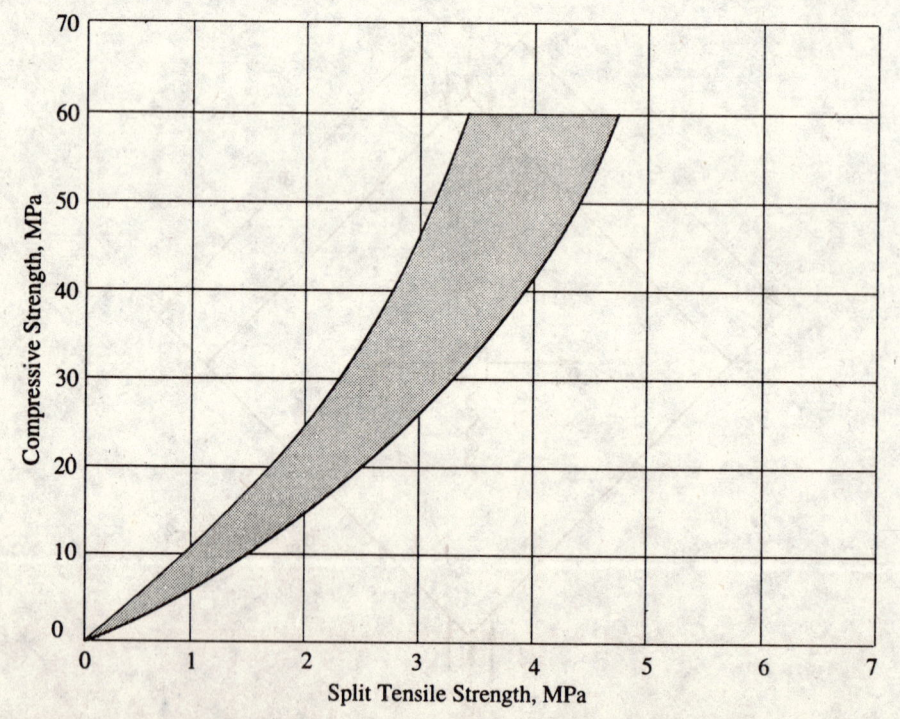

Fig. 8.6 Relationship between compressive strength and split tensile strength

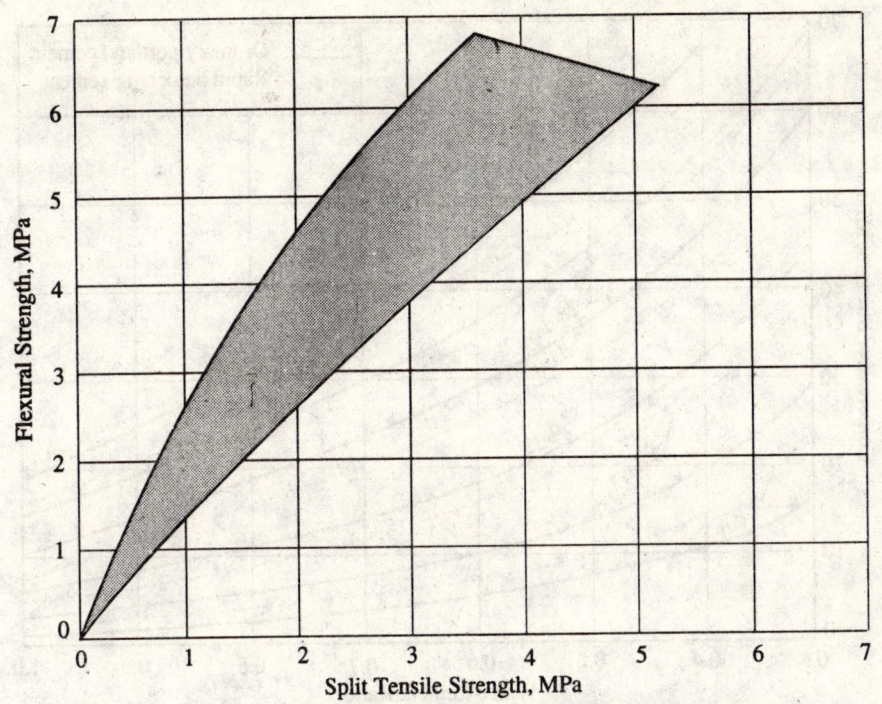

Properties of Hardened Concrete 89

Fig. 8.7 Relationship between flexural strength and splitting strength

Advantages of the splitting test for determining the tensile strength are as follows.
(i) The test is simple to perform and gives more uniform results than other tension tests.
(ii) The strength determined is closer to the actual tensile strength of the concrete than that given by the modulus of rupture test.
(iii) The same moulds can be used for casting specimens for both compression and tension tests.

The splitting tests have also been performed on prisms, i.e. on one-half of the specimen left after performing the modulus of rupture test. Splitting-type tests have also been done on the ring specimens to determine tensile strength. Mortar and concrete rings have been tested by subjecting them to internal pressure. The double punch test is another test performed on concrete cylinders to determine the tensile strength.

8.2.4 Factors Influencing the Strength of Concrete

The factors influencing the strength of concrete can be grouped into two categories:
(i) Factors depending on testing method: (a) size of test specimen, (b) size of specimen in relation to the size of aggregate, (c) support conditions of specimen, (d) moisture conditions of the specimen, (e) type of loading adopted, (f) rate of loading of the specimen, (g) type of testing machine, and (h) the assumptions made in the analysis relating stress to failure load.
(ii) Factors independent of the type of test: (a) the type of cement, age (Fig. 8.8), type of aggregate and admixture, (b) degree of compaction (Fig. 8.9), (c) concrete

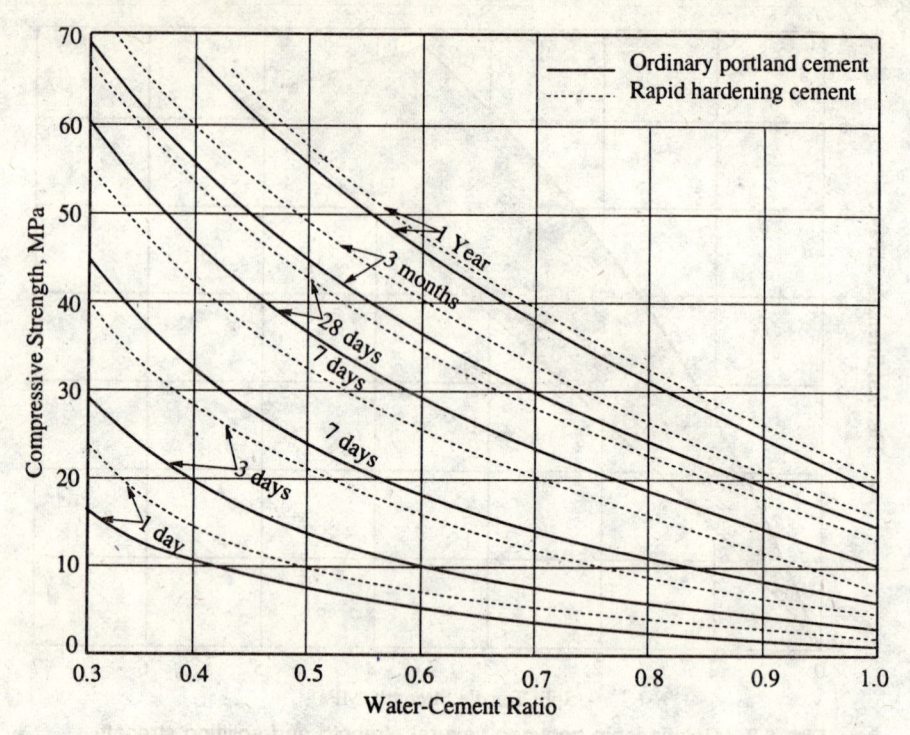

Fig. 8.8 Effect of water-cement ratio on compressive strength at different ages

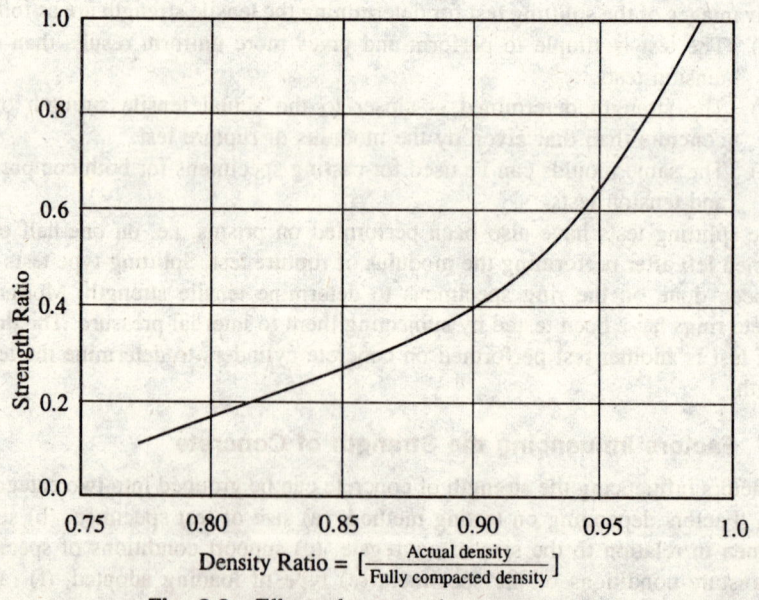

Fig. 8.9 Effect of compaction on strength

mix proportions, i.e. cement content (Fig. 8.10), aggregate-cement ratio (Fig. 8.11), amount of air-voids (Fig. 8.12) and water-cement ratio, (d) type of curing and temperature of curing, (e) nature of loading to which the specimen is subjected, i.e. static, sustained, dynamic etc., and (f) type of stress situation that may exist, viz. uniaxial, biaxial and triaxial.

The above factors have been described in different sections.

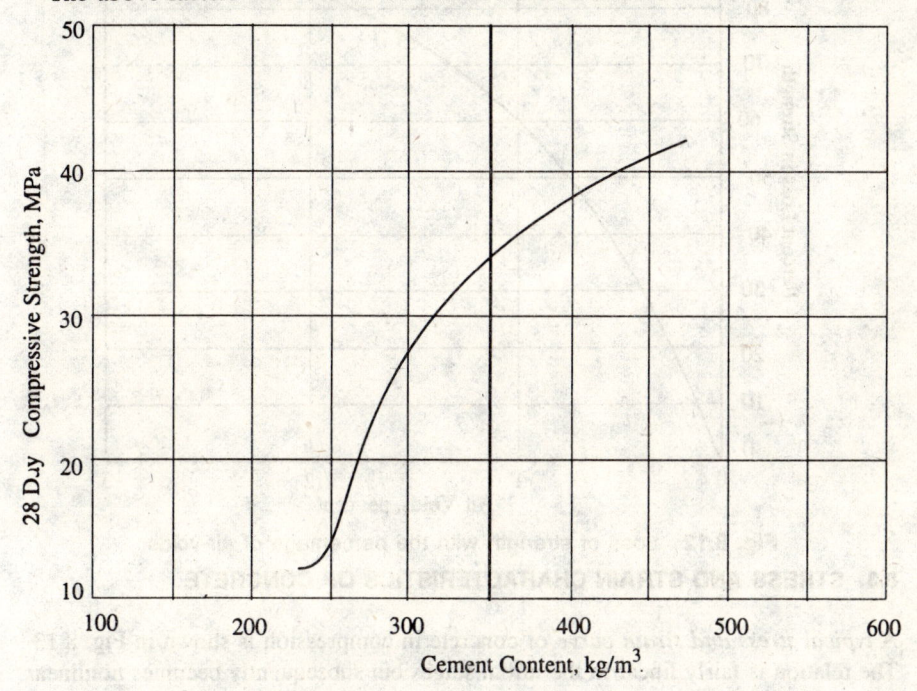

Fig. 8.10 Effect of cement content on strength

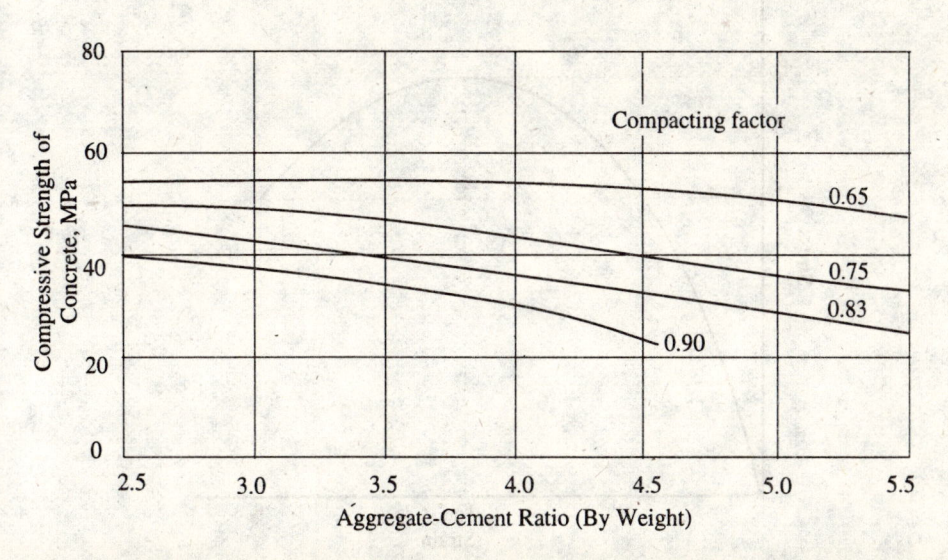

Fig. 8.11 Effect of aggregate-cement ratio on compressive strength

92 Concrete Technology

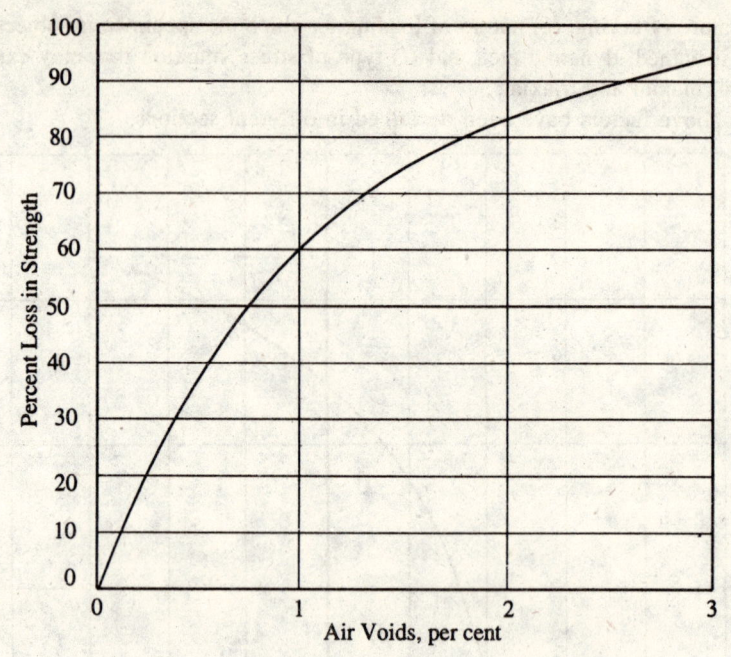

Fig. 8.12 Loss of strength with the percentage of air-voids

8.3 STRESS AND STRAIN CHARACTERISTICS OF CONCRETE

A *typical stress and strain curve* of concrete in compression is shown in Fig. 8.13. The relation is fairly linear in the initial stages but subsequently becomes nonlinear

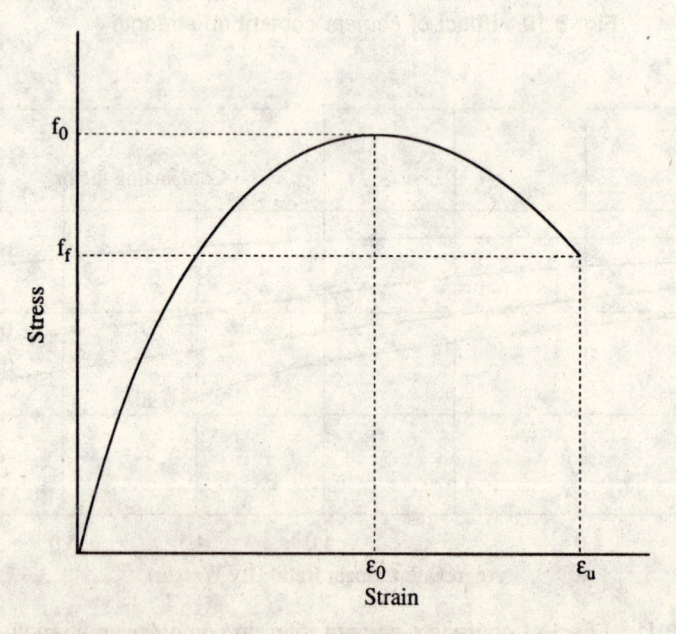

Fig. 8.13 Stress-strain relationship for concrete

reaching a maximum value and then a descending portion is obtained before concrete finally fails. The curve is usually obtained by testing a cylinder with a height-to-lateral dimension ratio of at least 2, the test being conducted under uniform rate of strain. If a uniform rate of stress is adopted, it will not be possible to obtain the descending portion of stress and strain curve beyond the maximum stress.

An equation representing the stress and strain curve completely should satisfy the following conditions:

(i) at $f = 0, \epsilon = 0$ $\quad\quad \dfrac{df}{d\epsilon} = E_t$

(ii) at $f = f_0, \epsilon = \epsilon_0$ $\quad\quad \dfrac{df}{d\epsilon} = 0$

(iii) at $f = f_f, \epsilon = \epsilon_u$

The equation satisfying all these conditions is used in the limit state design method. In another case, some simplifying assumptions are made. One of the major assumptions is made in approximating the stress-strain curve to a straight line, i.e. treating the concrete as linearly elastic material. This approximation is used in *working stress method* of design of structural concrete without much loss of accuracy up to about 50 per cent of f_0. Concrete is not strictly elastic in the sense that if it is unloaded after being stressed to $0.5 f_0$ or less, a *permanent set* is noticed (Fig. 8.14). However, the magnitude of the permanent set gradually decreases with more cycles of loading and unloading (within $0.5 f_0$) and the stress-strain curve tend to become a straight line. The *creep deformation* of concrete also varies linearly with the sustained stress up to a value of $0.5 f_0$. Hence, for all practical purposes, the concrete could be considered as a linear elastic material when stress does not exceed $0.5 f_0$.

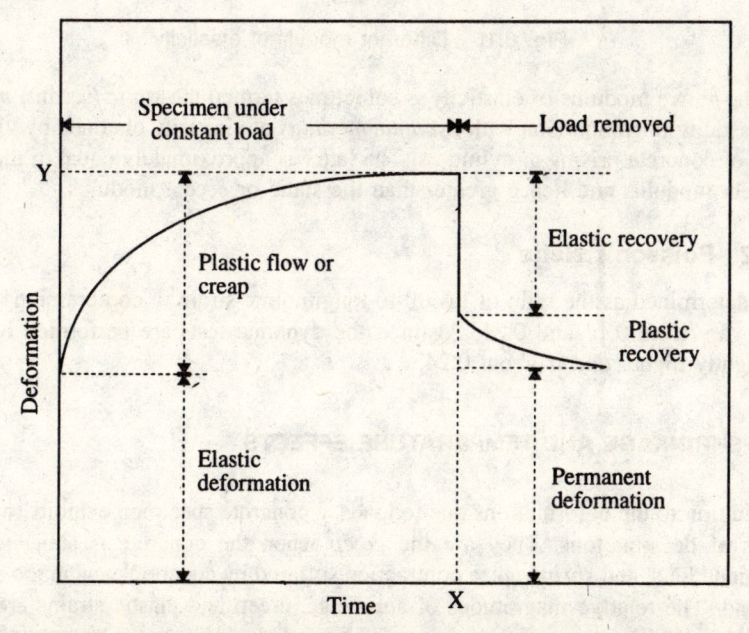

Fig. 8.14 Deformation of hardened concrete under load

8.3.1 Modulus of Elasticity

The *modulus of elasticity* of concrete would be a property for the case when the material is treated as elastic. If we consider the stress-strain curve of the first cycle, the modulus could be defined as *initial tangent modulus, secant modulus, tangent modulus* or *chord modulus*, as shown in Fig. 8.15. In the laboratory determination of the modulus of elasticity of concrete, a cylinder is loaded and unloaded (stress not exceeding one-third of f_0) for three or four cycles, the stress-strain curve is plotted after residual strain has become almost negligible and the average slope of stress-strain curve is taken.

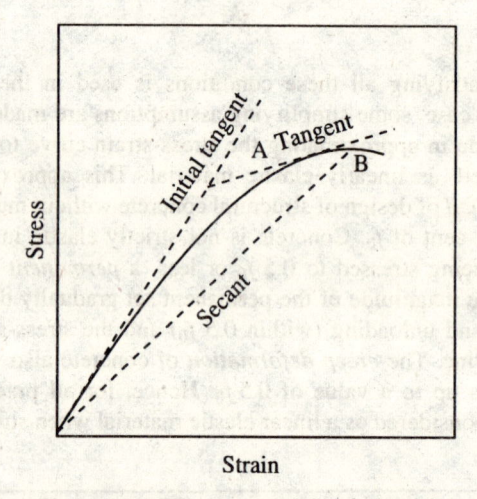

Fig. 8.15 Different modulii of elasticity

The above modulus of elasticity is sometimes termed the *static* (secant) *modulus* of elasticity in comparison with *dynamic modulus* of elasticity obtained by vibration tests of concrete prisms or cylinders. The latter is approximately equal to the initial tangent modulus and hence greater than the static or secant modulus.

8.3.2 Poisson's Ratio

It is determined as the ratio of lateral to longitudinal strain in compression test and may vary from 0.11 and 0.21. Again if the dynamic tests are performed its value is slightly higher and is about 0.24.

8.4 SHRINKAGE AND TEMPERATURE EFFECTS

In addition to the deformations due to loads, a concrete specimen exhibits two other types of deformations. They are the *creep* when the concrete is subjected to a sustained load, and *shrinkage*, a contraction suffered by concrete even in the absence of load. The relative magnitudes of shrinkage, creep and elastic strains are of the similar order. The term *volume change* is often used to refer to the change in volume that occurs due to the shrinkage, creep, temperature and, possibly, chemical disintegration. Two types of shrinkage strains are recognized, namely, plastic and drying shrinkage.

8.4.1 Plastic Shrinkage

The hydration of cement causes a reduction in the volume of the system of cement plus water to an extent of about 1 per cent of the volume of dry cement. This contraction is *plastic strain* and is aggravated due to loss of water by evaporation from the surface of concrete, particularly under hot climates and highwinds. This can result in surface cracking.

8.4.2 Drying Shrinkage

The shrinkage that takes place after the concrete has set and hardened is called *drying shrinkage* and most of it takes place in the first few months (it also coincides with the period of active creep and thus the two are inextricably related). Withdrawal of water from concrete stored in unsaturated air voids causes *drying shrinkage*. A part of this shrinkage is recovered on immersion of concrete in water (Fig. 8.16). It is termed moisture movement. In the absnece of other reliable data, the shrinkage can be estimated from Schorer's formula:

$$\epsilon_s = 0.00125(0.90 - h)$$

where ϵ_s is *shrinkage strain* and h represents *relative humidity* expressed as a fraction. In an environment of avergae humidity of 50 per cent $h = 0.5$, $\epsilon_s = 0.0005$ and it may be noticed that in fully saturated condition ($h = 1.0$), $\epsilon_s = -0.000125$ which indicates *swelling*.

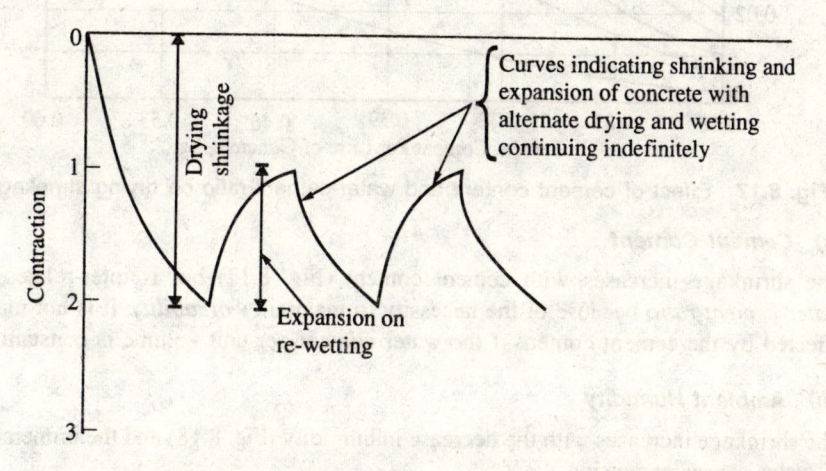

Fig. 8.16 Variation of drying shrinkage and moisture movement with alternate drying and wetting

The rate of shrinkage decreases with time. The tests indicate that 14 to 34 per cent of 20-years shrinkage occurs in two weeks, 40 to 70 per cent in 3 months and 66 to 80 per cent in one year.

The shrinkage is affected by:

(i) Water-cement Ratio

The shrinkage increases with the increase in the water-cement ratio as shown in Fig. 8.17.

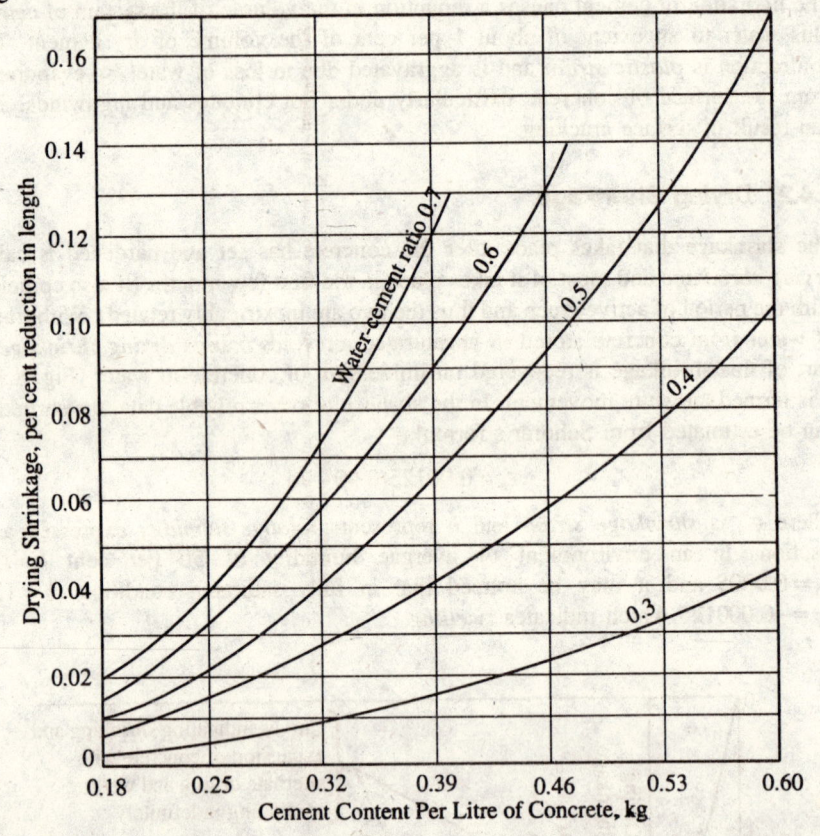

Fig. 8.17 Effect of cement content and water-cement ratio on drying shrinkage

(ii) Cement Content

The shrinkage increases with cement content (Fig. 8.17) but is inter-related to *water-cement ratio* because of the necessity to maintain *workability*. It is not much affected by the cement content if the water content per unit volume is constant.

(iii) Ambient Humidity

The shrinkage increases with the decrease in humidity (Fig. 8.18) and the immersion in water causes expansion.

(iv) Type of Aggregate

The aggregate which exhibit moisture movement themselves and have low elastic modulus cause large shrinkage. A concrete using sandstone may shrink twice as much as one using limestone. An increase in maximum size decreases the shrinkage.

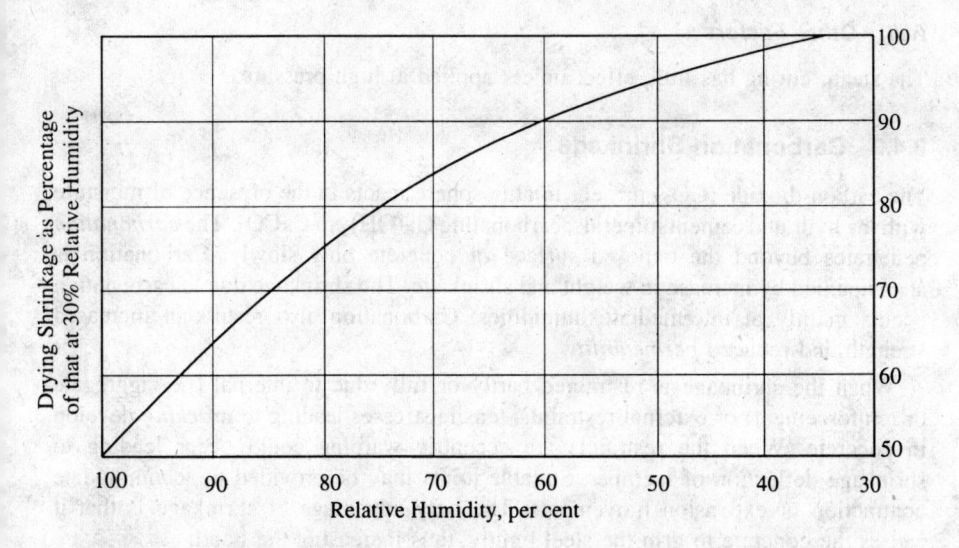

Fig. 8.18 Effect of relative humidity on drying shrinkage of concrete

The grading and shape has little effect on shrinkage. The effect of type of aggregate on the shrinkage is shown in Fig. 8.19.

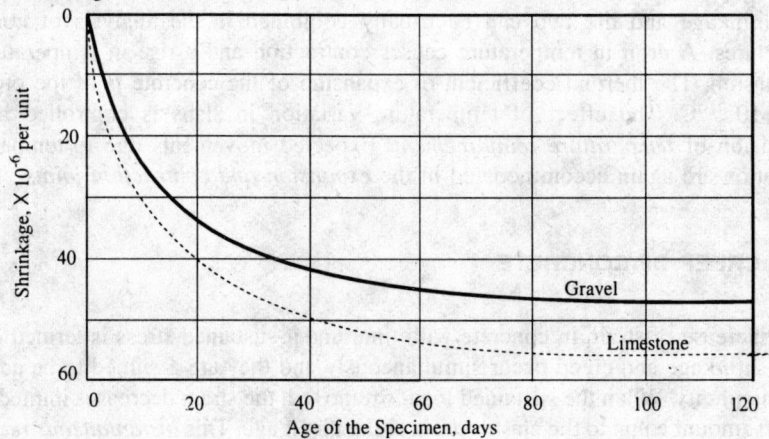

Fig. 8.19 Effect of the type of aggregate on the drying shrinkage of concrete

(v) Size and Shape of Specimen

Both rate and ultimate magnitude decrease with surface/volume ratio of the specimen.

(vi) Type of Cement

The rapid-hardening cement shrinks somewhat more than the others.

(vii) Admixtures

The shrinkage increases with the addition of calcium chloride and reduces with lime replacement.

(viii) Other Factors

The steam curing has little effect unless applied at high pressure.

8.4.3 Carbonation Shrinkage

The carbon dioxide (CO_2) present in atmosphere reacts in the presence of moisture with the hydrated cement minerals, carbonating $Ca(OH)_2$ to $CaCO_3$. The *carbonation* penetrates beyond the exposed surface of concrete only slowly. Carbonation is accompanied by increase in weight and shrinkage. The shrinkage due to carbonation occurs mainly at intermediate humidities. Carbonation also results in increased strength and reduced *permeability*.

When the shrinkage is restrained partly or fully due to internal (i.e. aggregate or reinforcement) or external restraints, tensile stresses leading to *cracking* develop in concrete. When the restraints are eccentric warping could occur leading to shrinkage deflection of member. Suitable joints may be provided to accommodate contraction or expansion movements. The only advantage of shrinkage is that it causes the concrete to grip the steel tightly, thus increasing the bond.

8.4.4 Temperature Variation

Effects of temperature variation on concrete could be considered similar to those of shrinkage and the two can be usually combined in the analysis of concrete structures. A drop in temperature causes contraction and a rise in temperature an expansion. The thermal coefficient of expansion of the concrete is of the order of $10 \times 10^{-6}/°C$. The effect of temperature variation in slabs is controlled by the provision of *temperature reinforcement*. Expected movements due to temperature variation are again accommodated in the *expansion and contraction joints*.

8.5 CREEP OF CONCRETE

The increase of strain in concrete with time under sustained stress is termed *creep*. The shrinkage and creep occur simultaneously and they are assumed to be additive for simplicity. When the sustained load is removed, the strain decreases immediately by an amount equal to the elastic strain at the given age. This *instantaneous recovery* is then followed by a gradual decrease in strain, called *creep recovery* which is a part of total creep strain suffered by the concrete. If a loaded concrete specimen is viewed as being subjected to a constant strain, the creep decreases the stress progressively with time. This is called *relaxation*.

The rate of creep decreases with time and the creep strains attained at a period of five years are usually taken as terminal values. While 80 to 85 per cent shrinkage strains occur in six months, only about 75 per cent of creep strains occur in twelve months. All the factors which influence shrinkage influence creep also in similar way. Types of aggregate, cement, and admixtures, entrained air, mix proportions, mixing time and consolidation, age of concrete, level of sustained stress, ambient humidity, temperature, and the size of the specimen are among the important factors influencing *creep*.

8.6 PERMEABILITY OF CONCRETE

The study of *permeability* of concrete is important for the following reasons:
  (i) The penetration by materials in solution may adversely affect the durability of concrete, e.g., $Ca(OH)_2$ leaches out, and the aggressive liquids attack the concrete.
  (ii) In case of reinforced concrete, ingress of moisture and air will result in corrosion of steel which leads to an increase in the volume of steel, and to cracking and spalling of concrete cover.
  (iii) The moisture penetration depends on permeability and if the concrete can become saturated with water it is more vulnerable to frost action.
  (iv) The permeability is also of interest in connection with water-tightness of liquid retaining structures and the problem of hydrostatic pressure in the interior of the dams.

The flow of water through concrete is similar to flow through any porous body. The *pores* in cement paste consist of *gel pores* and *capillary pores*. The pores in concrete due to incomplete compaction are voids of larger size which give a *honeycomb structure* leading to concrete of low strength, are not considered here. Since the capillary pores are larger in size than gel pores, and the cement paste is 20 to 100 times more permeable than the gel itself, the *permeability* of cement paste is controlled by the capillary porosity of the paste. In rocks the pores are fewer in number, but being of large size they lead to higher permeability.

The permeability of cement paste also varies with the age of concrete or with the degree of hydration. With age the permeability decreases because gel gradually fills the original water filled space. For the pastes hydrated to the same degree, the *permeability* is lower with lower *water-cement ratio* or higher cement content. For the same water-cement ratio the permeability of paste with coarser cement particles is higher than that with finer cement. In general, the higher the strength of cement paste, the lower will be the permeability. A durable concrete should be relatively impervious. Permeability can be measured by a simple test by measuring the quantity of water flowing through a given thickness of concrete in a given time. The drop in the hydraulic head using Darcy's equation is:

$$\frac{dq}{dt} \cdot \frac{1}{A} = k \cdot \frac{\Delta h}{L}$$

where $\frac{dq}{dt}$ is the rate of flow (ml/s), A the cross-sectional area (mm^2), Δh the drop in hydraulic head (mm), L the thickness of the sample in millimetres and k the coefficient of permeability (mm/s).

8.7 DURABILITY OF CONCRETE

A durable concrete is one which can withstand the conditions for which it has been designed, without deterioration over a period of years. The factors affecting the durability may be external or internal causes.

External causes may be grouped in the following categories:
  (i) physical, chemical or mechanical,
  (ii) environmental, such as occurrence of extreme temperatures, abrasion and electrostatic action, and
  (iii) attack by natural or industrial liquids and gases.

Internal causes include:
  (i) alkali-aggregate reaction,
  (ii) volume changes due to difference in thermal properties of the aggregate and cement paste, and
  (iii) permeability of concrete.

The common forms of chemical attack are: (a) leaching out of cement, (b) actions of sulphates, sea water and natural slightly acidic waters. The resistance to these attacks varies with the type of cement used and increases in the order: OPC and RHC; portland blast-furnace cement or low-heat cement; sulphate- resisting portland cement; pozzolanic cement; and supersulphated cement.

8.8 SULPHATE ATTACK

The ground water in clayey soils containing alkali, magnesium and calcium sulphates constitute the *sulphate solution* which reacts with $Ca(OH)_2$, and with calcium aluminate hydrate. The products of reaction are gypsum and calcium sulphoaluminate which have greater volume and thus lead to expansion and disruption of concrete. The resistance to sulphate attack can be improved by:
  (i) adding or partial replacement of cement by pozzolana,
  (ii) using high pressure steam curing, and
  (iii) reduced permeability.

The sulphates in *sea water* react chemically with concrete. In addition, the salts crystallize in the pores of concrete which after the evaporation of water cause disruption. Also the frost, wave impact and abrasion tend to aggravate the sea-water damage. The corrosion of reinforcement may lead to rupture of cover concrete. Use of low *water-cement ratio*, well compacted concrete, good workmanship, reduced porosity, sufficient cover over reinforcement along with the use of aluminous sulphate resisting cement, portland blast-furnace or portland pozzolana cements are recommended.

The resistance of concrete to sulphate attack can be tested by storing the specimen in a solution of sodium or magnesium sulphate or in a mixture of the two.

8.9 ACID ATTACK

Concrete structures are also used for storing the liquids some of which are harmful to the concrete. In industrial plants, concrete floors come in contact with liquids which damage the floor. In damp conditions SO_2 and CO_2 and other acid fumes present in the atmosphere affect concrete by dissolving and removing part of the set cement. In fact, no portland cement is acid resistant. Concrete is also attacked by water containing free CO_2. Sewerage water also very slowly causes deterioration of concrete.

8.10 EFFLORESCENCE

The water leaking through cracks or faulty joints or through the areas of poorly compacted porous concrete, dissolves some of the readily soluble calcium hydroxide and other solids, and after evaporation leaves on the surface the calcium carbonate as white deposit. These deposits on the surface of concrete resulting from the leaching of calcium hydroxide and subsequent carbonation and evaporation, are termed *efflorescence*. Unwashed seashore aggregate, gypsum and alkaline aggregate also cause efflorescence.

8.11 FIRE RESISTANCE

In general, the concrete has good properties with respect to fire resistance, i.e. the period of time under fire during which concrete continues to perform satisfactorily is relatively high and no toxic fumes are emitted. The length of time over which the structural concrete preserves structural action is known as fire-rating. Under sustained exposure to temperature in excess of 35°C under the condition that a considerable loss of moisture from concrete is allowed leads to decrease in strength and in modulus of elasticity. The loss of strength at higher temperatures is greater in saturated than in dry concrete. Excessive moisture at the time of fire is the primary cause of spalling. In general, moisture content of concrete is the most important factor determining the structural behaviour at higher temperature.

Leaner mixes appear to suffer a relatively lower loss of strength than rich ones. Flexural strength is affected more than compressive strength. The loss of strength is considerably lower when the aggregate does not contain silica, e.g. concrete made with limestone, crushed brick and blast furnace slag aggregate. Low *conductivity* of concrete improves its fire resistance, and hence a light-weight concrete is more fire resistant than ordinary concrete. The calcined material aggregate having a low density leads to a good fire resistance of concrete. Due to *endothermic* nature of carbonate aggregate during calcination at high temperature, heat is absorbed and further temperature rise is delayed. For example, dolomite gravel leads to a good fire resistance of concrete.

TABLE 8.1 Coefficient of reduction in compressive strength γ_c of dense concrete on heating

Type of concrete aggregate	Coefficient γ_c at the temperature, °C										
	20	100	200	300	400	500	600	700	800	900	1000
Limestone	1.0	1.0	1.0	1.0	1.0	1.0	0.90	0.67	0.45	0.22	0
Granite	1.0	1.0	1.0	1.0	1.0	0.92	0.70	0.46	0.25	0	0

The data on the variation of strength of concrete upon heating obtained experimentally are generally conditional. The data obtained by generalizing the results from the fire resistance tests on actual reinforced concrete structures are given in the Table 8.1. The variation in the strength with temperature is shown in Fig. 8.20.

102 Concrete Technology

The modulus of elasticity of concrete is considerably reduced and thermal creep increases considerably at high temperature.

(a) Compressive Strength (% Amb) — γ_c vs Temperature, °C

(b) Initial Modulus (% Amb) — E_c vs Temperature, °C

(c) Coefficient of Thermal Expansion × 10^{-6} — α_c vs Temperature, °C

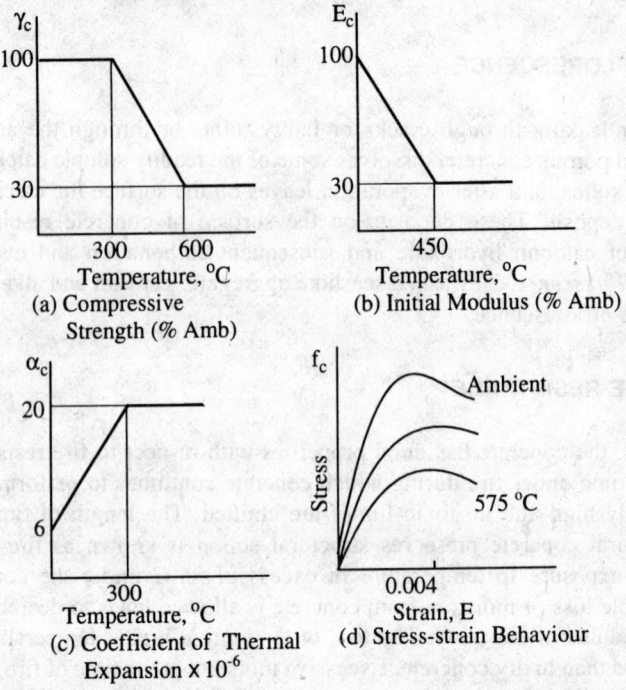

(d) Stress-strain Behaviour — f_c (Stress) vs Strain, E; curves for Ambient and 575 °C

Fig. 8.20 Thermomechanical properties of concrete at high temperature

The coefficient of thermal expansion of concrete using different types of aggregate given in Table 8.2 are valid for temperature up to 100°C. At higher temperature the values may differ considerably. The variation of thermo-mechanical properties of the concrete with temperature is shown in Fig. 8.21. The values of

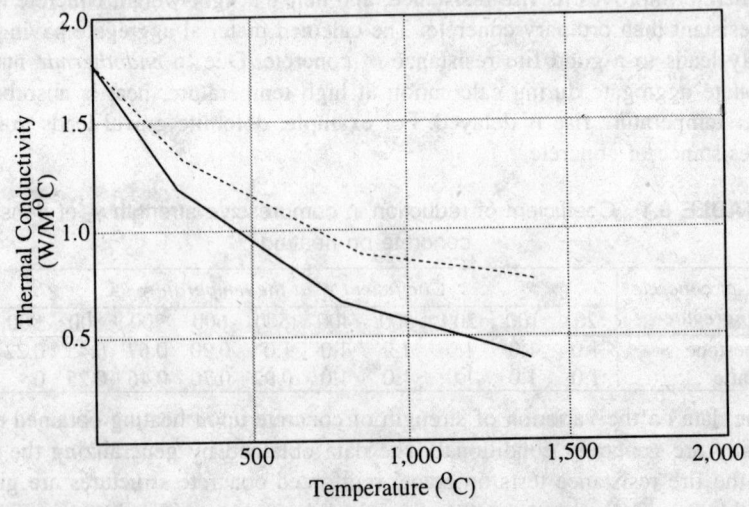

Fig. 8.21 Thermal conductivity of concrete

coefficient of thermal expansion of concrete prepared with different types of aggregates are given in the Table 8.2.

TABLE 8.2 Coefficient of thermal expansion of concrete

Type of aggregate	$\alpha_c'/°C$	Type of aggregate	$\alpha_c'/°C$
Quartz	11.9×10^{-6}	Granite	9.5×10^{-6}
Sandstone	11.7×10^{-6}	Basalt	8.6×10^{-6}
Gravel	10.8×10^{-6}	Limestone	6.8×10^{-6}

8.12 THERMAL PROPERTIES OF CONCRETE

The important thermal properties required for the design of structures are thermal conductivity, thermal diffusivity, specific heat, and coefficient of thermal expansion.

Thermal conductivity is a measure of the ability of the concrete to conduct heat and is measured in British Thermal Units per hour per square foot area of the body when the temperature difference is 1°F per foot thickness of the body. Thermal conductivity depends upon the composition of concrete. The structural concrete containing normal aggregate conducts heat more readily than light-weight concrete. Lower the *water-content* of the mix, the higher the conductivity of the hardened concrete. The density of the concrete does not appreciably affect the conductivity of ordinary concrete. The variation of thermal conductivity of concrete with temperature is shown in Fig. 8.21.

Thermal diffusivity is a measure of the rate at which temperature change within the mass takes place. Diffusivity can be determined by:

$$D = \frac{k}{Sd}$$

where D, k, S and d are the thermal diffusivity, thermal conductivity, specific heat and density of concrete, respectively.

The *specific heat* gives the heat capacity of concrete. It increases with the moisture content of concrete. The specific heat values of ordinary concrete are between 0.2 to 0.28 BTU/lb/°F.

The *coefficient of thermal expansion* of concrete depends on the composition of the mix and on the values of the coefficient of expansion of cement paste and aggregate. For ordinary cured concrete the coefficient decreases slightly with age but this is not the case in the concrete cured under high pressure steam. For ordinary concrete the value of coefficient of thermal expansion varies from 9×10^{-6} per °C to 12×10^{-6} per °C.

8.13 MICROCRACKING OF CONCRETE

Cracking of concrete can be defined as a separation of the individual components of concrete resulting in a discontinuous material. Depending upon the extent of cracking the cracks can be classified as macrocracks, microcracks and semi-microcracks. According to the location, the cracks can be classified as bond cracks, mortar cracks and aggregate cracks. The bond cracks are formed at the interface of the aggregate and mortar, whereas the mortar cracks and aggregate cracks are formed through the mortar and the aggregate, respectively.

A knowledge of the microcracking of concrete contributes considerably to the understanding of its properties, such as its inelastic nature, the descending portion of the stress-strain curve, the strength under combined, repeated and sustained loading, etc. The stress-strain curve is related to the internal cracking. The non-linearity of the stress-strain relation is due to propagation of bond and mortar cracks.

The progressive cracking in concrete with increasing strain has been indirectly determined by measuring the lateral expansion/contraction, surface cracking and by sonic methods. The exact nature of cracks and the strains at which they occur give precise information pertinent to the mechanism of cracking. Direct microscopic observations help in studying the extent of cracking both quantitatively and qualitatively.

The bond between aggregate and mortar plays an important role in controlling the strength characteristics of concrete. The existence of bond cracks prior to loading constitutes one of the weakest links in the heterogeneous concrete system. Due to the settlement of fresh concrete, hydration of cement, and shrinkage of concrete, the bond cracks exist in the hardened concrete near the large aggregates. It has been seen that the bond strength between the aggregate and the mortar is less than the tensile strength of the mortar. Hence it can be deduced that the bond between the aggregate and mortar controls the failure of concrete under uniaxial tensile loading whereas the tensile strength of mortar controls the strength of concrete under uniaxial compressive loading.

The failure process in plain concrete is a continuous one and proceeds in two ways. Bond cracks by themselves cannot cause failure, as they are isolated from each other. Failure occurs only when there are sufficient interconnected bond cracks with mortar cracks. The development of a continuous crack pattern does not lead to immediate loss of the load carrying capacity because concrete at this stage behaves as a highly redundant structure. As successive load paths become inoperative through bond cracking, alternative load paths (either entirely through mortar, or partly through mortar and partly through aggregate) continue to be available for carrying additional load. As the number of load paths decrease the intensity of stress and hence the magnitude of strain on remaining paths increases at a faster rate than external load. When an extensive continuous crack pattern has developed and the load paths have been reduced considerably, the carrying capacity of concrete decreases, and from this stage the stress-strain curve begins to descend.

9

Quality Control of Concrete

9.1 INTRODUCTION

Concrete is generally produced in batches at the site with the locally available materials of variable characteristics. It is, therefore, likely to be variable from one batch to another. The magnitude of this variation depends upon several factors, such as variation in the quality of constituent materials; variation in mix proportions due to batching process; variation in the quality of batching and mixing equipment available; the quality of overall workmanship and supervision at the site. Moreover, concrete undergoes a number of operations, such as transportation, placing, compacting and curing. During these operations considerable variations occur partly due to quality of plant available and partly due to differences in the efficiency of techniques used. Thus there are no unique attributes to define the quality of concrete in its entirety. Under such a situation concrete is generally referred to as being of *good, fair* or *poor* quality. This interpretation is subjective. It is, therefore, necessary to define the quality in terms of desired performance characteristics, economics, aesthetics, safety and other factors. Due to the large number of variables influencing the performance of concrete, quality control is an involved task. However, it should be appreciated that concrete has mainly to serve the dual needs of safety (under ultimate loads) and *serviceability* (under working loads) including *durability*. These needs vary from one situation and type of construction to another. Therefore, uniform standards valid for general application to all the works may not be practical.

Therefore, the aim of *quality control* is to reduce the above variations and produce uniform material providing the characteristics desirable for the job envisaged. Thus quality control is a corporate, dynamic programme to assure that all aspects of materials, equipment and workmanship are well looked after. The tasks and goals in these areas are properly set and defined in the *specifications* and *control requirements*. The specifications have to state clearly and explicitly the steps and requirements, adherence to which would result in a construction of acceptable quality. Except for compressive strength and appearance there is no early measure of construction performance. Each step in construction procedure is therefore to be specified. The probability based specifications containing allowable tolerances on its attributes is more rational and is preferred. *Quality control is thus conformity to the specifications, no more no less.* The most practical method of effective quality control is to check what is done in totality to conform to the specifications. An owner will have no right to expect anything more than what is in the specifications. The builder, on the other hand, knows that anything less than what is in the specifications will not be acceptable to the owner.

In view of the different processes involved in the manufacture of concrete, the problems of quality control are diversified and their solution elaborated. The factors involved are the personnel, the materials and equipment, the workmanship in all stages of concreting, i.e. batching of materials, mixing, transportation, placing, compaction, curing, and finally testing and inspection. It is therefore necessary to analyse the different factors causing variations in the quality and the manner in which they can be controlled.

9.2 FACTORS CAUSING VARIATIONS IN THE QUALITY OF CONCRETE

The main factors causing variation in concrete quality are as follows.

Personnel

The basic requirement for the success of any quality control plan is the availability of experienced, knowledgeable and trained personnel at all levels. The designer and the specification-writer should have the knowledge of construction operations as well. The site engineer should be able to comprehend the specification stipulations. Everything in quality control cannot be codified or specified and much depends upon the attitude and orientation of people involved. In fact, quality must be a discipline imbibed in the mind and there should be strong motivation to do every thing right the first time.

Material, Equipment and Workmanship

For uniform quality of concrete, the ingredients (particularly the cement) should preferably be used from a single source. When ingredients from different sources are used, the strength and other characteristics of the materials are likely to change and, therefore, they should only be used after proper evaluation and testing. The same type of cement from different sources and at different times from the same source exhibit variations in properties, especially in compressive strength. This variation in the strength of cement is related to the composition of raw materials as well as variations in the manufacturing process. The cement should be tested initially once from each source of supply and, subsequently, once every two months. Adequate storage under cover is necessary for protection from moisture. Set cement with hard lumps is to be rejected.

Grading, maximum size, shape, and moisture content of the aggregate are the major sources of variability. Aggregate should be separately stock piled in single sizes. The graded aggregate should not be allowed to segregate. The simple rule of grading is that:

(i) for fine aggregate, long continuous gradings are preferred and there should be minimum material passing through 300 micron and 150 micron sieves,

(ii) for fine aggregate, the gradings that are at the coarser end of the range are more suitable for rich mixes and those at the fine end of range should be suitable for lean mixes.

(iii) a coarser aggregate consistent with the size of the member and the spacing of reinforcement is more suitable, and

(iv) the aggregate sizes should be so selected that one size fits into the voids left by the next higher size.

The *aggregate* should be free from impurities and deleterious materials; since for every 1 per cent of clay in sand, there could be as much as 5 per cent reduction in the strength of the concrete. The moisture content of aggregates should be taken into account while arriving at the quantity of mixing water. Bulking of sand is important in several ways. It gives erroneous results, when volume batching is adopted, besides increasing the *water-cement ratio* which, in turn, enhances the workability but reduces the strength. The aggregates are required to be tested once initially for the approval of each sources of supply. Subsequently, tests should be conducted daily at the site for grading and moisture content.

The water used for mixing concrete should be free from silt, organic matter, alkali, and suspended impurities. Sulphates and chlorides in water should not exceed the permissible limits. Generally, water fit for drinking may be used for mixing concrete.

The equipment used for batching, mixing and vibration should be of the right capacity. Weight-batchers should be frequently checked for their accuracy. Weight-batching of materials is always preferred to volume batching. When weight-batching is not possible and the aggregates are batched by volume, such volume measures should be frequently checked for the weight-volume ratio. Mixer's performance should be checked for conformity to the requirements of the relevant standards. Concrete should be mixed for the required time, both under mixing and overmixing should be avoided. The vibrators should have the required *frequency* and *amplitude of vibration*.

The green concrete should be handled, transported and placed in such a manner that it does not get segregated. The time interval between mixing and placing the concrete should be reduced to the minimum possible. Anticipated targets of *strength, impermeability* and *durability of concrete* can be achieved only by thorough and adequate *compaction*. One per cent of the air voids left in concrete due to incomplete compaction can lower the compressive strength by nearly five per cent. Adequate *curing* is essential for handling and development of strength of concrete. The curing period depends upon the shape and size of member, ambient temperature and humidity conditions, type of cement, and the mix proportions. Nevertheless, the first week or ten days are the most critical, as any drying out during this young age can cause irreparable loss in the quality of concrete. Generally, the long-term compressive strength of concrete moist cured for only 3 days or 7 days will be about 60 per cent and 80 per cent, respectively, of the one moist cured for 28 days or more.

9.3 FIELD CONTROL

The field control, i.e. inspection and testing, play a vital role in the overall quality control plan. Inspection could be of two types, quality control inspection and acceptance inspection. For repeated operations early inspection is vital, and once the plant has stabilized, occasional checks may be sufficient to ensure continued satisfactory results. The operations which are not of repetitive type would require, on the other hand, more constant scrutiny.

Apart from the tests on concrete materials, concrete can be tested both in the fresh and hardened stages. Of these two, the tests on fresh concrete offer some opportunity for necessary corrective actions to be taken before it is too late. These include test on *workability, unit weight* or *air content* (where air-entrained concrete is used), etc. *Accelerated strength tests* by which a reliable idea about the potential 28-day strength can be obtained within few hours, are effective quality control tools. In contrast to this, the usual 28-day strength test is in fact a post mortem of concrete which has become history by then. It is, therefore, only *acceptance tests,* which help the decision-maker decide whether to accept or reject the concrete.

9.4 ADVANTAGES OF QUALITY CONTROL

The general feeling that quality control means extra cost is not correct, the advantages due to quality control offset the extra-cost. Some of the advantages of quality concrete are:

(i) Quality control means a rational use of the available resources after testing their characteristics and reduction in the materials costs.

(ii) In the absence of quality control there is no guarantee that over-spending in one-area will compensate for the weakness in another, e.g. an extra bag of cement will not compensate for incomplete compaction or inadequate curing. Proper control at all the stages is the only guarantee.

(iii) In the absence of quality control at the site, the designer is tempted to overdesign, so as to minimize the risks. This adds to the overall cost.

(iv) Checks at every stage of the production of concrete and rectification of the faults at the right time expedites completion and reduces delay.

(v) Quality control reduces the maintenance costs.

It should be realized that if the good quality concrete is made with cement, aggregates and water, the ingredients of bad concrete are exactly the same. The difference lies in the few *essential steps* collectively known as *quality control.*

9.5 STATISTICAL QUALITY CONTROL

Probability-based guidelines or specifications are usually laid down to ensure that the concrete attains its desired properties with the minimum expenditure. The specifications allow a certain *limits of variability* between individual samples. There is little gain in narrowing down the *tolerance limits* unless the process is capable of operating within these limits. The process of ensuring compliance to specifications which take into account the actual variability of concrete is termed *quality control.* The statistical quality control procedures are used to ascertain the range of values that can be expected under the existing conditions.

In the production of concrete the compliance to specifications requires that the *mix ingredients, size of aggregate, water-cement ratio, cement content, workability* as well as *methods of mixing, compaction* and *curing,* to be adopted for a particular work are specified such that they are easy to follow. It should be noted that the usual 28 day cube tests are not quality control measures in the strict sense, they are, in fact, acceptance tests. In situations of site production and placing, the quality of the concrete is to be controlled way ahead of the stage of testig cubes at 28 days. Moreover, the compressive strength, although taken as an index of the quality of concrete, does not satisfy the requirements of durability where impermeability and homogeneity are more important parameters. However, the acceptance criteria of the quality of the finished product can be based on the compressive strength of a specified number of 150 mm-cube specimens after 28-day moist-curing.

The basic parameters of statistical quality control are explained below.

Sampling

Since the quality of a larger mass of the materials or product is based on a few limited samples, it is necessary that samples be as representative as possible of the entire population. A sample should be chosen at random and not in a selective manner, i.e., obviously good or bad samples should not be purposely chosen.

Distribution of Results

The compressive strength test results of cubes from random sampling of a mix, although exhibit variations, when plotted on a histogram are found to follow a bell-shaped curve termed the *Normal* or *Gaussian distribution curve*. The results are said to follow a normal distribution as shown in Fig. 9.1, if they are equally

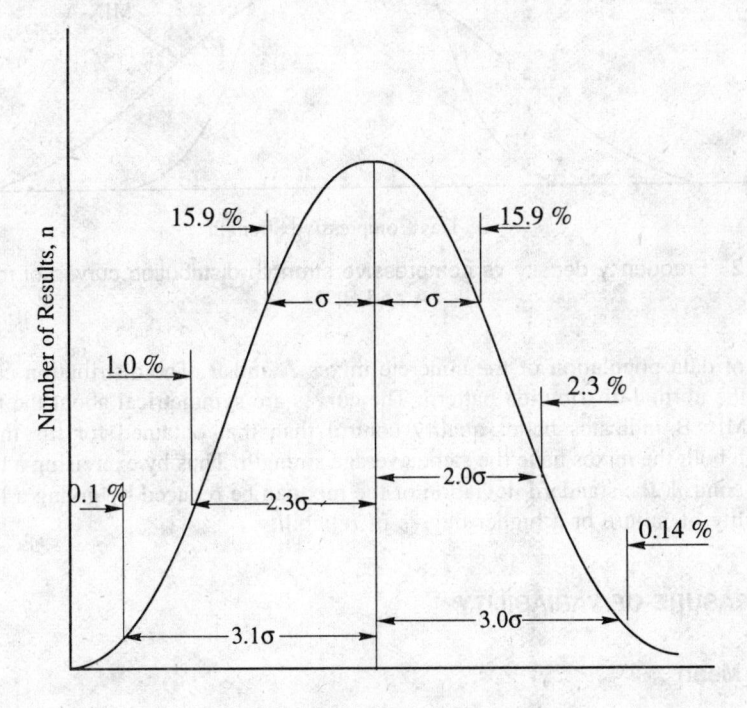

Fig. 9.1 Normal distribution of compressive strength results

110 Concrete Technology

spaced about the mean value. However, some divergence from the smooth curve is only to be expected, particularly if the number of results available is relatively small. The normal distribution curve can be used to ascertain the variation of strength from the mean. The area beneath the curve represents the total number of test results. The proportion of results less than the specified value is represented by the area beneath the curve to the left-hand side of the vertical line drawn through the specified value.

A *normal distribution* curve can be defined by two parameters, namely, the *mean strength* and the *standard deviation*. The mean strength is defined as the arithmetic mean of the set of actual test results. The standard deviation S is a measure of the spread of the results and the formula for computing the standard deviation is given in IS: 456-1978 as explained below.

Figure 9.2 shows the frequency density vs compressive strength distribution

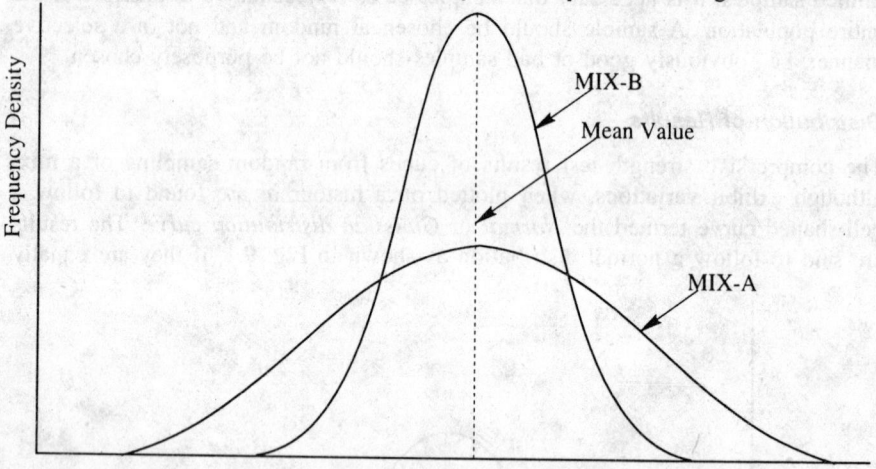

Fig. 9.2 Frequency density vs compressive strength distribution curves of mixes A and B.

curves of data population of the concrete mixes A and B. The distribution curves follow the normal distribution pattern. The curves are symmetrical about the mean value. Mix B indicates better quality control than that obtained for the mix A although both the mixes have the same average strength. Thus by exercising a better quality control, the standard deviation of the mix can be reduced by giving a lower probability of failure or a higher degree of reliability.

9.6 MEASURE OF VARIABILITY

9.6.1 Mean

The average or a mean $\bar{x}$ for a set of n observations $x_1, x_2 \ldots x_n$, is expressed as:

$$\bar{x} = \frac{\sum_{i=1}^{n} x_i}{n}$$

As the sample size n increases, $\bar{x}$ approaches the mean of the entire population.

9.6.2 Range

The range is the difference between the largest and the smallest values in a set of observations.

9.6.3 Standard Deviation

The *root mean square* (rms) *deviation* of the whole consignment from the mean $\bar{x}$ is termed as the *standard deviation* and is defined numerically as:

$$S = \sqrt{\frac{\sum_{i=1}^{n}(x_i - \bar{x})^2}{n-1}}$$

where
S = standard deviation of the set of observations
x_i = any value in the set of observations
$\bar{x}$ = arithmetic mean of the values
n = total number of observations

S has the same units as the quantity x. The square of standard deviation is called *variance*.

Standard deviation increases with increasing variability. It may be appreciated that the value of S is minimum for very good control and progressively increases as the level of control slackened as indicated by Table 9.1. An important property of standard deviation relating it to the proportions of all the results falling within or outside certain limits, can generally be assumed in the case of concrete work without serious loss of accuracy as long as techniques of random sampling are followed.

The spread of the normal distribution curve along the horizontal scale is governed by the standard deviation, while the position of the curve along the vertical scale is fixed by the average value, the limit below or above which the proportion of the results can be expected to fall are set out as $(x \pm kS)$, where k is the *probability factor*. For different values of k the percentage of results falling above and below a particular value is illustrated in Fig. 9.3, in relation to the area bounded by the normal probability curve. The values of k are given in Table 9.3.

Alternatively, the variation of results about the mean can be expressed by *coefficient of variation* which is a nondimensional measure of variation obtained by dividing the standard deviation by the average and is expressed as:

$$v = \frac{S}{\bar{x}} \times 100$$

With constant coefficient of variation, the standard deviation increases with strength and is larger for high strength concrete. The values of coefficient of variation suggested by Himsworth for different degrees of control are given in Table 9.1.

9.7 APPLICATIONS

The standard deviation and the coefficient of variation are useful in the design and quality control of concrete. As the strength test results follow normal distribution,

112 Concrete Technology

there is always the probability that some results may fall below the specified strength. Recognizing this fact IS: 456-1978 has brought in the concept of characteristic strength. The term *characteristic strength* indicates that value of the strength of material below which not more than 5 per cent of the test results are expected to fall. In the design of concrete mixes, the average strength to be aimed, i.e. the target mean strength, should be appreciably higher than the minimum or characteristic strength if the quality of concrete is to comply with the requirements of the specifications. If, from previous experience, the expected variation in compressive strength is represented by a certain standard deviation or coefficient of variation, it is possible to compute the target mean strength of the mix, which would carry with it a predetermined chance of results falling below a specified minimum strength. The target mean strength is obtained by using the following relation.

$$f_t = f_{ck} + kS$$

where
f_t = target mean strength
f_{ck} = characteristic strength
k = probability factor
S = standard deviation

The value of k where not more than 5 per cent (1 in 20) of test results are expected to fall below characteristic strength is 1.65 as obtained from the Fig. 9.3 or Table 9.3 and the above relation reduces to

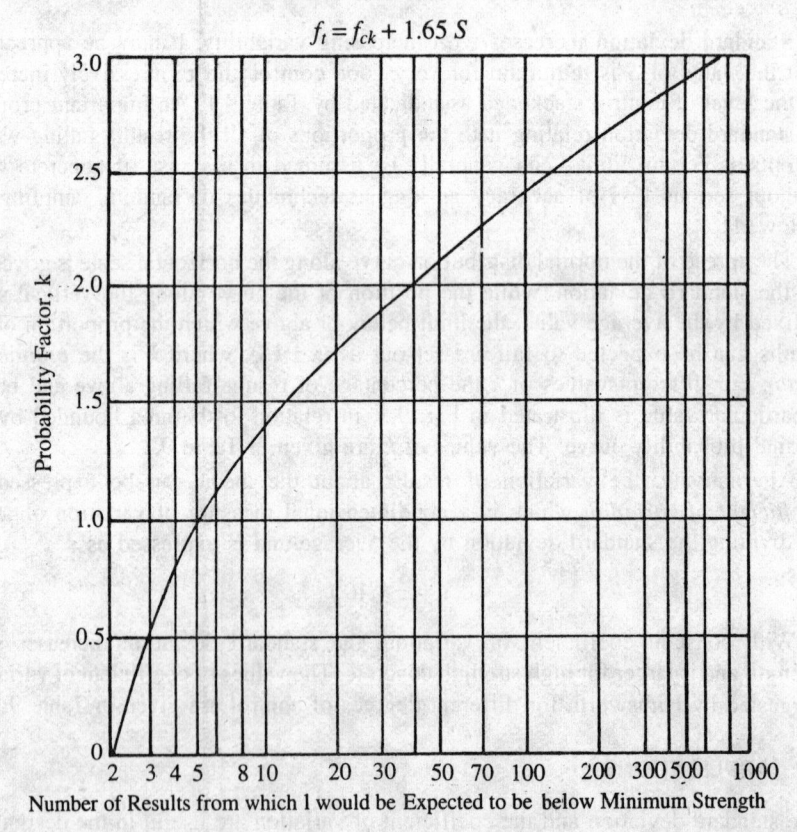

$$f_t = f_{ck} + 1.65\,S$$

Fig. 9.3 Probability factor k and proportion of results expected to be below the minimum strength

However, it should be noted that for a given degree of control, the standard deviation method yields higher target mean strengths than the coefficient of variation method for low-and medium-strength concretes. For high-strength concretes, the coefficient-of-variation method yields higher values of target mean strength. The cost of production being dependent on the target mean strength of concrete, the method of evaluation should be consistent with the observed trend of results for different ranges of strength. However, the use of the coefficient of variation is not envisaged in IS: 546-1978.

To keep a control on the quality of concrete produced, it is required to cast a number of specimens from random samples and test them at suitable intervals to obtain results as quickly as possible to enable the level of control to be established with reasonable accuracy in a short time. IS: 456-1978 stipulates that *random samples* from fresh concrete shall be taken as specified in IS: 1199-1959 and the cubes shall be made, cured and tested as described in IS: 516-1959. The code prescribes sampling every 150 m^3 of concrete or part thereof, the sample being drawn on each day for first four days of concreting and thereafter at least once in seven days of concreting. As far as the requirements of specifications with regard to the *acceptance criteria* for concrete is concerned, IS: 456-1978 stipulates that the concrete shall be deemed to satisfy the strength requirements provided the following specifications are satisfied.

(i) every sample has a test strength not less than the characteristic value; or

(ii) the strength of one or more samples though less than the characteristic value, is in each case not less than the greater of:

(a) the characteristic strength minus 1.35 times the standard deviation; and

(b) 0.80 times the characteristic strength; and

(iii) the average strength of all samples is not less than characteristic strength plus $1.65\left(1 - \dfrac{1}{\sqrt{n}}\right)$ times the standard deviation, where n is the number of samples.

According to IS: 456-1978, the concrete shall be deemed not to satisfy the strength requirements in the case of the following.

(i) The strength of any sample is less than the greater of:
   (a) $f_{ck} - 1.35\,S$
   (b) $0.80\,f_{ck}$

(ii) The average strength of all samples is less than

$$\left[f_{ck} + \left(1.65 - \dfrac{3}{\sqrt{n}}\right)S\right]$$

Concrete which does not satisfy the strength requirement specified in case (i) above but has strength greater than the requirements of case (ii) may be accepted as being structurally adequate without further testing at the discretion of the designer.

Concrete of each grade shall be assessed separately. The porous or honeycombed concrete may be rejected under certain conditions specified by the code.

As all the main variations of a job as in batching, proportions of ingredients, characteristics of aggregates, etc., are reflected in the fluctutions of the *water-cement ratio* and this ratio is, in itself, closely related to compressive strength, a *control-ratio* can be applied to reduce the water-cement ratio to take into account the observed variations in the strength. The control ratio is defined as:

$$\frac{\text{Water–cement ratio required to produce average strength}}{\text{Water–cement ratio required to produce minimum strength}}$$

Table 9.4 gives suggested values of control ratios for various probabilities of results falling below minimum with four different degrees of control.

TABLE 9.1 Standard deviation for different types of controls (according to Himsworth)

Type of control	Excellent	Very Good	Good	Fair	Poor	Uncontrolled
Standard Deviation, MPa	2.8	3.5	4.2	5.6	7.0	8.4
Coefficient of variation, per cent	5	12	15	18	20	25

TABLE 9.2 Standard deviation for different grades of concrete

Grades of Concrete	M10	M15	M20	M25	M30	M35	M40
Assumed standard deviation, MPa	2.3	3.5	4.6	5.3	6.0	6.3	6.6

TABLE 9.3 Probability factor for various tolerances

Percentage of results below characteristic strength	50 (1 in 2)	16 (1 in 6)	10 (1 in 10)	5 (1 in 20)	2.5 (1 in 40)	1.0 (1 in 100)	0.5 (1 in 200)	0.0
Probability factor k	0	1.0	1.28	1.65	1.96	2.33	2.58	infinity

TABLE 9.4 Control ratio for difference degrees of control

Degree of Control	Control Ratio for probabilities of			Remarks
	1 in 25	1 in 40	1 in 100	
A	0.82	0.80	0.76	Weight-batching of cement and aggregate by servo-operation
B	0.79	0.76	0.72	Weight-batching of cement and aggregate by manual operation
C	0.77	0.74	0.69	Weight-batching of cement and volume batching of aggregate
D	0.75	0.72	0.67	Volume batching for both cement and aggregate

Illustrative Example

In a construction where the concreting has been completed in three stages, a series of tests are conducted for a given grade of concrete. The specimens are tested at 28 days in each case and the results are represented in the Table 9.5. Establish the standard deviation for the grade of concrete.

TABLE 9.5 28-day compressive strengths of sets of cube specimens

Stage I		Stage II		Stage III	
Sample number	Concrete strength MPa	Sample number	Concrete strength MPa	Sample number	Concrete strength MPa
1	28.3	25	27.2	49	35.3
2	28.1	26	27.6	50	35.1
3	27.6	27	24.9	51	33.9
4	26.7	28	26.8	52	33.2
5	29.2	29	26.4	53	31.3
6	27.4	30	30.0	54	35.7
7	26.1	31	29.4	55	34.6
8	31.2	32	27.1	56	31.3
9	30.0	33	27.8	57	30.4
10	25.7	34	30.1	58	32.2
11	28.6	35	26.8	59	27.3
12	27.1	36	27.2	60	28.8
13	28.7	37	27.6	61	31.3
14	33.6	38	32.7	62	29.0
15	24.0	39	31.8	63	33.0
16	30.6	40	30.0	64	32.7
17	30.5	41	31.3	65	30.8
18	23.8	42	26.4	66	33.9
19	29.0	43	37.5	67	28.1
20	28.0	44	23.3	68	30.1
21	25.0	45	30.6	69	27.6
22	29.7	46	26.4	70	29.0
23	28.1	47	25.3	71	28.8
24	29.4	48	25.0	72	36.7
				73	29.2
				74	33.4
				75	27.6
				76	29.7
				77	35.0
				78	33.9

The standard deviation of the concrete produced up to the end of stage I, i.e. for samples 1 to 24,

$n = 24$
$\Sigma x = 676.4$
$\bar{x} = \Sigma x/n = 28.18$ MPa
$\Sigma(x - \bar{x})^2 = 120.61$

$$S = \sqrt{\frac{\Sigma(x - \bar{x})^2}{n - 1}} = 2.30 \text{ MPa}$$

For stage II (samples 25 to 48),

$n = 24$
$\Sigma x = 679.2$
$\bar{x} = 28.30$ MPa

$\Sigma(x-\bar{x})^2 = 217.64$
$S = 3.08$ MPa

For the standard deviation of concrete produced up to the end of second stage,
$n = 24 + 24 = 48$
$\Sigma x = 1355.6$
$\bar{x} = \Sigma x/n = 28.24$ MPa
$\Sigma(x-\bar{x})^2 = 338.42$
$S = 2.68$ MPa

For the samples (49 to 78) of stage III,
$n = 30$
$\Sigma x = 948.9$
$\bar{x} = 31.63$ MPa
$\Sigma(x-\bar{x})^2 = 220.40$
$S = 2.76$ MPa

To obtain the standard deviation of the concrete produced to date, it is necessary to combine the standard deviations from different stages.
$n = 24 + 24 + 30 = 78$
$\Sigma x = 2304.5$
$\bar{x} = 29.55$ MPa
$\Sigma(x-\bar{x})^2 = 770.77$

$$S = \sqrt{\frac{770.77}{77}} = 3.16 \text{ MPa}$$

If in the above construction work, the grade of concrete used is M30, apply the acceptance criteria of IS: 456-1978 to the following results each representing a day's production (average strength of 3 specimens tested at 28 days) expressed in MPa:

28.00, 27.77, 29.10, 27.13, 28.77, 28.30, 27.33, 29.07, 26.57, 27.73, 28.10, 28.03, 30.70, 29.23, 30.47, 25.57, 34.77, 33.40, 32.10, 29.43, 31.10, 32.47, 28.60, 31.50, 30.07, 32.87.

Arranging the data in the descending order:

34.77, 33.40, 32.87, 32.47, 32.10, 31.50, 31.10, 30.70, 30.47, 30.07, 29.43, 29.23, 29.10, 29.07, 28.77, 28.60, 28.30, 28.10, 28.03, 28.00, 27.77, 27.73, 27.33, 27.13, 26.57 and 25.57 MPa.

Applying the criteria of IS: 456-1978, it may be noted that:

(i) The first ten values are straightaway accepted, the sample strength being greater than the characteristic strength (30 MPa) in each case.

(ii) The 11th and the following results are less than the characteristic strength (30 MPa) and are compared with:

(a) 0.8x characteristic strength, i.e., 24 MPa, and

(b) Characteristic strength $-1.35 \times S$, i.e., 25.95 MPa (taking $S = 3$).

Althouth all these samples except the last one have values greater than 25.95 MPa. But none of them comply with the requirements as the average strength becomes less than $30 + 3 \times \left(1.65 - \frac{3}{\sqrt{n}}\right)$. Hence none of them is acceptable.

9.8 QUALITY MANAGEMENT IN CONCRETE CONSTRUCTION

As explained earlier the quality, meant to measure the degree of excellence, does in fact measure the degree of fulfilment. The quality is thus a philosophy rather than a mere attribute. It is from this philosophy the distinctive culture emanates guiding the society to attain targets set by it. The presence or absence of this culture makes all the difference which determines the level of acceptability. The constant awareness of this culture amongst other endowments have led many nations where they exist today.

In the industrial climate particularly in manufacturing and process industry, the concept of quality management is age old and is extensively used, whereas it is recent in concrete construction industry. Every piece of equipment or product is subjected to quality management in the industrial production as a matter of routine. The quality management ensures that every piece of product keeps on performing over a period of time without heavy maintenance and upkeep. Unfortunately in concrete construction even if rigid quality management measures are not followed, it performs, at least for reasonable period of time. On account of this co-operative property of the material, the concrete construction industry has been operating under the misconception that rigid quality management measure which are essential for an industrial product are not that essential for concrete. Thus in concrete industry of most of the developing countries, inspite of best efforts a great deal is yet to be achieved to derive maximum benefit out of this culture. Measures have been devised to elongate serviceable, maintenance and rehabilitation free life of the material and minimize, if not completely eliminate the possibilities of failure. The measures thought of are all related to *Quality Management*. Due to well co-ordinated efforts, a quantum jump has taken place in the design of reinforced concrete. The present day design methods are no longer limited to the earlier deterministic approaches such as working stress methods, but the limit state methods based on semi-probalistic approaches are now being extensively practised.

Today we are interested not only in 28-day cube strength, but also in its variability. The word *characteristic* has now come to stay in the codes of practice. The characteristic value approach gives insight and underlines the importance of quality assurance.

Apart from the strength of concrete, the other important area of concern is the *durability* of concrete. A great deal of attention has been focussed on this and concrete technologists have come up with many effective suggestions. Some of them are: (i) use of minimum quantities of cement, (ii) drastic reduction in *water-cement ratio* maintaining the workability by use of *plasticizers,* (iii) use of pozzolanas, (iv) use of low-heat cement, and (v) the most important of all is a good *quality control* in design, testing and production of concrete. During past decade and so, a good number of concrete structures have shown signs of distress much within their design life and most of these are due to poor durability considerations. The repair of such damages are highly expensive which could have been avoided with the application of quality control measures. The ever increasing use of concrete in engineering structures, has made a demand of very high order to fulfil the targets or engineering excellence. In some structures the design is not limited to ensure structural integrity, but is based on the axiom that the probability of failure of such structures must be as low as possible and lower than a predetermined value of extremely small order.

9.8.1 Management of Uncertainties

(i) Primary Uncertainties

All the structures have probabilities of failure inspite of being designed to carry the loads safely because in the probabilistic design approach, the design variables such as loads, material strength, etc., are considered as random variables. Hence the probability of occurrence of a very large or a very small value of variable is never zero; the probability of such occurence, may however, be very small. Thus whenever, the load variable exceeds the strength variable a failure situation occurs. If by applying a better quality control the standard deviation of mix is reduced, then the probability of failure will be reduced.

(ii) Secondary Uncertainties

The secondary uncertainties are introduced during both the design and construction phases. Selection of inappropriate design conditions, use of inapplicable site data, injudicious assumptions regarding boundary conditions and other data in design introduces secondary uncertainties. During construction more secondary uncertainties are introduced e.g. use of inappropriate materials, violation of design conditions and incorrect interpretation of designers requirements, etc. Thus the level of confidence which may be viewed as a measure of closeness of the behaviour of the actual constructed structure to that of analytical model influences the probability of failure.

Although the odds of primary uncertainties can be taken care of by allowing for the randomness of the design variables, no proven analytical approach is available within the present state-of-the-art to increase the level of confidence against the effect of secondary uncertainties. It is therefore, imperative that a systematic implementation of quality management system in design, manufacture and construction is a must as to produce a safe and reliable structure.

9.8.2 Quality Management System (QMS)

QMS is the management and control system document having three elements: Quality Assurance (QA) plans, implementation of *Quality Control* (QC) process and *Quality Audit* (QA) system of tracking and documentation of quality assurance and quality control programmes. QMS ensures that the intended degree of excellence is attained. The owner or his representative formulates the policy, determines the scope of quality planning and quality management, establishes the relationship between the various participating agencies and delegates responsibilities and authorities to them so that the quality objectives as set by owner are achieved. It must be understood that QMS cannot be developed in totality at the inception. QMS has to undergo stages of development as various project phases such as design, procurement of materials, construction, inspection, erection and commissioning are entered into with more and more agencies being involved and interfaces take place. The various stages of development of QMS are:

Stages	QMS Elements
Planning	: Owner formulates QA policy and develops QA plan.
Engineering	: The consultant develops his own design QA program and that of prospective vendors and contractors.
Procurement	: Suppliers develop and submit their own QA programs and QC methods.
Construction	: Contractor develop and submit their QA programs and QC methods.
Inspection	: The testing agencies develop their QA programs.

The stage-wise development of QA program based of owner's QA plan are required to be reviewed and approved by the owner or by his consultant as the case may be.

Quality Assurance (QA)

It is planned and systematic pattern of all actions necessary to provide adequate confidence that a product will conform to established requirements. It is a system of procedures for selecting the levels of quality required for a project or a portion there of to perform the functions intended and assuring that those levels are obtained. QA is thus the responsibility of the owner/user to ensure that consultants follow codes and sound engineering practices and that contractors and suppliers of materials comply with the contract requirements. QA program developed by each agency responsible to the extent of its contractual obligation must contain the policies, practices, procedures and method to be followed such that the quality objectives laid down by the owner in his QA plan are fully met. The QA program must be addressed fully (to the extent applicable) to the following aspects:

   (i) Organization set-up.
  (ii) Responsibilities and authorities of various personnel involved.
 (iii) Identification of co-ordinating personnel.
 (iv) Quality control measures in design including field changes.
  (v) Establishment of control norms, acceptance and rejection criteria for materials.
 (vi) Inspection program for verification of contractual compliance including acceptance and rejection criteria.
(vii) Sampling, testing, documentation and material qualifications.
(viii) Corrective measures during non-complying conditions and non-conformance.
 (ix) Resolution of technical differences/disputes.
  (x) Preparation, submission and maintenance of records at all stages.

The quality assurance activity has to start right at the planning and design stage. Development of a QA program for design activities is an art by itself and is beyond the scope of this book. Apart from organizational and administrative aspects, it has to cover procedures of design, conformance to codes, proper detailing and attention to durability and constructability. One important part of quality assurance is *Peer Review*. It is review of the project including its design, drawings and specifications by an *Independent Professional* or an agency, with equal or more experience and qualifications than of the professionals engaged for the design of the project.

Quality Control (QC)

It implements the quality plan by those actions necessary for conformance to the established requirements. It is the system of procedure and standards by which a contractor, product manufacturer, material processor or the like monitors the properties of finished work.

QC is the responsibility of the contracting organization. The contracting organization is also responsible for QC activities related to its sub-contractors. Quality control starts with the construction. The constructing organization prepares the QA program manual describing and establishing the QA and control system to be used by it in performing design, purchasing, fabrication, production of concrete and other construction activities for the contractual responsibilities assigned to it. Application area, indentification of agencies and personnel responsible for implementing, managing and documenting the QC programs, their responsibilities and authorities must be well established in the document. The detailed steps in these procedures depend upon the scope and type of work and owner's policy decision.

Quality Audit (QA)

This is a system of tracking and documentation of Quality Assurance and Quality Control programs. Quality Audit is the responsibility of the owner, and has to be performed at regular intervals through the tenure of the project. Quality Audit covers both the design as well as the construction phases. Thus the concept of Quality Management encompasses a total project and each element of that project. The systems on methodology of implementing concept of Quality Management depend on the available materials and construction technology. As the concrete technology changes, these systems also change. As such the systems of implementing concepts of Quality Management are not universal but regional and not static but dynamic, and ever changing.

An integrated systematic implementation of QMS is extremely beneficial, but any attempt to make its piece-meal use will defeat the very purpose for which it is extended. In other words in order to produce a safe, reliable and durable structure, Quality Culture must begin at the beginning and be carried through all the stages of design, procurement, construction and be continued further into the inservice regime. It is only a matter of systematic cultivation and a desire towards increased perfection that can make a complete metamorphosis of a developing construction industry.

9.8.3 Cost Effectiveness of Quality Management

It has been the general experience that whether it is the owner who has to cover the cost of *Quality Assurance, Quality Audit* and *Peer Review* or the contractor who has to cover the cost of *Quality Control*, the expenditure is met out of savings which accrue from the project due to implementation of *Quality Management Systems* (QMS). On the part of owner the Quality Management ensures a product of assured quality, strength, reliability and maintenance free durable life cycle. This is achieved by eliminating chances of mistakes in planning, over design or under design and ensures proper detailing and constructability. Any of these items if over-looked can later cost heavily to the owner. It is universally accepted that every project has a *Quality Cost Component*. Every contractor has a choice as to when he will pay the cost. He can pay the controlled cost of Quality Control during contsruction, or he can pay the uncontrolled cost of correcting the defective workmanship and materials later. Patched up work, dismantling and re-doing unacceptable work, maintenance and up-keep during performance guarantee period may cost a contractor up to 20 to 25 per cent of his gross income. The unwilling contractors may be motivated to introduce *Quality Control* within their organizations, by fixing the criteria of acceptability of concrete based on statistical control of strength which has a small range with a provision of a bonus for better quality control than stipulated. For example, if the criteria satisified 100 per cent, the contractor receives 100 per cent payment. If it satisfies the lower limit (say 90 per cent), he gets paid 90 per cent. But if he satisfies the criteria by more than 100 per cent he gets a bonus up to a maximum of 2 per cent. An intensive dialogue between consultants and contractors on concrete specifications, acceptability criteria, testing procedures, field controls, inspection systems etc., with the common objective of updating these documents and procedures for definitely attaining the desired quality may be extremely helpful.

Since the development of concrete technology is closely linked to general construction industry, the passing away of period of shortage and variance have forced the construction industry to change and modernize. The changed situation is bound to give an impetus to concrete technology to update itself. An alternate criteria for acceptance of concrete based on its durability instead of its load carrying ability which helps in prolonging the serviceability life of concrete may be designed. The later is based on passivity of concrete which can be evaluated by its minimum strength at 28 days. The former is based on active concrete that functions under changing conditions and respond to varying environments and abuses. Its performance is mainly based on *water-cement ratio*. Thus the designs must not be linked just to the strength of concrete but also to the durability of concrete.

The introduction and implementation of Quality Management systems can be successful if the concrete industry directs its efforts towards increasing reliability, durability, economy, energy efficiency, versatility, capability, adaptability and aesthetics, as well as towards improvements in materials, material handling, quality control, education of users, construction methods, codes and specifications, disposal and recycling of waste and extension of the environment under which concrete can be used and placed. In addition efforts should be directed to the development of accurate non-destructive testing procedures, continuous batching and new placing methods, immediate quality control tests, simplified forming methods, simplified reinforcing procedures, simplified methods of joining structural members, new design concepts, performance codes and improved cold and hot weather construction practices. The QMS need be updated to keep pace with advancement in concrete technology.

10

Proportioning of Concrete Mixes

10.1 INTRODUCTION

Concrete of different qualities can be obtained by using its constituents namely, cement, water, fine and coarse aggregates, in different proportions. Also, the ingredients of widely varying characteristics can be used to produce concrete of acceptable quality. The common method of expressing the proportions of the materials in a concrete mix is in the form of parts, of ratios of cement, the fine and coarse aggregates with cement being taken as unity. For example, a 1 : 2 : 4 mix contains one part of cement, two parts of fine aggregate and four parts of coarse aggregate. The amount of water, entrained air and admixtures, if any, are expressed separately. The proportion should indicate whether it is *by volume* or *by mass*. The *water-cement ratio* is generally expressed by mass. The amount of *entrained air* in concrete is expressed as a percentage of the volume of concrete. The amounts of *admixtures* are expressed relative to the weight of cement. Other forms of expressing the proportions are by ratio of cement to the sum of fine and coarse aggregates, i.e., *cement-aggregate*, ratio and by *cement factor* or number of bags of cement per cubic metre of concrete.

The wide use of concrete as construction material has led to the use of mixes of fixed proportions, which ensure adequate strength. These mixes are known as *nominal mixes*. These offer simplicity and, under normal circumstances, have a margin of strength above that specified. However, these do not account for the varying characteristics of the constituents and may result in under- or over-rich mixes. Generally, a *nominal mix* is expressed in terms of *aggregate-cement ratio* by volume. If weigh batching is adopted, bulk density has to be considered. The nominal mixes commonly used are given in the Table 10.1.

TABLE 10.1 Proportions of nominal mix concrete

Grade of Concrete	Total quantity of dry aggregate per bag of cement of 50 kg, kg	Maximum water content per bag of cement of 50 kg, litres	Proportions of fine aggregate to coarse aggregate by mass
M5	800	60	Generally 1 : 2 with upper limit as 1 : 1.5 and lower limit as 1 : 2.5
M7.5	625	45	
M10	480	34	
M15	350	32	
M20	250	30	

To avoid the ambiguities of nominal mixes, IS: 456-1978 has introduced a set of mix proportions by volume for the concretes of commonly used strengths. These mixes called *standard mixes* are by definition conservative, but are useful as *off the shelf sets of proportion* that allow the desired concrete to be produced with minimum preparatory work. For example, for M15 grade concrete the proportion is 1 : 2 : 4. For the ordinary concrete from which quite *undemanding performance* is expected, the nominal or standard mixes may be used.

The concrete making materials being essentially variable result in the production of mixes of variable quality. In such a situation, for *high performance* concrete, the most rational approach of mix proportioning is to select proportions with specific materials in mind which possess more or less unique characteristics. This will ensure the concrete with the appropriate properties to be produced, most economically. Other factors like *workability, durability, compaction equipment available, curing methods* adopted, etc., also influence the choice of the mix proportion. The mix proportion so arrived at is called the *designed mix*. However, the method does not guarantee the correct mix for the desired strength, thereby necessitating the use of *trial mixes*. In the process of mix proportioning, a number of *subjective decisions* are required on which hinge the important ramifications for the concrete. The *designed mix* serves only as a guide. For many works it is desirable to go through the process of mix design, for example, where a large volume of concrete is required, a minimization of the cement content may reduce the cost appreciably, or where for technical reasons the type of concrete required necessitates careful selection and proportioning of ingredients.

10.2 BASIC CONSIDERATIONS FOR CONCRETE MIX DESIGN

The concrete mix design is a process of selecting suitable ingredients for concrete and determining their proportions which would produce, as economically as possible, concrete that satisfies the job requirements, i.e., concrete having a certain minimum *compressive strength, workability* and *durability*. The proportioning of the ingredients of concrete is an important phase of concrete technology as it ensures quality and economy.

The proportioning of concrete mixes is accomplished by the use of certain empirical relations which afford a reasonably accurate guide to select the best combination of the ingredients so as to achieve the desired properties. The design of plastic concretes of medium strengths can be based on the following assumptions.

(i) The *compressive strength* of concrete is governed by its *water-cement ratio*.

(ii) For the given aggregate characteristics, the *workability* of concrete is governed by its *water content*.

For *high-strength concrete* mixes of *low workability*, considerable interaction occurs between the above two criteria and the validity of such assumptions may become limited. Moreover, there are various factors which affect the properties of concrete, e.g., the quality and quantity of cement, water and aggregates; techniques used for batching, mixing, placing, compaction and curing, etc. Therefore, the specific relationships used in the proportioning of a concrete mix should be considered only as a basis for making an *initial guess at the optimum combination of the ingredients* and the *final mix proportion* is obtained only on the basis of further *trial mixes*.

10.3 FACTORS INFLUENCING THE CHOICE OF MIX PROPORTIONS

According to IS: 456–1978 and IS: 1343–1980, the design of concrete mix should be based on the following factors.
 (i) Grade designation
 (ii) Type of cement
 (iii) Maximum nominal size of aggregates
 (iv) Grading of combined aggregates
 (v) Water-cement ratio
 (vi) Workability
 (vii) Durability
 (viii) Quality control

10.3.1 Grade Designation

The *grade designation* gives *characteristic compressive strength* requirements of the concrete. As per IS: 456–1978, the *characteristic compressive strength* is defined as that value below which not more than 5 per cent of the test results are expected to fall. It is the major factor influencing the mix design. Depending upon the *degree of control* available at the site, the concrete mix has to be designed for *a target mean compressive strength* which is somewhat higher than the characteristic strength.

10.3.2 Type of Cement

The type of cement is important mainly through its influence on the *rate of development of compressive strength* of concrete. The choice of the type of cement depends upon the requirements of performance at hand. Where very high compressive strength is required, for example, in prestressed concrete railway sleepers high strength Portland cement conforming to IS: 8112-1976 will be found suitable. In situations where an early strength development is required rapid-hardening Portland cement conforming to IS:8041-1978 and for *mass concrete* construction, low-heat Portland cement conforming to IS:269-1979 is preferable. The *blended cements* such as Portland pozzolana cement and Portland slag cement are permitted for use in reinforced concrete construction; while Portland slag cement is also permitted for prestressed concrete construction. The rate of development of early strength may be somewhat slower with *blended cements*.

10.3.3 Maxmium Nominal Size of Coarse Aggregate

The maximum *nominal size* of the coarse aggregate is determined by sieve analysis and is designated by the sieve size higher than the largest size on which 15 per cent or more of the aggregate is retained. The maximum nominal size of the aggregate to be used in concrete is governed by the size of the section and the spacing of the reinforcement. According to IS: 456–1978 and IS: 1343–1980, the maximum nominal size of the aggregate should not be more than one-fourth of the minimum thickness of the member, and it should be restricted to 5 mm less than the minimum clear distance between the main bars or 5 mm less than the minimum cover to the reinforcement or 5 mm less than the spacing between the prestressing cables. Within these limits, the nominal maximum size of aggregate may be as large as possible, because larger the maximum size of aggregate smaller is the cement requirement for a particular *water-cement ratio*. The *workability* also increases with an increase in the maximum size of aggregate. However, the smaller size aggregates provide larger *surface area* for bonding with the *mortar matrix* which increases the *compressive strength* and reduces the stress *concentration* in the *mortar-aggregate interface*. For the concrete with higher water-cement ratio, the larger maximum size of aggregate may be beneficial whereas for *high strength concrete*, 10 or 20 mm size of aggregate is preferable.

10.3.4 Grading of Combined Aggregate

The relative proportions of the fine and coarse aggregates in a concrete mix is one of the important factors affecting the strength of concrete. For dense concrete it is essential that the coarse and fine aggregates be well graded. Generally, the locally available aggregates do not conform to the *standard gradings*. In such cases the aggregates need to be combined in suitable proportions so that the resultant (combined) grading approximates to a *continuous grading* close to the desired (or standard) grading. The process of combining aggregates is aimed at obtaining a grading close to the *coarsest grading* of standard grading curves, the most economical mix having highest permissible *aggregate-cement ratio*. IS: 383–1963 has recommended limits to the coarsest and *finest gradings* which are listed in Tables 10.2 to 10.5. The aggregates can be combined either by analytical calculations or graphically using the method suggested in Road Note No. 4.

(i) Analytical Method

The method is easy to understand and calculations are trivial. Let two aggregates (designated as aggregate-I and aggregate-II) are to be combined. Let α, β and γ represent the percentages of the combined (resultant) aggregate, aggregate-I and aggregate-II, respectively, passing the seive corresponding to the point on the *standard grading curve* taken as criterion, i.e., the point to which the combined aggregate is required to approximate. If x and y are the proportions of two aggregates in the combined state, then to satisfy the condition that α per cent of combined aggregate passes the criterion sieve:

$$\beta x + \gamma y = \alpha(x + y)$$

126 Concrete Technology

or
$$\frac{x}{y} = \frac{\alpha - \gamma}{\beta - \alpha} = \frac{1}{k}$$

i.e.
$$x : y = 1 : k \qquad (10.1)$$

where $k = (\beta - \alpha)/(\alpha - \gamma)$

Hence the two aggregates have to be combined in the proportions of $1 : k$. The grading of the resulting combined aggregate is determined by first multiplying the gradings of aggregate-I and aggregate-II by 1 and k, respectively. The sum of corresponding products of the percentages passing the sieve sizes is then divided by $(1 + k)$, the values being rounded off to the nearest percentage.

Example 10.1 The gradings of fine and coarse aggregates available at a construction site are listed in columns b and c of Table 10.2. These aggregates are to be combined in suitable proportions so as to obtain the specified grading chosen from standard grading curves which is listed in the column d of the Table 10.2.

Solution Let one kilogram of fine aggregate be combined to x kilogram of coarse aggregate to obtain the desired grading. Suppose the percentage passing IS: 4.75-mm sieve is selected as criterion. In the standard grading 42 per cent of the total aggregate passes the IS: 4.75 mm sieve. Hence using Table 10.2,

$$96(1) + 3(x) = 42(1 + x)$$

or
$$x = (96 - 42)/(42 - 3) = 1.3846$$

Therefore the fine and coarse aggregates must be combined in the proportion 1.0 : 1.3846. The grading of resulting combined aggregate is obtained by multiplying columns b and c of Table 10.2 by 1.0 and 1.3846, respectively and dividing the sum of these products by 1.0 + 1.3846 (= 2.3846). The resulting combined grading is listed in column e of Table 10.2. Comparing column e with column d, it can be noted that the percentage passing IS: 4.75-mm sieve is same and combined grading is close to the desired grading.

TABLE 10.2 Combining fine and coarse aggregates to a stipulated grading

IS sieve	Percentage passing Fine aggregate	Percentage passing Coarse aggregate	Specified grading	Combined grading
(a)	(b)	(c)	(d)	(e)
40-mm	100	100	100	100
20-mm	100	98	100	99
10-mm	100	43	65	67
4.75-mm	96	03	42	42
2.36-mm	89	0	35	37
1.18-mm	73	0	28	31
600-μm	48	0	20	20
300-μm	20	0	07	08
150-μm	02	0	0	01

In the above problem there is only one point on the grading curve to which the aggregate is required to approximate. Comparing the grading of resulting combined curve with the selected standard grading curve, the percentage passing the *criterion sieve* necessarily agree but the other values may not. In some cases variation is very small which may be ignored. If, however, the discrepancies are large, the proportions may be changed by adopting another criterion point. It should be realized that mix proportioning is approximate, and it is extremely doubtful that the result would be better if the grading is further made closer.

The method can also be applied if three or more aggregates are to be combined. The following example will illustrate the procedure.

Example 10.2 The gradings of fine and two coarse aggregates available at a project site are listed in columns b, c and d, respectively, of Table 10.3. These aggregates are to be combined so as to approximate to the curve 1 of Fig. 10.4. The grading of the selected curve is listed in column e of the Table.

Solution It is required to determine fractions x and y of the two coarse aggregates to be combined with unit weight of fine aggregate so as to obtain the specified grading. Two unknowns need two equations for solution. Let the criterion sieve sizes be 10 mm and 2.36 mm.

According to the specified grading, the combined aggregate passing the IS: 10 mm sieve is 45 per cent, hence using Table 10.3.

$$100(1) + 94x + 18y = 45(1 + x + y)$$

or
$$49x - 27y = -55$$

The combined aggregate passing IS: 2.36-mm sieve is 23 per cent, hence using Table 10.3.

$$84(1) + 2(x) + 0(y) = 23(1 + x + y)$$

or
$$21x + 23y = 61$$

Solving the two equations

$$x = 0.2255 \quad \text{and} \quad y = 2.4463$$

Hence, the fine aggregate, coarse aggregate-I and coarse aggregate-II must be combined in the proportions 1.000 : 0.2255 : 2.4463, respectively. The grading of combined aggregate is obtained by multiplying columns b, c and d of Table 10.3 by 1.0, 0.2255 and 2.4463, respectively, and dividing sum of these products by 1 + 0.2255 + 2.4463 (= 3.6718). The resulting combined grading is listed in the column f of Table 10.3. On comparing the resulting grading with the specified grading, it is noticed that the percentage of combined aggregate passing 10-mm and 2.36-mm sieve are the same as in the specified grading. The error is mainly in the percentages passing IS: 1.18-mm and IS: 600-μm sieves. However, since the mix proportioning is only a guide for trial mixes any further effort in approximating it more accurately is not necessary.

128 Concrete Technology

TABLE 10.3 Combining two coarse aggregates with the fine aggregate to the stipulated grading

IS sieve	Percentage passing			Specified grading, per cent	Combined grading, percent (f)=[(b) + 0.2255(c) + 2.4463 (d)]/ 3.6718
	Fine aggregate	Coarse aggregate-I	Coarse aggregate-II		
(a)	(b)	(c)	(d)	(e)	(f)
20-mm	100	100	95	100	97
10-mm	100	94	18	45	45
4.75-mm	100	12	2	30	29
2.36-mm	84	2	0	23	23
1.18-mm	75	0	0	16	20
600-μm	51	0	0	9	14
300-μm	11	0	0	2	3
150-μm	02	0	0	0	0.5

(ii) Graphical Method

The graphical method as suggested in Road Note No. 4 consists of marking a square paper with the percentage scales on three sides as shown in Fig. 10.1. The procedure for combining a fine aggregate with a coarse aggregate with gradings shown in Fig. 10.1 is described below.

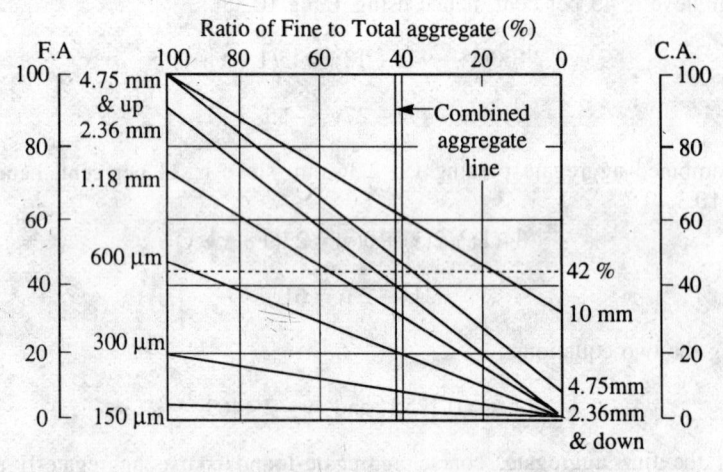

Material	Percentage passing the IS sieve							
	20-mm	10-mm	4.75-mm	2.36-mm	1.18-mm	600-μm	300-μm	150-μm
Coarse aggregate	100	31	7	-	-	-	-	-
Fine aggregate	-	-	100	92	76	48	20	3

Fig. 10.1 Graphical method for combining the aggregates

On the left-hand vertical axis are marked the points showing the *grading of fine aggregate*. These points are numbered with sieve sizes such that the ordinate of each point represents the percentage of material passing the corresponding sieve. The *grading of the coarse aggregate* is marked along the right-hand axis in a similar manner. The corresponding points representing the sieve sizes on the left-and right-hand axes are joined by straight lines. A vertical line called the *combined aggregate line* is drawn passing through the point where the sloping line representing IS: 4.75-mm sieve (the *criterion sieve*) intersects the horizontal line representing the percentage of material passing the IS: 4.75-mm sieve required in the combined grading. The ordinates of the intersections of the combined aggregate line with the sloping lines represent the grading of the combined aggregate.

Comparing the combined grading curve with the standard grading curve selected, the percentage passing the sieve corresponding to the criterion point (usually 4.75 mm) necessarily agree but the other values may not. In some cases the variation is very small which may be ignored. If, however, the discrepancies are large, the proportions may be changed by shifting the combined aggregate line to the right or to the left so that the discrepancy is reduced.

Where three aggregates are to be combined, two should be combined first and the resulting grading is combined with the third. For example, if two coarse aggregates are to be combined with a sand, the two coarse sizes should be combined, using the percentage passing a 20-mm sieve as a criterion. If two sands are to be combined, the criterion should be the amount passing IS: 600-µm sieve.

10.3.5 Water-cement Ratio

The *compressive strength* of concrete at a given age and under normal temperature depends primarily on the *water-cement ratio*, lower the water-cement ratio greater

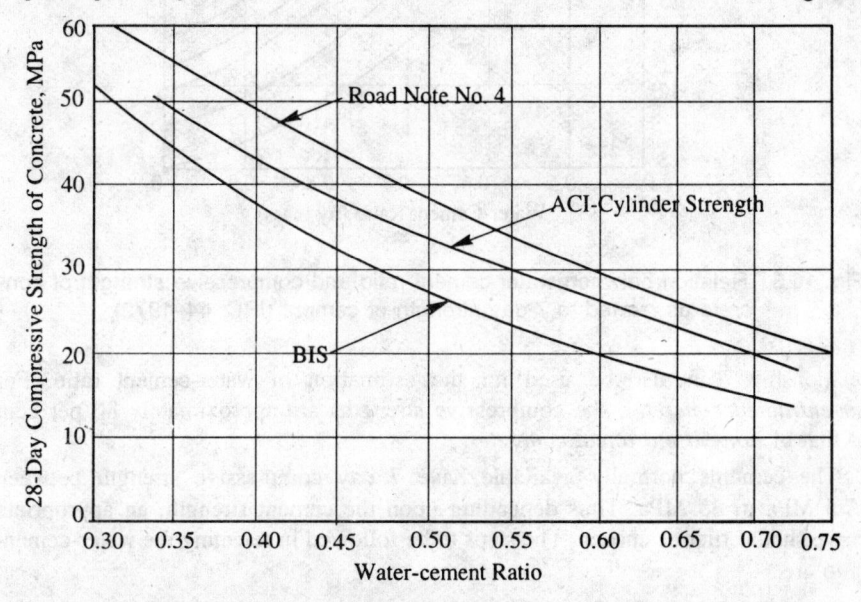

Fig. 10.2 Generalized relationship between water-cement ratio and compressive strength of concrete

is the compressive strength and vice-versa. A number of relationships between compressive strength and water-cement ratio are available which are supposed to be valid for a wide range of conditions. In so far as the selection of the *water-cement ratio* for the *target compressive strength* at 28 days is concerned, Fig. 10.2 is applicable for both ordinary Portland and Portland pozzolana cements with comparable validity. However, the 28-day compressive strength of concrete is related to the 7-day compressive strength of cement mortar as shown in Fig. 10.3. These

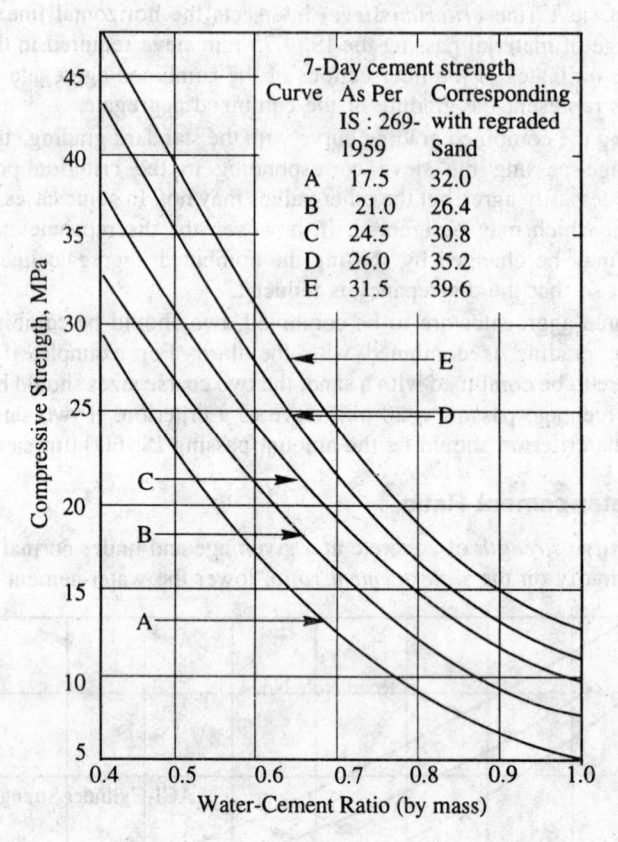

Fig. 10.3 Relation between water-cement ratio and compressive strength of concrete as related to 7-day strength of cement (IRC: 44–1972)

Curve	7-Day cement strength As Per IS : 269-1959	Corresponding with regraded Sand
A	17.5	32.0
B	21.0	26.4
C	24.5	30.8
D	26.0	35.2
E	31.5	39.6

relationships can also be used for the estimation of water-cement ratio. For *air-entrained concretes,* the compressive strengths are approximately 80 per cent of that of *non-air-entrained concretes.*

The cements normally available have 7 day compressive strength between 17.5 MPa to 35 MPa. Thus depending upon the cement strength, an appropriate curve should first be chosen. The steps to be followed in selecting the water-cement ratio are:

1. The strength of cement to be used is determined. In India only one type of cement is officially recognized, the one which gives 7-day strength of 22 MPa.

2. When cement strength data are available, corresponding curve is chosen for the determination of water-cement ratio. In the absence of such data, the curve corresponding to cement strength of 22 MPa, the minimum permissible as per the Indian Standards may be used.

10.3.6 Workability

The workability of concrete for satisfactory placing and compaction is controlled by the size and shape of the section to be concreted, the quantity and spacing of reinforcement, and the methods to be employed for transportation, placing and compaction of concrete. The situation should be properly assessed to arrive at the desired workability. The aim should be to have the minimum possible workability consistent with satisfactory placing and compaction of concrete. It should be kept in mind that insufficient workability resulting in *incomplete compaction* may severely affect the *strength, durability* and *surface finish* of concrete and may thus prove to be uneconomical in the long run.

There is no rigid correlation between workabilities of concrete as measured by different test methods. It is desirable that for a given concrete, the test method be identified before hand and workability be measured accordingly. The workability measured by different test methods for comparable concretes are given in Table 10.4.

TABLE 10.4 Recommended workability values

Degree of workability	Values of workability in terms of			
	Compacting factor	Slump, mm	Vee-Bee time, sec	Drop table revolutions
Extremely low (very stiff)	≤0.70*	—	30–20	96–48
Very low (stiff)	0.75–0.80	0–25	20–10	48–24
Low (stiff plastic)	0.80–0.85	25–50	10–5	24–12
Medium (plastic)	0.85–0.92	50–75	5–2	12–6
High (flowing)	>0.92	75–150	2–0	6–0

*Compacting factor test is not used for concrete with aggregate having maximum nominal size of 40 mm and higher.

10.3.7 Durability

The durability of concrete can be defined and interpreted to mean its *resistance to deteriorating influences* which may reside inside the concrete itself, or to the aggressive environments. The requirements of *durability* are achieved by restricting the *minimum cement content* and the *maximum water-cement ratio* to the values given in Table 10.5, and the type of cement. The *permeability* of cement paste increases exponentially with increase in *water-cement ratio* above 0.45 or so. Thus from considerations of permeability, the water-cement ratio is usually restricted to 0.45 to 0.55, except in mild environments. For a given water-cement ratio, the cement content in the concrete mix should correspond to the required workability keeping in view the placing conditions and the concentration of reinforcement. In addition, the *cement content* is chosen to ensure sufficient *alkalinity* to provide a passive environment against *corrosion of steel*, e.g,. in concrete for marine environment or sea water a minimum cement content of 350 kg/m^3 or more is required.

TABLE 10.5 Minimum cement content (kg/m^3) required in the portland cement concrete to ensure durability under specified conditions of exposure

Exposure	Plain concrete — Nominal size of aggregate, mm — 40	20	12.5	10	Maximum free water-cement ratio	Reinforced concrete — Nominal size of aggregate, mm — 40	20	12.5	10	Maximum free water-cement ratio	Prestressed concrete — Nominal size of aggregate, mm — 40	20	12.5	10	Maximum free water-cement ratio
Mild: Completely protected against weather or aggressive conditions, except for a brief period of exposure to normal weather conditions during construction.	200	220	260	270	0.70	220	250	280	290	0.65	—	300	300	300	0.65
Moderate: Sheltered from severe rain and against freezing whilst saturated with water. Buried concrete and concrete continuously under water.	220	250	290	300	0.60	260	290	330	340	0.55	300	300	330	340	0.55
Severe: Exposed to sea water, moorland water, driving rain, alternate wetting and drying and to freezing whilst wet. Subject to heavy condensation or corrosive fumes.	270	310	350	360	0.50	320	360	400	410	0.45	320	360	400	410	0.45
Subject to salt used for de-icing.	240	280	320	330	0.55	260	290	330	340	0.55	300	300	330	340	0.55

Moreover, the cement content and water-cement ratio are so chosen as to provide a sufficient volume of cement paste to overfill the *voids* in the compacted aggregates. The *blended cements* like Portland pozzolana cement and Portland slag cement render greater *durability* to the concrete in *sulphatic environments* and sea water. Resistance to *alternate freezing and thawing* is not so important for Indian conditions, but wherever situations demand, *air-entrained concrete* could be employed using an *air-entraining admixture*. Air-entrainment lowers the compressive strength but increases *workability* which may permit certain reduction in the water content to make up the loss in *compressive strength*.

10.3.8 Quality Control

The strength of concrete varies from batch to batch over a period of time. The sources of variability in the strength of concrete may be considered due to variation in the quality of the constituent materials, variations in mix proportions due to batching process, variations in the quality of batching and mixing equipment available, the quality of supervision and workmanship. These variations are inevitable during production to varying degrees. Controlling these variations is important in lowering the difference between the *minimum strength* and *characteristic mean strength* of the mix and hence reducing the *cement content*. The factor controlling this difference is *quality control*. The *degree of control* is ultimately evaluated by the variation in test results usually expressed in terms of the *coefficient of variation*.

It can be summarized that the aim of mix design is to obtain a most practical and economical combination of materials that will produce a concrete mix of necessary *plasticity (workability)* and, at the same time, produce *hardened concrete* of required *strength* and *durability*. Most of the mix design procedures are primarily based on the *water-cement ratio law* and *absolute volume system* of calculating the amount of materials. As explained earlier, according to the water-cement ratio law, the strength of hardened concrete is approximately inversely proportional to the water content per cubic metre of cement. The calculation of the quantities of the aggregates to be used with a given cement paste is based on the absolute volume method. The absolute volume of loose material is the actual volume of the solid matter in all the particles ignoring the space occupied by the *voids* between the particles. The absolute volume is calculated as follows.

$$\text{Absolute volume} = \frac{\text{Mass of loose dry material}}{\text{Specific gravity} \times \text{Mass of unit volume of water}}$$

The process of mix design is outlined in the flowchart given in Fig. 10.4.

10.4 METHODS OF CONCRETE MIX DESIGN FOR MEDIUM STRENGTH CONCRETES

Most of the available mix design methods are based on empirical relationships, charts and graphs developed from extensive experimental investigations. Basically they follow the same principles enunciated in the preceding section and only minor variations exist in different mix design methods in the process of selecting the mix

134 Concrete Technology

```
┌─────────────────────────────────────────┐
│ DEGREE OF QUALITY CONTROL ENVISAGED     │
│ STIPULATED CHARACTERISTIC STRENGTH      │
└─────────────────────────────────────────┘
               │
               ▼
   ┌───────────────────────┐      ┌──────────────────┐
   │ MEAN TARGET STRENGTH  │◄─────│ TYPE OF CEMENT   │
   └───────────────────────┘      │ TYPE OF EXPOSURE │
               │                  │ (DURABILITY)     │
               ▼                  └──────────────────┘
   ┌───────────────────────┐
   │   WATER-CEMENT RATIO  │           • MAXIMUM SIZE OF
   └───────────────────────┘             AGGREGATE
               │                       • TYPE AND SHAPE
               ▼                         OF AGGREGATE
   ┌───────────────────────────────┐   • GRADING OF
   │      WATER CONTENT            │◄──  FINE AGGREGATE
   │ FINE AGGREGATE AS PERCENT OF  │   • REQUIRED
   │ TOTAL AGGREGATE BY ABSOLUTE   │     WORKABILITY
   │ VOLUME                        │
   └───────────────────────────────┘
               │
               ▼
   ┌───────────────────────┐
   │ CONCRETE MIX PROPORTIONS │◄──┐
   └───────────────────────┘     │
               │                 │
               ▼                 │
          ╱ IS       ╲           │
         ╱ TRIAL MIXES╲  NO      │
        ╱  STRENGTH    ╲─────────┘
        ╲  ADEQUATE?   ╱
         ╲            ╱
          ╲   YES    ╱
               │                ┌───────────────────┐
               ▼◄───────────────│ CAPACITY OF       │
   ┌───────────────────────┐    │ CONCRETE MIXER    │
   │ WEIGHT OF INGREDIENTS │    └───────────────────┘
   │ PER BATCH             │
   └───────────────────────┘
```

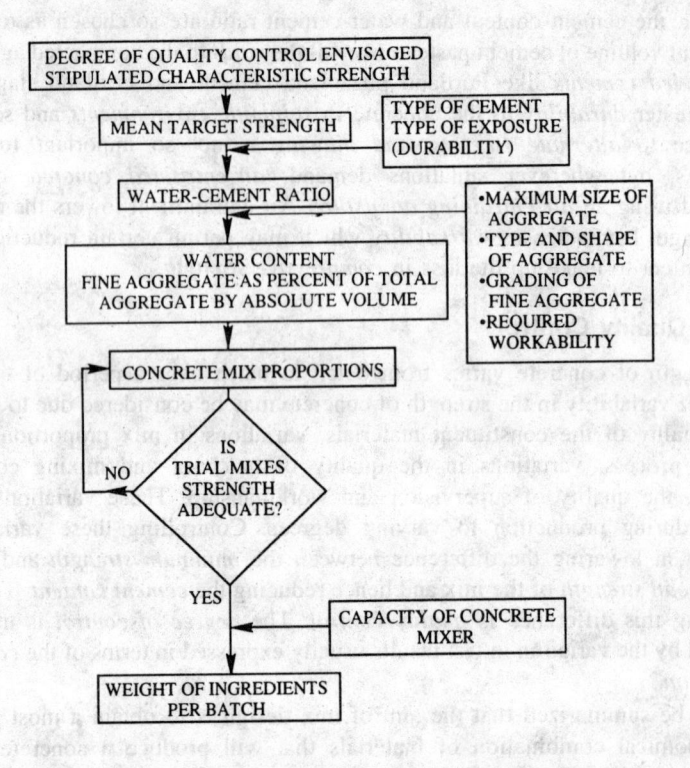

Fig. 10.4 Steps involved in mix proportioning

proportions. The *requirements* of the concrete mix are usually dictated by the general experience with regard to the *structural design conditions, durability* and *conditions of placing.* Some of the commonly used mix design methods for medium strength concrete are:

(i) Trial and adjustment method of mix design
(ii) Road note No. 4 method of mix design
(iii) DoE (British) mix design method
(iv) ACI mix design method
(v) Mix design according to Indian Standard Recommended Guidelines
(vi) Rapid method for mix design.

The step-by-step procedure for concrete mix design can be summarized as follows:

(i) The *mean target strength* is estimated from the specified *characteristic strength* and the *level of quality control.*

(ii) The *water-cement ratio* is selected for the mean target strength and is checked for the *requirements of durability.*

(iii) The degree of *workability* required in terms of slump, compacting factor or Vee-Bee time is selected.

(iv) The percentage of fine aggregate in the total aggregate is determined from the characteristics of coarse and fine aggregates. Alternatively, the *aggregate-cement ratio* may be determined.

(v) The water content for the required workability is arrived at.

(vi) The *cement content* is calculated and its quantity is checked for the *requirements of durability.*

(vii) The concrete mix proportions for the first *trial mix* are computed and concrete cubes are cast in the laboratory following the standard procedure. After the required *period of curing,* the cubes are tested for the compressive strength of the mix.

(viii) The *trial batches* obtained by making suitable adjustment in water-cement ratio or aggregate-cement ratio are tested till the *final mix composition* is arrived at.

(ix) The final proportions are expressed either on mass or volume basis.

Most of the available mix design methods are essentially based on the above procedure and due consideration should be given for the *moisture content of aggregate* and the *entrained air.*

10.4.1 Trial and Adjustment Method of Mix Design

The method is based on experimental approach and aims at producing a concrete mix which has *minimum voids* and hence, *maximum density.* The fine aggregate is mixed in sufficient quantity to fill the voids in coarse aggregate; and cement paste is used in sufficient quantity to fill the voids in the mixed aggregate. The proportion of fine to coarse aggregate which gives maximum mass of combined aggregate can be obtained by trials. The process consists of filling a container of known volume with the two materials in thin layers, the fine being placed over the coarse aggregate and lightly rammed after each layer. If the container is shaken too much, the coarse aggregate will try to come on the top and the fine aggregate will deposit at the bottom without filling the voids of the coarse aggregate. Since the density of the particles of fine and coarse aggregates is nearly the same, the mixture giving maximum weight will have maximum solid matter and hence *least voids.* Such a combination will need the least amount of cement per cubic metre of concrete and will be most economical for a given *water-cement ratio* and *slump.*

In an alternate trial mix method, the sand is combined with the coarse aggregate in several proportions, such as 20 : 80, 30 : 70, 40 : 60, 50 : 50 and 60 : 40, and for each such mixture, the quantity of cement paste of a certain *water-cement ratio* per unit volume of concrete is determined to give the required *workability* (expressed in terms of slump). The percentage of sand corresponding to the ratio requiring minimum cement, is termed *optimum percentage.* If the quantity of sand used is more than the optimum, more cement will be needed to have the same consistency. On the other hand, a smaller quantity of sand will make the mix harsh unless more cement is used for proper consistency. The optimum percentage of sand is lower for a low water-cement ratio. The step-by-step procedure of mix proportioning is as follows:

(i) The target mean compressive strength is determined from the characteristic strength.

(ii) The water-cement ratio is chosen for the target mean strength computed in step (i). The water-cement ratio so chosen is checked against limiting water-cement ratio for the requirements of durability and the lower of the two values is adopted.

(iii) The workability is determined in terms of the slump required for a particular job.

(iv) The maximum nominal size of the coarse aggregate that is available or desired to be used, is determined.

(v) The fine and coarse aggregates are so mixed that either the weight per litre of mixed aggregate is maximum or the sand percentage corresponds to the optimum value.

(vi) By actual trials the quantity of cement (in the form of cement paste) required per unit volume of aggregate to give the desired slump is determined.

(vii) The proportions of cement, fine aggregate, coarse aggregate and water to meet the requirements of strength, durability, workability and economy are computed and concrete cubes are cast and tested after the required period of curing for the compressive strength.

(viii) The trial mix is adjusted, if necessary, by varying the water-cement ratio or the aggregate-cement ratio to suit the actual requirements of the job.

10.4.2 Mix Design According to Road Note No. 4

The Road Research Laboratory of the Department of Scientific and Industrial Research, London, has prepared a series of grading charts which are useful for the design of concrete mixes. The design procedure has been compiled as Road Note No. 4. The method is based on the fact that a designed concrete must be satisfactory both in the plastic as well as in the hardened state and therefore, the choice of mix proportions is governed by both these conditions.

The method is based on the use of specific grading curves of combined aggregate. The *parameters* involved in the mix design using this method are: *the nominal maximum size of aggregate, type of aggregate* and *its grading, workability of concrete mix* and its *characteristic strength*. The method recommends the use of aggregates with nominal sizes as 10 mm, 20 mm and 40 mm for general reinforced concrete works. The method is applicable to any of the three types of aggregates such as rounded, irregular or crushed rock. For each of the three sizes of aggregate the method recommends four types of gradings. For 20 mm and 10 mm maximum nominal size aggregates the grading curves are shown in Figs 10.5 and 10.6, respectively. These are not the ideal curves but represent gradings used in road research laboratory testing. Higher the number of the grading curve, the larger will be the proportion of fine particles. The *coarsest grading curve* No. 1 is suitable for *harsh mixes*, i.e., the most economical mix having the highest permissible aggregate-cement ratio. The *finest grading curve* No. 4 is suitable for *lean mixes* where a high workability is required. The change from one extreme to the other is progressive. The outer curves 1 and 4 represent the limits for the normal *continuous gradings*. The saving in cement affected by using a coarse grading can be considerable. The method classifies the *workability* in five categories: *extremely low, very low, low, medium,* and *high* (refer to Table 10.9). The target mean strength, as usual, is obtained from characteristic strength and standard deviation. The water-cement ratio is determined to satisfy both strength and durability requirements. Design tables correlating water-cement ratio, aggregate-cement ratio, maximum size of aggregate,

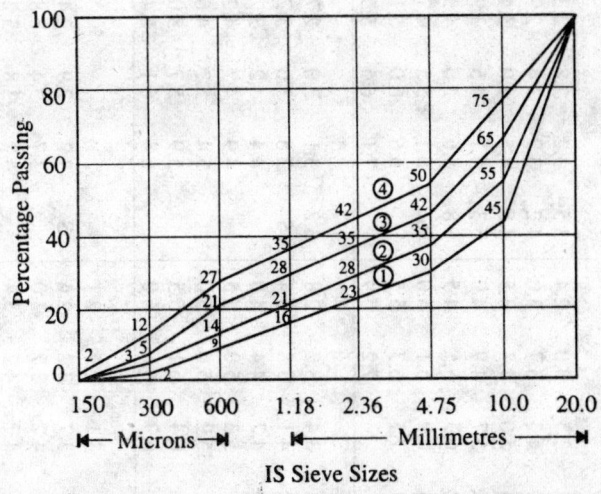

Fig. 10.5 Grading curves for 20 mm maximum nominal size aggregate

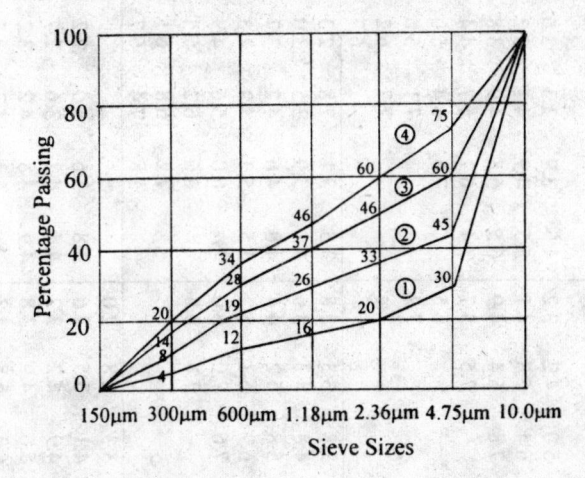

Fig. 10.6 Grading curves for 10 mm maximum nominal size aggregate

type of aggregate (differing in shape as rounded, angular and irregular), degree of workability and overall grading (curve) of combined aggregates, are given.

These tables are based on the assumption that fine and coarse aggregates are of the same type. The values given in the design tables have been obtained by the extrapolation of other data and not based directly on the results of trial mixes. The data given in Table 10.7 is based on the water in excess of the absorbed by the aggregate, while those in Table 10.6 refers to the total water added to the air-dry-aggregate. Therefore, the adjustments should be made for water contained in the aggregate. If the locally available aggregate does not conform to the standard grading, the finer and coarser fractions of the aggregate can be suitably combined to obtain the desired standard grading. This can be achieved by analytical calculations or graphically as explained in Sec. 10.3.

138 Concrete Technology

TABLE 10.6 Aggregate-cement ratio for different workabilities with 10-mm nominal size aggregate

Aggregate-cement ratio by mass

Degree of workability (Table 10.9)		Very low				Low				Medium				High			
Grading number (Fig. 10.6)		1	2	3	4	1	2	3	4	1	2	3	4	1	2	3	4
Water-cement ratio by mass																	
Rounded gravel aggregate	0.40	5.6	5.0	4.2	3.2	4.5	3.9	3.3	2.6	3.9	3.5	3.0	2.4	3.5	3.2	2.8	2.3
	0.45	7.2	6.4	5.3	4.1	5.5	4.9	4.1	3.2	4.7	4.3	3.7	3.0	4.2	3.9	3.4	2.9
	0.50		7.8	6.4	4.9	6.5	5.8	4.9	3.8	5.4	5.0	4.3	3.5	4.8	4.5	4.0	3.4
	0.55			7.5	5.7	7.4	6.7	5.7	4.4	6.1	5.7	4.9	4.0	5.3	5.1	4.5	3.9
	0.60				6.5		7.5	6.4	5.0	6.7	6.3	5.5	4.5	5.8	5.6	5.0	4.3
	0.65				7.2			7.1	5.6	7.3	6.9	6.1	5.0	S	6.1	5.5	4.7
	0.70							7.7	6.2	7.9	7.5	6.7	5.5		6.6	6.0	5.1
	0.75								6.7			7.2	5.9		7.1	6.5	5.5
Irregular gravel aggregate	0.40	4.1	3.8	3.3	2.8	3.3	3.1	2.8	2.3								
	0.45	5.1	4.8	4.3	3.6	4.1	3.9	3.5	3.0	3.5	3.4	3.2	2.8	3.2	3.1	3.0	2.7
	0.50	6.1	5.8	5.2	4.4	4.8	4.6	4.2	3.7	4.2	4.1	3.8	3.4	S	3.8	3.6	3.2
	0.55	7.0	6.7	6.1	5.2	5.5	5.3	4.9	4.3	S	4.7	4.4	4.0		4.4	4.2	3.7
	0.60	7.9	7.6	7.0	6.0	S	6.0	5.6	4.9		5.3	5.0	4.5		4.9	4.7	4.2
	0.65			7.8	6.8		6.6	6.2	5.5		5.9	5.6	5.0		5.4	5.2	4.6
	0.70						7.2	6.8	6.1		6.4	6.1	5.5		5.9	5.7	5.0
	0.75						7.8	7.4	6.7		6.9	6.6	6.0		6.4	6.1	5.4
Crushed gravel aggregate	0.40	3.7	3.3	2.8	2.0	3.8	3.6	3.0	2.2	3.3	3.1	2.7	2.1				
	0.45	4.5	4.1	3.5	2.6	4.4	4.2	3.6	2.7	3.8	3.7	3.2	2.6	S	3.2	2.9	2.4
	0.50	5.2	4.9	4.2	3.2	4.9	4.8	4.2	3.2	S	4.2	3.7	3.0		3.7	3.4	2.8
	0.55	5.9	5.6	4.9	3.8	S	5.3	4.7	3.7		4.7	4.2	3.4		4.2	3.8	3.2
	0.60	6.6	6.3	5.5	4.3		5.8	5.2	4.2		5.1	4.6	3.8		4.6	4.2	3.6
	0.65	7.3	7.0	6.1	4.8		6.3	5.7	4.6		5.6	5.1	4.2		5.0	4.6	4.0
	0.70	7.9	7.6	6.7	5.3		6.8	6.2	5.0		6.0	5.5	4.6		5.4	5.0	4.4
	0.75			7.3	5.8												

With crushed aggregate of poorer shape than the tested, segregation may occur at a lower aggregate-cement ratio. S indicates that the mix would segregate.

Proportioning of Concrete Mixes 139

TABLE 10.7 Aggregate-cement ratio for different workabilities with 20-mm nominal size aggregate

Aggregate-cement ratio by mass

Degree of workability (Table 10.9)		Very low				Low				Medium				High			
Grading number (Fig. 10.5)		1	2	3	4	1	2	3	4	1	2	3	4	1	2	3	4
Water-cement ratio by mass																	
Rounded gravel aggregate	0.35	4.5	4.5	3.5	3.2	3.8	3.6	3.2	3.1	3.1	3.0	2.8	2.7	2.8	2.8	2.6	2.5
	0.40	6.6	6.3	5.3	4.5	5.3	5.1	4.5	4.1	4.2	4.2	3.9	3.7	3.6	3.7	3.5	3.3
	0.45	8.0	7.7	6.7	5.8	6.9	6.6	5.9	5.1	5.3	5.3	5.0	4.5	4.6	4.8	4.5	4.1
	0.50			8.0	7.0	8.2	8.0	7.0	6.0	6.3	6.3	5.9	5.4	5.5	5.7	5.3	4.8
	0.55				8.1			8.2	6.9	7.3	7.3	7.4	6.4	6.3	6.5	6.1	5.5
	0.60								7.7			8.0	7.2	S	7.2	6.8	6.1
	0.65								8.5				7.8	S	7.7	7.4	6.6
	0.70															7.9	7.2
Irregular gravel aggregate	0.35	3.7	3.7	3.5	3.0	3.0	3.0	3.0	2.7	2.6	2.6	2.7	2.4	2.4	2.5	2.5	2.2
	0.40	4.8	4.7	4.7	4.0	3.9	3.9	3.8	3.5	3.3	3.4	3.5	3.2	3.1	3.2	3.2	2.9
	0.45	6.0	5.8	5.7	5.0	4.8	4.8	4.6	4.3	4.0	4.1	4.2	3.9	S	3.9	3.9	3.5
	0.50	7.2	6.8	6.5	5.9	5.5	5.5	5.4	5.0	4.6	4.8	4.8	4.5	S	4.4	4.4	4.1
	0.55	8.3	7.8	7.3	6.7	6.2	6.2	6.0	5.7	S	5.4	5.5	5.1	S	4.8	4.9	4.7
	0.60	9.4	8.6	8.0	7.4	6.8	6.9	6.7	6.2	S	6.0	6.0	5.6	S	S	5.4	5.2
	0.65				8.0	7.4	7.5	7.3	6.8	S	S	6.4	6.1	S	S	5.8	5.6
	0.70					8.0	8.0	7.7	7.4	S	S	6.8	6.6	S	S	6.2	6.1
Crushed gravel aggregate	0.35	3.2	3.0	2.9	2.7	2.7	2.7	2.5	2.5	2.4	2.4	2.3	2.2	2.2	2.3	2.1	2.1
	0.40	4.5	4.2	3.7	3.5	3.5	3.5	3.2	3.0	3.1	3.1	2.9	2.7	2.9	2.9	2.8	2.6
	0.45	5.5	5.0	4.6	4.3	4.3	4.2	3.9	3.7	3.7	3.7	3.4	3.3	3.5	3.5	3.2	3.1
	0.50	6.5	5.8	5.4	5.0	5.0	4.9	4.5	4.3	4.2	4.2	3.9	3.8	S	3.9	3.8	3.5
	0.55	7.2	6.6	6.0	5.6	5.7	5.4	5.0	4.8	4.7	4.7	4.5	4.3	S	S	4.3	4.0
	0.60	7.8	7.2	6.6	6.3	6.3	6.0	5.6	5.3	S	5.2	4.9	4.8	S	S	4.7	4.4
	0.65	8.3	7.8	7.2	6.9	6.9	6.5	6.1	5.8	S	5.7	5.4	5.2	S	S	5.1	4.9
	0.70	8.7	8.3	7.7	7.5	7.4	7.0	6.5	6.3	S	6.2	5.8	5.7	S	S	5.5	5.3

S indicates that the mix would segregate. These proportions are based on specific gravities of approximately 2.5 for the coarse aggregate and 2.6 for the fine aggregate.

The procedure used in the design of a concrete mix using Road Note No. 4 method is as follows:

(i) The *maximum nominal size of the aggregate* which is economically available is determined as per the specified requirements. The degree of workability is selected as per job requirements.

(ii) The target mean strength based on 28-day characteristic strength and standard deviation is determined.

(iii) A suitable water-cement ratio to obtain a concrete mix of desired strength is selected from Fig. 10.2. The water-cement ratio so chosen is compared with that required for durability, the lower value is adopted.

(iv) The gradings of different size aggregates are determined.

(v) The proportions of different size aggregates to obtain a combined grading matching any of the grading curves given in Figs. 10.5 and 10.6 are selected.

(vi) The aggregate-cement ratio based on the water-cement ratio and aggregate characteristics is obtained from the relevant table.

(vii) The mix proportions per cubic metre of concrete are calculated.

(viii) The properties of the trial batch are checked by testing in accordance with relevant code specifications and suitable adjustment are made, if necessary.

The design tables of *aggregate-cement ratio* covered in Road Note No. 4 are for limited shapes of aggregates and types of gradings, but in practice many aggregates having different shapes, sizes and properties are to be used. At best, the method should be considered as a guide to select the mix proportions since it is strictly applicable only to the actual aggregates used in their derivation. The Road Note No. 4 data cannot be used directly for design of *air-entrained concrete*. However, the nominal mix designed by using the data can be suitably adjusted for the desired air content in the mix.

Example 10.3 Using Road Note No. 4 method of mix design, determine aggregate-cement ratio for a concrete mix having a characteristic strength of 30 MPa. The other design stipulations are:

Type of cement	ordinary Portland cement
Type of construction	reinforced concrete work using high frequency vibrator (i.e. low workability)
Aggregate type	irregular
maximum nominal size:	10 mm
grading:	similar to curve No. 2
Type of exposure	moderate

Solution For the given characteristic strength

$$\text{Target mean strength, } f_t = f_{ck} + kS$$

$$= 30 + 1.65 \times 6 \approx 40 \text{ MPa.}$$

The free water-cement ratio, for a 28 day compressive strength of 30 MPa from Fig. 10.2 is 0.47. From Table 10.6 for irregular aggregate with low workability, the aggregate-cement ratio for grading curve No. 2 is 3.9 for water-cement ratio of 0.45, and 4.6 for water-cement ratio of 0.50. By linear interpolation, for a water-cement ratio of 0.47.

$$\text{Aggregate-cement ratio} = 3.9 + \frac{0.47 - 0.45}{0.50 - 0.45} \times (4.6 - 3.9) = 4.18$$

The Road Note No. 4 method has been superseded by a new method known as the DoE or British mix design method for normal concrete mixes. The new method is discussed in detail in the following sections.

10.4.3 The DoE (British) Mix Design Method

The traditional British mix design method of Road Note No. 4 has been replaced by the Department of the Environment's Design of normal concrete mixes. The method uses the relationship between *water-cement ratio* and *compressive strength* of concrete depending on the type of cement and the type of aggregate used. The water contents required to give various *levels of workability*, namely, *very low, low, medium* and *high* (expressed in terms of slump or Vee-Bee time) are determined for the two types of aggregates, viz., crushed (angular) and uncrushed (gravel). The method gives mix proportions in terms of quantities of materials per unit volume of concrete.

The method is suitable for the design of normal concrete mixes having 28-day cube compressive strength as high as 75 MPa for *non-air-entrained concrete*. The step-by-step procedure of mix proportioning is as follows.

(i) Determination of Free Water-cement Ratio

(a) For the stipulated characteristic strength at a specified age, the *target mean compressive strength* at that age is first determined.

(b) For the given types of cement and aggregate, the compressive strength at the specified age corresponding to a free water-cement ratio of 0.50 is obtained from the Table 10.8. For example, when ordinary portland cement and uncrushed aggregate are used, the compressive strength is 40 MPa at 28 days. Adopting this pair of data (40 MPa and water-cement ratio = 0.50) representing a *controlling point*, the appropriate strength versus water-cement ratio curve is located in Fig.10.7. In this particular case it is the fourth curve from the top of the figure passing the point. Using this curve the water-cement ratio corresponding to computed target mean strength is determined.

In case an existing curve is not available which passes through the *controlling point*, it is necessary to interpolate between two curves in the figure.

TABLE 10.8 Approximate compressive strength of concrete mixes with water-cement ratio of 0.5

Type of cement	Type of coarse aggregate	Compressive strength, MPa Age (days)			
		3	7	28	91
Ordinary or sulphate- resisting portland cement	Uncrushed	18	27	40	48
	Crushed	23	33	47	55
Rapid-hardening portland cement	Uncrushed	25	34	46	53
	Crushed	30	40	53	60

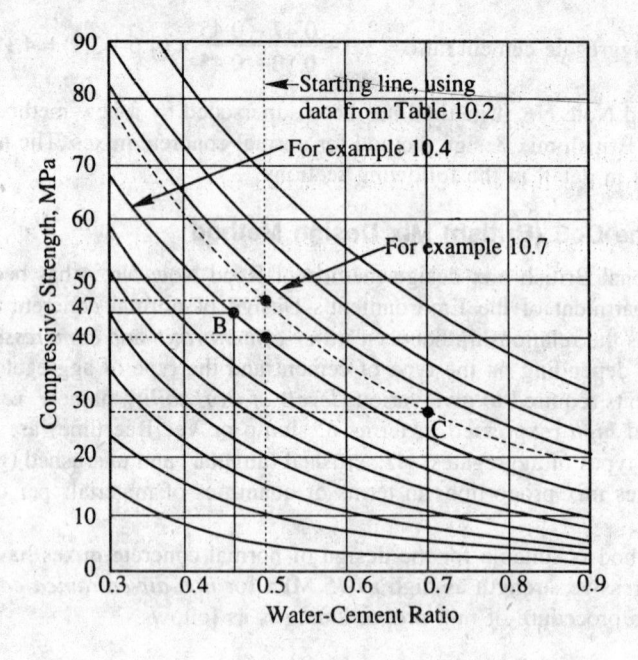

Fig. 10.7 Variation of compressive strength with water-cement ratio (DoE 185)

(c) Compare the *water-cement ratio* obtained in step i(b) with the *maximum water-cement ratio* specified for the durability and lower of the two values is adopted.

(ii) Determination of Water Content

Depending upon the type and maximum nominal size of aggregate, and workability (specified in terms of slump or Vee-Bee time) the water content is determined from Table 10.9.

(iii) Determination of Cement Content

The *cement content* of the mix is calculated using the selected *water-cement ratio* (from step i) and the *water-content* (from step ii).

$$\text{Cement content (kg/m}^3) = \frac{\text{water content}}{\text{water-cement ratio}} \quad (10.3)$$

The above value of cement content is checked against any maximum or minimum cement content that may have been specified for durability. If the computed cement content is below the specified minimum, this minimum must be used which results in a reduced water-cement ratio and hence in a higher strength. However, if the calculated *cement content* is higher than the *specified maximum,* then the specified strength and workability cannot simultaneously be met with selected materials. Try by changing the type of cement, the type and maximum size of aggregate.

TABLE 10.9 Approximate free water content required to give various levels of workability

	Description	Extremely low	Very low	Low	Medium	High
Level of workability	Slump, mm	0	0–10	10–30	30–60	60–180
	Vee-Bee, s	>20	20–12	12–6	6–3	3–0
	Compacting factor	0.65–0.75	0.75–0.85	0.85–0.90	0.90–0.93	>0.93

Maximum size of aggregate, mm	Type of aggregate*	Water content, kg/m³				
10	Uncrushed	—	150	180	205	225
	Crushed	—	180	205	230	250
20	Uncrushed	—	135	160	180	195
	Crushed	—	170	190	210	225
40	Uncrushed	—	115	140	160	175
	Crushed	—	155	175	190	205

*When the coarse and fine aggregates used are of different types, the water content is estimated by the following expression:

$$W = \frac{2}{3}W_f + \frac{1}{3}W_c \qquad (10.2)$$

where W_f = Water content appropriate to the type of fine aggregate
W_c = Water content appropriate to the type of coarse aggregate

(iv) Determination of Aggregate-cement Ratio

The total *aggregate content* (in saturated surface dry condition) is obtained by subtracting the cement and water contents from the wet density of concrete obtained from Fig. 10.8 which depends on the *specific gravity of combined aggregates* (in the saturated surface dry condition). Alternatively, it can also be calculated easily from first principles as follows:

$$\text{Absolute volume occupied by the aggregate} = 1 - \frac{\text{cement content (kg)}}{1000\,S_c} - \frac{\text{water content (kg)}}{1000} \qquad (10.4)$$

where $S_c (= 3.15)$ is the specific gravity of cement particles.

Therefore total aggregate content (kg/m³)
$$= 1000\,S_a \times (\text{absolute volume occupied by the aggregates})$$

where S_a is the specific gravity of aggregate particles. If no information is available S_a should be taken as 2.6 for uncrushed aggregate and 2.7 for crushed aggregate.

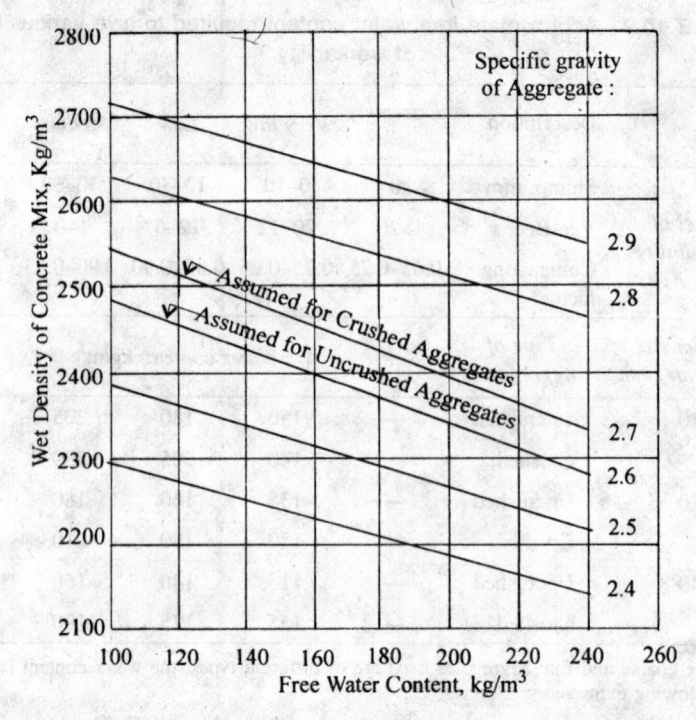

Fig. 10.8 Estimated wet density of fully compacted concrete (DoE 185)

(v) Determination of the Fine and Coarse Aggregate Contents

Depending on the water-cement ratio, the nominal maximum size of coarse aggregate, the workability and grading zone of fine aggregate, the proportion of fine aggregate is determined from Fig. 10.9. Having calculated the proportion of the fine aggregate expressed as a percentage of total aggregate, the content of coarse aggregate is calculated.

Coarse aggregate content (per cent) = 100 − content of fine aggregate (per cent).

(vi) Determination of Final Proportions

The trial mixes of the proportions are prepared and checked for their specified strength and suitable adjustments are made to obtain the final proportions satisfying the design stipulations.

Example 10.4 Using the DoE mix design method design a concrete mix for an application requiring a characteristic strength of 35 MPa at 28 days and a slump of 10–30 mm. The materials available are ordinary portland cement, and uncrushed fine and coarse aggregates of specific gravity of 2.65. The maximum nominal size of the coarse aggregate is 10 mm and fine aggregate conforms to the grading zone III. The concrete is to be used in a structure likely to be exposed to a moderate climate. The standard deviation is 6.3 and defective rate is 5 per cent, i.e., probability factor $k = 1.65$.

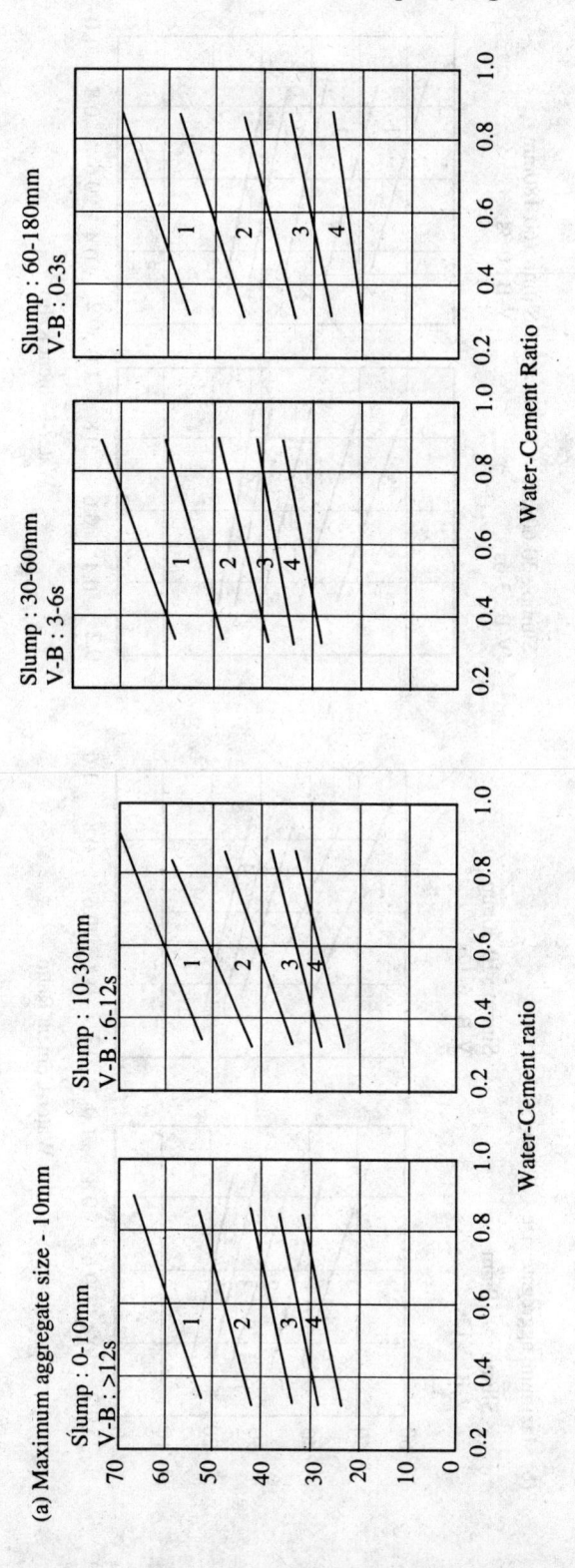

Proportioning of Concrete Mixes 145

(a) Maximum aggregate size - 10mm

146 Concrete Technology

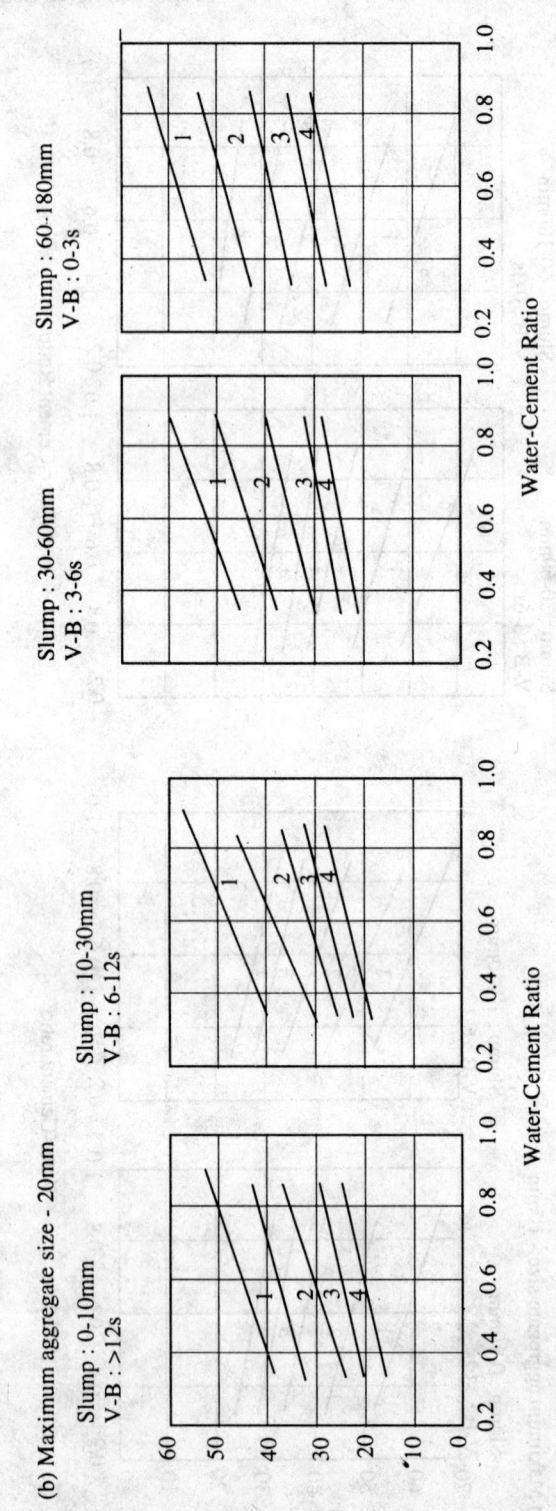

(b) Maximum aggregate size - 20mm

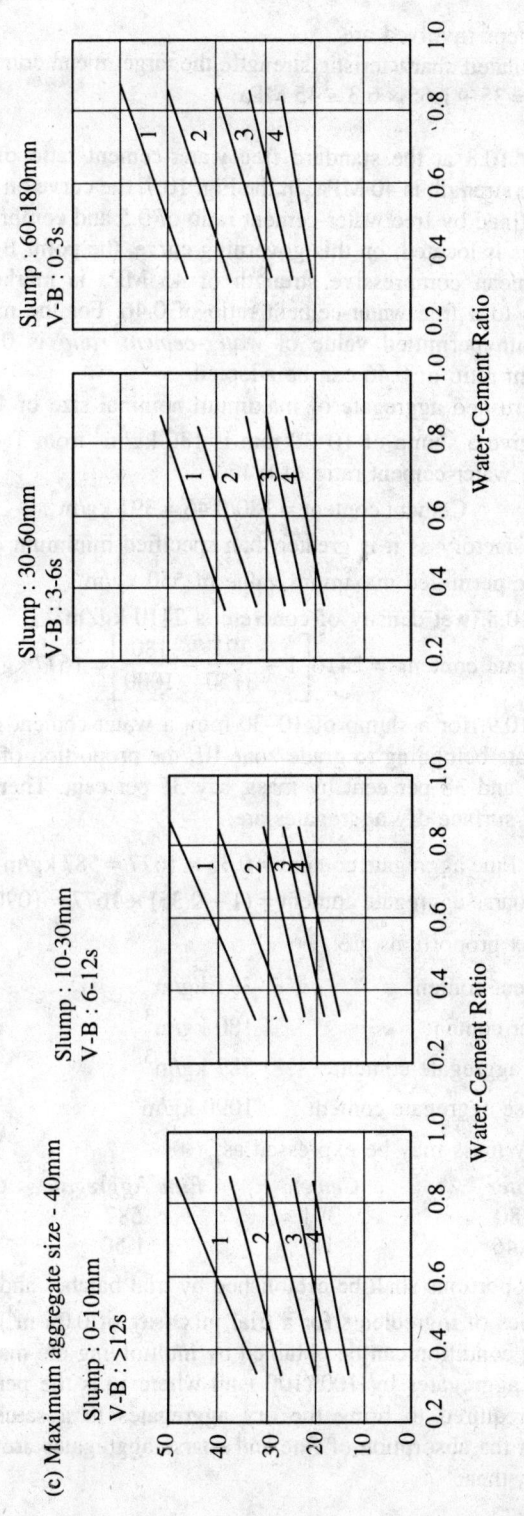

Fig. 10.9 Recommended proportions of fine aggregate for different grading zones (DoE 185)

148 Concrete Technology

Solution The steps involved are:
1. For the stipulated characteristic strength, the target mean compressive strength,
$f_t = f_{ck} + kS = 35 + 1.65 \times 6.3 \approx 45$ MPa

2. From Table 10.8 at the standard free water-cement ratio of 0.5, the 28 day compressive strength is 40 MPa. In the Fig. 10.7, the curve on which the control point A (defined by free water-cement ratio of 0.5 and compressive strength of 40 MPa) lies is located; on this governing curve, the point B corresponding to the target mean compressive strength of 45 MPa is marked. This point B corresponds to a free water-cement ratio of 0.46. For the moderate exposure the maximum permitted value of *water-cement ratio* is 0.55. Therefore, a water-cement ratio of 0.46 can be adopted.

3. For the uncrushed aggregate of maximum nominal size of 10 mm, the water content to give a slump of 10–30 mm is 180 kg/m^3 from Table 10.9.

4. For the free water-cement ratio of 0.46
$$\text{Cement content} = 180/0.46 = 391 \text{ kg/m}^3.$$
Thus is satisfactory as it is greater than specified minimum of 300 kg/m^3 and less than the permitted maximum value of 550 kg/m^3.

5. From Fig. 10.8, wet density of concrete is 2410 kg/m^3.

6. Total aggregate content $= 2410 \left[1 - \dfrac{391}{3150} - \dfrac{180}{1000} \right] = 1677$ kg/m^3.

7. From Fig. 10.9, for a slump of 10–30 mm, a water-cement ratio of 0.46, and fine aggregate belogning to grade zone III, the proportion of fine aggregate is between 32 and 38 per cent by mass, say 35 per cent. Therefore proportions of saturated surface dry aggregates are:

$$\text{Fine aggregate content} = 0.35 \times 1677 = 587 \text{ kg/m}^3$$
$$\text{Coarse aggregate content} = (1 - 0.35) \times 1677 = 1090 \text{ kg/m}^3$$

The required mix proportions are:

cement content	391 kg/m^3
water content	180 kg/m^3
fine aggregate content	587 kg/m^3
coarse aggregate content	1090 kg/m^3

The mix ratio by mass may be expressed as:

	Water	Cement	Fine Aggregate	Coarse Aggregate
	180 :	391 :	587 :	1090
or	0.46 :	1.0 :	1.50 :	2.79

The final proportions shall be established by trial batches and site adjustments.

8. The quantities of ingredients for a trial mix (say of 0.05 m^3) to be batched in an oven-dry condition can be obtained by multiplying the masses of saturated-surface-dry aggregates by $100/(100 + w)$ where w is the percentage of water (by mass) required to bring the dry aggregates to a saturated surface dry condition. If the absorption of fine and coarse aggregates are 2 and 1 per cent, respectively, then:

Mass of oven-dry fine aggregate = 29.35 × 100/102 = 28.77 kg
Mass of oven-dry coarse aggregate = 54.50 × 100/101 = 53.96 kg
Water required for absorption = (29.35 − 28.77) + (54.50 − 53.96) = 1.12 kg.

Add 10 per cent to allow for any under estimation and waste. Thus the quantities for trial batch are: cement 19.55 kg, water 10.12 kg, fine aggregate 28.77 kg (oven-dry) and coarse aggregate 53.96 kg (oven-dry).

10.4.4 The ACI Method of Mix Proportioning

The ACI method is based on the fact that for a given *maximum size of well shaped aggregate*, the *water-content* (kg/m^3) determines the *workability* of mix, i.e. it is largely independent of mix proportions. The method further assumes that the *optimum ratio* of the *bulk volume of coarse aggregate* to the total volume of concrete depends only on the maximum size of aggregate and on the *grading of fine aggregate*. The optimum volume of coarse aggregate when used with fine aggregates of different fineness moduli is given in Table 10.12. Having determined the maximum size and type of available aggregate, the water content for a specified workability is selected from the Table 10.11 and bulk volume of coarse aggregate from the Table 10.12. The *water-cement ratio* is determined as in other methods to satisfy both strength and durability requirements. The *air content* in concrete is taken into account for calculating the volume of fine aggregate. The step-by-step procedure adopted for the selection of mix proportions is as follows.

(i) The water-cement ratio is selected from Table 10.10 for the average strength.

(ii) The maximum size of the coarse aggregate to be used is determined by sieve analysis. The degree of workability is decided depending upon the placing conditions, etc.

(iii) The water content is selected from Table 10.11 for the desired workability and maximum size of aggregate.

(iv) The cement content is calculated from the water content and water-cement ratio required for strength and durability.

(v) The coarse aggregate content is estimated from Table 10.12 for maximum size of aggregate and fineness modulus of sand.

(vi) The content of fine aggregate is determined by subtracting the sum of absolute volumes of the coarse aggregate, cement, water and entrained air from unit volume of concrete. Trial batches are tested and final proportions are obtained by adjustments.

TABLE 10.10 Relationship between water-cement ratio and average compressive strength (aci manual of concrete practice, part i, 1979)

Compressive Strength at 28-days, MPa	Water-cement Ratio by Mass	
	Non-air-entrained concrete	Air-entrained concrete
45	0.38	—
40	0.43	—
35	0.48	0.40
30	0.55	0.46
25	0.62	0.53
20	0.70	0.61
15	0.80	0.71

TABLE 10.11 Approximate water requirements for different slumps and maximum size coarse aggregate (ACI manual of concrete practice, part I, 1979)

Slump, mm	Mixing water (kg/m^3 of concrete) for maximum sizes of aggregate, mm							
	10	12.5	20	25	40	50	70	150
Non-air-entrained concrete									
30–50	205	200	185	180	160	155	145	125
80–100	225	215	200	195	175	170	160	140
150–180	240	230	210	205	185	180	170	—
Approx. percentage of entrained air	3.0	2.5	2.0	1.5	1.0	0.5	0.3	0.2
Air-entrained concrete									
30–50	180	175	165	160	145	140	135	120
80–100	200	190	180	175	160	155	150	135
150–180	215	205	190	185	170	165	160	—
Recommended average total air content, per cent	8.0	7.0	6.0	5.0	4.5	4.0	3.5	3.0

TABLE 10.12 Bulk volume of coarse aggregate (ACI manual of concrete practice, part I, 1979)

Maximum size of aggregate, mm	Bulk volume of dry-rodded coarse aggregate per unit volume of concrete, for fineness modulus of fine aggregate			
	2.40	2.60	2.80	3.00
10	0.50	0.48	0.46	0.44
12.5	0.59	0.57	0.55	0.53
20	0.66	0.64	0.62	0.60
25	0.71	0.69	0.67	0.65
40	0.76	0.74	0.72	0.70
50	0.78	0.76	0.74	0.72
70	0.81	0.79	0.77	0.75
150	0.87	0.85	0.83	0.81

Limitation of the ACI Method

The ACI mix design method is suitable for *normal* and *heavy weight concretes* in the workability range of 25 to 100 mm slump and having maximum 28-day *cylinder compressive strength* of 45 MPa. The method is recommended for well-shaped aggregates within the range of generally acceptable specifications. Suitable adjustments are required to maintain the *consistency* of the mix in case of increased angularity number, and for the departure from the standard grading. The volume of *dry-rodded coarse aggregate* recommended in Table 10.12 applies to aggregates of specific gravity $S_{ca} = 2.68$. For an aggregate having a specific gravity of S'_{ca}, the volume of coarse aggregate obtained from the table should be multiplied by the ratio S'_{ca}/S_{ca}.

Example 10.5 It is required to design a non-air-entrained concrete mix with mean cylinder compressive strength of 25 MPa at 28 days and a slump of the order of 75–100 mm. The coarse aggregate available at the site is well graded having a maximum nominal size of 40 mm, and its specific gravity of 2.64. The available fine aggregate has a fineness modulus of 2.60 and a specific gravity of 2.68. The bulk density of coarse aggregate is 1600 kg/m^3. The type of cement available is ordinary portland cement. The exposure condition may be assumed to be moderate.

Design of Mix

(i) *Coarse aggregate content* Bulk volume of dry-rodded coarse aggregate per cubic metre of concrete from Table 10.12 is 0.74 m^3. Thus the mass of coarse aggregate per cubic metre of concrete = $0.74 \times 1600 = 1184$ kg/m^3.

(ii) *Water and cement contents* The water-cement ratio from Table 10.10 is 0.62. However, from durability considerations it is limited to 0.50. Hence adopt a water-cement ratio of 0.50.

Water content per cubic metre of concrete for a slump of 75–100 mm and maximum nominal size of aggregate of 40 mm as obtained from Table 10.11 is 175 kg. Therefore the cement content is $175/0.50 = 350$ kg/m^3. For a maximum size of aggregate of 40 mm the entrapped air content is 1 per cent.

(iii) *Fine aggregate content* The absolute volumes of mix ingredients per cubic metre of concrete are therefore:

$$\text{Cement} = \frac{C}{S_c} = \frac{350}{3.15 \times 1000} = 0.111 \text{ m}^3$$

$$\text{Water} = \frac{W}{1000} = \frac{175}{1000} = 0.175 \text{ m}^3$$

$$\text{Coarse aggregate} = \frac{C_a}{S_{ca}} = \frac{1184}{2.64 \times 1000} = 0.448 \text{ m}^3$$

$$\text{Entrapped air} = 0.01 \times 1 = 0.010 \text{ m}^3$$

$$\text{Total} = 0.745 \text{ m}^3$$

Hence, volume of fine aggregate required = $1 - 0.745 = 0.255$ m^3

mass of fine aggregate = $0.255 \times 2.68 \times 1000 = 684$ kg

(iv) *Trial Mix Proportions* The mix proportions (by mass) per cubic metre of concrete are:

Cement	350 kg/m^3
Fine aggregate	684 kg/m^3
Coarse aggregate	1184 kg/m^3
Water	175 kg/m^3
Total	2393 kg/m^3

Hence the unit weight of fully compacted fresh concrete is 2393 kg/m^3.

Mix ratio by mass:

Water		Cement		Fine Aggregate		Coarse Aggregate
175	:	350	:	684	:	1184 (kg)
0.5	:	1.0	:	1.95	:	3.38

10.4.5 Mix Design by Indian Standard Recommended Guidelines

The Indian Standard recommended guidelines for mix design include the design of normal concrete mixes (non-air-entrained), for both medium and high strength concretes. This method of mix design consists of determining the *water content* and *percentage of fine aggregate* corresponding to the *maximum nominal size of aggregate* for the *reference values of workability, water-cement ratio* and *grading of fine aggregate*. The water content and percentage of fine aggregate are then adjusted for any difference in workability, water-cement ratio and grading of fine aggregate from the reference values. Finally, the mass of materials per unit volume of concrete is calculated by *absolute volume method*. The guidelines are applicable to both ordinary portland and portland pozzolana cements. The final mix proportions, selected after trial mixes, may entail some minor changes in each case. In case of flyash-cement concrete the water content can be reduced by about 3 to 5 per cent and proportion of fine aggregate can be reduced by 2 to 4 per cent.

The basic data required for the design of concrete mix are: *characteristic compressive strength* at 28 days (f_{ck}); *degree of workability; limitations on water-cement ratio* and *minimum cement content* to ensure adequate durability for the type of exposure; *type and maximum nominal size of aggregate*; and *standard deviation* (S) for compressive strength of concrete. According to IS: 456–1978, the characteristic strength is defined as that value below which not more than 5 per cent of test results are expected to fall. The procedure adopted for the selection of mix proportions is as follows:

(i) The target mean strength (f_t) is determined by using the relation

$$f_t = f_{ck} + kS$$
$$= f_{ck} + 1.65\ S \qquad (10.5)$$

where f_t is the *target mean compressive strength* at 28 days, f_{ck} is the characteristic compressive strength at 28 days, S is standard deviation and k is a *statistical coefficient*. For the definition of characteristic strength given in IS: 456–1978, $k = 1.65$.

(ii) The water cement ratio for the target mean strength is determined from Fig. 10.2. The water cement ratio so chosen is compared with the maximum water-cement ratio specified for durability, and the lower of the two values is used.

(iii) Approximate air content is estimated from Table 10.14 for the maximum nominal size of coarse aggregate used.

(iv) The water-content and percentage of sand in total aggregate by absolute volume are next selected from Table 10.14 for the medium and high strength concretes for the following standard reference conditions:

(a) Crushed (angular) coarse aggregate,
(b) Fine aggregate consisting of natural sand conforming to grading Zone II of Table 4 of IS: 383–1973, in saturated surface dry condition,
(c) Water-cement ratio of 0.60 and 0.35 for medium and high strength concretes, respectively, and
(d) Workability corresponding to compacting factor of 0.80.

TABLE 10.13 Suggested values of standard deviation

Degree of control	Condition of production	Standard deviation, MPa — Grade of concrete								
		M10	M15	M20	M25	M30	M35	M40	M45	M50
Very good	Weigh batching of all materials, control of aggregate grading, and moisture content, frequent supervision, field laboratory facilities	2.0	2.5	3.6	4.3	5.0	5.3	5.6	6.0	6.4
Good	Weigh batching of all materials, graded aggregate, periodic tests, intermittent supervision, experienced workers	2.3	3.5	4.6	5.3	6.0	6.3	6.6	7.0	7.4
Fair	Volume batching, occasional supervision and tests	3.3	4.5	5.6	6.3	7.0	7.3	7.6	8.0	8.4

TABLE 10.14 Approximate sand and water contents per cubic metre of concrete

Stipulated conditions of the mix	Nominal maximum size of aggregate, mm	Water content per cubic metre of concrete, kg	Proportion of sand of total aggregate by absolute volume, per cent	Entrapped air, as per cent of volume of concrete
Water-cement ratio = 0.60	10	208	40	3.0
Workability = 0.8 CF	20	186	35	2.0
	40	165	30	1.0
Water-cement ratio = 0.35	10	200	28	3.0
Workability = 0.80 CF	20	180	25	2.0

(v) The water content and percentage of sand in total aggregate are adjusted as per Table 10.15 for any difference in workability, water-cement ratio, grading of fine aggregate and for rounded aggregate for the particular case from the reference values.

TABLE 10.15 Adjustment of values in water content and sand percentage for other conditions

Change in conditions stipulated for Table 10.13	Adjustment required in	
	Water content	Per cent sand in total aggregate
For Sand conforming to Grade Zones, I, III and IV Table 4, IS: 383–1970	0	+1.5 for Zone I −1.5 for Zone III −3.0 for Zone IV
Increase or decrease in the value of compacting factor by 0.1	±3 per cent	0
Each 0.05 increase or decrease in water-cement ratio	0	±1 per cent
For rounded aggregate	−15 kg/m^3	−7 per cent

(vi) The cement content is calculated using the selected water-cement ratio and the final water content of the mix obtained after adjustment. The cement content so calculated is compared with minimum cement content from the requirements of durability and the greater of the two values is used.

(vii) The total aggregate content (saturated surface dry condition) per unit volume of concrete is determined by subtracting the air, cement and water quantities per unit volume of concrete. With the percentage of sand in total aggregate already determined, the coarse and fine aggregate contents per unit volume of concrete are calculated.

If W be the mass (kg) of water, C be the mass (kg) of cement and v be the air content (m^3) per cubic metre of concrete, the absolute volume of total aggregate (V_a) per unit volume of concrete is given by:

$$V_a = 1.0 - \left(W + \frac{C}{S_c}\right)\frac{1}{1000} - v \qquad (10.6)$$

where S_c is the specific gravity of cement. For the ratio of fine aggregate to total aggregate by absolute volume of p, the absolute volumes of coarse and fine aggregates per unit volume of concrete may be calculated as:

$$V_{fa} = pV_a$$

and

$$V_{ca} = (1-p)V_a \qquad (10.7)$$

where V_{fa} and V_{ca} are absolute volumes of fine and coarse aggregates (m^3), respectively. The mix proportions by mass (kg) become:

Water	Cement	Fine Aggregate	Coarse Aggregate	
W :	C :	$V_{fa}S_{fa}(1000)$:	$V_{ca}S_{ca}(1000)$	(10.8)

where S_{fa} and S_{ca} are specific gravities of saturated surface dry fine and coarse aggregates, respectively, in kilograms per litre.

The mix proportions by volume (m^3) can be expressed as:

Water	Cement	Fine Aggregate	Coarse Aggregate	
$\dfrac{W}{1000}$:	$\dfrac{C}{\gamma_c}$:	$\dfrac{V_{fa}S_{fa}(1000)}{\gamma_{fa}}$:	$\dfrac{V_{ca}S_{ca}(1000)}{\gamma_{ca}}$	(10.9)

where γ_c, γ_{fa} and γ_{ca} are bulk densities (kg/m^3) of cement, fine and coarse aggregates, respectively.

Proportioning of Concrete Mixes 155

(viii) The actual mix proportions are arrived at by means of tests on a number of trial mixes.

Example 10.6 Using Indian Standard recommended guidelines design a concrete mix for a structure to be subjected to the mild exposure conditions for the following requirements:

(a) *Design stipulations*

Characteristic strength at 28 days:	15 MPa
Maximum nominal size of the aggregate:	20 mm
Type of aggregate:	Angular (crushed)
Degree of workability:	Medium (0.9 CF or 75 mm slump)
Degree of quality control:	good
Grading zone of sand:	III

(b) *Characteristics of the materials*

Cement

Type of cement	Ordinary portland cement
Specific gravity	3.15
Bulk density of cement	1450 kg/m^3

Aggregate	Fine aggregate	Coarse aggregate
Specific gravity	2.60	2.75
Bulk density (kg/m^3)	1700	1800
Free surface moisture	2.0	1.0
Fineness modulus	2.2	6.0

(c) *Mix design*

Target mean strength
$f_t = f_{ck} + kS = 15 + (1.65)(3.5)$ 21 MPa

Water-cement ratio

From compressive strength consideration (from Fig. 10.2)	0.59
From durability consideration	0.65
Hence water-cement ratio adopted	0.59

Water and cement contents

Water content (from Table 10.14)	186 kg/m^3
Sand content as percentage of total aggregate by absolute volume	35

Adjustment for the change in conditions from the reference values specified in Table 10.14:

	Water	Sand
(i) For decrease in water-cement-ratio of 0.01	0.0	–0.2
(ii) For increase in compacting factor by 0.10	+3.0	0.0
(iii) For sand conforming to zone III	0.0	–1.5
Total	+3.0	–1.7

Therefore,

water content = 186 × 1.03	191.6 kg/m^3
sand proportion = 35 − 1.7	33.3 per cent
air content	2 per cent
cement content = (191.6)/(0.59)	324.75 kg/m^3

Coarse and fine aggregates

Total absolute volume of aggregates

$$= 1 - \left(0.02 + \frac{324.75}{3.15 \times 1000} + \frac{191.6}{1000}\right) \quad 0.6853 \text{ m}^3$$

Absolute volume of fine aggregate 0.228 m^3

Absolute volume of coarse aggregate 0.457 m^3

Mix proportions

(i) *By mass*

Water	Cement	Fine Aggregate (dry)	Coarse Aggregate (dry)
191.6 :	324.75 :	0.228 × 2600 :	0.457 × 2750 (kg)
0.59 :	1.00 :	1.83 :	3.87

(ii) *By volume of dry aggregates*

Water	Cement	Fine Aggregate	Coarse Aggregate
— :	$\frac{324.75}{1450}$:	$\frac{592.8}{1700}$:	$\frac{1256.75}{1800}$ (m)^3
0.59 :	1.00 :	1.56 :	3.12

(by mass)

Quantities of ingredients

Free water in coarse aggregate = 1256.75 × (0.01)	12.57 kg
Free water in fine aggregate = 592.8 × (0.02)	11.86 kg
Total water present in the aggregate	24.43 kg
Mass of wet coarse aggregate in mix = 1256 + 12.57	1269.32 kg
Mass of wet fine aggregate in mix = 592.8 + 11.86	604.66 kg
Mixing water required = 191.6 − 24.43	167.17 kg

Actual mix proportions by mass

Water	Cement	Fine aggregate	Coarse Aggregate
167.17 :	324.75 :	604.66 :	1269.32
0.51 :	1.00 :	1.86 :	3.91

10.4.6 Rapid Method for Mix Design

A more realistic approach to estimate the preliminary *water-cement ratio* corresponding to the *target mean strength* is to correlate it with the 28-day *compressive strength of cement*. In contrast to the usual 7-day strength, the cement is characterized by its 28-day strength because characteristic strength is found to be better related to 28-day strength of cement rather than at earlier ages particularly so for *blended cements*. However, this approach will need 28 days for determining the strength characteristics of cement and at least another 28 days for trial mixes of concrete.

The 28 or 56 days time is too long a period for a contractor to wait for trial mix results. There is a tendency of straight a way using the mix without waiting for *trial mixes*. In order to cut down the time required for trials, the Cement Research Institute of India (CRI) has developed an alternate rapid method where the compressive strengths of cement and concrete are obtained by using *accelerated curing* method as described in IS: 9013–1978.

The *28-day compressive strength of concrete* is found to be statistically significantly related to its *accelerated strength*, therefore the trial mixes are correlated to the *target mean accelerated strength* rather than to target mean 28-day strength. This correlation is found to be independent of the type or characteristics of cement, presumably because they affect both the *accelerated* and *normal strengths of concrete* in a proportionate manner. On the other hand, results of accelerated compressive strength tests on standard cement mortar (IS: 4932–1968) have been found to be unreliable. In the method suggested by CRI, this problem has been overcome by using *accelerated strength of standard* or *reference concrete mix* having *water-cement ratio* of 0.35 and *workability* of 0.80 CF (compacting factor) using cement at hand. The strength is determined by using accelerated curing in accordance with IS: 9013–1978. The nominal maximum size of coarse aggregate of reference concrete should be 10 mm and the fine aggregate should conform to Zone II given in IS: 383–1970. The mix proportion of reference concrete is 1 : 0.81 : 2.07 with a water-cement ratio of 0.35.

Using the above proportions, 150 mm cube specimens of reference concrete are made and the *accelerated strength* is determined by using accelerated *curing* by the

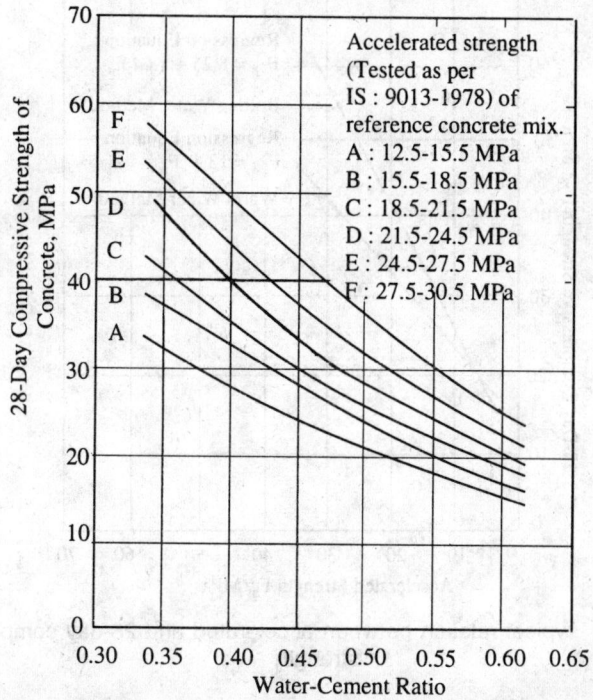

Fig. 10.10 Water-cement ratio vs compressive strength of concrete for different reference strengths (boiling water method)

boiling-water method. Corresponding to the accelerated strength of the reference concrete, the water-cement ratio for the required target mean strength of normal concrete is determined from Fig. 10.10.

The *accelerated strength* of the *trial mix* using this water-cement ratio is checked against the characteristic target strength using the correlation of accelerated and normal 28-day strengths of concrete. The step-by-step procedure of mix design is as follows:

(i) The accelerated strength of reference or standard concrete using the cement at hand is determined by testing 150 mm cubes cured by the boiling water method in accordance with IS: 9013–1978.

(ii) The water-cement ratio for the required target mean strength of normal concrete is determined by using the corresponding accelerated strength of standard concrete obtained in step (i).

(iii) The mix proportions are determined by any of the accepted methods of mix design and checked for workability of fresh concrete against the desired value.

(iv) The accelerated compressive strength of the trial mix is determined on 150 mm cubes cured by the boiling-water method as specified in IS: 9013–1978.

(v) The 28-day compressive strength of normal concrete is estimated from its accelerated strength obtained in step (iv), by using the correlation of accelerated test results to the 28-day strengths of normally cured specimens given in Fig. 10.11.

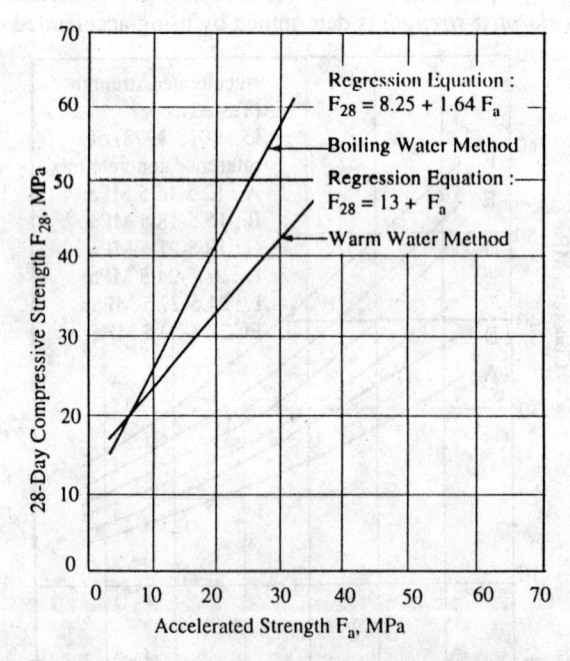

Fig. 10.11 Typical relation between accelerated and 28-day compressive strength

Regression Equation: $F_{28} = 8.25 + 1.64 F_a$ — Boiling Water Method

Regression Equation: $F_{28} = 13 + F_a$ — Warm Water Method

The compressive strength is checked against the target mean strength to judge the suitability of the trial mix.

The significant *reduction in the time* required for the *trial mixes* will help in the adoption of *designed mix concrete* and curb the tendency of using the trial mix without waiting for the strength results. The *accelerated curing cycles* given by IS: 9013–1978 are as follows:

(i) *Boiling water method*
   (a) Specimens are cured for $23 \pm \frac{1}{4}$ h under standard moist conditions at $27 \pm 2°C$.
   (b) At the end of this period, the specimens are cured in boiling water (100°C) for $3\frac{1}{2}$ h $\pm$ 5 min.
   (c) Specimens are cooled to a normal temperature of $27 \pm 2°C$ in two hours before the testing.

(ii) *Warm water method*
   (a) One-and-a-half to three-and-half hours after casting, the specimens are immersed in water maintained at $55 \pm 1°C$ and cured for 20 h $\pm$ 10 min.
   (b) Demould the specimen and cool at $27 \pm 2°C$ for one hour before testing.

Either method may be adopted as a standard for the prediction of *accelerated strength*. These are applicable to most test specimens and give results of *low*

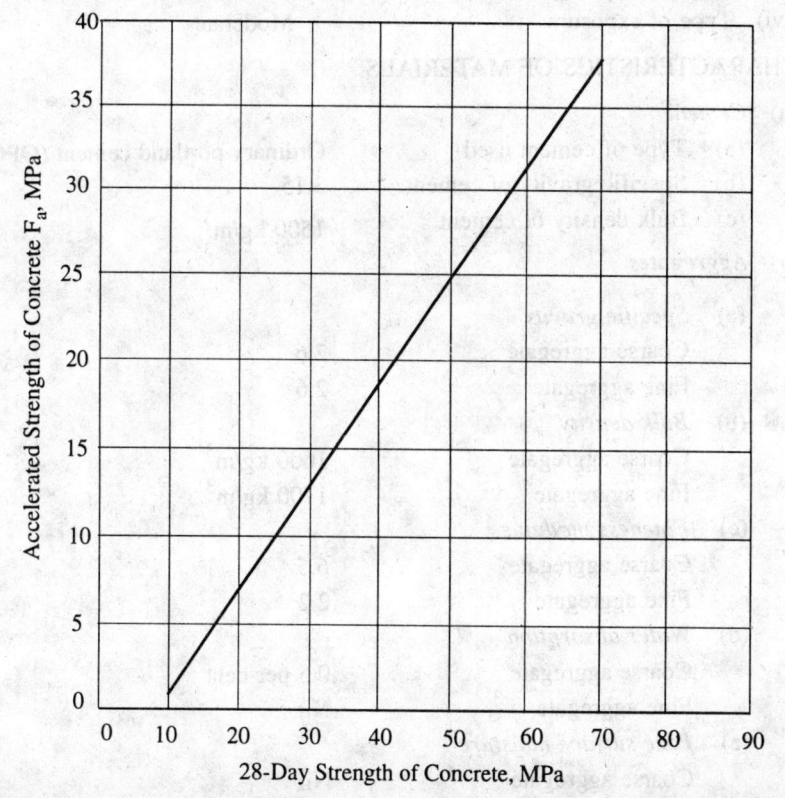

Fig. 10.12 Relationship between accelerated and 28-day strength of concrete

160 Concrete Technology

variability. The actual correlation of accelerated strength to 28-day strength of the normally cured specimen depends upon the *curing cycle* adopted, the chemical composition of cement, and the concrete mix proportions. The average correlation shown in Fig. 10.12 is generally used for different concretes, and in the absence of any past records concerning local materials they can be used to predict the 28-day compressive strength within ±15 per cent.

10.5 CONCRETE MIX DESIGN

The methods discussed in the preceding section are compared by means of following concrete mix proportioning problem:

Example 10.7

1. DESIGN STIPULATIONS
   - (i) Characteristic compressive cube strength at 28 days — 20 MPa
   - (ii) Maximum size of aggregate — 20 mm
   - (iii) Type of aggregate — Crushed rock (angular)
   - (iv) Degree of workability — 0.90 CF
   - (v) Degree of quality control — Good
   - (vi) Type of exposure — Moderate

2. CHARACTERISTICS OF MATERIALS
   - (i) *Cement*
     - (a) Type of cement used — Ordinary portland cement (OPC)
     - (b) Specific gravity of cement — 3.15
     - (c) Bulk density of cement — 1500 kg/m^3
   - (ii) *Aggregates*
     - (a) *Specific gravity*
       - Coarse aggregate — 2.6
       - Fine aggregate — 2.6
     - (b) *Bulk density*
       - Coarse aggregate — 1600 kg/m^3
       - Fine aggregate — 1700 kg/m^3
     - (c) *Fineness modulus*
       - Coarse aggregate — 6.5
       - Fine aggregate — 2.2
     - (d) *Water absorption*
       - Coarse aggregate — 0.5 per cent
       - Fine aggregate — Nil
     - (e) *Free surface moisture*
       - Coarse aggregate — Nil
       - Fine aggregate — 2.0 per cent

(f) *Grading of aggregate*

Type of aggregate	Percentage passing the IS sieve							
	20-mm	*10-mm*	*4.75-mm*	*2.36-mm*	*1.18-mm*	*600-μm*	*300-μm*	*150-μm*
Coarse	100	49	1	0	0	0	0	0
Fine	100	100	100	98	82	63	30	6.5

3. TARGET MEAN STRENGTH

Target mean compressive strength = $(20.0 + 1.65 \times 4.6) = 27.6$ MPa

4. MAXIMUM WATER-CEMENT RATIO

Water-cement ratio for durability under moderate
exposure conditions   0.55

5. WATER CONTENT

Maximum size of aggregate   20 mm
Workability   medium
Water content per cubic metre of concrete   (as per the method used).

6. MIX PROPORTIONS

10.5.1 Mix Design as per Road Note No. 4

(i) Proportioning of Aggregates to Obtain Standard Grading

The gradings of the available fine and coarse aggregates are to be combined in a suitable proportion so as to obtain the desired standard grading. Adopt for trial the standard grading curve No. 3 of Fig. 10.5 for this example. Suppose the percentage passing IS: 4.75-mm sieve is selected as criterion. Let one kg of fine aggregate is combined with x kg of coarse aggregate to obtain desired standard grading and assume that 38 per cent of combined aggregate passes the criterion sieve. The percentage passing IS: 4.75-mm sieve individually must be equal to the total aggregate passing the same sieve, i.e.,

$$100 \times (1) + 1 \times (x) = 38(1 + x) \quad \text{or} \quad x = 1.63$$

Hence the fine and coarse aggregates must be combined in the proportion 1 : 1.63. The combined grading obtained is compared with the standard grading in the Table 10.16.

TABLE 10.16 Comparison of combined grading with standard grading

Grading	Percentage passing							
	20-mm	*10-mm*	*4.75-mm*	*2.36-mm*	*1.18-mm*	*600-μm*	*300-μm*	*150-μm*
Combined aggregate	100	68	38	37	31	24	11	2
Standard grading curve 3	100	65	42	35	28	21	5	1

(ii) Aggregate Cement Ratio

Water-cement ratio for target mean
strength: 0.62 > 0.55
Water-cement ratio (adopted)   0.55

Workability   medium

Maximum size of aggregate   20 mm

Type of aggregate   Angular (Crushed rock)

From Table 10.7 corresponding to
curve 3 aggregate-cement ratio   4.5

Cement content:

Let C be the mass of cement required per cubic metre of concrete then,

$$\frac{C}{3.15 \times 1000} + \frac{4.5\,C}{2.6 \times 1000} + \frac{0.55\,C}{1000} = 1.00$$

$$\therefore \quad C = 385 \text{ kg/m}^3$$

(iii) Mix Proportions

By mass

Water	Cement	Fine aggregate	Coarse aggregate
0.55 × 385 :	385 :	(385 × 4.5 × 1)/2.63 :	(385 × 4.5 × 1.63)/2.63
0.55 :	1.00 :	1.71 :	2.79

10.5.2 DoE Method of Mix Proportioning

(i) Free Water-cement Ratio

From Table 10.8, at the standard free water-cement ratio of 0.50 the 28-day compressive strength is 47 MPa. This pair of data provides a controlling point on Fig. 10.7. By visual interpolation, plot the curve passing through the controlling point. Locate the point C corresponding to compressive strength of 28 MPa on the interpolated curve which is seen to correspond to a free water-cement ratio of 0.7. This value is more than the specified maximum value of 0.55 from durability considerations. Hence adopt a free water-cement ratio of 0.55.

(ii) Cement Content

From Table 10.9,

$$\text{Free water content} = 210 \text{ kg/m}^3$$

Hence,   cement content = 210/0.55 = 382 kg/m^2

This is satisfactory as it is greater than the specified minimum of 290 kg/m^3.

(iii) Aggregate Content

From Fig. 10.8, wet density of concrete = 2345 kg/m^3

Total aggregate content = 2345[1 − (382/3150) − (210/1000)] = 1568 kg/m^3

From Fig. 10.9, for a slump of 30–60 mm, a water-cement ratio of 0.55 and a fine aggregate belonging to grading zone III, a proportion of fine aggregate of 30 per cent may be adopted.

The aggregate content are:

$$\text{Fine aggregate} = 0.30 \times 1568 = 471 \text{ kg/m}^3$$
$$\text{Coarse aggregate} = (1 - 0.30) \times 1568 = 1097 \text{ kg/m}^3$$

(iv) Mix Proportions

Thus the trial mix proportions (by mass) are:

	Water	:	Cement	:	Fine aggregate	:	Coarse aggregate
	210	:	382	:	471	:	1097 (kg/m³)
or	0.55	:	1.00	:	1.23	:	2.87

10.5.3 Mix Design by ACI Method of Mix Proportioning

(i) Coarse Aggregate Content

Maximum size of coarse aggregate	20 mm
Fineness modulus of fine aggregate	2.2
Bulk volume of dry-rodded coarse aggregate per cubic metre of concrete from Table 10.12 (by extrapolation)	0.68 m³
Dry mass of coarse aggregate = 0.68 × 1600	1088 kg

(ii) Cement Content

(a) Equivalent cylinder strength = 0.8×27.6 22.08 MPa
Water-cement ratio:
From strength considerations from Table 10.10   0.67
From durability consideration   0.55
Water-cement ratio adopted   0.55
Water content per cubic metre of concrete from workability considerations from Table 10.11   185 kg
Thus, cement content   185/0.55 = 366.5 kg
(b) From durability considerations the cement content   290 kg
Cement content adopted   336.5 kg

(iii) Air Content

Air content for a maximum size of aggregate of 20 mm from Table 10.11   2 per cent

(iv) Estimation of Mass of Fresh Concrete

Mass of fully compacted fresh concrete per cubic metre volume (in kilograms)

$$= \{1000 - \{v + (C/1000S_c) + (W/1000)\}\}S_a + C + W$$

where C and W are the cement and water contents in kilograms per cubic metre of concrete, respectively, S_c is the specific gravity of cement, S_a the weighted average specific gravity of fine and coarse aggregate combined. The v is air content given in terms of percentage of volume of concrete.

Mass per cubic metre of fresh concrete (kg)

$$= 1000\left[1 - \left(0.02 + \frac{336.5}{1000 \times 3.15} + \frac{185}{1000}\right)\right] \times 2.6 + 336.5 + 185$$

$$= 2311 \text{ kg}$$

Total mass of aggregate per cubic metre of concrete
$$= 2311 - 336.5 - 185 = 1789 \text{ kg}$$

Mass of fine aggregate 1789 − 1088 = 701 kg i.e., the percentage of sand is 39.

(v) Trial Mix Proportions

(a) *By mass*

Water	Cement	Fine aggregate	Coarse aggregate
185 :	336.5 :	701 :	1088
0.55 :	1 :	2.08 :	3.23

(b) *By Volume*

Cement	Fine aggregate	Coarse aggregate
1 :	1.83 :	3.03

Water-cement ratio = 0.55.

10.5.4 Mix Design as per Indian Standard Recommended Guidelines

Water content   186 kg/m^3
Percentage of fine aggregate from Table 10.14   35
Adjustments due to change in conditions from those specified in Table 10.14

		In water (per cent)	In Sand (per cent)
(i)	For decrease in water cement ratio by 0.10	0	−2
(ii)	For increase in compacting factor by 0.10	+3	0
(iii)	For sand conforming to zone III (IS: 383–1970)	0	−1.5
	Total	+3	−3.5

Water-content = 186 × 1.03   191.6 kg
Sand content   31.5 per cent
Air content   2.0 per cent
Concrete volume = 1.00 − 0.02   0.98 m^3
Cement content = (191.6/0.5)   383 kg > 290 kg (adequate)

$$\text{Total volume of aggregates} = 0.98 - \left(\frac{383}{3.15} + 191.6\right)\frac{1}{1000} = 0.67 \text{ m}^3$$

Total mass of aggregates = 0.67 × (2.6 × 1000) = 1742 kg

Mix Proportions

(a) *By mass*

Water	Cement	Fine aggregate	Coarse aggregate
191.6 :	383 :	549 :	1193
0.50 :	1.00 :	1.43 :	3.11

(b) *By volume*

Cement	Fine aggregate	Coarse aggregate
1 :	1.26 :	2.92

Water-cement ratio = 0.5.

The mix proportions arrived at by different methods are summarized in the Table 10.17.

TABLE 10.17 Summary of mix proportions obtained by various methods

Method of mix design	Mix proportions by mass			Aggregate-cement ratio
	Cement	Fine aggregate	Coarse aggregate	
1. Road Note No. 4	1	1.71	2.79	4.50
2. DoE method	1	1.23	2.87	4.10
3. ACI method	1	2.08	3.23	5.31
4. Indian Standard Recommended Guidelines	1	1.43	3.11	4.54

7. ACTUAL QUANTITIES OF MATERIAL REQUIRED PER BAG OF CEMENT

The weights of materials must be adjusted for moisture content. As an illustration, adopt the mix proportions obtained by using the Indian Standard Recommended Guidelines for mix design. Thus for one bag of cement:

Cement content	=	50 kg
Fine aggregate (dry)	=	71.5 kg
Coarse aggregate (dry)	=	155 kg
Weight of wet fine aggregate = 71.5 × 1.02 =		72.93 kg

The free moisture present in the aggregate must be deducted from the water to be added and extra water is needed to provide for absorption. Surface moisture contributed by:

Fine aggregate = 71.5 × 0.02	= 1.43 kg
Coarse aggregate	= Nil

Extra water to be added to provide for absorption in:

Fine aggregate	= Nil
Coarse aggregate = 155 × 0.005	= 0.78 kg

Therefore,

   Estimated requirement of water = 25.00 − 1.43 + 0.78 = 24.35 kg (or litres)

Batch mass per bag of cement

Water content	=	24.35 kg
Cement content	=	50.00 kg
Fine aggregate (wet)	=	72.93 kg
Coarse aggregate	=	155.00 kg
Total	=	302.28 kg

Therefore, the net mix proportion by mass is 1.00 : 1.46 : 3.10 with free water cement ratio as 0.487.

8. TRIAL MIXES

Trial mixes should be prepared using these proportions as explained in Sec. 10.6 and tested to check if the mix meets the design stipulations. Otherwise, suitable adjustments should be made till it satisfies the design stipulations.

The final mix proportions were found to be 1 : 1.55 : 2.95 by mass (or 1 : 1.37 : 2.77 by dry volume) and the water-cement ratio was 0.49.

10.6 COMPUTER-AIDED CONCRETE MIX DESIGN

10.6.1 Conventional Design

Based on the procedure described in sec. 10.4.5., a computer code CADCOM has been developed. The design parameters considered in the code are grade of concrete, water-cement ratio, degree of workability, degree of quality control, type of exposure, type of aggregate, maximum size of aggregate, and grade of fine aggregate. The flow chart is shown in the Fig. 10.13. The proportions of materials required for various grades of concrete are listed in Tables 10.18 to 10.35. The proportions listed in the tables assume specific gravity of cement as 3.15, and that of fine and coarse aggregates as 2.6.

10.6.2 Optimum Design

In case of big projects a comparative analysis of cost of concretes produced using materials from different sources can help to identify the appropriate source. The *optimum mix proportions* of the ingredients giving the least cost of concrete should satisfy the criteria of *strength* and *durability of hardened concrete* and *workability of fresh concrete*. Since the cost of cement in general is very high in comparison with other ingredients, a *leanest mix* satisfying the requirements of durability in terms of cement content, and workability is desirable.

The process of estimating the optimum relative proportions of ingredients to achieve the most economical mix satisfying the design stipulations regarding strength, durability and workability, can be formulated into a *mathematical optimization problem*. The formulation consists in minimization of an *objective function* expressing production cost of unit volume of concrete subject to the *constraints* of compressive strength, durability, workability and aggregate characteristics.

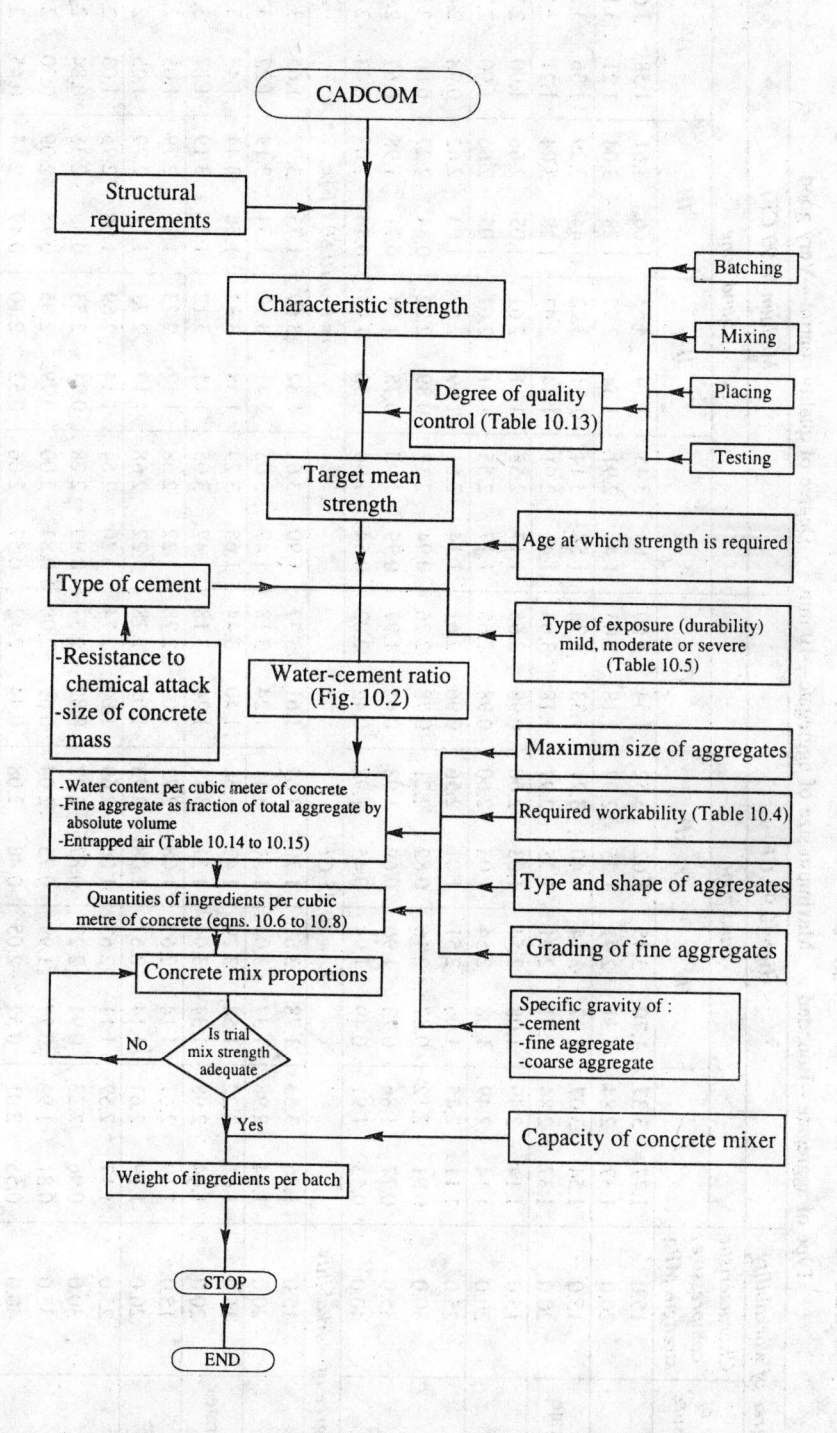

Fig. 10.13 Flow chart of computer program CADCOM

TABLE 10.18 Fine, and coarse aggregate-cement ratios (by mass) for different characteristic compressive strengths

Type of aggregate—Rounded Maximum size of aggregate—10 mm Degree of quality control—Very good

Degree of workability		*High (0.95 CF)*								*Medium (0.90 CF)*							
Type of exposure	*Characteristic compressive strength MPa*	*Sand zone*								*Sand zone*							
		I		*II*		*III*		*IV*		*I*		*II*		*III*		*IV*	
Mild	15.0	1.77	3.37	1.70	3.45	1.62	3.52	1.54	3.60	1.81	3.45	1.74	3.53	1.66	3.61	1.58	3.68
	20.0	1.37	2.84	1.31	2.91	1.25	2.97	1.18	3.03	1.41	2.91	1.34	2.97	1.28	3.04	1.21	3.10
Moderate	15.0	1.54	3.07	1.47	3.14	1.40	3.21	1.33	3.28	1.58	3.14	1.51	3.22	1.44	3.29	1.36	3.36
	20.0	1.37	2.84	1.31	2.91	1.25	2.97	1.18	3.03	1.41	2.91	1.34	2.97	1.28	3.04	1.21	3.10
Severe	15.0	1.14	2.49	1.08	2.54	1.03	2.60	0.98	2.65	1.17	2.55	1.11	2.61	1.05	2.66	1.00	2.72
	20.0	1.14	2.49	1.08	2.54	1.03	2.60	0.98	2.65	1.17	2.55	1.11	2.61	1.05	2.66	1.00	2.72
All	25.0	1.11	2.45	1.06	2.51	1.01	2.56	0.96	2.61	1.14	2.51	1.09	2.57	1.03	2.62	0.98	2.68
	30.0	0.91	2.12	0.87	2.16	0.82	2.21	0.78	2.26	0.94	2.17	0.89	2.22	0.84	2.27	0.80	2.31
	35.0	0.77	1.86	0.73	1.90	0.69	1.93	0.65	1.97	0.79	1.90	0.75	1.94	0.71	1.98	0.67	2.02
	40.0	0.53	1.91	0.49	1.94	0.45	1.98	0.42	2.02	0.54	1.96	0.50	1.99	0.46	2.03	0.43	2.07
Degree of workability		*Low (0.85 CF)*								*Very low (0.80 CF)*							
Mild	15.0	1.86	3.53	1.78	3.61	1.70	3.69	1.61	3.77	1.90	3.61	1.82	3.69	1.73	3.77	1.65	3.86
	20.0	1.44	2.98	1.37	3.04	1.31	3.11	1.24	3.18	1.47	3.05	1.41	3.12	1.34	3.19	1.27	3.25
Moderate	15.0	1.61	3.22	1.54	3.29	1.47	3.36	1.40	3.44	1.65	3.29	1.58	3.37	1.56	3.44	1.43	3.52
	20.0	1.44	2.98	1.37	3.04	1.31	3.11	1.24	3.18	1.47	3.05	1.41	3.12	1.34	3.19	1.27	3.25
Severe	15.0	1.19	2.61	1.14	2.67	1.08	2.73	1.02	2.78	1.22	2.68	1.17	2.73	1.11	2.79	1.05	2.85
	20.0	1.19	2.61	1.14	2.67	1.08	2.73	1.02	2.78	1.22	2.68	1.17	2.73	1.11	2.79	1.05	2.85
All	25.0	1.17	2.57	1.11	2.63	1.06	2.69	1.00	2.74	1.20	2.64	1.14	2.69	1.08	2.75	1.03	2.81
	30.0	0.96	2.23	0.91	2.27	0.87	2.32	0.82	2.37	0.99	2.28	0.94	2.33	0.89	2.38	0.84	2.43
	35.0	0.81	1.95	0.77	1.99	0.73	2.04	0.69	2.08	0.83	2.00	0.79	2.05	0.75	2.09	0.70	2.13
	40.0	0.55	2.01	0.52	2.05	0.48	2.08	0.44	2.12	0.57	2.06	0.53	2.10	0.49	2.14	0.45	2.18

TABLE 10.19 Fine, and coarse aggregate-cement ratios (by mass) for different characteristic compressive strengths

Type of aggregate—Rounded   Maximum size of aggregate—10 mm

			High (0.95 CF)						Medium (0.90 CF)			
			Sand zone						Sand zone			
Type of exposure	Characteristic compressive strength MPa	I	II	III	IV		I	II	III	IV

Degree of quality control—Good

Degree of workability

Type of exposure	Char. strength MPa	I	II	III	IV	I	II	III	IV
Mild	15.0	1.66 / 1.30	1.59 / 1.24	1.52 / 1.18	1.44 / 1.12	1.70 / 1.33	1.63 / 1.27	1.55 / 1.20	1.48 / 1.14
	20.0	3.23 / 2.73	3.30 / 2.79	3.38 / 2.85	3.45 / 2.91	3.31 / 2.80	3.38 / 2.86	3.46 / 2.92	3.53 / 2.98
Moderate	15.0	1.54 / 1.30	1.47 / 1.24	1.40 / 1.18	1.33 / 1.12	1.58 / 1.33	1.51 / 1.27	1.44 / 1.20	1.36 / 1.14
	20.0	3.07 / 2.73	3.14 / 2.79	3.21 / 2.85	3.28 / 2.91	3.14 / 2.80	3.22 / 2.86	3.29 / 2.92	3.36 / 2.98
Severe	15.0	1.14 / 1.14	1.08 / 1.08	1.03 / 1.03	0.98 / 0.98	1.17 / 1.17	1.11 / 1.11	1.05 / 1.05	1.00 / 1.00
	20.0	2.49 / 2.49	2.54 / 2.54	2.60 / 2.60	2.65 / 2.65	2.55 / 2.55	2.61 / 2.61	2.66 / 2.66	2.72 / 2.72
All	25.0	1.06 / 0.87	1.01 / 0.82	0.95 / 0.78	0.90 / 0.74	1.08 / 0.89	1.03 / 0.85	0.98 / 0.80	0.93 / 0.76
	30.0	2.36 / 2.04	2.41 / 2.08	2.46 / 2.12	2.51 / 2.17	2.42 / 2.09	2.47 / 2.13	2.52 / 2.18	2.57 / 2.22
	35.0	0.73 / 1.78	0.69 / 1.82	0.66 / 1.86	0.62 / 1.89	0.75 / 1.83	0.71 / 1.87	0.67 / 1.91	0.63 / 1.94
	40.0	0.50 / 1.83	0.46 / 1.86	0.43 / 1.90	0.39 / 1.93	0.51 / 1.88	0.48 / 1.91	0.44 / 1.95	0.40 / 1.98

Degree of workability

Low (0.85 CF) | Very low (0.80 CF)

Type of exposure	Char. strength MPa	I	II	III	IV	I	II	III	IV
Mild	15.0	1.74 / 1.36	1.66 / 1.30	1.59 / 1.23	1.51 / 1.17	1.78 / 1.39	1.70 / 1.33	1.62 / 1.26	1.55 / 1.20
	20.0	3.38 / 2.86	3.46 / 2.93	3.54 / 2.99	3.61 / 3.05	3.46 / 2.93	3.54 / 3.00	3.62 / 3.06	3.70 / 3.13
Moderate	15.0	1.61 / 1.36	1.54 / 1.30	1.47 / 1.23	1.40 / 1.17	1.65 / 1.39	1.58 / 1.33	1.50 / 1.26	1.43 / 1.20
	20.0	3.22 / 2.86	3.29 / 2.93	3.36 / 2.99	3.44 / 3.05	3.29 / 2.93	3.37 / 3.00	3.44 / 3.06	3.52 / 3.13
Severe	15.0	1.19 / 1.19	1.14 / 1.14	1.08 / 1.08	1.02 / 1.02	1.22 / 1.22	1.17 / 1.17	1.11 / 1.11	1.05 / 1.05
	20.0	2.61 / 2.61	2.67 / 2.67	2.73 / 2.73	2.78 / 2.78	2.68 / 2.68	2.73 / 2.73	2.79 / 2.79	2.85 / 2.85
All	25.0	1.11 / 0.91	1.06 / 0.87	1.00 / 0.82	0.95 / 0.78	1.14 / 0.94	1.08 / 0.89	1.03 / 0.84	0.97 / 0.80
	30.0	2.48 / 2.14	2.53 / 2.19	2.58 / 2.23	2.64 / 2.28	2.54 / 2.20	2.59 / 2.24	2.65 / 2.29	2.70 / 2.34
	35.0	0.77 / 1.88	0.73 / 1.92	0.69 / 1.96	0.65 / 2.00	0.79 / 1.93	0.75 / 1.97	0.71 / 2.01	0.67 / 2.05
	40.0	0.52 / 1.93	0.49 / 1.96	0.45 / 2.00	0.42 / 2.04	0.54 / 1.98	0.50 / 2.02	0.46 / 2.05	0.43 / 2.09

TABLE 10.20 Fine, and coarse aggregate–cement ratios (by mass) for different characteristics compressive strengths

Type of aggregate—Rounded Maximum size of aggregate—10 mm Degree of quality control—Fair

Type of exposure	Characteristic compressive strength MPa	High (0.95 CF) Sand zone I	II	III	IV	Medium (0.90 CF) Sand zone I	II	III	IV
Degree of workability									
Mild	15.0	1.56 / 1.23	1.49 / 1.17	1.42 / 1.11	1.35 / 1.05	1.60 / 1.26	1.53 / 1.20	1.45 / 1.14	1.38 / 1.08
	20.0	3.10 / 2.63	3.17 / 2.68	3.24 / 2.74	3.31 / 2.80	3.17 / 2.69	3.24 / 2.75	3.31 / 2.81	3.39 / 2.87
Moderate	15.0	1.54 / 1.23	1.47 / 1.17	1.40 / 1.11	1.33 / 1.05	1.58 / 1.26	1.51 / 1.20	1.44 / 1.14	1.36 / 1.08
	20.0	3.07 / 2.63	3.14 / 2.68	3.21 / 2.74	3.28 / 2.80	3.14 / 2.69	3.22 / 2.75	3.29 / 2.81	3.36 / 2.87
Severe	15.0	1.14 / 1.14	1.08 / 1.08	1.03 / 1.03	0.98 / 0.98	1.17 / 1.17	1.11 / 1.11	1.05 / 1.05	1.00 / 1.00
	20.0	2.49 / 2.49	2.54 / 2.54	2.60 / 2.60	2.60 / 2.60	2.55 / 2.55	2.61 / 2.61	2.66 / 2.66	2.72 / 2.72
All	25.0	1.00 / 0.82	0.95 / 0.78	0.90 / 0.74	0.86 / 0.70	1.03 / 0.85	0.98 / 0.80	0.93 / 0.76	0.88 / 0.72
	30.0	2.27 / 1.96	2.32 / 2.00	2.37 / 2.04	2.41 / 2.08	2.32 / 2.01	2.37 / 2.05	2.42 / 2.09	2.47 / 2.14
	35.0	0.69 / 0.47	0.66 / 0.44	0.62 / 0.41	0.79 / 0.37	0.71 / 0.48	0.67 / 0.45	0.64 / 0.42	0.60 / 0.38
	40.0	1.71 / 1.75	1.75 / 1.79	1.78 / 1.82	1.82 / 1.85	1.76 / 1.80	1.79 / 1.83	1.83 / 1.87	1.87 / 1.90

Type of exposure	Characteristic compressive strength MPa	Low (0.85 CF) Sand zone I	II	III	IV	Very Low (0.80 CF) Sand zone I	II	III	IV
Degree of workability									
Mild	15.0	1.63 / 1.29	1.56 / 1.23	1.49 / 1.17	1.42 / 1.10	1.67 / 1.32	1.60 / 1.26	1.52 / 1.19	1.45 / 1.13
	20.0	3.25 / 2.75	3.32 / 2.81	3.39 / 2.87	3.46 / 2.93	3.32 / 2.82	3.40 / 2.88	3.47 / 2.94	3.55 / 3.01
Moderate	15.0	1.61 / 1.29	1.54 / 1.23	1.47 / 1.17	1.40 / 1.10	1.65 / 1.32	1.58 / 1.26	1.50 / 1.19	1.43 / 1.13
	20.0	3.22 / 2.75	3.29 / 2.81	3.36 / 2.87	3.44 / 2.93	3.29 / 2.82	3.37 / 2.88	3.44 / 2.94	3.52 / 3.01
Severe	15.0	1.19 / 1.19	1.14 / 1.14	1.08 / 1.08	1.02 / 1.02	1.22 / 1.22	1.17 / 1.17	1.11 / 1.11	1.05 / 1.05
	20.0	2.61 / 2.61	2.67 / 2.67	2.73 / 2.73	2.78 / 2.78	2.68 / 2.68	2.73 / 2.73	2.79 / 2.79	2.85 / 2.85
All	25.0	1.05 / 0.87	1.00 / 0.82	0.95 / 0.78	0.90 / 0.74	1.08 / 0.89	1.03 / 0.84	0.97 / 0.80	0.92 / 0.76
	30.0	2.38 / 2.06	2.43 / 2.10	2.48 / 2.15	2.54 / 2.19	2.44 / 2.11	2.49 / 2.16	2.55 / 2.20	2.60 / 2.25
	35.0	0.73 / 0.50	0.69 / 0.46	0.65 / 0.43	0.62 / 0.39	0.75 / 0.51	0.71 / 0.47	0.67 / 0.44	0.63 / 0.40
	40.0	1.80 / 1.85	1.84 / 1.88	1.88 / 1.92	1.92 / 1.95	1.85 / 1.90	1.89 / 1.93	1.93 / 1.97	1.97 / 2.01

TABLE 10.21 Fine, and coarse aggregate–cement ratios (by mass) for different characteristic compressive strengths

Type of aggregate—Rounded   Maximum size of aggregate—20 mm   Degree of quality control—Very good

High (0.95 CF) — Sand zone

Degree of workability	Type of exposure	Characteristic compressive strength MPa	I	II	III	IV
Mild		15.0	1.83 / 4.37	1.74 / 4.46	1.64 / 4.55	1.55 / 4.65
		20.0	1.41 / 3.70	1.33 / 3.77	1.26 / 3.85	1.18 / 3.92
Moderate		15.0	1.58 / 3.99	1.50 / 4.07	1.42 / 4.16	1.33 / 4.24
		20.0	1.41 / 3.70	1.33 / 3.77	1.26 / 3.85	1.18 / 3.92
Severe		15.0	1.17 / 3.25	1.10 / 3.31	1.03 / 3.38	0.97 / 3.44
		20.0	1.17 / 3.25	1.10 / 3.31	1.03 / 3.38	0.97 / 3.44
All		25.0	1.14 / 3.20	1.08 / 3.27	1.01 / 3.33	0.95 / 3.40
		30.0	0.94 / 2.78	0.88 / 2.83	0.82 / 2.89	0.77 / 2.94
		35.0	0.79 / 2.45	0.74 / 2.49	0.69 / 2.54	0.64 / 2.59
		40.0	0.50 / 2.20	0.46 / 2.24	0.42 / 2.28	0.38 / 2.32

Low (0.85 CF)

Degree of workability	Type of exposure	Characteristic compressive strength MPa	I	II	III	IV
Mild		15.0	1.91 / 4.56	1.81 / 4.66	1.72 / 4.76	1.62 / 4.85
		20.0	1.47 / 3.86	1.39 / 3.94	1.31 / 4.02	1.23 / 4.10
Moderate		15.0	1.66 / 4.17	1.57 / 4.26	1.48 / 4.34	1.39 / 4.43
		20.0	1.47 / 3.86	1.39 / 3.94	1.31 / 4.02	1.23 / 4.10
Severe		15.0	1.22 / 3.40	1.15 / 3.47	1.08 / 3.54	1.01 / 3.60
		20.0	1.22 / 3.40	1.15 / 3.47	1.08 / 3.54	1.01 / 3.60
All		25.0	1.19 / 3.35	1.13 / 3.42	1.06 / 3.49	0.99 / 3.55
		30.0	0.98 / 2.91	0.92 / 2.97	0.86 / 3.03	0.81 / 3.08
		35.0	0.83 / 2.57	0.78 / 2.62	0.72 / 2.67	0.67 / 2.72
		40.0	0.53 / 2.32	0.49 / 2.36	0.45 / 2.40	0.40 / 2.44

Medium (0.90 CF) — Sand zone

Degree of workability	Type of exposure	Characteristic compressive strength MPa	I	II	III	IV
Mild		15.0	1.87 / 4.46	1.77 / 4.56	1.68 / 4.65	1.58 / 4.75
		20.0	1.44 / 3.78	1.36 / 3.86	1.29 / 3.93	1.21 / 4.01
Moderate		15.0	1.62 / 4.08	1.53 / 4.16	1.45 / 4.25	1.36 / 4.33
		20.0	1.44 / 3.78	1.36 / 3.86	1.29 / 3.93	1.21 / 4.01
Severe		15.0	1.19 / 3.32	1.12 / 3.39	1.06 / 3.46	0.99 / 3.52
		20.0	1.19 / 3.32	1.12 / 3.39	1.06 / 3.46	0.99 / 3.52
All		25.0	1.17 / 3.27	1.10 / 3.34	1.03 / 3.41	0.97 / 3.47
		30.0	0.96 / 2.84	0.90 / 2.90	0.84 / 2.96	0.79 / 3.01
		35.0	0.81 / 2.50	0.76 / 2.55	0.71 / 2.60	0.66 / 2.65
		40.0	0.52 / 2.26	0.48 / 2.30	0.43 / 2.34	0.39 / 2.38

Very low (0.80 CF)

Degree of workability	Type of exposure	Characteristic compressive strength MPa	I	II	III	IV
Mild		15.0	1.95 / 4.66	1.85 / 4.76	1.75 / 4.86	1.66 / 4.96
		20.0	1.51 / 3.95	1.43 / 4.03	1.34 / 4.11	1.26 / 4.20
Moderate		15.0	1.69 / 4.26	1.60 / 4.35	1.51 / 4.44	1.43 / 4.53
		20.0	1.51 / 3.95	1.43 / 4.03	1.34 / 4.11	1.26 / 4.20
Severe		15.0	1.25 / 3.48	1.18 / 3.55	1.11 / 3.62	1.04 / 3.69
		20.0	1.25 / 3.48	1.18 / 3.55	1.11 / 3.62	1.04 / 3.69
All		25.0	1.22 / 3.43	1.15 / 3.50	1.08 / 3.57	1.01 / 3.64
		30.0	1.00 / 2.98	0.94 / 3.04	0.88 / 3.10	0.83 / 3.16
		35.0	0.85 / 2.63	0.79 / 2.68	0.74 / 2.73	0.69 / 2.78
		40.0	0.54 / 2.37	0.50 / 2.42	0.46 / 2.46	0.41 / 2.51

172 Concrete Technology

TABLE 10.22 Fine, and coarse aggregage—cement ratios (by mass) for different characteristic compressive strengths

Type of aggregate—Rounded Maximum size of aggregate—20 mm

Degree of workability		High (0.95 CF)				Medium (0.90 CE)			
		Degree of quality control—Good							
		Sand zone				Sand zone			
Type of exposure	Characteristic compressive strength MPa	I	II	III	IV	I	II	III	IV
Mild	15.0	1.71 / 1.33	1.62 / 1.26	1.53 / 1.18	1.45 / 1.11	1.75 / 1.36	1.66 / 1.29	1.57 / 1.21	1.48 / 1.14
	20.0	4.19 / 3.55	4.28 / 3.63	4.37 / 3.70	4.45 / 3.77	4.28 / 3.63	4.37 / 3.71	4.46 / 3.78	4.55 / 3.86
Moderate	15.0	1.58 / 1.33	1.50 / 1.26	1.42 / 1.18	1.33 / 1.11	1.62 / 1.36	1.53 / 1.29	1.45 / 1.21	1.36 / 1.14
	20.0	3.99 / 3.55	4.07 / 3.63	4.16 / 3.70	4.24 / 3.77	4.08 / 3.63	4.16 / 3.71	4.25 / 3.78	4.33 / 3.86
Severe	15.0	1.17 / 1.17	1.10 / 1.10	1.03 / 1.03	0.97 / 0.97	1.19 / 1.19	1.12 / 1.12	1.06 / 1.06	0.99 / 0.99
	20.0	3.25 / 3.25	3.31 / 3.31	3.38 / 3.38	3.44 / 3.44	3.32 / 3.32	3.39 / 3.39	3.46 / 3.46	3.52 / 3.52
All	25.0	1.08 / 0.89	1.02 / 0.83	0.96 / 0.78	0.89 / 0.73	1.11 / 0.91	1.04 / 0.85	0.98 / 0.80	0.92 / 0.75
	30.0	3.08 / 2.67	3.14 / 2.73	3.21 / 2.78	3.27 / 2.83	3.15 / 2.74	3.22 / 2.79	3.28 / 2.85	3.34 / 2.90
	35.0	0.75 / 0.48	0.70 / 0.44	0.66 / 0.40	0.61 / 0.36	0.77 / 0.49	0.72 / 0.45	0.67 / 0.41	0.62 / 0.37
	40.0	2.35 / 2.11	2.40 / 2.15	2.44 / 2.19	2.49 / 2.23	2.41 / 2.17	2.46 / 2.21	2.50 / 2.25	2.55 / 2.29

Degree of workability		Low (0.85 CF)				Very low (0.80 CF)			
		I	II	III	IV	I	II	III	IV
Mild	15.0	1.79 / 1.39	1.69 / 1.31	1.60 / 1.24	1.51 / 1.16	1.83 / 1.42	1.73 / 1.34	1.64 / 1.27	1.54 / 1.19
	20.0	4.38 / 3.72	4.47 / 3.79	4.56 / 3.87	4.65 / 3.95	4.47 / 3.80	4.57 / 3.88	4.66 / 3.96	4.76 / 4.04
Moderate	15.0	1.66 / 1.39	1.57 / 1.31	1.48 / 1.24	1.39 / 1.16	1.69 / 1.42	1.60 / 1.34	1.51 / 1.27	1.43 / 1.19
	20.0	4.17 / 3.72	4.26 / 3.79	4.34 / 3.87	4.43 / 3.95	4.26 / 3.80	4.35 / 3.88	4.44 / 3.96	4.53 / 4.04
Severe	15.0	1.22 / 1.22	1.15 / 1.15	1.08 / 1.08	1.01 / 1.01	1.25 / 1.25	1.18 / 1.18	1.11 / 1.11	1.04 / 1.04
	20.0	3.40 / 3.40	3.47 / 3.47	3.54 / 3.54	3.60 / 3.60	3.48 / 3.48	3.55 / 3.55	3.62 / 3.62	3.69 / 3.69
All	25.0	1.13 / 0.93	1.07 / 0.88	1.00 / 0.82	0.94 / 0.76	1.16 / 0.95	1.09 / 0.90	1.03 / 0.84	0.96 / 0.78
	30.0	3.23 / 2.80	3.29 / 2.86	3.36 / 2.91	3.42 / 2.97	3.30 / 2.87	3.37 / 2.93	3.43 / 2.98	3.50 / 3.04
	35.0	0.78 / 0.50	0.74 / 0.46	0.69 / 0.42	0.64 / 0.38	0.80 / 0.52	0.75 / 0.47	0.70 / 0.43	0.66 / 0.39
	40.0	2.47 / 2.22	2.52 / 2.26	2.57 / 2.31	2.61 / 2.35	3.53 / 2.28	2.58 / 2.32	2.63 / 2.37	2.68 / 2.41

TABLE 10.23 Fine, and coarse aggregate–cement ratios (by mass) for different characteristic compressive strengths

Type of aggregate—Rounded Maximum size of aggregate—20 mm Degree of quality control—Fair

Degree of workability		High (0.95 CF) Sand zone				Medium (0.90 CF) Sand zone			
Type of exposure	Characteristic compressive strength MPa	I	II	III	IV	I	II	III	IV
Mild	15.0	1.60 / 4.02	1.52 / 4.11	1.44 / 4.19	1.35 / 4.27	1.64 / 4.11	1.55 / 4.20	1.47 / 4.28	1.38 / 4.37
	20.0	1.26 / 3.42	1.19 / 3.49	1.12 / 3.56	1.05 / 3.63	1.28 / 3.50	1.21 / 3.57	1.14 / 3.64	1.07 / 3.71
Moderate	15.0	1.58 / 3.99	1.50 / 4.07	1.42 / 4.16	1.33 / 4.24	1.62 / 4.08	1.53 / 4.16	1.45 / 4.25	1.36 / 4.33
	20.0	1.26 / 3.42	1.19 / 3.49	1.12 / 3.56	1.05 / 3.63	1.28 / 3.50	1.21 / 3.57	1.14 / 3.64	1.07 / 3.71
Severe	15.0	1.17 / 3.25	1.10 / 3.31	1.03 / 3.38	0.97 / 3.44	1.19 / 3.32	1.12 / 3.39	1.06 / 3.46	0.99 / 3.52
	20.0	1.17 / 3.25	1.10 / 3.31	1.03 / 3.38	0.97 / 3.44	1.19 / 3.32	1.12 / 3.39	1.06 / 3.46	0.99 / 3.52
All	25.0	1.02 / 2.97	0.97 / 3.03	0.91 / 3.09	0.85 / 3.15	1.05 / 3.04	0.99 / 3.10	0.93 / 3.16	0.87 / 3.22
	30.0	0.84 / 2.57	0.79 / 2.62	0.74 / 2.68	0.69 / 2.73	0.86 / 2.64	0.81 / 2.69	0.76 / 2.74	0.71 / 2.79
	35.0	0.71 / 2.26	0.67 / 2.31	0.62 / 2.35	0.58 / 2.39	0.73 / 2.32	0.68 / 2.36	0.64 / 2.41	0.59 / 2.45
	40.0	0.45 / 2.03	0.42 / 2.07	0.38 / 2.10	0.34 / 2.14	0.46 / 2.08	0.43 / 2.12	0.39 / 2.16	0.35 / 2.20

Degree of workability		Low (0.85 CF)				Very low (0.80 CF)			
Mild	15.0	1.68 / 4.20	1.59 / 4.29	1.50 / 4.38	1.41 / 4.47	1.71 / 4.30	1.62 / 4.39	1.53 / 4.48	1.44 / 4.57
	20.0	1.31 / 3.58	1.24 / 3.65	1.17 / 3.72	1.09 / 3.80	1.34 / 3.66	1.27 / 3.73	1.20 / 3.81	1.12 / 3.88
Moderate	15.0	1.66 / 4.17	1.57 / 4.26	1.48 / 4.34	1.39 / 4.43	1.69 / 4.26	1.60 / 4.35	1.51 / 4.44	1.43 / 4.53
	20.0	1.31 / 3.58	1.24 / 3.65	1.17 / 3.72	1.09 / 3.80	1.34 / 3.66	1.27 / 3.73	1.20 / 3.81	1.12 / 3.88
Severe	15.0	1.22 / 3.40	1.15 / 3.47	1.08 / 3.54	1.01 / 3.60	1.25 / 3.48	1.18 / 3.55	1.11 / 3.62	1.04 / 3.69
	20.0	1.22 / 3.40	1.15 / 3.47	1.08 / 3.54	1.01 / 3.60	1.25 / 3.48	1.18 / 3.55	1.11 / 3.62	1.04 / 3.69
All	25.0	1.07 / 3.11	1.01 / 3.17	0.95 / 3.23	0.89 / 3.29	1.10 / 3.18	1.04 / 3.24	0.97 / 3.31	0.91 / 3.37
	30.0	0.88 / 2.70	0.83 / 2.57	0.78 / 2.81	0.72 / 2.86	0.91 / 2.76	0.85 / 2.82	0.80 / 2.87	0.74 / 2.93
	35.0	0.75 / 2.37	0.70 / 2.42	0.65 / 2.47	0.61 / 2.51	0.76 / 2.43	0.72 / 2.48	0.67 / 2.53	0.62 / 2.58
	40.0	0.48 / 2.14	0.44 / 2.17	0.40 / 2.21	0.36 / 2.25	0.49 / 2.19	0.45 / 2.23	0.41 / 2.27	0.37 / 2.31

174 Concrete Technology

TABLE 10.24 Fine, and coarse aggregate–cement ratios (by mass) for different characteristic compressive strengths

Type of aggregate—Rounded   Maximum size of aggregate—40 mm   Degree of quality control—Very good

Degree of workability		High (0.95 CF)				Medium (0.90 CF)			
		Sand zone				Sand zone			
Type of Exposure	Characteristic compressive strength MPa	I	II	III	IV	I	II	III	IV
Mild	15.0	1.84	1.73	1.61	1.50	1.88	1.76	1.65	1.53
		5.65	5.77	5.88	5.99	5.77	5.89	6.00	6.12
	20.0	1.40	1.31	1.22	1.12	1.43	1.34	1.24	1.15
		4.79	4.89	4.98	5.07	4.90	4.99	5.09	5.18
Moderate	15.0	1.56	1.46	1.36	1.26	1.57	1.47	1.37	1.27
		5.09	5.19	5.29	5.39	5.11	5.21	5.31	5.41
	20.0	1.40	1.31	1.22	1.12	1.43	1.34	1.24	1.15
		4.79	4.89	4.98	5.07	4.90	4.99	5.09	5.18
Severe	15.0	1.12	1.04	1.96	0.88	1.12	1.04	0.96	0.88
		4.09	4.17	4.24	4.32	4.10	4.18	4.26	4.34
	20.0	1.12	1.04	1.96	0.88	1.12	1.04	0.96	0.88
		4.09	4.17	4.24	4.32	4.10	4.18	4.26	4.34
All	25.0	1.11	1.03	1.95	0.88	1.11	1.03	0.96	0.88
		4.09	4.17	4.24	4.32	4.11	4.18	4.26	4.34
	30.0	0.92	0.85	0.78	0.72	0.94	0.87	0.80	0.73
		3.63	3.70	3.77	3.83	3.71	3.78	3.85	3.92
	35.0	0.77	0.71	0.65	0.59	0.79	0.73	0.67	0.61
		3.21	3.27	3.33	3.39	3.28	3.34	3.40	3.46
	40.0	—	—	—	—	—	—	—	—

Degree of workability		Low (0.85)				Very low (0.80 CF)			
		I	II	III	IV	I	II	III	IV
Mild	15.0	1.92	1.80	1.68	1.57	1.94	1.83	1.71	1.59
		5.90	6.01	6.13	6.25	5.98	6.10	6.22	6.34
	20.0	1.46	1.37	1.27	1.17	1.50	1.40	1.30	1.20
		5.00	5.10	5.20	5.29	5.11	5.21	5.31	5.41
Moderate	15.0	1.57	1.47	1.37	1.27	1.58	1.48	1.37	1.27
		5.13	5.23	5.33	5.43	5.14	5.25	5.35	5.45
	20.0	1.46	1.37	1.27	1.17	1.50	1.40	1.30	1.20
		5.00	5.10	5.20	5.29	5.11	5.21	5.31	5.41
Severe	15.0	1.12	1.04	0.96	0.88	1.13	1.05	0.97	0.89
		4.10	4.18	4.26	4.34	4.13	4.21	4.29	4.37
	20.0	1.12	1.04	0.96	0.88	1.13	1.05	0.97	0.89
		4.10	4.18	4.26	4.34	4.13	4.21	4.29	4.37
All	25.0	1.12	1.04	0.96	0.88	1.12	1.04	0.96	0.88
		4.10	4.18	4.26	4.34	4.13	4.21	4.29	4.37
	30.0	0.96	0.89	0.82	0.75	0.98	0.91	0.84	0.77
		3.79	3.87	3.94	4.01	3.88	3.95	4.03	4.10
	35.0	0.81	0.75	0.68	0.62	0.83	0.76	0.70	0.64
		3.36	3.42	3.68	3.54	3.43	3.50	3.56	3.63
	40.0	—	—	—	—	—	—	—	—

TABLE 10.25 Fine, and coarse aggregate–cement ratios (by mass) for different characteristic compressive strengths

Type of aggregate—Rounded   Maximum size of aggregate—40 mm   Degree of quality control—Good

High (0.95 CF) Sand zone / Medium (0.90 CF) Sand zone

Type of exposure	Characteristic compressive strength MPa	I	II	III	IV	I	II	III	IV
Mild	15.0	1.71	1.61	1.50	1.39	1.75	1.64	1.53	1.42
		5.43	5.53	5.64	5.75	5.54	5.65	5.76	5.87
	20.0	1.32	1.23	1.14	1.05	1.35	1.26	1.17	1.08
		4.62	4.70	4.79	4.88	4.71	4.81	4.90	4.99
Moderate	15.0	1.56	1.46	1.36	1.26	1.57	1.47	1.37	1.27
		5.09	5.19	5.29	5.39	5.11	5.21	5.31	5.41
	20.0	1.32	1.23	1.14	1.05	1.35	1.26	1.17	1.08
		4.62	4.70	4.79	4.88	4.71	4.81	4.90	4.99
Severe	15.0	1.12	1.04	0.96	0.88	1.12	1.04	0.96	0.88
		4.09	4.17	4.24	4.32	4.10	4.18	4.26	4.34
	20.0	1.12	1.04	0.96	0.88	1.12	1.04	0.96	0.88
		4.09	4.17	4.24	4.32	4.10	4.18	4.26	4.34
All	25.0	1.07	0.99	0.92	0.84	1.09	1.01	0.94	0.86
		4.01	4.09	4.17	4.24	4.10	4.18	4.26	4.34
	30.0	0.87	0.81	0.74	0.68	0.89	0.83	0.76	0.69
		3.50	3.56	3.63	3.69	3.58	3.64	3.71	3.78
	35.0	0.73	0.68	0.62	0.56	0.75	0.69	0.63	0.57
		3.09	3.15	3.20	3.26	3.16	3.22	3.28	3.34
	40.0	—	—	—	—	—	—	—	—

Degree of workability — Low (0.85 CF) / Very low (0.80 CF)

Type of exposure	Characteristic compressive strength MPa	I	II	III	IV	I	II	III	IV
Mild	15.0	1.79	1.68	1.57	1.45	1.83	1.71	1.60	1.49
		5.66	5.77	5.88	5.99	5.78	5.90	6.01	6.12
	20.0	1.38	1.29	1.19	1.10	1.41	1.32	1.22	1.13
		4.82	4.91	5.00	5.10	4.92	5.02	5.11	5.21
Moderate	15.0	1.57	1.47	1.37	1.27	1.58	1.48	1.37	1.27
		5.13	5.23	5.33	5.43	5.14	5.25	5.35	5.45
	20.0	1.38	1.29	1.19	1.10	1.41	1.32	1.22	1.13
		4.82	4.91	5.00	5.10	4.92	5.02	5.11	5.21
Severe	15.0	1.12	1.04	0.97	0.89	1.13	1.05	0.97	0.89
		4.12	4.19	4.27	4.35	4.13	4.21	4.29	4.37
	20.0	1.12	1.04	0.97	0.89	1.13	1.05	0.97	0.89
		4.12	4.19	4.27	4.35	4.13	4.21	4.29	4.37
All	25.0	1.10	1.02	0.94	0.87	1.10	1.03	0.95	0.87
		4.14	4.22	4.29	4.37	4.15	4.23	4.31	4.39
	30.0	0.91	0.84	0.78	0.71	0.93	0.86	0.79	0.72
		3.66	3.73	3.79	3.86	3.74	3.81	3.88	3.95
	35.0	0.77	0.71	0.65	0.59	0.78	0.72	0.66	0.60
		3.32	3.29	3.35	3.41	3.31	3.37	3.43	3.49
	40.0	—	—	—	—	—	—	—	—

TABLE 10.26 Fine, and coarse aggregate–cement ratios (by mass) for different characteristic compressive strengths

Type of aggregate—Rounded Maximum size of aggregate—40 mm Degree of quality control—Fair

Degree of workability		High (0.95 CF) Sand zone				Medium (0.90 CF) Sand zone			
Type of exposure	Characteristic compressive strength MPa	I	II	III	IV	I	II	III	IV
Mild	15.0	1.60 / 5.21	1.50 / 5.31	1.40 / 5.42	1.30 / 5.52	1.64 / 5.32	1.53 / 5.43	1.43 / 5.53	1.33 / 5.63
	20.0	1.25 / 4.44	1.16 / 4.53	1.08 / 4.61	0.99 / 4.70	1.27 / 4.54	1.19 / 4.63	1.10 / 4.71	1.01 / 4.80
Moderate	15.0	1.56 / 5.09	1.46 / 5.19	1.36 / 5.29	1.26 / 5.39	1.57 / 5.11	1.47 / 5.21	1.37 / 5.31	1.27 / 5.41
	20.0	1.25 / 4.44	1.16 / 4.53	1.08 / 4.61	0.99 / 4.70	1.27 / 4.54	1.19 / 4.63	1.10 / 4.71	1.01 / 4.80
Severe	15.0	1.12 / 4.09	1.04 / 4.17	0.96 / 4.24	0.88 / 4.32	1.12 / 4.10	1.04 / 4.18	0.96 / 4.26	0.88 / 4.34
	20.0	1.12 / 4.09	1.04 / 4.17	0.96 / 4.24	0.88 / 4.32	1.12 / 4.10	1.04 / 4.18	0.96 / 4.26	0.88 / 4.34
All	25.0	1.01 / 3.87	0.94 / 3.94	0.86 / 4.01	0.79 / 4.09	1.03 / 3.95	0.96 / 4.03	0.88 / 4.10	0.81 / 4.18
	30.0	0.83 / 3.37	0.76 / 3.43	0.70 / 3.50	0.64 / 3.56	0.85 / 3.45	0.78 / 3.51	0.72 / 3.58	0.65 / 3.64
	35.0	0.70 / 2.98	0.64 / 3.03	0.59 / 3.09	0.53 / 3.14	0.71 / 3.04	0.66 / 3.10	0.60 / 3.16	0.54 / 3.21
	40.0	—	—	—	—	—	—	—	—

Degree of workability		Low (0.85 CF)				Very low (0.80 CF)			
Mild	15.0	1.67 / 5.44	1.57 / 5.54	1.46 / 5.65	1.35 / 5.76	1.71 / 5.55	1.60 / 5.66	1.49 / 5.77	1.38 / 5.88
	20.0	1.30 / 4.64	1.21 / 4.73	1.12 / 4.82	1.03 / 4.91	1.33 / 4.74	1.24 / 4.83	1.15 / 4.92	1.06 / 5.02
Moderate	15.0	1.57 / 5.13	1.47 / 5.23	1.37 / 5.33	1.27 / 5.43	1.58 / 5.14	1.48 / 5.25	1.37 / 5.35	1.27 / 5.45
	20.0	1.30 / 4.64	1.21 / 4.73	1.12 / 4.82	1.03 / 4.91	1.33 / 4.74	1.24 / 4.83	1.15 / 4.92	1.06 / 5.02
Severe	15.0	1.12 / 4.12	1.04 / 4.19	0.97 / 4.27	0.89 / 4.35	1.13 / 4.13	1.05 / 4.21	0.97 / 4.29	0.89 / 4.37
	20.0	1.12 / 4.12	1.04 / 4.19	0.97 / 4.27	0.89 / 4.35	1.13 / 4.13	1.05 / 4.21	0.97 / 4.29	0.89 / 4.37
All	25.0	1.06 / 4.04	0.98 / 4.12	0.90 / 4.19	0.83 / 4.27	1.08 / 4.13	1.00 / 4.21	0.92 / 4.29	0.85 / 4.37
	30.0	0.87 / 3.53	0.80 / 3.59	0.73 / 3.66	0.67 / 3.72	0.89 / 3.61	0.82 / 3.67	0.75 / 3.74	0.68 / 3.81
	35.0	0.73 / 3.12	0.67 / 3.17	0.61 / 3.23	0.56 / 3.29	0.75 / 3.19	0.69 / 3.25	0.63 / 3.31	0.57 / 3.37
	40.0	—	—	—	—	—	—	—	—

TABLE 10.27 Fine, and coarse aggregate–cement ratios (by mass) for different characteristic compressive strengths

Type of aggregate—Angular   Maximum size of aggregate—10 mm

Degree of workability		High (0.95 CF)				Medium (0.90 CF)				
Degree of quality control—Very good										
Type of exposure	Characteristic compressive strength MPa	Sand zone				Sand zone				
		I	II	III	IV	I	II	III	IV	
Mild	15.0	1.92 2.71	1.85 2.78	1.78 2.85	1.71 2.92	1.96 2.77	1.89 2.84	1.82 2.91	1.75 2.98	
	20.0	1.49 2.28	1.44 2.34	1.38 2.40	1.32 2.45	1.53 2.34	1.47 2.39	1.41 2.45	1.35 2.51	
Moderate	15.0	1.67 2.47	1.61 2.53	1.55 2.59	1.49 2.66	1.71 2.53	1.65 2.59	1.58 2.65	1.52 2.72	
	20.0	1.49 2.28	1.44 2.34	1.38 2.40	1.32 2.45	1.53 2.34	1.47 2.39	1.41 2.45	1.35 2.51	
Severe	15.0	1.24 2.00	1.19 2.05	1.15 2.10	1.10 2.14	1.27 2.05	1.22 2.10	1.17 2.14	1.12 2.19	
	20.0	1.24 2.00	1.19 2.05	1.15 2.10	1.10 2.14	1.27 2.05	1.22 2.10	1.17 2.14	1.12 2.19	
All	25.0	1.22 1.97	1.17 2.02	1.12 2.06	1.07 2.11	1.25 2.02	1.20 2.06	1.15 2.11	1.10 2.16	
	30.0	1.00 1.70	0.96 1.74	0.92 1.78	0.88 1.82	1.03 1.74	0.98 1.78	0.94 1.82	0.90 1.86	
	35.0	0.84 1.48	0.81 1.52	0.77 1.55	0.74 1.59	0.86 1.52	0.83 1.56	0.79 1.59	0.76 1.63	
	40.0	0.61 1.53	0.58 1.56	0.55 1.59	0.52 1.63	0.63 1.57	0.59 1.60	0.56 1.63	0.53 1.67	

Degree of workability		Low (0.85 CF)				Very low (0.80 CF)				
Mild	15.0	2.00 2.83	1.93 2.90	1.86 2.97	1.79 3.05	2.05 2.89	1.97 2.97	1.90 3.04	1.83 3.11	
	20.0	1.56 2.39	1.50 2.45	1.44 2.51	1.39 2.57	1.60 2.44	1.54 2.51	1.48 2.57	1.42 2.63	
Moderate	15.0	1.75 2.58	1.68 2.65	1.62 2.71	1.55 2.78	1.79 2.64	1.72 2.71	1.66 2.77	1.59 2.84	
	20.0	1.56 2.39	1.50 2.45	1.44 2.51	1.39 2.57	1.60 2.44	1.54 2.51	1.48 2.57	1.42 2.63	
Severe	15.0	1.30 2.09	1.25 2.14	1.20 2.20	1.15 2.25	1.33 2.14	1.28 2.20	1.23 2.25	1.18 2.30	
	20.0	1.30 2.09	1.25 2.14	1.20 2.20	1.15 2.25	1.33 2.14	1.28 2.20	1.23 2.25	1.18 2.30	
All	25.0	1.28 2.06	1.23 2.11	1.18 2.16	1.13 2.21	1.31 2.11	1.26 2.16	1.20 2.21	1.15 2.27	
	30.0	1.05 1.78	1.01 1.82	0.97 1.87	0.92 1.91	1.08 1.83	1.03 1.87	0.99 1.91	0.95 1.96	
	35.0	0.89 1.56	0.85 1.60	0.81 1.63	0.78 1.67	0.91 1.60	0.87 1.64	0.83 1.67	0.80 1.71	
	40.0	0.64 1.61	0.61 1.64	0.58 1.68	0.54 1.71	0.66 1.65	0.63 1.69	0.59 1.72	0.56 1.75	

TABLE 10.28 Fine, and coarse aggregate–cement ratios (by mass) for different characteristic compressive strengths

Type of aggregate—Angular   Maximum size of aggregate—10 mm   Degree of quality control—Good

Degree of workability		High (0.95 CF)				Medium (0.90 CF)			
		Sand zone				Sand zone			
Type of exposure	Characteristic compressive strength MPa	I	II	III	IV	I	II	III	IV
Mild	15.0	1.80 / 2.60	1.73 / 2.66	1.67 / 2.73	1.60 / 2.79	1.84 / 2.65	1.77 / 2.72	1.70 / 2.79	1.64 / 2.86
	20.0	1.41 / 2.19	1.36 / 2.25	1.30 / 2.30	1.25 / 2.36	1.44 / 2.25	1.39 / 2.30	1.33 / 2.36	1.28 / 2.41
Moderate	15.0	1.67 / 2.47	1.61 / 2.53	1.55 / 2.59	1.49 / 2.66	1.71 / 2.53	1.65 / 2.59	1.58 / 2.65	1.52 / 2.72
	20.0	1.41 / 2.19	1.36 / 2.25	1.30 / 2.30	1.25 / 2.36	1.44 / 2.25	1.39 / 2.30	1.33 / 2.36	1.28 / 2.41
Severe	15.0	1.24 / 2.00	1.19 / 2.05	1.15 / 2.10	1.10 / 2.14	1.27 / 2.05	1.22 / 2.10	1.17 / 2.14	1.12 / 2.19
	20.0	1.24 / 2.00	1.19 / 2.05	1.15 / 2.10	1.10 / 2.14	1.27 / 2.05	1.22 / 2.10	1.17 / 2.14	1.12 / 2.19
All	25.0	1.15 / 1.89	1.11 / 1.94	1.06 / 1.98	1.02 / 2.03	1.18 / 1.94	1.14 / 1.98	1.09 / 2.03	1.04 / 2.08
	30.0	0.95 / 1.63	0.91 / 1.67	0.87 / 1.71	0.83 / 1.75	0.97 / 1.67	0.93 / 1.71	0.90 / 1.75	0.86 / 1.79
	35.0	0.80 / 1.42	0.77 / 1.46	0.73 / 1.49	0.70 / 1.52	0.82 / 1.46	0.79 / 1.49	0.75 / 1.53	0.76 / 1.56
	40.0	0.58 / 1.47	0.55 / 1.50	0.52 / 1.53	0.49 / 1.56	0.59 / 1.50	0.56 / 1.54	0.53 / 1.57	0.50 / 1.60

Degree of workability		Low (0.85 CF)				Very low (0.80 CF)			
Mild	15.0	1.88 / 2.71	1.81 / 2.78	1.74 / 2.85	1.67 / 2.92	1.92 / 2.77	1.85 / 2.84	1.78 / 2.91	1.71 / 2.99
	20.0	1.48 / 2.30	1.42 / 2.35	1.37 / 2.41	1.31 / 2.47	1.51 / 2.35	1.46 / 2.41	1.40 / 2.47	1.34 / 2.52
Moderate	15.0	1.75 / 2.58	1.68 / 2.65	1.62 / 2.71	1.55 / 2.78	1.79 / 2.64	1.72 / 2.71	1.66 / 2.77	1.59 / 2.84
	20.0	1.48 / 2.30	1.42 / 2.35	1.37 / 2.41	1.31 / 2.47	1.51 / 2.35	1.46 / 2.41	1.40 / 2.47	1.34 / 2.52
Severe	15.0	1.30 / 2.09	1.25 / 2.14	1.20 / 2.20	1.15 / 2.25	1.33 / 2.14	1.28 / 2.20	1.23 / 2.25	1.18 / 2.30
	20.0	1.30 / 2.09	1.25 / 2.14	1.20 / 2.20	1.15 / 2.25	1.33 / 2.14	1.28 / 2.20	1.23 / 2.25	1.18 / 2.30
All	25.0	1.21 / 1.98	1.16 / 2.03	1.11 / 2.08	1.07 / 2.13	1.24 / 2.03	1.19 / 2.08	1.14 / 2.13	1.09 / 2.18
	30.0	1.00 / 1.71	0.96 / 1.75	0.92 / 1.79	0.88 / 1.83	1.02 / 1.75	0.98 / 1.80	0.94 / 1.84	0.90 / 1.88
	35.0	0.84 / 1.50	0.81 / 1.53	0.77 / 1.57	0.74 / 1.60	0.86 / 1.53	0.83 / 1.57	0.79 / 1.61	0.76 / 1.64
	40.0	0.61 / 1.54	0.58 / 1.57	0.55 / 1.61	0.51 / 1.64	0.63 / 1.58	0.59 / 1.62	0.56 / 1.65	0.53 / 1.68

TABLE 10.29 Fine, and coarse aggregate–cement ratios (by mass) for different characteristic compressive strengths

Type of aggregate—Angular   Maximum size of aggregate—10 mm   Degree of quality control—Fair

Degree of workability		High (0.95 CF) Sand zone				Medium (0.90 CF) Sand zone			
Type of exposure	Characteristic compressive strength MPa	I	II	III	IV	I	II	III	IV
Mild	15.0	1.69 2.49	1.63 2.55	1.57 2.62	1.50 2.68	1.73 2.55	1.67 2.61	1.60 2.67	1.54 2.74
	20.0	1.34 2.11	1.28 2.16	1.23 2.21	1.18 2.16	1.37 2.16	1.31 2.21	1.26 2.26	1.21 2.32
Moderate	15.0	1.67 2.47	1.61 2.53	1.55 2.59	1.49 2.66	1.71 2.53	1.65 2.59	1.58 2.65	1.52 2.72
	20.0	1.34 2.11	1.28 2.16	1.23 2.21	1.18 2.26	1.37 2.16	1.31 2.21	1.26 2.26	1.21 2.32
Severe	15.0	1.24 2.00	1.19 2.05	1.15 2.10	1.10 2.14	1.27 2.05	1.22 2.10	1.17 2.14	1.12 2.19
	20.0	1.24 2.00	1.19 2.05	1.15 2.10	1.10 2.14	1.27 2.05	1.22 2.10	1.17 2.14	1.12 2.19
All	25.0	1.10 1.82	1.05 1.86	1.01 1.91	0.96 1.95	1.12 1.86	1.08 1.91	1.03 1.95	0.99 2.00
	30.0	0.90 1.57	0.87 1.60	0.83 1.64	0.79 1.68	0.93 1.61	0.89 1.64	0.85 1.68	0.81 1.72
	35.0	0.76 1.36	0.73 1.40	0.70 1.43	0.66 1.46	0.78 1.40	0.75 1.43	0.71 1.46	0.68 1.50
	40.0	0.55 1.40	0.52 1.43	0.49 1.46	0.46 1.49	0.56 1.44	0.53 1.47	0.50 1.50	0.47 1.53

Degree of workability		Low (0.85 CF)				Very low (0.80 CF)			
		I	II	III	IV	I	II	III	IV
Mild	15.0	1.77 2.60	1.70 2.67	1.64 2.73	1.57 2.80	1.81 2.66	1.74 2.73	1.68 2.80	1.61 2.86
	20.0	1.40 2.21	1.35 2.26	1.29 2.32	1.24 2.37	1.43 2.26	1.38 2.31	1.32 2.37	1.27 2.43
Moderate	15.0	1.75 2.58	1.68 2.65	1.62 2.71	1.55 2.78	1.79 2.64	1.72 2.71	1.66 2.77	1.59 2.84
	20.0	1.40 2.21	1.35 2.26	1.29 2.32	1.24 2.37	1.43 2.26	1.38 2.31	1.32 2.37	1.27 2.43
Severe	15.0	1.30 2.09	1.25 2.14	1.20 2.20	1.15 2.25	1.33 2.14	1.28 2.20	1.23 2.25	1.18 2.30
	20.0	1.30 2.09	1.25 2.14	1.20 2.20	1.15 2.25	1.33 2.14	1.28 2.20	1.23 2.25	1.18 2.30
All	25.0	1.15 1.91	1.10 1.95	1.06 2.00	1.01 2.04	1.18 1.95	1.13 2.00	1.08 2.05	1.04 2.09
	30.0	0.95 1.65	0.91 1.68	0.87 1.72	0.83 1.76	0.97 1.69	0.93 1.73	0.89 1.77	0.85 1.81
	35.0	0.80 1.44	0.77 1.47	0.73 1.50	0.70 1.54	0.82 1.47	0.79 1.51	0.75 1.54	0.72 1.58
	40.0	0.58 1.48	0.55 1.51	0.52 1.54	0.49 1.57	0.59 1.52	0.56 1.55	0.53 1.58	0.50 1.61

180 Concrete Technology

TABLE 10.30 Fine, and coarse aggregate–cement ratios (by mass) for different characteristic compressive strengths

Type of aggregate—Angular   Maximum size of aggregate—20 mm   Degree of quality control—Very good

Type of exposure	Characteristic compressive strength MPa	High (0.95 CF) Sand zone I	II	III	IV	Medium (0.90 CF) Sand zone I	II	III	IV
Mild	15.0	2.02 / 3.52	1.94 / 3.60	1.85 / 3.68	1.77 / 3.76	2.06 / 3.59	1.98 / 3.67	1.89 / 3.76	1.81 / 3.84
	20.0	1.57 / 2.97	1.50 / 3.04	1.44 / 3.11	1.37 / 3.18	1.61 / 3.04	1.54 / 3.11	1.47 / 3.18	1.40 / 3.25
Moderate	15.0	1.76 / 3.21	1.68 / 3.29	1.61 / 3.36	1.54 / 3.43	1.80 / 3.28	1.72 / 3.35	1.64 / 3.43	1.57 / 3.51
	20.0	1.57 / 2.97	1.50 / 3.04	1.44 / 3.11	1.37 / 3.18	1.61 / 3.04	1.54 / 3.11	1.47 / 3.18	1.40 / 3.25
Severe	15.0	1.31 / 2.61	1.25 / 2.67	1.19 / 2.73	1.13 / 2.79	1.34 / 2.67	1.28 / 2.73	1.22 / 2.79	1.16 / 2.85
	20.0	1.31 / 2.61	1.25 / 2.67	1.19 / 2.73	1.13 / 2.79	1.34 / 2.67	1.28 / 2.73	1.22 / 2.79	1.16 / 2.85
All	25.0	1.28 / 2.57	1.22 / 2.63	1.17 / 2.66	1.11 / 2.75	1.31 / 2.63	1.25 / 2.69	1.19 / 2.75	1.13 / 2.81
	30.0	1.06 / 2.23	1.01 / 2.28	0.96 / 2.33	0.91 / 2.38	1.08 / 2.28	1.03 / 2.33	0.98 / 2.38	0.93 / 2.43
	35.0	0.89 / 1.96	0.85 / 2.00	0.81 / 2.04	0.76 / 2.09	0.91 / 2.00	0.87 / 2.05	0.83 / 2.09	0.78 / 2.13
	40.0	0.61 / 1.77	0.57 / 1.81	0.54 / 1.84	0.50 / 1.88	0.62 / 1.81	0.59 / 1.85	0.55 / 1.89	0.51 / 1.92

Degree of workability: Low (0.85 CF) / Very low (0.80 CF)

Type of exposure	Characteristic compressive strength MPa	Low (0.85 CF) I	II	III	IV	Very low (0.80 CF) I	II	III	IV
Mild	15.0	2.10 / 3.66	2.02 / 3.75	1.93 / 3.84	1.85 / 3.92	2.15 / 3.74	2.06 / 3.83	1.97 / 3.92	1.88 / 4.01
	20.0	1.64 / 3.10	1.57 / 3.17	1.50 / 3.24	1.43 / 3.32	1.68 / 3.17	1.60 / 3.24	1.53 / 3.32	1.46 / 3.39
Moderate	15.0	1.83 / 3.35	1.76 / 3.43	1.68 / 3.50	1.60 / 3.58	1.87 / 3.42	1.79 / 3.50	1.72 / 3.58	1.64 / 3.66
	20.0	1.64 / 3.10	1.57 / 3.17	1.50 / 3.24	1.43 / 3.32	1.68 / 3.17	1.60 / 3.24	1.53 / 3.32	1.46 / 3.39
Severe	15.0	1.37 / 2.73	1.30 / 2.79	1.24 / 2.85	1.18 / 2.91	1.40 / 2.79	1.33 / 2.85	1.27 / 2.91	1.21 / 2.98
	20.0	1.37 / 2.73	1.30 / 2.79	1.24 / 2.85	1.18 / 2.91	1.40 / 2.79	1.33 / 2.85	1.27 / 2.91	1.21 / 2.98
All	25.0	1.34 / 2.69	1.28 / 2.75	1.22 / 2.81	1.16 / 2.87	1.37 / 2.75	1.31 / 2.81	1.25 / 2.87	1.18 / 2.93
	30.0	1.11 / 2.33	1.05 / 2.38	1.00 / 2.43	0.95 / 2.49	1.13 / 2.39	1.08 / 2.44	1.03 / 2.49	0.97 / 2.54
	35.0	0.93 / 2.05	0.89 / 2.10	0.85 / 2.14	0.80 / 2.18	0.96 / 2.10	0.91 / 2.14	0.87 / 2.19	0.82 / 2.24
	40.0	0.64 / 1.86	0.60 / 1.90	0.56 / 1.93	0.53 / 1.97	0.66 / 1.91	0.62 / 1.94	0.58 / 1.98	0.54 / 2.02

TABLE 10.31 Fine, and coarse aggregate–cement ratios (by mass) for different characteristic compressive strengths

Type of aggregate—Angular   Maximum size of aggregate—20 mm

Degree of quality control									
			Good						

Degree of workability		High (0.95 CF)				Medium (0.90 CF)			
Type of exposure	Characteristic compressive strength MPa	Sand zone				Sand zone			
		I	II	III	IV	I	II	III	IV
Mild	15.0	1.89 / 3.37	1.82 / 3.45	1.74 / 3.53	1.66 / 3.61	1.93 / 3.44	1.85 / 3.52	1.77 / 3.60	1.69 / 3.68
	20.0	1.49 / 2.86	1.42 / 2.92	1.36 / 2.99	1.29 / 3.05	1.52 / 2.92	1.45 / 2.99	1.39 / 3.05	1.32 / 3.12
Moderate	15.0	1.76 / 3.21	1.68 / 3.29	1.61 / 3.36	1.54 / 3.43	1.80 / 3.28	1.72 / 3.35	1.64 / 3.43	1.57 / 3.51
	20.0	1.49 / 2.86	1.42 / 2.92	1.36 / 2.99	1.29 / 3.05	1.52 / 2.92	1.45 / 2.99	1.39 / 3.05	1.32 / 3.12
Severe	15.0	1.31 / 2.61	1.25 / 2.67	1.19 / 2.73	1.13 / 2.79	1.34 / 2.67	1.28 / 2.73	1.22 / 2.79	1.16 / 2.85
	20.0	1.31 / 2.61	1.25 / 2.67	1.19 / 2.73	1.13 / 2.79	1.34 / 2.67	1.28 / 2.73	1.22 / 2.79	1.16 / 2.85
All	25.0	1.22 / 2.48	1.16 / 2.53	1.11 / 2.59	1.05 / 2.64	1.24 / 2.53	1.19 / 2.59	1.13 / 2.64	1.07 / 2.70
	30.0	1.00 / 2.14	0.96 / 2.19	0.91 / 2.24	0.86 / 2.29	1.03 / 2.19	0.98 / 2.24	0.93 / 2.29	0.88 / 2.34
	35.0	0.85 / 1.88	0.81 / 1.92	0.77 / 1.96	0.73 / 2.00	0.87 / 1.93	0.83 / 1.97	0.78 / 2.01	0.74 / 2.05
	40.0	0.58 / 1.70	0.54 / 1.73	0.51 / 1.77	0.48 / 1.80	0.59 / 1.74	0.56 / 1.78	0.52 / 1.81	0.49 / 1.85

Degree of workability		Low (0.85 CF)				Very low (0.80 CF)			
Mild	15.0	1.97 / 3.52	1.89 / 3.60	1.81 / 3.68	1.73 / 3.76	2.02 / 3.59	1.93 / 3.67	1.85 / 3.76	1.77 / 3.84
	20.0	1.55 / 2.98	1.48 / 3.05	1.42 / 3.12	1.35 / 3.19	1.59 / 3.05	1.52 / 3.12	1.45 / 3.19	1.38 / 3.26
Moderate	15.0	1.83 / 3.35	1.76 / 3.43	1.68 / 3.50	1.60 / 3.58	1.87 / 3.42	1.79 / 3.50	1.72 / 3.58	1.64 / 3.66
	20.0	1.55 / 2.98	1.48 / 3.05	1.42 / 3.12	1.35 / 3.19	1.59 / 3.05	1.52 / 3.12	1.45 / 3.19	1.38 / 3.26
Severe	15.0	1.37 / 2.73	1.30 / 2.79	1.24 / 2.85	1.18 / 2.91	1.40 / 2.79	1.33 / 2.85	1.27 / 2.91	1.21 / 2.98
	20.0	1.37 / 2.73	1.30 / 2.79	1.24 / 2.85	1.18 / 2.91	1.40 / 2.79	1.33 / 2.85	1.27 / 2.91	1.21 / 2.98
All	25.0	1.27 / 2.59	1.21 / 2.65	1.16 / 2.70	1.10 / 2.76	1.30 / 2.65	1.24 / 2.70	1.18 / 2.76	1.12 / 2.82
	30.0	1.05 / 2.24	1.00 / 2.29	0.95 / 2.34	0.90 / 2.39	1.08 / 2.30	1.02 / 2.35	0.97 / 2.40	0.92 / 2.45
	35.0	0.89 / 1.97	0.85 / 2.01	0.80 / 2.06	0.76 / 2.10	0.91 / 2.02	0.87 / 2.06	0.82 / 2.11	0.78 / 2.15
	40.0	0.61 / 1.78	0.57 / 1.82	0.54 / 1.86	0.50 / 1.89	0.62 / 1.83	0.59 / 1.87	0.55 / 1.90	0.51 / 1.94

TABLE 10.32 Fine, and coarse aggregate–cement ratios (by mass) for different characteristic compressive strengths

Type of aggregate—Angular Maximum size of aggregate—20 mm

Degree of exposure	Degree of workability	Characteristic compressive strength MPa	High (0.95 CF) Sand zone I	II	III	IV	Medium (0.90 CF) Sand zone I	II	III	IV
							Degree of quality control—Fair			
Mild		15.0	1.78 3.24	1.71 3.31	1.63 3.39	1.56 3.46	1.82 3.31	1.74 3.38	1.67 3.46	1.59 3.54
		20.0	1.41 2.75	1.34 2.81	1.28 2.88	1.22 2.94	1.44 2.81	1.37 2.87	1.31 2.94	1.25 3.00
Moderate		15.0	1.76 3.21	1.68 3.29	1.61 3.36	1.54 3.43	1.80 3.28	1.72 3.35	1.64 3.43	1.57 3.51
		20.0	1.41 2.75	1.34 2.81	1.28 2.88	1.22 2.94	1.44 2.81	1.37 2.87	1.31 2.94	1.25 3.00
Severe		15.0	1.31 2.61	1.25 2.67	1.19 2.73	1.13 2.79	1.34 2.67	1.28 2.73	1.22 2.79	1.16 2.85
		20.0	1.31 2.61	1.25 2.67	1.19 2.73	1.13 2.79	1.34 2.67	1.28 2.73	1.22 2.79	1.16 2.85
All		25.0	1.15 2.38	1.10 2.44	1.05 2.49	1.00 2.54	1.18 2.44	1.13 2.49	1.07 2.54	1.02 2.60
		30.0	0.95 2.06	0.91 2.11	0.86 2.15	0.82 2.20	0.98 2.11	0.93 2.16	0.88 2.20	0.84 2.25
		35.0	0.81 1.81	0.77 1.85	0.73 1.89	0.69 1.93	0.83 1.85	0.79 1.89	0.75 1.93	0.71 1.97
		40.0	0.55 1.63	0.51 1.66	0.48 1.69	0.45 1.73	0.56 1.67	0.53 1.70	0.49 1.74	0.46 1.77

Degree of workability			Low (0.85 CF)				Very low (0.80 CF)			
Mild		15.0	1.86 3.38	1.78 3.45	1.70 3.53	1.62 3.61	1.90 3.45	1.82 3.53	1.74 3.61	1.66 3.69
		20.0	1.47 2.87	1.40 2.94	1.34 3.00	1.27 3.07	1.50 2.94	1.43 3.00	1.37 3.07	1.30 3.13
Moderate		15.0	1.83 3.35	1.76 3.43	1.68 3.50	1.60 3.58	1.87 3.42	1.79 3.50	1.72 3.58	1.64 3.66
		20.0	1.47 2.87	1.40 2.94	1.34 3.00	1.27 3.07	1.50 2.94	1.43 3.00	1.37 3.07	1.30 3.13
Severe		15.0	1.37 2.73	1.30 2.79	1.24 2.85	1.18 2.91	1.40 2.79	1.33 2.85	1.27 2.91	1.21 2.98
		20.0	1.37 2.73	1.30 2.79	1.24 2.85	1.18 2.91	1.40 2.79	1.33 2.85	1.27 2.91	1.21 2.98
All		25.0	1.21 2.49	1.15 2.55	1.10 2.60	1.04 2.66	1.23 2.55	1.18 2.60	1.12 2.66	1.07 2.72
		30.0	1.00 2.16	0.95 2.21	0.90 2.25	0.86 2.30	1.02 2.21	0.97 2.26	0.93 2.31	0.88 2.35
		35.0	0.85 1.90	0.80 1.94	0.76 1.98	0.72 2.02	0.87 1.94	0.82 1.98	0.78 2.02	0.74 2.07
		40.0	0.58 1.71	0.54 1.74	0.51 1.78	0.47 1.81	0.59 1.75	0.56 1.79	0.52 1.82	0.49 1.86

TABLE 10.33 Fine, and coarse aggregate–cement ratios (by mass) for different characteristic compressive strengths

Type of aggregate—Angular   Maximum size of aggregate—40 mm   Degree of quality control—Very good

High (0.95 CF) — Sand zone

Type of exposure	Characteristic compressive strength MPa	I	II	III	IV	I	II	III	IV
Mild	15.0	2.09 / 4.54	1.99 / 4.64	1.89 / 4.74	1.79 / 4.84	2.13 / 4.63	2.03 / 4.73	1.93 / 4.84	1.83 / 4.94
	20.0	1.62 / 3.85	1.54 / 3.93	1.46 / 4.02	1.38 / 4.10	1.65 / 3.93	1.57 / 4.01	1.49 / 4.10	1.40 / 4.18
Moderate	15.0	1.82 / 4.15	1.73 / 4.24	1.64 / 4.33	1.55 / 4.42	1.85 / 4.24	1.76 / 4.33	1.67 / 4.42	1.58 / 4.51
	20.0	1.62 / 3.85	1.54 / 3.93	1.46 / 4.02	1.38 / 4.10	1.65 / 3.93	1.57 / 4.01	1.49 / 4.10	1.40 / 4.18
Severe	15.0	1.35 / 3.39	1.28 / 3.46	1.20 / 3.53	1.13 / 3.60	1.37 / 3.46	1.30 / 3.53	1.23 / 3.61	1.16 / 3.68
	20.0	1.35 / 3.39	1.28 / 3.46	1.20 / 3.53	1.13 / 3.60	1.37 / 3.46	1.30 / 3.53	1.23 / 3.61	1.16 / 3.68
All	25.0	1.32 / 3.34	1.25 / 3.41	1.18 / 3.48	1.11 / 3.55	1.35 / 3.41	1.28 / 3.48	1.20 / 3.56	1.13 / 3.63
	30.0	1.09 / 2.91	1.03 / 2.97	0.97 / 3.03	0.91 / 3.09	1.11 / 2.97	1.05 / 3.03	0.99 / 3.09	0.93 / 3.15
	35.0	0.92 / 0.87	0.87 / 2.62	0.81 / 2.67	0.76 / 2.72	0.94 / 2.62	0.89 / 2.67	0.83 / 2.73	0.78 / 2.78
	40.0	—	—	—	—	—	—	—	—

Low (0.85 CF) — Sand zone / Very low (0.80 CF)

Type of exposure	Characteristic compressive strength MPa	I	II	III	IV	I	II	III	IV
Mild	15.0	2.17 / 4.73	2.07 / 4.83	1.97 / 4.93	1.86 / 5.04	2.22 / 4.82	2.11 / 4.93	2.01 / 5.03	1.90 / 5.14
	20.0	1.69 / 4.01	1.60 / 4.10	1.52 / 4.18	1.43 / 4.27	1.72 / 4.09	1.64 / 4.18	1.55 / 4.27	1.46 / 4.35
Moderate	15.0	1.89 / 4.32	1.80 / 4.42	1.70 / 4.51	1.61 / 4.60	1.93 / 4.41	1.83 / 4.51	1.74 / 4.60	1.64 / 4.70
	20.0	1.69 / 4.01	1.60 / 4.10	1.52 / 4.18	1.43 / 4.27	1.72 / 4.09	1.64 / 4.18	1.55 / 4.27	1.46 / 4.35
Severe	15.0	1.40 / 3.53	1.33 / 3.61	1.25 / 3.68	1.18 / 3.75	1.43 / 3.61	1.36 / 3.68	1.28 / 3.76	1.21 / 3.83
	20.0	1.40 / 3.53	1.33 / 3.61	1.25 / 3.68	1.18 / 3.75	1.43 / 3.61	1.36 / 3.68	1.28 / 3.76	1.21 / 3.83
All	25.0	1.37 / 3.48	1.30 / 3.56	1.23 / 3.63	1.16 / 3.70	1.40 / 3.56	1.33 / 3.63	1.26 / 3.71	1.18 / 3.78
	30.0	1.13 / 3.03	1.07 / 3.10	1.01 / 3.16	0.95 / 3.22	1.16 / 3.10	1.09 / 3.16	1.03 / 3.23	0.97 / 3.29
	35.0	0.96 / 2.68	0.90 / 2.73	0.85 / 2.79	0.80 / 2.84	0.98 / 2.74	0.92 / 2.79	0.87 / 2.85	0.81 / 2.90
	40.0	—	—	—	—	—	—	—	—

TABLE 10.34 Fine, and coarse aggregate–cement ratios (by mass) for different characteristic compressive strengths

Type of aggregate—Angular   Maximum size of aggregate—40 mm   Degree of quality control—Good

High (0.95 CF) — Sand zone

Type of exposure	Characteristic compressive strength MPa	I		II		III		IV	
Mild	15.0	1.96	4.36	1.86	4.45	1.77	4.55	1.67	4.64
	20.0	1.53	3.71	1.45	3.79	1.37	3.86	1.30	3.94
Moderate	15.0	1.82	4.15	1.73	4.24	1.64	4.33	1.55	4.42
	20.0	1.53	3.71	1.45	3.79	1.37	3.86	1.30	3.94
Severe	15.0	1.35	3.39	1.28	3.46	1.20	3.53	1.13	3.60
	20.0	1.35	3.39	1.28	3.46	1.20	3.53	1.13	3.60
All	25.0	1.25	3.22	1.18	3.29	1.12	3.35	1.05	3.42
	30.0	1.03	2.80	0.98	2.86	0.92	2.92	0.86	2.97
	35.0	0.87	2.47	0.82	2.52	0.77	2.57	0.72	2.62
	40.0	—	—	—	—	—	—	—	—

Medium (0.90 CF) — Sand zone

Type of exposure	Characteristic compressive strength MPa	I		II		III		IV	
Mild	15.0	2.00	4.45	1.90	4.54	1.80	4.64	1.71	4.74
	20.0	1.56	3.78	1.48	3.86	1.40	3.94	1.32	4.02
Moderate	15.0	1.85	4.24	1.76	4.33	1.67	4.42	1.58	4.51
	20.0	1.56	3.78	1.48	3.86	1.40	3.94	1.32	4.02
Severe	15.0	1.37	3.46	1.30	3.53	1.23	3.61	1.16	3.68
	20.0	1.37	3.46	1.30	3.53	1.23	3.61	1.16	3.68
All	25.0	1.28	3.29	1.21	3.36	1.14	3.42	1.07	3.49
	30.0	1.05	2.86	1.00	2.92	0.94	2.98	0.88	3.04
	35.0	0.89	2.52	0.84	2.57	0.79	2.63	0.74	2.68
	40.0	—	—	—	—	—	—	—	—

Low (0.85 CF)

Type of exposure	Characteristic compressive strength MPa	I		II		III		IV	
Mild	15.0	2.04	4.54	1.94	4.63	1.84	4.73	1.74	4.83
	20.0	1.59	3.86	1.51	3.94	1.43	4.02	1.35	4.11
Moderate	15.0	1.89	4.32	1.80	4.42	1.70	4.51	1.61	4.60
	20.0	1.59	3.86	1.51	3.94	1.43	4.02	1.35	4.11
Severe	15.0	1.40	3.53	1.33	3.61	1.25	3.68	1.18	3.75
	20.0	1.40	3.53	1.33	3.61	1.25	3.68	1.18	3.75
All	25.0	1.30	3.36	1.23	3.43	1.16	3.50	1.10	3.57
	30.0	1.08	2.92	1.02	2.98	0.96	3.04	0.90	3.10
	35.0	0.91	2.58	0.86	2.63	0.81	2.68	0.76	2.74
	40.0	—	—	—	—	—	—	—	—

Very low (0.80 CF)

Type of exposure	Characteristic compressive strength MPa	I		II		III		IV	
Mild	15.0	2.08	4.63	1.98	4.73	1.88	4.83	1.78	4.93
	20.0	1.63	3.94	1.54	4.02	1.46	4.11	1.38	4.19
Moderate	15.0	1.93	4.41	1.83	4.51	1.74	4.60	1.64	4.70
	20.0	1.63	3.94	1.54	4.02	1.46	411	1.38	4.19
Severe	15.0	1.43	3.61	1.36	3.68	1.28	3.76	1.21	3.83
	20.0	1.43	3.61	1.36	3.68	1.28	3.76	1.21	3.83
All	25.0	1.33	3.43	1.26	3.50	1.19	3.57	1.12	3.64
	30.0	1.10	2.99	1.04	3.05	0.98	3.11	0.92	3.17
	35.0	0.93	2.64	0.88	2.69	0.83	2.77	0.77	2.80
	40.0	—	—	—	—	—	—	—	—

TABLE 10.35 Fine, and coarse aggregate–cement ratios (by mass) for different characteristic compressive strengths

Type of aggregate—Angular   Maximum size of aggregate—40 mm   Degree of quality control—Fair

Degree of workability			High (0.95 CF) Sand zone							Medium (0.90 CF) Sand zone						
Type of exposure	Characteristic compressive strength MPa		I		II		III		IV		I		II		III		IV
Mild	15.0		1.84	4.19	1.75	4.28	1.66	4.37	1.57	4.46	1.88	4.27	1.78	4.36	1.69	4.46	1.60	4.55
	20.0		1.45	3.57	1.37	3.64	1.30	3.72	1.22	3.79	1.48	3.64	1.40	3.72	1.32	3.80	1.25	3.87
Moderate	15.0		1.82	4.15	1.73	4.24	1.64	4.33	1.55	4.42	1.85	4.24	1.76	4.33	1.67	4.42	1.58	4.51
	20.0		1.45	3.57	1.37	3.64	1.30	3.72	1.22	3.79	1.48	3.64	1.40	3.72	1.32	3.80	1.25	3.87
Severe	15.0		1.35	3.39	1.28	3.46	1.20	3.53	1.13	3.60	1.37	3.46	1.30	3.53	1.23	3.61	1.16	3.68
	20.0		1.35	3.39	1.28	3.46	1.20	3.53	1.13	3.60	1.37	3.46	1.30	3.53	1.23	3.61	1.16	3.68
All	25.0		1.19	3.10	1.12	3.17	1.06	3.23	1.00	3.30	1.21	3.17	1.15	3.23	1.08	3.30	1.02	3.36
	30.0		0.98	2.70	0.95	2.75	0.87	2.81	0.82	2.86	1.00	2.76	0.95	2.81	0.89	2.87	0.83	2.92
	35.0		0.83	2.38	0.78	2.42	0.73	2.47	0.69	2.52	0.85	2.43	0.80	2.48	0.75	2.53	0.70	2.58
	40.0		—		—		—		—		—		—		—		—	

Degree of workability			Low (0.85 CF)								Very low (0.80 CF)							
			I		II		III		IV		I		II		III		IV		
Mild	15.0		1.91	4.36	1.82	4.45	1.73	4.55	1.63	4.64	1.95	4.45	1.86	4.54	1.76	4.64	1.67	4.73
	20.0		1.51	3.72	1.43	3.80	1.35	3.87	1.27	3.95	1.54	3.79	1.46	3.87	1.38	3.95	1.30	4.03
Moderate	15.0		1.89	4.32	1.80	4.42	1.70	4.51	1.61	4.60	1.93	4.41	1.83	4.51	1.74	4.60	1.64	4.70
	20.0		1.51	3.72	1.43	3.80	1.35	3.87	1.27	3.95	1.54	3.79	1.46	3.87	1.38	3.95	1.30	4.03
Severe	15.0		1.40	3.53	1.33	3.61	1.25	3.68	1.18	3.75	1.43	3.61	1.36	3.68	1.28	3.76	1.21	3.83
	20.0		1.40	3.53	1.33	3.61	1.25	3.68	1.18	3.75	1.43	3.61	1.36	3.68	1.28	3.76	1.21	3.83
All	25.0		1.24	3.23	1.17	3.30	1.10	3.37	1.04	3.44	1.26	3.30	1.20	3.37	1.13	3.44	1.06	3.51
	30.0		1.02	2.82	0.97	2.87	0.91	2.93	0.85	2.99	1.05	2.88	0.99	2.94	0.93	2.99	0.87	3.05
	35.0		0.87	2.48	0.82	2.53	0.77	2.58	0.72	2.63	0.89	2.54	0.84	2.59	0.79	2.64	0.73	2.69
	40.0		—		—		—		—		—		—		—		—	

Design Variables and Constraints

The aggregates of different maximum sizes having different properties and unit costs are to be combined in such a way that the grading of resulting aggregate lies within the preselected grading limits and promotes workability. The constraints generally considered are:

(i) Compressive strength
(ii) Durability
(iii) Workability
(iv) Aggregate characteristics

The objective function to be minimized comprises the total cost of production of unit volume of concrete expressed as the sum of the cost of individual materials.

1. *Compressive strength* In addition to the water-cement ratio, the other important factor governing the strength of concrete is the *maximum nominal size* of the aggregate.

2. *Durability* The durability criterion is generally satisfied by limiting the *water-cement ratio* and *minimum cement content*. Depending on the type of environmental conditions to which a structure is likely to be exposed, a maximum permissible water-cement ratio constraint is introduced. Sometimes a *minimum water-cement ratio* constraint is also imposed from practical considerations.

3. *Workability* The degree of workability designated as *very low, low, medium* and *high* can be expressed in terms of range of slump, compaction factor and Vee-Bee time values. For the purpose of formulating constraints in the optimization problem, the workability is generally related to the water-cement ratio, aggregate-cement ratio and standard consistency of cement.

4. *Aggregate characteristics* The type of aggregate, its maximum size and grading influence the water content to produce a workable concrete mix. Generally, the standard grading curves of road note No. 4 are well suited for constraint formulation, since the maximum and minimum permissible grading limits of the combined aggregates are specified for different maximum nominal sizes of aggregates ranging from 10 to 40 mm. If ten standard sieves are 40 mm down to 75 µm are used, there will be 20 constraints. Corresponding to the maximum and minimum grading limits.

Solution of optimization problem If the objective function is a linear function of various constraints which themselves are linear, then the resulting formulation is called the *linear programming problem*. Its solution can be obtained by *simplex method*. The *nonlinear programming problem* can be transformed into a form which permits application of Simplex Alogrithm. Alternatively, the nonlinear constrained optimization problem can be solved directly by using the available methods.

10.7 DESIGN OF HIGH STRENGTH CONCRETE MIXES

The properties of high strength concrete with a compressive strength above 40 MPa or 50 MPa are highly influenced by the properties of aggregate in addition to that of the water-cement ratio.

To achieve high strength, it is necessary to use the lowest possible water-cement ratio with high cement content which invariably affects the workability of the mix and necessitates the use of special vibration techniques for proper compaction. It should be kept in mind that high cement content may liberate large heat of hydration causing rise in temperature which may affect setting and may result in excessive shrinkage. In the present state of art, concrete which has a desired 28-day compressive strength up to 70 MPa can be made by suitably proportioning the ingredients and using normal vibration techniques for compacting the mix. A number of methods for designing high strength concrete mixes are available. The method of Erntroy and Shacklock is described below along with design examples.

10.7.1 Erntroy and Shacklock's Empirical Method

Erntroy and Shacklock have suggested empirical graphs relating *compressive strength* to an arbitrary *reference number* for concretes made with crushed granite coarse aggregates and irregular gravel. These graphs are shown in Figs 10.14 and

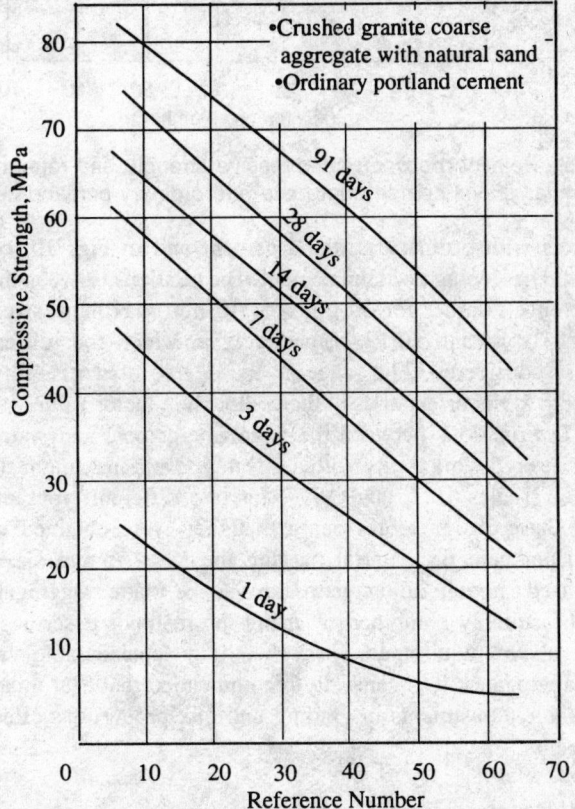

Fig. 10.14 Relation between compressive strength and reference number (crushed coarse aggregate and ordinary portland cement)

188 Concrete Technology

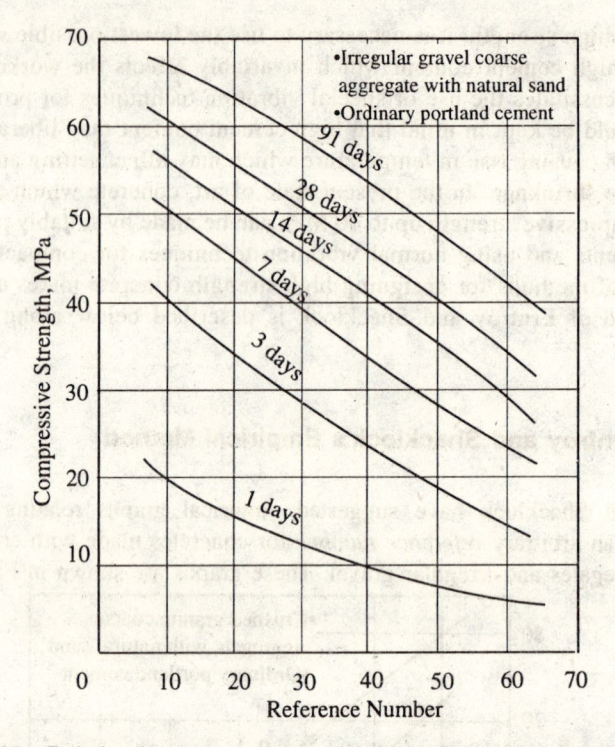

Fig. 10.15 Relation between compressive strength and reference number (irregular gravel coarse aggregate and ordinary portland cement)

10.15 for mixes with ordinary portland cement and in Figs 10.16 and 10.17 for mixes with rapid hardening portland cement. The relations between the water-cement ratio and reference number for 20 mm and 10 mm maximum size aggregates are shown in Figs. 10.18 and 10.19, respectively in which four different degrees of workability are considered. The range of the degree of workability varying from extremely low to high corresponds to the compacting factor values of 0.65 and 0.95, respectively. The relations between the aggregate-cement and water-cement ratios to achieve the desired workability with a given type and maximum size of aggregate are compiled in Tables 10.36 and 10.37 for two different types of cements. The limitations of these design tables being that they were obtained with aggregates containing 30 per cent of material passing the IS : 4.75 mm sieve, and if other gradings are used suitable adjustments have to be made. Aggregates available at the site may be suitably combined by using the methods described in Sec. 10.3.4 to satisfy the above requirement. In view of the considerable variations in the properties of aggregates, it is generally recommended that trial mixes must first be made and suitable adjustments in grading and mix proportions effected to achieve the desired results.

Mix Design Procedure
1. The mean design strength is obtained by applying suitable control factors to the specified minimum strength.

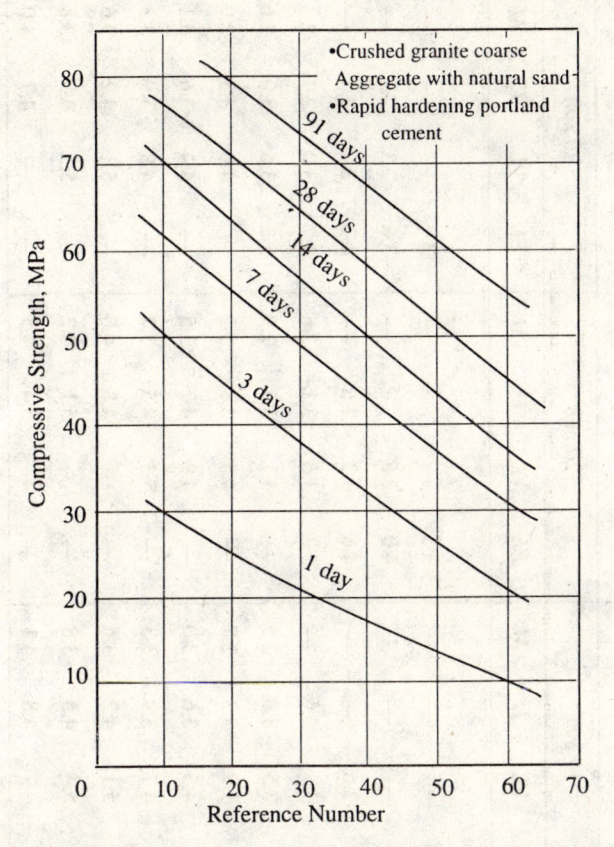

Fig. 10.16 Relation between compressive strength and reference number (crushed coarse aggregate and rapid hardening cement)

2. For the given types of cement and coarse aggregate to be used, the reference number corresponding to the design strength at the particular age is interpolated from Figs. 10.14 to 10.17.

3. Corresponding to the reference number the water-cement ratio to achieve the required workability is obtained from Figs. 10.18 to 10.19 for aggregates with a maximum nominal size of 20 mm and 10 mm, respectively.

4. The aggregate-cement ratio to give the desired workability with the known water-cement ratio is obtained from Tables 10.36 and 10.37.

5. Having obtained the water-cement and aggregate-cement ratios and knowing the specific gravities of the ingredients of the mix, the cement content is obtained by the absolute volume method.

6. Batch quantities are worked out after adjustment for moisture content in the aggregates.

TABLE 10.36 Aggregate-cement ratio (by mass) required to give four degrees of workability with different water-cement ratios using ordinary portland cement

Maximum size of aggregate	Irregular gravel									Crushed granite								
	20 mm				10 mm					20 mm				10 mm			
Degree of workability**	EL	VL	L	M	EL	VL	L	M		EL	VL	L	M	EL	VL	L	M
Water-cement ratio by mass																	
0.30	3.0	—	—	—	2.4	—	—	—		3.3	—	—	—	2.9	—	—	—
0.32	3.8	2.5	—	—	3.2	—	—	—		4.0	2.6	—	—	3.6	2.3	—	—
0.34	4.5	3.0	2.5	—	3.9	2.6	—	—		4.6	3.2	2.6	—	4.2	2.8	2.3	—
0.36	5.2	3.5	3.0	2.5	4.6	3.1	2.6	—		5.2	3.6	3.1	2.6	4.7	3.2	2.7	2.3
0.38	—	4.0	3.4	2.9	5.2	3.5	3.0	2.5		—	4.1	3.5	2.9	5.2	3.6	3.0	2.6
0.40	—	4.4	3.8	3.2	—	3.9	3.3	2.7		—	4.5	3.8	3.2	—	4.0	3.3	2.9
0.42	—	4.9	4.1	3.5	—	4.3	3.6	3.0		—	4.9	4.2	3.5	—	4.4	3.6	3.1
0.44	—	5.3	4.5	3.8	—	4.7	3.9	3.3		—	5.3	4.5	3.7	—	4.8	3.9	3.3
0.46	—	—	4.8	4.0	—	5.1	4.2	3.6		—	—	4.8	4.0	—	5.1	4.2	3.6
0.48	—	—	5.2	4.4	—	5.4	4.5	3.8		—	—	5.1	4.2	—	5.5	4.5	3.8
0.50	—	—	5.5	4.7	—	—	4.8	4.1		—	—	5.4	4.5	—	—	4.7	4.0

*Natural sand is used in combination with both types of coarse aggregate
**EL = extremely low, VL = very low, L = low, and M = Medium.

TABLE 10.37 Aggregate-cement ratio (by mass) required to give four degrees of workability with different water-cement ratios using rapid-hardening portland cement

Type of coarse aggregate*	Irregular gravel								Crushed granite							
Maximum size of aggregate	20 mm				10 mm				20 mm				10 mm			
Degree of workability**	EL	VL	L	M	EL	VL	L	M	EL	VL	L	M	EL	VL	L	M
Water-cement ratio by mass																
0.32	2.6	—	—	—	—	—	—	—	2.9	—	—	—	3.5	—	—	—
0.34	3.4	2.2	—	—	2.8	—	—	—	3.6	2.4	—	—	2.2	—	—	—
0.36	4.1	2.7	2.3	—	3.5	2.4	—	—	4.3	2.9	2.4	—	3.9	2.5	—	—
0.38	4.8	3.2	2.8	2.3	4.2	2.9	2.4	—	4.9	3.4	2.9	2.4	4.5	3.0	2.5	—
0.40	5.5	3.7	3.2	2.7	4.9	3.3	2.8	2.3	5.5	3.9	3.3	2.7	5.0	3.4	2.9	2.4
0.42	—	4.2	3.6	3.0	—	3.7	3.1	2.6	—	4.2	3.6	3.0	5.5	3.8	3.2	2.7
0.44	—	4.6	4.0	3.4	—	4.1	3.5	2.9	—	4.7	4.0	3.3	—	4.2	3.5	3.0
0.46	—	5.0	4.3	3.7	—	4.5	3.8	3.2	—	5.1	4.3	3.6	—	4.6	3.8	3.2
0.48	—	5.5	4.7	4.0	—	4.9	4.1	3.5	—	5.5	4.6	3.9	—	5.0	4.1	3.4
0.50	—	—	5.0	4.3	—	5.2	4.4	3.7	—	—	4.9	4.1	—	5.3	4.4	3.7

*Natural sand is used in combination with both types of coarse aggregate
**EL = extremely low, VL = very low, L = low, and M = Medium.

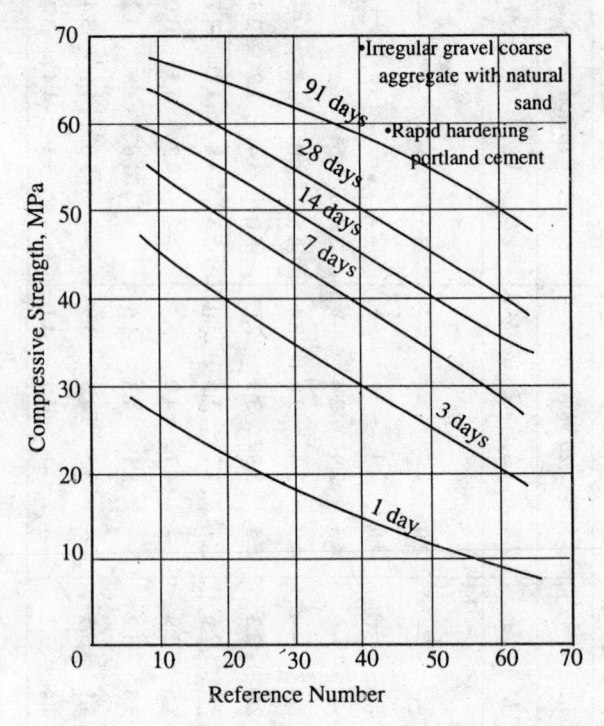

Fig. 10.17 Relation between compressive strength and reference number (irregular gravel coarse aggregate and rapid hardening portland cement)

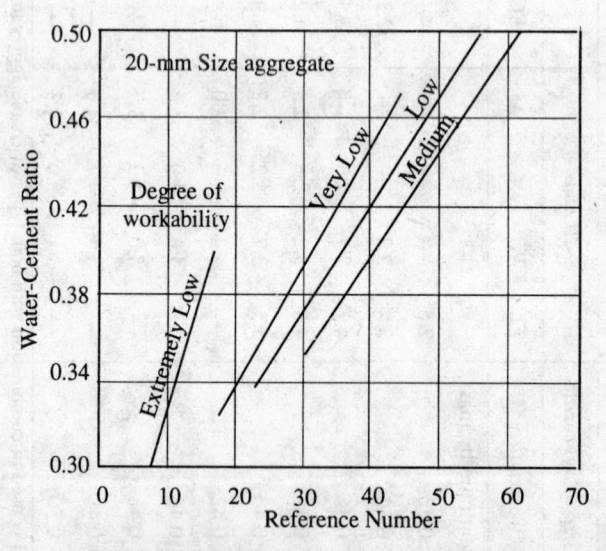

Fig. 10.18 Relation between water-cement ratio and reference number for 20mm maximum nominal size aggregates

Proportioning of Concrete Mixes 193

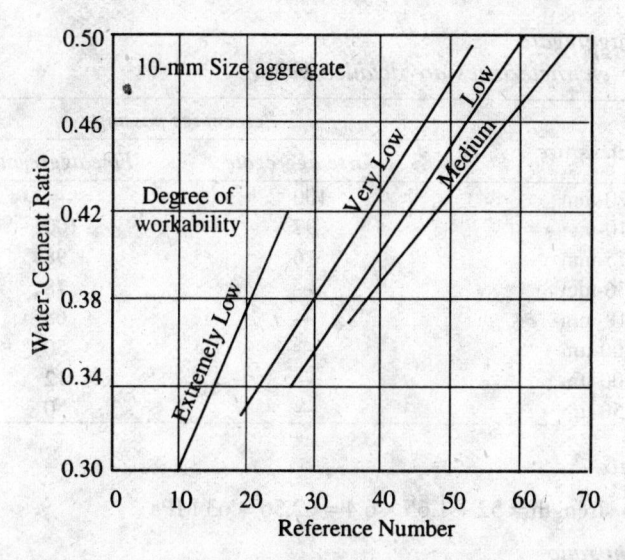

Fig. 10.19 Relation between water-cement ratio and reference number for 10 mm nominal size aggregates

10.7.2 Mix Design Examples

Example 10.8 Design a concrete mix for use in production of prestressed concrete elements to suit the following requirements:

Characteristic compressive strength at 28 days	52 MPa
Degree of quality control	Very good
Degree of workability	Very low
Type of cement	Ordinary Portland Cement (OPC)
Specific gravity of cement	3.15
Type of coarse aggregate	Crushed granite (angular)
Maximum nominal size of aggregate	10 mm
Type of fine aggregate	Natural sand
Specific gravity	
Sand	2.60
Coarse aggregate	2.50
Free surface moisture	
Fine aggregate	5 per cent
Coarse aggregate	1 per cent

Grading of aggregate
The grading characteristics are detailed below:

IS sieve size	Percentage passing	
	Coarse aggregate	Fine aggregate
20-mm	100	—
10-mm	97	100
4.75-mm	6	98
2.36-mm	—	78
1.18-mm	—	68
600-μm	—	50
300-μm	—	12
150-μm	—	0

Design of mix
Target mean strength = $52 + 1.65 \times 6.4 = 62.56 \approx 63$ MPa

Water-cement ratio
  Reference number (Fig. 10.14) 25
  Water-cement ratio (Fig. 10.19) 0.35

Aggregate-cement ratio
  For 10 mm maximum nominal size aggregate and very low workability:
  Aggregate-cement ratio (Table 10.36) 3.0

Mix proportions
The fine and coarse aggregates are combined by using the method described in Sec. 10.3.4 so that 30 per cent of the combined aggregate passes through the IS : 4.75 mm sieve. Let one kg of fine aggregate and x kg of coarse aggregate be combined to obtain the desired aggregate.

Then $\qquad 98 + 6x = 30(1 + x)$

giving $\qquad x = 2.83$

  Ratio of fine to total aggregate = 1.00 : 2.83

Therefore, fine and coarse aggregates are combined in ratio 1.00 : 2.83. Required proportions by mass of dry-aggregates are

Water	Cement	Fine aggregate	Coarse aggregate
0.35 :	1 :	$\left(\dfrac{1.00}{3.83} \times 3.0\right)$:	$\left(\dfrac{2.83}{3.83} \times 3.0\right)$
0.35 :	1 :	0.78 :	2.22

If C is the mass of cement required per m³ of concrete,

Then $\qquad \dfrac{C}{3.15 \times 1000} + \dfrac{0.78C}{2.6 \times 1000} + \dfrac{2.22C}{2.5 \times 1000} + \dfrac{0.35C}{1000} = 1.00$

Therefore $\qquad C = 539$ kg

Batch Quantities per Cubic Metre of Concrete

Ingredient	Batch quantity, kg	
	Dry-aggregate	Moist-aggregate
Cement	539	539
Water	189	147
Fine aggregate	420	441
Coarse aggregate	1196	1208

10.8 TRIAL MIXES

The mix proportions arrived at shall be checked by means of trial batches. The quantity of material for each trial batch shall be sufficient for at least three 150 mm concrete cube specimens and concrete required to carry out the workability test.

The mix proportions computed by a mix design method shall comprise trial mix no. 1. The workability of this trial batch in terms of slump or compaction factor shall be measured and the mix carefully observed for any tendency for segregation and bleeding, and for its finishing properties. If the measured workability of trial batch no. 1 is different from the stipulated value, the water content shall be adjusted according to Table 10.15 (use appropriate criterion of the method used) for the required change in compacting factor. For this adjusted water content, the mix proportions shall be recalculated keeping the free water-cement ratio at the preselected value, this will comprise trial mix no. 2. In addition, two more trial mixes no. 3 and 4 shall be made with the water content kept at the level of trial mix no. 2 but varying the free water-cement ratio by ± 10 per cent of the pre-selected value. The mix proportions for the trial mixes 3 and 4 shall be recalculated for the changed free water-cement ratio by making suitable adjustments in accordance with Table 10.15. The trial batches 2 and 4 will normally provide sufficient information to arrive at the field mix proportions.

10.9 CONVERSION OF MIX PROPORTIONS FROM MASS TO VOLUME BASIS

For volume batch mixing it is desirable to express concrete mix proportions by volume. The mix proportions by mass can be converted into volume proportions by dividing the mass proportions by the corresponding bulk densities. Let the contents of cement, fine aggregate and coarse aggregate per cubic metre of concrete are C, F_a and C_a, respectively, and γ_c, γ_{fa} and γ_{ca} represent the bulk densities of the corresponding materials.
Then mix proportion by mass are:
$$C : F_a : C_a \quad \text{(kg)}$$
and mix proportions by volume are:
$$\frac{C}{\gamma_c} : \frac{F_a}{\gamma_{fa}} : \frac{C_a}{\gamma_{ca}} \quad (\text{m}^3) \qquad (10.10)$$

The proportions obtained above are based on volume of dry aggregates. If fine aggregate contains moisture, suitable modifications for bulking shall be made.

10.10 QUANTITIES OF MATERIALS TO MAKE SPECIFIED VOLUME OF CONCRETE

When the mix proportions have been determined, the quantities of materials required to produce a specified quantity of concrete can be calculated by *absolute volume method*. The method is based on the principle that the volume of fully compacted concrete is equal to the absolute volume of all the ingredients. If W, C, F_a and C_a are the mass of water, cement, fine aggregate and coarse aggregate, respectively, used in making the concrete; S_c, S_{fa} and S_{ca} are the specific gravity of cement, fine aggregate and coarse aggregate, respectively; and v is the percentage of entrained air in the concrete. Then the absolute volume of fully compacted fresh concrete (ignoring air content) is given by:

$$V_c = \frac{W}{1000} + \frac{C}{1000 S_c} + \frac{F_a}{1000 S_{fa}} + \frac{C_a}{1000 S_{ca}} \qquad (10.14)$$

The method is illustrated in Example 10.7.

Example 10.7 Calculate the quantities of ingradients required to produce one cubic metre of structural concrete. The mix is to be used in proportions of 1 part of cement to 1.37 parts of sand to 2.77 parts of 20 mm nominal size crushed coarse aggregate by dry-volumes with a water-cement ratio of 0.49 (by mass). Assume the bulk densities of cement, sand and coarse aggregate to be 1500, 1700 and 1600 kg/m^3, respectively. The percentage of entrained air is 2.

Solution The mix proportions of 1 : 1.37 : 2.77 by dry-volume to be used in the production of structural concrete can be expressed in terms of masses as follows:

Water	Cement	Sand	Coarse aggregate
—	: 1 × 1500	: 1.37 × 1700	: 2.77 × 1600 (kg)
0.49	: 1	: 1.55	: 2.95

The absolute volume of concrete produced by one bag of cement of 50 kg is

$$V_c = \frac{0.49 \times 50}{1000} + \frac{1 \times 50}{100 \times 3.15} + \frac{1.55 \times 50}{1000 \times 2.6} + \frac{2.95 \times 50}{1000 \times 2.6} = 0.127 \text{ m}^3$$

With an entrained air of 2 per cent, the absolute volume of ingredients in one cubic metre of fully compacted fresh concrete is 1.0 − 0.02 = 0.98 m^3.
Therefore,

Cement content per m^3 of concrete, $C = (0.98)/(0.127) = 7.72$ bags

or

$$C = 386 \text{ kg}$$

Therefore, ingredient requirement are:

Cement	386 kg/m^3
Sand	598 kg/m^3
Coarse aggregate	1139 kg/m^3
Water	189 kg/m^3

10.11 ACCEPTANCE CRITERIA FOR CONCRETE

In order to ensure proper quality control, IS : 456–1978 requires that a minimum number of random samples from the fresh concrete of each grade should be taken as specified in IS : 1199–1959 and cubes should be made, cured and tested as described in IS : 516–1959. The minimum number of samples is given in Table 10.38. The average of the strengths of three specimens is the *test strength* of any sample. The acceptance criteria given in IS : 456–1978 stipulates that the strength requirement is satisfied if:

TABLE 10.38 Frequency of sampling of concrete

Quantity of concrete in the job, m^3	Number of samples
1–5	1
6–15	2
16–30	3
31–50	4
51 and above	4 + one additional sample for each additional 50 m^3 or part thereof

(i) Every sample has a *test strength* not less than the characteristic value f_{ck}.
Or
(ii) The strength of one or more samples though less than the characteristic value, is in each case not less than the greater of:

    (a) $f_{ck} - 1.35\,S$; and   (10.11)

    (b) $0.8\,f_{ck}$

Further, the average strength of all samples is not less than

$$f_{ck} + 1.65\left(1 - \frac{1}{\sqrt{n}}\right)S \tag{10.12}$$

where f_{ck} is characteristic strength,
          S is standard deviation
          n represents the number of samples

When n is sufficiently large, the term $\left[f_{ck} + 1.65\left(1 - \frac{1}{\sqrt{n}}\right)S\right]$ approaches the *mean* or *target strength*, as it should be.

The strength criteria shall be deemed not satisfied if:
(i) The strength of any sample is less than the greater of the following:

    (a) $f_{ck} - 1.35\,S$, and

    (b) $0.8\,f_{ck}$. Or

(ii) The average strength of all sample is less than

$$f_{ck} + \left(1.65 - \frac{3}{\sqrt{n}}\right)S$$

According to IS : 456–1978, the individual variation in strength of the specimens of a sample should not be more than ± 15 per cent of the average strength. The code has recommended values of standard deviation for different concrete mixes based on a minimum of 30 test results. For a given concrete mix and corresponding *standard deviation*, the *target mean strength* can be determined by using the stipulation that the average strength of all samples should not be less than

$$f_{ck} + 1.65\left(1 - \frac{1}{\sqrt{n}}\right) S \qquad (10.13)$$

where f_{ck} and S are the characteristic strength and standard deviation, respectively, and n is the number of samples. When n is sufficiently large, this approaches to $(f_{ck} + 1.65\, S)$ and its termed *mean* or *characteristic strength*.

To take into account the variation of individual sample, the laboratory design strength can be obtained by increasing the *average field strength* or the *target mean strength* by 10 to 15 per cent. The recommended values of *laboratory design strength* of samples for different concrete mixes are given in Table 10.39.

TABLE 10.39 Laboratory design strength of concrete and corresponding water-cement radio

Grade of concrete	Assumed standard deviation*, MPa	Target mean strength, MPa	Laboratory design strength, MPa	Water-cement ratio
M15	3.5	21	23	0.55
M20	4.6	27	30	0.47
M25	5.3	34	37	0.41
M30	6.0	40	44	0.36
M35	6.3	45	50	0.32
M40	6.6	51	56	0.30

*The values are for good control.

10.12 FIELD ADJUSTMENTS

In a concrete mix if W, C, F_a and C_a are the required quantities of water, cement, fine aggregate, and coarse aggregate, respectively, to produce one cubic metre of fully compacted concrete, then based on concept that volume of compacted concrete is equal to the sum of the absolute volumes of all ingredients, the following relation is obtained:

$$\frac{W}{1000} + \frac{C}{1000 S_c} + \frac{F_a}{1000 S_{fa}} + \frac{C_a}{1000 S_{ca}} = 1.0 \qquad (10.15)$$

When entrained air is also present and its content is v per cent of the volume of concrete, the right-hand side of above equation would read: $(1.0 - 0.01v)$.

If the specific gravities of fine and coarse aggregates are assumed to be same say S_a, then for a given type of cement the above relation can be written as:

$$\frac{W}{C} + \frac{1}{S_a}\left(\frac{F_a + C_a}{C}\right) = \frac{1000}{C} - \frac{1}{S_c}$$

i.e. (Water-cement ratio) $+ \dfrac{\text{(Aggregate-cement ratio)}}{S_a} = \dfrac{1000}{C} - \dfrac{1}{S_c} \qquad (10.16)$

This relation can be used to convert the *aggregate-cement ratio* into *cement content* for the given *water- cement* ratio or vice versa. Figure 10.20 renders such conversion quite simple.

If the aggregate contains free surface moisture whose content is, say, w per cent of the mass of saturated surface dry aggregate then the masses of added water W

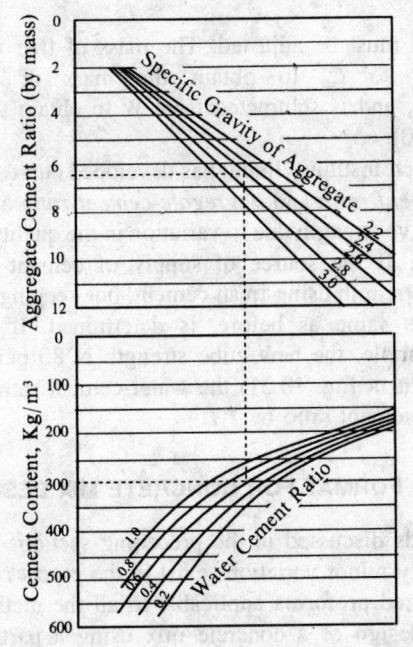

Fig. 10.20 Relationship between aggregate-cement ratio and cement content

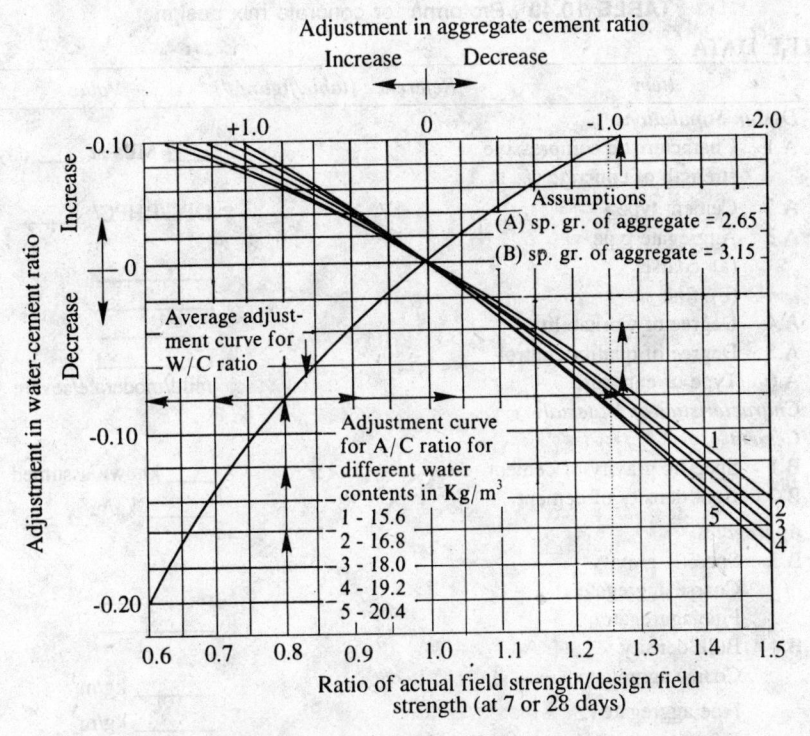

Fig. 10.21 Field adjustment for variation in cement quality (in terms of compressive strength)

200 Concrete Technology

and of (wet) aggregate must be adjusted. The mass of free water is C_a (w/100). This mass is added to C_a to obtain the mass of wet aggregate required, $C_a[1 + (w/100)]$, and is subtracted from W to obtain the mass of water to be added, $W - C_a(w/100)$.

Central Road Research Institute (India) has developed curves shown in Fig. 10.21 to adjust the *water-cement ratio* and *aggregate-cement ratio* at site to take care of the change in compressive strength due to variation in the quality of cement obtained from different sources. If the source of supply of cement changes during the construction, concrete strength using fresh cement, but keeping the mix proportions and water-cement ratio same as before, is determined. If there is substantial difference, say for example, the new cube strength is 80 per cent of the design strength, then (as shown in Fig. 10.21) the water-cement ratio should be reduced by 0.09 and aggregate-cement ratio by 1.1.

10.13 GENERALIZED FORMAT FOR CONCRETE MIX DESIGN

The mix design methods discussed in the preceding sections basically follow the same principles and only minor variations exist in the process of selecting the mix proportions. A generalized proforma applicable to all the methods is suggested in Table 10.40. For the design of a concrete mix using a particular method, only relevant items need to be filled up.

TABLE 10.40 Proforma for concrete mix design

PART-I: DATA

		Item	Reference (table/figure #)	Value
A.		*Design Stipulations*		
	A.1	Characteristic compressive strength of concrete (f_{ck}):		_____ MPa at _____ days
	A.2	Cement type		OPC/RHPC/ _____
	A.3	Aggregate type		
		(a) coarse		_____
		(b) fine		_____
	A.4	Degree of workability		_____
	A.5	Degree of quality control		_____
	A.6	Type of exposure		mild/moderate/severe
B.		*Characteristics of Material*		
		Cement		
	B.1	Specific gravity of cement		_____ known/assumed
	B.2	Bulk density of cement		_____ kg/m^3
		Aggregates		
	B.3	Specific gravity		
		Coarse aggregate		_____
		Fine aggregate		_____
	B.4	Bulk density		
		Coarse aggregate		_____ kg/m^3
		Fine aggregate		_____ kg/m^3
	B.5	Water absorption		
		Coarse aggregate		_____ per cent

	Fine aggregate	_____ per cent
B.6	Free surface moisture	
	Coarse aggregate	_____ per cent
	Fine aggregate	_____ per cent
B.7	Grading of aggregate	

Type of aggregate	Percentage passing the IS sieve
	40-mm 20-mm 10-mm 4.75-mm 2.36-mm 1.18-mm 600-μm 300-μm 150-μm
Coarse	
Fine	

B.8	Maximum size of coarse aggregate	_____ mm
B.9	Grading zone of fine aggregate	_____
B.10	Fineness modulus	
	Coarse aggregate	_____
	Fine aggregate	_____

PART-II: MIX DESIGN

Stage		Item	Reference of calculation	Value
I	1.1	Characteristic compressive strength (f_{ck})	Specified (Part-I)	_____ at _____ days Proportion of defective specimens _____ per cent
	1.2	Standard deviations (S)	Fig./Table	_____ MPa
	1.3	Probability factor (k)		_____
	1.4	Target mean strength (f_t)	$f_t = f_{ck} + kS$	__ + __ × __ = __ MPa
	1.5	Free water-cement ratio	Table/Fig.	__ ⎤
	1.6	Maximum free water-cement ratio	Specified	__ ⎦ — use the lower value
II	2.1	Compacting factor or Slump or V-B	Specified	C.F. _____ Slump _____ mm or V-B _____ s
	2.2	Maximum aggregate size	Specified (Part-I)	_____ mm
	2.3	Free water content	Table	_____ kg/m^3
III	3.1	Cement content		___ / ___ = ___ kg/m^3
	3.2	Maximum cement content	Specified	_____ kg/m^3
	3.3	Minimum cement content	Specified	_____ kg/m^3 Use if greater than Item 3.1 and calculate Item 3.4
	3.4	Modified free water-cement ratio		_____
IV	4.1	Relative density of aggregate (SSD)	Specified (Part-I)	_____
	4.2	Concrete density		_____ kg/m^3
	4.3	Total aggregate content		__ − __ − __ = __ kg/m^3
V	5.1	Grading of fine aggregate	Specified (Part-I)	_____
	5.2	Proportion of fine aggregate	Fig.	_____ per cent
	5.3	Fine aggregate content		__ × __ = __ kg/m^3
	5.4	Coarse aggregate content		__ − __ = __ kg/m^3

VI	Ingredients	Water (kg)	Cement (kg or 1)	Fine aggregate (kg)	Coarse aggregate (kg)
	Quantity per m^3 (to nearest 5 kg)	———	———	———	———
	Quantity per trial mix of ____ m^3	———	———	———	———
	Ratio	———	———	———	———

10.14 DESIGN OF CONCRETE MIX AS A SYSTEM

The foregoing procedures of concrete mix design can be expressed almost without qualification in terms of design methodology of *System Engineering*. The term 'System Engineering' is currently the popular name for engineering processes of planning and design used in the creation of a system or project. In the most general sense, a system may be defined as a collection of various structural and non-structural (e.g. human) components which are so interconnected and organised to achieve a specified objective by the control and distribution of material resources, information and energy. The fundamental characteristic of a properly designed and operated system is that the performance achieved by the whole is beyond the total capability of the separate components operating in isolation.

The purpose of this section is to set the scene for development of mathematical model for an efficient and economical concrete mix design. In system engineering the physical quantities and processes are represented by mathematical models; performance is analyzed and objective measures of costs and benefits are obtained by mathematical operations; and the influence of uncertainty particularly where humans are involved, is modelled by probability distributions where necessary. This principle of system engineering can be applied to select a most efficient or optimal proposal from a large number of feasible alternatives which may be imperfect to some degree. Sometimes the model as modified and improved in light of preliminary design, tested for its feasibility and optimized for main system parameters may help to reach a reasonably firm decision concerning acceptance or rejection of the proposal.

In the mix design problems the mix proportion, water-cement ratio or water content may be selected as decision variables, since effectiveness of the concrete (system) can be evaluated directly or indirectly in terms of these variables. The object of the analysis is to determine the best possible set of values with respect to system effectiveness. This is called optimal proposal. The objective function which is the measure of effectiveness of a particular proposal is expressed as a function of these decision variables.

The conditions which a mathematical model must satisfy before the decision variable values can represent a feasible solution are termed constraints. The process of mix design may be summarized by the following five sequential activities:

(i) Selection of decision variables
(ii) Definition of objectives and identification of design criteria
(iii) Generation of design alternatives
(iv) Testing of feasibility of proposals
(v) Optimization and refinement of design to maximise the effectiveness.

The *objectives* should be stated in the most basic and general terms possible. The information provided in the preceding sections may help in building up a picture of the problem environment. In concrete mix design problems, the economy of end product i.e. the concrete may be the objective.

Once the objectives have been determined, design criteria must be identified. In mix design problems the workability, the 28-day compressive strength, and durability are generally taken as the design criteria.

Testing of Feasibility

For each alternative proposal, the first test must be that of feasibility. Technical constraints are normally carried out routinely in the course of preliminary design. The other constraints of economic and inter-disciplinary nature also exist, they must be identified and quantified at this stage. As the design proceeds any constraint which is violated will result in the proposal being modified or rejected.

Measure of Effectiveness

The most important factor influencing the nature of the final solution is the definition of objective and selection of appropriate measures of effectiveness. In its simplest form the effectiveness of a mix design may be measured in terms of cost of final product, i.e. the concrete.

To illustrate the formulation of objective function and constraints consider the following example.

Example 10.8 A concrete mixing plant has to supply M15 grade concrete in large quantity to a dam project. The mix proportions have been estimated as 1 : 1.91 : 4.46 (by mass). This concrete requires sand and gravel (C.A.) mixture of 30 per cent sand and 70 per cent gravel by mass. The natural deposits at five pits near the dam site are found to have different compositions and their cost including transportation to the site also varies as shown in Table 10.41. However the constituents satisfy the specifications. Determine the quantities of deposit to be obtained from each source in order to minimize the cost per cubic of concrete.

Let x_i be the fraction taken from pit i, the cost per cubic metre of concrete can be expressed as:
$$Z = 2.0x_1 + 3.0x_2 + 1.5x_3 + 1.0x_4 + 2.5x_5$$
The fractions x_i should be of the magnitudes such that ratio of sand and gravel in concrete should be 30 and 70 per cent, respectively.

Thus
$$0.45x_1 + 0.4x_2 + 0.5x_3 + 0.55x_4 + 0.20x_5 = 0.3$$
$$0.55x_1 + 0.6x_2 + 0.5x_3 + 0.45x_4 + 0.80x_5 = 0.7$$

TABLE 10.41 Sand and gravel mixture

Aggregate type	Mixture composition, per cent Pit No.				
	1	2	3	4	5
Sand	45	40	50	55	20
Gravel	55	60	50	45	80
Relative cost per cubic metre of mixture	2.0	3.0	1.5	1.0	2.5

The mathematical optimization problem of the system can be stated as:
Minimize:
$$Z = 2.0x_1 + 3.0x_2 + 1.5x_3 + 1.0x_4 + 2.5x_5$$
Subject to:
$$0.45x_1 + 0.4x_2 + 0.5x_3 + 0.55x_4 + 0.20x_5 = 0.3$$
$$0.55x_1 + 0.6x_2 + 0.5x_3 + 0.45x_4 + 0.80x_5 = 0.7$$

This linear programming problem can be solved easily by SIMPLEX METHOD. The values of fractions obtained by this method are:
$$x_1 = 0.0, \quad x_2 = 0.0, \quad x_3 = 0.0, \quad x_4 = 2/7, \quad x_5 = 5/7$$

Thus the deposits from the pits 4 and 5 when mixed in the ratio 2 : 5 will result in a most economical concrete satisfying all the stipulations. Any deviation from these values will result in an increase of cost of the material.

11

Production of Concrete

11.1 INTRODUCTION

The design of a satisfactory mix proportion is by itself no guarantee of having achieved the objective of quality concrete work. The batching, mixing, transportation, placing, compaction, finishing and curing are very complimentary operations to obtain desired good quality concrete. The good quality concrete is a homogeneous mixture of water, cement, aggregates and other admixtures. It is not just a matter of mixing these ingredients to obtain some kind of plastic mass, but it is scientific process which is based on some well established principles and governs the properties of concrete mixes in fresh as well as in hardened state. The aim of *quality control* is to ensure the production of concrete of uniform strength in such a way that there is a continuous supply of concrete delivered to the place of deposition, each batch of which is as nearly like the other batches as possible. The production of concrete of uniform quality involves five definable phases:

  (i) Batching or measurement of materials,
  (ii) Mixing of concrete,
  (iii) Transportation,
  (iv) Placing, compaction and finishing of concrete, and
  (v) Curing.

11.2 BATCHING OF MATERIALS

A proper and accurate measurement of all the materials used in the production of concrete is essential to ensure uniformity of proportions and aggregate grading in successive batches. All the materials should be measured to the tolerances indicated in Table 11.1.

TABLE 11.1 Batching tolerances

Material	Accuracy of measurement
Aggregates, cement and water	± 3 per cent of batch quantity
Admixtures	± 5 per cent of batch quantity

For most of the large and important jobs the batching of materials is usually done by weighing. In weigh batching, the weight of surface water carried by the wet aggregate must be taken into account. The factors affecting the choice of proper batching system are: (i) size of job, (ii) required production rate, and (iii) required standards of batching performance. The production capacity of a plant is determined by the material handling system, the bin size, the batcher size, and the plant mixer size, and their number available. The batching equipment falls into three general categories, namely, manual, semiautomatic, and fully automatic systems.

Manual Batching

In this sort of batching all operations of weighing and batching of concrete ingredients are done manually. Manual batching is acceptable for small jobs having low batching rates. Attempt to increase the capacity of manual plants by rapid batching often results in excessive weighing inaccuracies. The weighing may also be done by an ordinary platform scale.

Semiautomatic Batching

This batching is one in which the aggregate bin gates for charging batchers are opened by manually operated switches. Gates are closed automatically when the designated weight of material has been delivered. The system contains interlocks which prevent batcher charging and discharging occuring simultaneously. Provision is made for the visual inspection of the scale reading for each material being weighed. All the weighing hoppers should be constructed in a manner facilitating their easy inspection.

Automatic Batching

Automatic batching is one in which all scales for the materials are electrically activated by a single switch and complete autographic records are made of the weight of each material in each batch. However, interlocks interrupt the batching cycle when preset weighing tolerances are exceeded.

The *batching plant* generally compris two, three, four or six compartment bins of several capacities together with a supporting system. Below the bins are provided the weight batchers discharging over the conveyor belts. The use of separate hoppers as shown in Fig. 11.1 is preferable as it accomplishes some mixing of materials before they enter the mixer.

The *mobile plant* consisting of batching equipment mounted on pneumatic tyred wheels has the advantage that the plant can be kept close to the site where concreting is required. The mobile plant is particularly useful where concrete is required over a very large area, e.g. an aerodrome; the runways and road construction work where the plant can follow the progess of work.

In addition to accurate batching of mixing water, the amount of moisture present in the aggregate (particularly in the sand) as it is batched should be taken into account.

For most of the small jobs, *volume batching* is adopted, i.e. the amount of each solid ingredient is measured by loose volume using measuring boxes, wheel barrows, etc. In batching by volume, allowance has to be made for the moisture present in sand which results in its bulking. The proportions by volume are generally specified in terms of the dry-rodded condition of the aggregate; the batch quantities must also be specified in the damp and loose condition.

In volume batching it is generally advisable to set the volumes in terms of whole bags of cement. Fractioned bags lead to variable proportions, resulting in concrete of non-uniform strength in successive batches. Before the batching operations are started, the engineer-in-charge should check the batch box volumes. When filling the boxes, the material should be thrown loosely into the box and struck off, and no compaction is to be allowed. At the end of each day's work the boxes should be stacked upside down to prevent any accumulation of rain water.

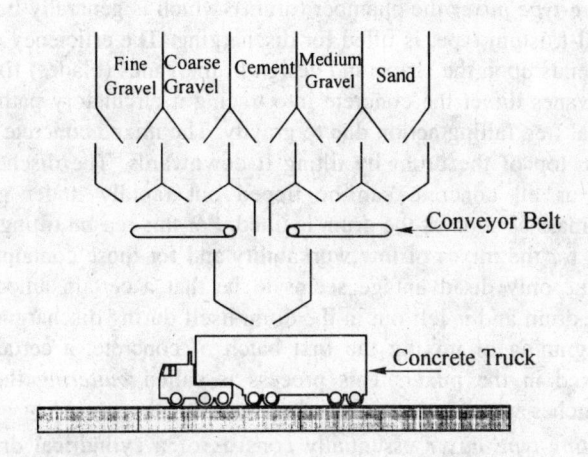

Fig. 11.1 A typical scheme for a batching plant

11.3 MIXING OF CONCRETE MATERIALS

The object of *mixing* is to coat the surface of all aggregate particles with cement paste, and to blend all the ingredients of concrete into a uniform mass. The mixing action of concrete thus involve two operations: (i) a general blending of different particle sizes of the ingredients to be uniformly distributed through out the concrete mass, and (ii) a vigorous rubbing action of cement paste on to the surface of the inert aggregate particles. Concrete mixing is normally done by mechanical means called *mixer*, but sometimes the *mixing* of concrete is done by hand. Machine mixing is more efficient and economical compared to hand mixing.

In the mixing process the cement paste is formed first with simultaneous absorption of water in the aggregates. In the second stage the cement paste coats the aggregate particles. The mixing process should be continued till a thoroughly and properly mixed concrete is obtained. At the end of this stage the concrete appears to be of uniform colour and grading. The uniformity must be maintained while discharging the concrete from the mixer. As a matter of fact the classification of the mixers is based on the technique of discharging the mixed concrete as follows:

   (i)   the tilting type mixer,
  (ii)   the nontilting type, and
 (iii)   the pan or stirring mixer.

The size of a mixer is designated by a number representing its nominal mix batch capacity in litres, i.e. the total volume of mixed concrete in litres which can be obtained from the mixer per batch. The capacity of a mixer for a particular job should be such that the required volume of concrete per hour is obtained without speeding up the mixer or reducing the mixing time below the specified period and without overloading the mixer above its rated capacity. The standardized sizes of the mixers given in IS: 1971–1968 are given in Table 11.2. Most of the mixers can handle a ten-per cent overload satisfactorily. If the quantity mixed is much less than the rated capacity of the mixer the resulting mix may not be uniform, and the mixing operation becomes uneconomical.

In the *tilting-type mixer* the chamber (drum), which is generally bowl-shaped or double-conical-frustum type, is tilted for discharging. The efficiency of the mixing operation depends upon the shape and design of the vanes (blades) fixed inside the drum. These vanes direct the concrete into tracing a circulatory path. In addition, there is vertical free falling action due to gravity. The mixed concrete is discharged from the open top of the drum by tilting it downwards. The discharge action is always good as all concrete can be tipped out rapidly under gravity in an unsegregated mass as soon as the drum is tilted. For this reason tilting drum mixers are preferable for the mixes of low workability and for those containing large-size aggregates. The only disadvantage seems to be that a certain amount of mortar adheres to the drum and is left out in the drum itself during discharging. Therefore, before the beginning of mixing the first batch of concrete, a certain amount of mortar is mixed in the mixer. This process is called *buttering* the mixer. The subsequent batches will be as desired.

The *nontilting-type mixer* essentially consists of a cylindrical drum with two circular openings at the ends and blades fixed inside the drum. The drum rotates about a horizontal axis and cannot be tilted. The mixer is loaded through a central opening at one end of drum and, after mixing, the concrete is discharged through the opening at the other end by a chute. Owing to a rather slow rate of discharge, the concrete is sometimes susceptible to *segregation*. In particular, the larger size aggregate may tend to stay in the mixer, when the other constituents are being discharged. Hence the discharge may initially consists of mortar and then as a collection of a large size coated aggregate. However, it is worthwhile to check the performance of the mixer for a particular type of mix, before it is actually used.

The *pan mixer* consists essentially of a circular pan rotating about a vertical axis. One or two stars of paddles also rotate in the pan about a vertical axis not coincident with the axis of the pan. In some types, the pan is static and the axis of star travels along a circular path about the axis of the pan. In other types, the paddles are stationary and the pan rotates about the vertical axis. In either case, the relative movement between the paddles and the concrete is the same, and concrete in every part of the pan is thoroughly mixed. There is another set of blades called scrapper blades, which prevent the sticking of the mortar to the pan sides by continuous scrapping. The paddle height can be adjusted to prevent a permanent coating of mortar forming on the bottom of pan. The mix is discharged through a central hole at the bottom of the pan.

The pan mixer is generally not mobile and is, therefore, used either as a central mixing plant on a large concrete project or at a precast concrete factory. This mixer is particularly efficient with stiff and cohesive mixes. Pan mixers are extensively used in the laboratories for mixing small quantities of concrete of consistent quality, because of the efficient scrapping arrangement. A pan mixer consisting of a bowl-and-stirrer working on the principle of the cake mixer is sometimes used for mixing the mortar.

Apart from the three types of mixers given above, another type called *dual drum mixer* is extensively used for mixing concrete for road or pavement construction. The dual drum mixer consists of a long drum divided into two parts by a central diaphram. Both parts are operated in series. The concrete is intially mixed up to a certain time in the first compartment of the drum and then transferred to the second compartment for the remaining operation of mixing. In the meanwhile, the first compartment is recharged with the constituents of the mix. The dual drum mixers are useful as the mixing capacity can be doubled with the same batching equipment and supervisory staff. The mobile or truck mixers consisting of mixer drum mounted on a conventional truck chassis are powered either from the truck engine or from a separate diesel engine. These mixers are used in *ready mixed concrete* industry.

TABLE 11.2 Standard sizes of mixer

Type of mixer	Nominal mixed batch capacity, litres
Tilting (T)	85T, 100T, 140T, 200T
Nontilting (NT)	200NT, 280NT, 340NT, 400NT, 800NT
Reversing (R)	200R, 280R, 340R, 400R

11.3.1 Mixing Time

It is the time required to produce the uniform concrete. The mixing time is reckoned from the time when all the solid materials have been put in the mixer, and it is usual to specify that all water has to be added not later than after one quarter of mixing time. The time varies with the type of mixer and depends on its size. Strictly speaking, it is not the mixing time but the number of revolutions of the mixer that are to be considered, because there is an optimum speed of rotation for the mixer. The number of revolutions and the time of mixing are independent of each other. In high-speed pan mixers, the mixing time can be as short as 35 s. On the other hand, when light weight aggregate is used, the mixing time should not be less than 5 minutes, sometimes divided into 2 minutes of mixing the aggregate with water followed by 3 minutes with cement added. In general, the length of mixing time required for sufficient uniformity of mix depends on the quality of blending of materials during charging of the mixer.

With machine mixing, there is an increase in strength with time of mixing up to about 5 minutes. The increase in strength is largest in first one minute and after 2 minutes the increase is very small. A mixing time of not less than one minute after all the materials have been added in the mixer drum is generally recognized as a satisfactory period for mixers up to a capacity of 750 litres. For mixers of larger capacity the mixing time should be increased at the rate of 20 s or more for each cubic metre or fraction thereof. The recommended minimum mixing times are given in Table 11.3.

The *order of feeding the ingredients* into the mixer depends on the properties of the mix and those of the mixer. Generally, a small amount of water should be fed first, followed by all the solid materials, preferably fed uniformly and simultaneously into the mixer. If possible, the greater part of the water should also be fed during the same time the remainder of water is added after the solids have been fed.

TABLE 11.3 Recommended minimum mixing time

Capacity of mixer, m^3	Mixing time, minutes
0.8	1.00
1.5	1.25
2.3	1.50
3.1	1.75
3.8	2.00
4.6	2.25
7.6	3.25

The *choice of mixer* depends upon the size, extent, and the nature of work. The choice between central and site mixing will be governed by local factors, such as accessibility, water supply, transport routes, availability of working space, etc.

11.3.2 Hand Mixing

There may be occasions when the concrete has to be mixed by hand, and because in this case uniformity is more difficult to achieve, particular care and effort are necessary. The aggregate should be spread in a uniform layer on a hard, clean and non-porous base; cement is then spread over the aggregate and the dry materials are mixed by turning over from one end of the heap to another and *cutting* with a shovel until the mix appears uniform. Turning three times is usually required. The water is gradually added to the trough formed by the uniform dry mix and the mix is turned over until a homogeneous mixture of uniform colour and consistency is obtained.

11.4 TRANSPORTATION OF CONCRETE

Concrete from the mixer should be transported to the point where it has to be placed as rapidly as possible by a method which prevents the *segregation* or loss of ingredients. The concrete has to be placed before *setting* has commenced. Attempts have been made to limit the time lapse between *mixing* and *compaction* within the forms. The specifications, however, permit a maximum of two hours between the introduction of mixing water to the cement and aggregates, and the discharge, if the concrete is transported in a truck mixer or agitator. In the absence of an agitator, this figure is reduced to one hour only. All these, however, presume that the temperature of concrete, when deposited, is not less than 5°C or more than 32°C. It has now been established that delays in placing concrete, after the so-called *initial* set has taken place, are not injurious and may give increased compressive strengths, provided the concrete retains adequate *workability* to allow *full compaction*.

The requirements to be fulfilled during transportation are:
  (i) no segregation or separation of materials in the concrete, and
  (ii) concrete delivered at the point of placing should be uniform and of proper consistency.

The prevention of *segregation* is the most important consideration in handling and transporting concrete. The segregation should be prevented and not corrected after its occurrence. The concrete being a non-homogeneous composite of materials of widely differing particle sizes and specific gravities, is subjected to internal and external forces during transportation and placing tending to separate the dissimilar constituents.

Segregation can be prevented by ensuring that the direction of fall during the dumping or dropping or concrete is vertical. When the discharge is at an angle, the larger aggregate is thrown to the far side of the container being charged and the mortar is collected at the near side, thus resulting in *segregation*.

The plant required for transporting the concrete varies according to the size of the job and the level at which the concrete is to be placed. The principal methods of transporting concrete from the mixer are:

(i) Barrows
  (a) Wheel barrows and handcarts
  (b) Power barrows or powered buggies or dumpers
(ii) Tippers and Lorries
(iii) Truck mixers and agitator lorries
(iv) Dump buckets
(v) The monorail system or trolley or rails.

The most commonly used method of transporting concrete by the hand pans passing from hand to hand is slow, wasteful and expensive. If concrete is to be placed at or below the mixer level, steel wheelbarrows are a better mode of transportation. Concrete can be discharged from the wheelbarrow to the required point. When concrete is to be placed much below the general ground level, as in basement slabs, foundations, etc., a wooden or steel chute may be used for chuting the concrete into place. The wheelbarrows are suitable for small jobs and where the length of transport is small, and over muddy ground. The average quantity that can be carried in one wheelbarrow is about 35 litres (80 kg). Sometimes, for relatively bigger jobs, *power arrows* which are motorized version of wheelbarrows are used.

Dumpers and ordinary open-steel body tipping lorries can be used economically for hauls of up to about 5 km. These lorries are suitable only for dry mixes to avoid difficulties caused by *segregation* and *consolidation*. The time of journey should be as short as possible. It is essential that the lorry body be watertight to prevent loss of fines. The concrete has to be covered with tarpaulins to prevent the concrete being exposed to sun, wind and rain. If the haul is long, agitators have to be used to prevent segregation. Steel buckets transported by rail or road may be used to transport the concrete for long distances and for large jobs like dams, bridges, etc. While using this method it is necessary to see that: (i) the entire mixer batch is placed in the bucket, and (ii) segregation is prevented while filling the bucket.

The monorail system is useful when the ground conditions are not suitable for normal wheeled traffic. In the monorail system, the rail can easily be provided at such a level that the concrete be tipped directly into the formwork. Basically, the system consists of a power wagon mounted on a single rail capable of a travelling speed of 90 m/min. The engine may be diesel or petrol powered, without a driver.

Conveyor belts have also been used for conveying fairly stiff concrete, but there is a tendency of segregation on steep inclines and at transfer points. When using conveyor belts, it is necessary that the flow of concrete be continuous to minimize the effects of *segregation*.

In the jobs where the concrete is to be lifted up to 5 m inclined runways with one or two landings for carrying the concrete up to the required level can be built. Another method of lifting concrete to greater heights is by using some sorts of hoists. The various types of hoists are chain hoists, platform hoists or skip hoists. In the chain hoisting, a chain sling is suspended from a pulley and is operated by a power winch at the ground level. The sling is attached to the container, which is then lifted bodily to the working level. Wheelbarrows and carts can be elevated by platform hoists operating on vertical steel guides. With some hoists, two platforms are provided, one descends while the other is being raised. The major types of hoists used in tall buildings are tip skip hoists, automatic skip discharge hoists, twin automatic skip, and passenger/concrete hoists.

The tip *skip hoists*, normally fed by direct discharge from the mixer are elevated and discharged into a receiving hopper at the working level; from that point wheelbarrows or other transport can deliver the concrete to the forms. In a modified tip skip model, bottom doors, which open automatically at the required level allowing the concrete discharge via drop chute into the floor hopper, are provided. In tall buildings where the time taken for the travelling of skips and discharging of concrete is large, skips with individual winch units are provided. When one skip travels, the other one is filled, thereby allowing greater efficiency of the whole operation.

In certain tall buildings where a passenger or passenger/goods hoist is essential due to mechanical and other permanent services involved, a combined passenger hoist with concrete carrying and discharging abilities can be used. When the hoist is not in use for concrete conveyance, the cage floor is left perfectly free for normal passenger duties.

In an arrangement consisting of an elevated tower and chutes, concrete is raised in buckets to the central tower and distributed through sloping chutes from the top of the central tower. This system is suitable for large dam jobs. There is a tendency of segregation for dryer mix, it may be necessary to fit vibrators to the chutes.

Pumping of concrete through steel pipelines is one of the successful methods of transporting concrete. *Pumped concrete* has largely been used in construction of multistoreyed buildings, tunnels, and bridges. The equipment consists of a heavy-duty, single-acting horizontal piston pump of special design. The concrete is fed from the hopper into the pump cylinder largely by gravity, but is assisted by the vacuum created on the suction stroke of the piston and forced into the pipeline on the pressure stroke. The pipeline is completely filled and concrete moves uniformly. The pump capacity can range from 15 m^3/h to 150 m^3/h. The normal distance to which the concrete can be pumped is about 400 m, horizontally, and 80 m vertically. Usually 1 m of vertical movement is equivalent to 10 m horizontally. Bends in the pipeline reduce the effective pumping distance by approximately 10 m for each 90° bend, 5 m for a 45° bend, and 3 m for a 22.5° bend.

Mix design for pumpable concrete needs special attention. In general, concrete should be very *cohesive* and *fatty* having a slump value of 50 mm to 100 mm or more. The mix proportions should be so chosen that the total quantity of fines passing 200-micron sieve should not be less than 350 kg/m^3. For obtaining high slump *flowing* concrete admixtures called superplasticizers are being used.

Although the method of transporting and placing concrete by pumps is fast and efficient, a small part of unpumpable mix in hopper can block the pump, leading to delay while the pump is stripped down. Great attention is required in the design of mix as a minor variation in the concrete mix is sufficient to make an otherwise pumpable mix completely unpumpable. At the end of the run, the pipeline must be cleared of concrete by inserting a plunger at the pump end and forcing it through under pressure. After the concrete is cleared, the pipeline is washed out to leave a smooth clean surface ready for next day's work.

A concrete whose constituents are weight-batched at a central batching plant and are mixed either at the plant itself or in truck mixers, and is transported to the construction site and delivered in a condition ready to use is termed *ready mixed concrete (RMC)*. This enables the places of manufacture and use of concrete being separated and linked by suitable transport operation. The technique is useful in congested sites or at diverse work places and saves the consumer from the botheration of procurement, storage and handling of concrete materials. Ready-mixed concrete is produced under factory conditions and permits a close control of all operations of manufacture and transportation of fresh concrete.

There are two principal categories of RMC. In the first, called the *central mixed*, the mixing is done at a central plant and the mixed concrete is delivered generally in an agitator truck which revolves slowly so as to prevent segregation and undue stiffening of the mix. In the second, called the *transit mix*, the materials batched at a central plant are mixed during the period of transit to the site or immediately prior to concrete being discharged. Transit mixing permits a longer haul. Sometimes the concrete is partially mixed at the central plant and the mixing is completely enroute; such concrete is known as *shrink-mixed concrete*. This enables better utilization of transporting trucks. The time of transit after water is added is generally limited from one to one-and-a-half hours. The total number of revolutions during both mixing and agitation are limited to 300.

11.5 PLACING OF CONCRETE

The methods used in placing concrete in its final position have an important effect on its homogeneity, density and behaviour in service. The same care which has been used to secure homogeneity in mixing and the avoidance of segregation in transporting must be exercised to preserve homogeneity in placing.

To secure good concrete it is necessary to make certain preparations before placing. The forms must be examined for correct alignment and adequate rigidity to withstand the weight of concrete, impact loads during construction without undue deformation. The forms must also be checked for tightness to avoid any loss of mortar which may result in honeycombing. Before placing the concrete, the inside of the forms are cleaned and treated with a release agent to facilitate their removal when concrete is set. Any coating of the hardened mortar on the forms should be removed. The reinforcement should be checked for tightness and clean surface. It should also be freed of all loose rust or scales by wire brushing or any other method. Coatings like paint, oil, grease, etc. are removed. The reinforcement should be checked for conformity with the detailing plans for size, spacing and location. It should be properly spliced, anchored and embedded to a given minimum distance from the surface. Anchor bolts, pipe sleeves, pipe conduits, wiring and other fixtures should, in general, be firmly fixed in position before the concrete is placed. Rubbish, such as sawdust shavings and wire, must be blown out with compressed air.

The concrete should be placed in its final position rapidly so that it is not too stiff to work. Water should not be added after the concrete has left the mixer. The concrete must be placed as closely as possible to its final position. It should never be moved by vibrating it and allowing it to flow, as this may result in segregation which will show on the surface of the finished work. When placing the concrete, care should be taken to drop the concrete vertically and from not too great height. *Segregation*, if it occurs, should be eliminated by taking remedial measures.

The surfaces against which the fresh concrete is to be placed must be examined as to their possible effect in absorbing mixing water. For example, subgrades should be compacted and thoroughly dampened to prevent loss of moisture from concrete.

Where fresh concrete is required to be placed on a previously placed and hardened concrete, special precautions must be taken to clean the surface of all foreign matter and remove the laitance or scum before the fresh concrete is placed. For securing a good bond and watertight joint, the receiving surface should be made rough and a rich mortar placed on it unless it has been poured just before. The mortar layer should be about 15 mm thick, and have the same water-cement ratio as the concrete to be placed. In all cases, the base course should be rough, clean, and moistened. The surface can be cleaned by a stiff or steel broom a few hours after placement when the concrete is still soft enough to allow removal of scum but hardened enough not to permit loosening of aggregate particles.

It is becoming increasingly more economical to place concrete in deep lifts. This technique saves time and reduces number of horizontal joints. For placing in deep lifts to be successful, the mix must be designed to have a low risk of segregation and bleeding. The concrete should be introduced into the forms through trunkling, as this reduces impact damage to the forms and reinforcement, and enables the layer of concrete to be built up evenly.

The actual procedure depends largely upon the type of structure, the quality of concrete and of the receiving surface. In mass concrete construction, as in dams, two principal methods are employed in preparing the surface to receive the fresh concrete. For the surface having excessive laitance, it has been common practice to remove all laitance and inferior surface concrete and to wash to mortar from the protuding aggregate by means of a high-velocity jet of air and water as soon as concrete has hardened sufficiently to prevent the jet revealing the concrete below the desired depth. Ordinarily, the surface is cut to a depth of about 3 mm. The time interval between placing and clean up operation may range from 4 to 12 hours depending upon the temperature, humidity, and the setting characteristics of concrete. This surface is thereafter protected and cured by covering it with a layer of about 40 mm wet sand until concreting is resumed, when it receives a final clean up. The final clean up is most effectively accomplished by wet sand blasting and washing.

While *concreting* in walls, footings and other thin sections of appreciable height, the concrete should be placed in horizontal layers not less than 150 mm in depth, unless some other thickness is specified. The concreting should start at the ends or corners of forms and continue towards the centre. In large openings, concrete should be placed first around the perimeter.

On a slope, the concreting should begin at the lower end of slope. To avoid cracking due to settlement, the concrete in columns and walls should be allowed to stand at least for two hours before concrete is placed in slabs or beams which they are to support. Haunches and columns capital are a part of the floor or roof and should be concreted integrally with them.

Concrete in cast-in-situ piles and deep caisson footings has to be dropped from a considerable height. The concreting should be as nearly continuous as possible, because the consolidation in the lower portion of footing depends upon the impact of succeeding increments of concrete. Plastic consistency of about 10 mm (slump) is adequate.

While concreting a slab, the batches of concrete should be placed against or towards preceding ones, not away from them. Batches should not be dumped in separate, individual piles.

11.5.1 Construction Joints

Construction joints are a potential source of weakness and should be located and formed with care and their number is kept to a minimum. As the construction proceeds, water sometimes collects on horizontal surfaces. If this occurs, a drier mix should be used for the layer to be poured to avoid the formation of *laitance*. Any laitance so formed should be removed by spraying the surface with water and brushing it to expose the coarse aggregate. Preferably this should be done an hour or so after the concrete has been placed. The best joints are obtained by light brushing soon after pouring.

Water bars are often installed across construction joints to provide a positive barrier against movement of water through the joint. Great care is needed when placing concrete around water bars because the space is often congested. If the concrete is not properly compacted and is honeycombed, water can pass round the water bar and its object is defeated. Insufficient care in placing may even displace the bars.

11.5.2 Effect of Delay in Placing

It is now generally recognized that there is a gain in compressive strength with delay in placing provided the concrete can still be adequately compacted. The limits imposed by the latter requirement varies with the type of mix. Only a short delay can be allowed for a dry mix in hot weather, a delay of several hours is possible with very wet mix in cold weather. According to the current specifications in general the delay between mixing and final placing of concrete is limited to between half and one hour. Brook has suggested a sliding scale of half an hour for ambient temperature exceeding 20°C, three-quarters of one hour for temperature between 15° and 20°, and one hour for temperature below 15°C.

The effect of delay in placement of concrete varies with the richness of the mix and the initial slump. A low slump concrete could be compacted satisfactorily for only up to one-and-a half hours, but high slump concrete could be compacted satisfactorily even after five hours in agitation.

According to the ASTM specification C-94-71, the environmental and other handling conditions are automatically taken into account by controlling the uniformity of the concrete as delivered for placement.

11.6 COMPACTION OF CONCRETE

During the manufacture of concrete a considerable quantity of air is entrapped and during its transportation there is a possibility of partial segregation taking place. If the entrapped air is not removed and the *segregation* of coarse aggregate not corrected, the concrete may be porous, nonhomogeneous and of reduced strength. The process of removal of *entrapped air* and of uniform placement of concrete to form a homogeneous dense mass is termed *compaction*. Compaction is accomplished by doing external work on the concrete. The density and, consequently, the strength and durability of concrete depend upon the quality of this compaction. Therefore, thorough compaction is necessary for successful concrete manufacture. The concrete mix is designed on the basis that after being placed in forms it may be thoroughly compacted with available equipment. The presence of even 5 per cent voids in hardened concrete left due to incomplete compaction may result in a decrease in compressive strength by about 35 per cent. The compaction is necessary for the following reasons.

(i) The internal friction between the particles forming the concrete, between concrete and reinforcement, and between concrete and formwork, makes it difficult to spread the concrete in the forms. The friction also prevents the concrete from coming in close contact with the reinforcement, thereby leading to poor bond between the reinforcement and surrounding concrete. The compaction helps to overcome the above frictional forces.

(ii) Friction can also be reduced by adding more water than can combine with cement. The water in excess to that required to hydrate the cement fully forms water voids which have as harmful an affect in reducing strength as air voids. Nevertheless, it is preferable to use slightly more water than run the risk of securing inadequate compaction. The compaction reduces the voids to minimum.

The voids due to inadequate compaction can be readily seen when they are at the surface. The patching done to hide surface honeycombing is regarded as a very poor substitute for properly compacted concrete since it can never improve concrete which may honeycombed right through. Furthermore, badly honeycombed concrete does not allow necessary bond to be developed between concrete and reinforcement, and over a period of time the moisture may penetrate to corrode the steel.

Compaction Methods

The compaction of the concrete can be achieved in four ways: (i) hand rodding, (ii) mechanical vibrations, (iii) centrifugation or spinning, and (iv) high pressure and shock.

The choice of a particular technique of compaction of concrete depends upon the following factors:

(i) the type of structural element,

(ii) the properties of the concrete mix, particularly its water-cement ratio,

(iii) the desired properties of the hardened concrete, i.e. strength, durability and watertightness etc., and

(iv) the duration of the production process and the rate of the output in the case of precast concrete products. The different methods are compared in Table 11.4.

11.6.1 Hand Rodding

Rodding is the process of ramming the concrete manually with a heavy flat-flaced tool in an effort to work it around the reinforcement, the embedded fixtures, and corners of the formwork. The rodding action is effective for a depth of concrete equal to five times the maximum size of aggregate and hence the depth of each layer has to be restricted to this value. The rod should penetrate to the full depth of the concrete layer and into underlying layer if it is still plastic to ensure proper bonding of the layers. The compaction should continue until the cement mortar spreads on the surface of the concrete. Fast rodding can be done by using rodding or tamping equipment. Fast rodding produces better compaction than hand rodding. The main disadvantage of rodding is that it produces large pressures on the formwork. However, such a system though better than no compaction, cannot assure a thoroughly dense and compacted concrete free of air pockets.

TABLE 11.4 Different methods of compaction

Method of compaction	Limiting characteristics of concrete		Typical applications
	Workability	Type of concrete	
Hand rodding	Mixes of all workabilities except very fluid and very plastic mixes	All grades including lightweight concrete	Flat elements like slab etc.
Mechanical vibration	All mixes except fluid and very plastic mixes	All grades of concrete	All elements
Centrifugation or spinning	Plastic mix	All grades of concrete; dense and rich mixes	Precast products having radial symmetry like poles and pipes
Other methods like high pressure and shock	All mixes	Only dense concrete	Precast elements

11.6.2 Mechanical Vibrations

Vibration is the commonly used method of compaction of concrete, which reduces the internal friction between the different particles of concrete by imparting oscillations to the particles and thus consolidates the concrete into a dense, and compact mass. The oscillations are in the form of simple harmonic motion. The mechanical vibrations can be imparted by means of vibrators which are operated with the help of an electric motor of diesel engine or pneumatic pressure.

The vibration, in general, is caused by the rotation of an eccentrically loaded shaft at high speed usually greater than 2800 rpm. The tendency at present is to use higher frequencies beyond 6000 rpm (up to 15000 rpm), such a vibration being termed as *high frequency vibration*. The lower frequencies cause oscillations mainly of coarse aggregate particles which transfer the oscillations to the other particles. On the other hand, the higher frequencies affect mainly the fine aggregate particles which in turn transfer the vibrations to the other particles. However, in both cases, the vibrations are communicated rapidly to the particles of concrete making it fluid and enabling it to flow around the reinforcement and enter into the corners. Any entrapped air is forced to the surface and the particles occupy a more stable position making the concrete considerably denser. The acceleration produced on the particles in the case of high frequency vibrations is of the order of $4g$ to $7g$, where g is the acceleration due to gravity. The amplitude of oscillation is very small, of the order of 0.5 mm. The kinetic energy imparted to the concrete to cause compaction is found to depend upon the square of the amplitude and the square of the frequency.

The optimum frequency of vibration of concrete depends on the size of the particles and on the mobility or stiffness of concrete. For the concrete mix containing relatively coarser fractions of aggregate, a lower frequency of vibration with greater amplitude, and for concrete containing finer fractions, a higher frequency with lower amplitude are necessary.

For all practical purposes, the vibration can be considered to be sufficient when the air bubbles cease to appear and sufficient mortar appears to close the surface interstices and facilitate easy finishing operations. The period of vibration required for a mix depends upon the workability of the mix. Plastic mixes need less time of vibration than harsh or dry mixes, since the latter need more compacting energy to form dense masses. Every mix has an optimum period of vibration depending upon the characteristics of the mix. This optimum period can be estimated by conducting trials with different periods of vibration to obtain compaction without segregation and then choosing the period which gives maximum strength of concrete cubes.

Choice of Vibrators

Since concrete contains particles of varying sizes, the most satisfactory compaction would perhaps be obtained by using vibrators with different speeds of vibration. Polyfrequency vibrators for compacting concrete of stiff consistency are being developed. The vibrators used in practice have frequency suitable for average particle size of concrete. Vibrators for compacting concrete are manufactured with frequencies of vibration from 2800 to 15000 rmp. The various types of vibrators used are described in the following subsections.

Immersion or needle vibrators Of the several types of vibrators, this is perhaps the most commonly used. It essentially consists of a steel tube (with one end closed and rounded) having an eccentric vibrating element inside it. This steel tube called poker is connected to an electric motor or a diesel engine through a flexible tube. They are available in sizes varying from 40 to 100 mm in diameter. The diameter of the poker is decided from the consideration of the spacing between the reinforcing bars in the form-work. The frequency of vibration varies up to 15000 rpm. However, a range between 3,000 to 6,000 rpm is suggested as a desirable minimum with an acceleration of $4g$ to $10g$. The normal radius of action of an immersion vibrator is

0.50 to 1.0 m. However, it would be preferable to immerse the vibrator into concrete at invervals of not more than 600 mm or 8 to 10 times the diameter of the poker. The period of vibration required may be of the order of 30 s to 2 min. The concrete should be placed in layers not more than 600 mm high. The vibrator can be placed vertically or at a small inclination of not more than 10° to the vertical to avoid flow of concrete due to vibration and consequent scope for segregation. The vibrator should be allowed to penetrate the concrete under its own weight during vibration. The vibrator should be removed while still running at a rate of 75 mm/s so that the hole left by the vibrator closes without any air being entrapped. The vibrator should be immersed through the entire depth of freshly placed concrete and into the layer below if this is still plastic or can be brought into plastic state (by revibration) to avoid the plane of weakness at the junction of the two layers.

Internal vibrators are comparatively more efficient since all energy is utilized to vibrate the concrete unlike other types of vibrators.

External or Shutter Vibrators

These vibrators called form vibrators are clamped rigidly to the form-work at the predetermined points so that both the form and concrete are vibrated. They consume more power for a given compaction effect than internal vibrators. These vibrators can compact up to 450 mm from the face but have to be moved from one place to another as concreting progresses. These vibrators operate at a frequency of 3,000 to 9,000 rpm at an acceleration of 4g.

If external vibrators are to be used, the shuttering must be stronger and more rigid than for other types of vibrators. The formwork should also be absolutely watertight. In case parallel forms are used for the casting of a structural element, the distance between parallel shutters should not be more than 750 mm. The use of an immersion vibrator along with the form vibrator can be considered for vibration of top layer concrete, if the spacing of the reinforcement allows the pocker. This will ensure more uniform compaction of concrete in the case of wide sections.

The external vibrators are more often used for the precasting of thin in-situ sections of such shape and thickness as cannot be compacted by internal vibrators.

Surface Vibrators

Surface vibrators are placed directly on the concrete mass. These are best suited for the compaction of shallow elements and should not be used when the depth of the concrete to be vibrated is more than 250 mm. For example these are used for compacting plain concrete or one-way-reinforced concrete floors, and road surfaces where immersion vibrators is impraticable. Surface vibrators can also be used as supplementary compacting equipment for vibrating the top layer of concrete when the concrete underneath is subjected to the action of immersion or form vibrators. Very dry mixes can be most effectively compacted with surface vibrators, since the vibration acts in the direction of gravity, thereby minimizing the tendency for segregation. Surface vibrators cause movement of finer material to the top and hence aid the finishing operation. However, the movement of a large amount of fine material in case of plastic mixes should be avoided. The surface vibrators commonly used are pan vibrators and vibrating screeds. The pan vibrator consists of a flat steel pan of approximate size of 400 mm × 600 mm on which an electric motor is mounted. The main application of this type of vibrator is in the compaction of small slabs, not exceeding 150 mm in thickness, and patching and repair work of pavement slabs. A vibrating screed on the other hand consists of a steel beam of 4 to 5 m length over which one or more vibrators are mounted. The operating frequency is about 4000 rpm at an acceleration of $6g$ to $9g$. The screeds are useful for compacting flat slabs or pavements whose depth is not more than 150 mm.

Vibrating Table

The vibrating table consists of a rigidly built steel platform mounted on flexible springs and is driven by an electric motor. The normal frequency of vibration is 4000 rpm at an acceleration of about $4g$ to $7g$. The springs are so designed that they cause resonance. The moulds are rigidly clamped on the platform to enable the system to vibrate in unison. Vibration is considered adequate when the concrete develops a smooth level surface.

Large vibrating tables are fitted with more than one vibrator. All vibrators should produce oscillations which are perfectly synchronized. The compaction is thorough since the vibrations are in the direction of gravity. The vibrating tables are very efficient in compacting stiff and harsh concrete mixes required for the manufacture of precast elements in the factories and test specimens in the laboratories.

11.6.3 Prolonged Vibration and Revibration

The vibration of concrete with low water-cement ratio can be continued beyond two minutes (the time required for the satisfactory compaction of concrete). Prolonged or overvibration has been found to increase the strength appreciably from 30 s to 3 min. and marginally after that for a vibration frequency of 5000 rpm. At higher frequencies of the order of 8000 rpm the strength is found to increase appreciably even after 12 min. The increase in strength due to prolonged vibration can be attributed to a decreased water-cement ratio of the concrete mass.

Generally, the concrete is vibrated immediately after placement to complete its consolidation before it has stiffened. However, in order to ensure a good bond between layers, the upper part of the underlying layer should be revibrated provided the layer can still regain the plastic state. If the concrete has not already set, the mass once again becomes plastic due to the revibration and any residual air is forced out. If the concrete is at the point of the initial set of cement, the revibration disrupts the setting mass slightly and causes reconsolidation of concrete with possible expulsion of free water. If the concrete has already set, it becomes so stiff it can not be revibrated.

Concrete can be successfully revibrated up to about four hours from the time of mixing. Revibration up to three hours after initial vibration is found to increase the 28-day compressive strength by as much as 25% and bond strength of plain reinforced bars by about 30 to 50 per cent. The bond strength at first slip is increased by almost 100 per cent compared to the unvibrated reinforced concrete. If retarders are used, the concrete can be revibrated up to 10 hours after placing. Revibration is preferred when watertightness is required. It can also be advantageously used for the manufacture of precast and pre-stressed elements.

The vibration technique is suitable only for properly graded or designed concretes. While using vibrators for compacting concrete mixes, the following general points should be kept in mind.

(i) Vibration should not be used as a means of spreading the concrete in forms, as this may result in the segregation of coarse aggregate.

(ii) The prolonged vibration of concrete mixes with a slump of more than 100 mm entails their being segregated causing smaller and lighter constituents of the mix to rise to the surface causing a layer of mortar or even laitance on the surface. The resulting concrete may have honeycoming at the bottom and a dusty surface at the top. This causes planes of weakness is succeeding layers. In the case of a single layer, the surface lacks resistance to abrasion.

(iii) Vibration may reduce the entrained air in the air-entrained concrete to about 50 per cent. Hence air-entrainment should be doubled if the mix is to be vibrated. Undervibration should not be resorted to for fear of expelling the entrained air.

(iv) When concrete is compacted by internal vibrators, the thickness of the layers placed should not exceed the depth of the operating part of the vibrator by more than 25 per cent.

(v) The period for which a vibrator is kept in one position should be such as to ensure adequate compaction taking into account the stiffness of the mix and the thickness of the compacted layer. The adequacy of the compaction is indicated by no further settlement of concrete, appearance of slurry on surface, and disapperance of rising bubbles.

(vi) On completion of compaction at one place, the vibrator is transferred to another place. The distance between successive positions of the vibrator must not exceed one-and-a-half times its radius of operation. When an internal vibrator is transferred, it must be removed slowly by switching off the meter.

(vii) The internal vibrator must be set up at a distance not exceeding 50 to 100 mm from the wall of the form. The reinforcing bars must not be touched by the operating vibrator as the bond between reinforcement and concrete may be disturbed by vibration.

(viii) The compaction of concrete with surface vibrator is carried out in straight continuous strokes with a 100 to 200 mm overlapping on the previously compacted area. The vibration time at a position may be approximately 30 to 60s depending upon the mobility of concrete and as verified by external indications discussed in (v).

(ix) The surface vibrator should be withdrawn by an upward jerk and not pulled slowly through the concrete.

(x) The external vibrator should be firmly clamped to the form as otherwise, its efficiency is reduced drastically. The time of vibration using external vibrators varies from 60 to 90s.

(xi) The vibrators must be switched off at regular intervals to allow cooling of the meters.

(xii) The total number of vibrators on the site should be about 30 to 50 per cent more than the calculated number.

In oversanded concrete mixes a phenomenon called *rotational instability* occurs during compaction using vibration technique. The coarse aggregate particles coated with cement-mortar form nearly ball-like particles. These particles, during compaction, do not consolidate and settle in a dense mass but continue to rotate about a horizontal axis passing through them. This phenomenon is found to occur in cases where low-frequency (below 6000 rpm) and high-amplitude (above 0.13 mm) vibrations are employed as a means of compaction. Under this condition the air is sucked into concrete and entrapped, causing reduction in strength. This phenomenon is found to occur particularly when vibrating low slump concrete in narrow sections on a vibrating table.

11.6.4 Centrifugation or Spinning

The method is used in the production of elements which are circular in cross-section, such as concrete pipes, concrete lamp posts, etc. It comprises feeding the concrete into the horizontal mould spinning at a low speed. After the predetermined amount of concrete is fed into mould, the spinning speed is increased to a high value. The water is forced out of the mix which flows out of the mould. At the end of the spinning process, the speed is slowly reduced and dry cement sprinkled in small quantity such that any free water on the surface does not increase the local *water-cement ratio*. A round rod is held against the two end-rings to finish the surface.

The initial water-cement ratio for effective compaction without segregation should be between 0.35 and 0.40. The final water-cement ratio after spinning reduces to about 0.30. High speed of rotation and prolonged centrifugation may cause segregation of concrete. The coarser particles, due to their higher mass, consolidate on the outer face of the product. The optimum speed and duration of spinning depend upon the diameter of the pipe and quality of mix. The segregation can be minimized by adopting a continuous grading curve for the aggregate. The centrifugation results in a watertight product and hence is used in the manufacture of both pressure pipes for water-supply and nonpressure pipes for sewerage disposal and storm water drains.

11.6.5 Vibropressing

The method comprises of applying external pressure from the top and vibration from below the mould. The vibration tables can be used for this purpose. The excess water added during the mixing is forced out due to large pressure. The *water-cement ratio* at the end of vibropressing can be reduced to a value as low as 0.30. The product obtained by this process is of extremely good quality and durability. The technique has been successfully used for mass manufacturing of concrete kerbs etc.

11.6.6 Other Methods

Jolting

This method of compaction consists of subjecting the mould containing the concrete to a series of jolting actions at a frequency of 100 to 150 jolts per minute. This jolting is in effect a vibrating action of a low frequency and high amplitude. The cams used for the purpose raise the mould by about 12 mm and then allow it to fall to its original position under gravity. The method is quite effective for dry mixes and is used for the manufacture of precast concrete products.

Rolling

It is a continuous pressing operation for compacting the soft and plastic concrete obtained by previbration. The previbrated concrete is fed continuously in between rubber rollers employing pressures up to 50 atmospheres which force out the excess water in the concrete. The continuity of production makes it best suited for automated factory production of very thin concrete products like concrete tiles.

11.7 CURING OF CONCRETE

The physical properties of concrete depend to a large extent on the extent of hydration of cement and the resultant *microstructure* of the hydrated cement. Upon coming in contact with water, the hydration of cement proceeds both inwards and outward in the sense that the hydration products get deposited on the outer periphery of cement grains, and the nucleus of unhydrated cement inside gets gradually diminished in volume. At any stage of hydration the cement paste consists of the product of hydration (called gel because of its large area), the remnant of unreacted cement, $Ca(OH)_2$ and water. The product of hydration forms a random three-dimensional network gradually filling the space originally occupied by the water. Accordingly, the hardened cement paste has a porous structure, the pore sizes varying from very small (4×10^{-10}m) to very large and are called *gel* pores and *capillary* pores. As the hydration proceeds, the deposit of hydration products on the original cement grains makes the diffusion of water to the unhydrated nucleus more and more difficult, and so the rate of hydration decreases with time. Therefore, the development of the strength of concrete, which starts immediately after setting is completed, continues for an indefinite period, though at a rate gradually diminishing with time. Eighty to eighty five per cent of the eventual strength is attained in the first 28 days and hence this 28-day strength is considered to be the criterion for the design and is called *characteristic strength*.

As mentioned above, the hydration of cement can take place only when the *capillary pores* remain saturated. Moreover, additional water available from on outside source is needed to fill the *gel pores* which will otherwise make the capillary empty. Thus, for complete and proper strength development, the loss of water in concrete from evaporation should be prevented, and the water consumed in hydration should be replenished. Thus the concrete continues gaining strength with time provided sufficient moisture is available for the hydration of cement which can be assured only by a creation of favourable conditions of temperature and humidity. This process of creation of an environment during a relatively short period immediately after the placing and compaction of the concrete, favourable to the setting and the hardening of concrete, is termed *curing*. The desirable conditions are; a suitable temperature, as it governs the rate at which the chemical reactions involving setting and hardening take place; the provision of ample moisture or the prevention of loss of moisture; and the avoidance of premature stressing or disturbance. All the care taken in the selection of materials, mixing, placing and compaction, etc. will be brought to nought if the curing is neglected. The curing increases compressive strength, improves durability, impermeability and abrasion resistance.

11.7.1 Curing Conditions

Proper curing practice is one of the important steps in making high quality concrete. A good mix design with low water-cement ratio alone cannot ensure good concrete. The favourable conditions to be set up at early hardening periods for best results are:

(i) adequate moisture within concrete to ensure sufficient water for continuing hydration process, and

(ii) warm temperature to help the chemical reaction.

In addition, the length of curing is also important. The first three days are most critical in the life of portland cement concrete. In this period the hardening concrete is susceptible to permanent damage. On an average, the one-year strength of continuously moist cured concrete is 40 per cent higher than that of 28-days moist cured concrete, while no moist-curing can lower the strength to about 40 per cent. Moist curing for the first 7 to 14 days may result in a compressive strength of 70 to 85 per cent of that of 28 day moist-curing as shown in Fig. 11.2.

It has been observed that the hydration takes place only when the vapour pressure in the capillaries is more than 80% of the saturation pressure. The rate of hydration

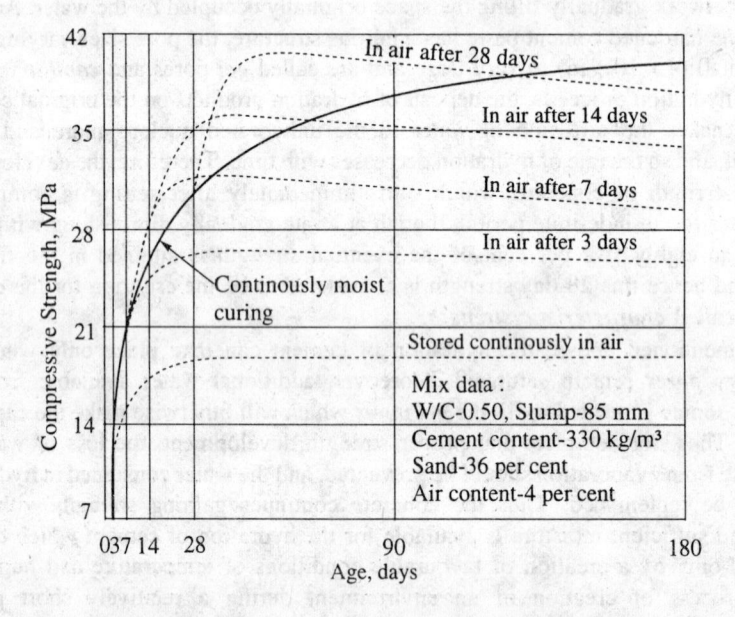

Fig. 11.2 Strength of concrete dried in air after preliminary moist curing

is maximum at the saturation pressure and is minimum at three times the saturation pressure. The vapour pressure in capillaries reduces with the passage of time resulting in a reduction of rate of hydration and hence of development of strength. It is necessary to prevent even a small loss of water during the process of hardening. If the concrete is left in air, *i.e.* without any method of curing being adopted, there is a continuous loss of moisture due to evaporation and self desiccation. The rate of evaporation depends upon the temperature and the relative humidity of the surrounding air and on the velocity of wind. An air-cured concrete develops considerably less strength compared to the moist-cured concrete as is seen in Fig. 11.2. The rate of development of strength with curing period is given in Fig. 11.3.

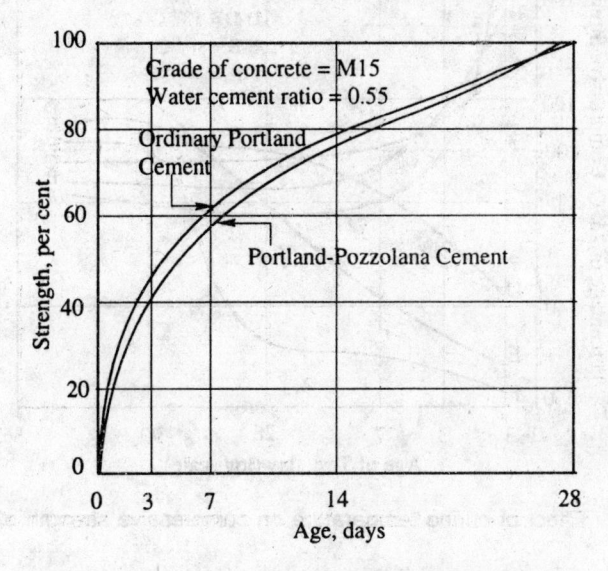

Fig. 11.3 Development of strength with curing period

The rate of development of strength not only depends on the period of curing but also on the temperature during the period of curing. The influence of temperature on the strength is shown in Fig. 11.4. It can be seen that the optimum temperature during the curing period is 15°C to 38°C. The ambient temperature in most parts of India provide warmth required for satisfactory hydration.

11.7.2 Maturity of Concrete

Since the strength of concrete depends on both the period of curing (*i.e.*, age) and temperature during curing, the strength can be visulized as a function of period and temperature of curing. The product (period × temperature) is called the *maturity of concrete*. Here the temperature is reckoned from —10°C which is a reasonable value of the lowest temperature at which an appreciable increase in strength can take place and the period in hours or days. The maturity of concrete is measured in °C hours or °C days. The strength of concrete is found to increase linearly with its maturity as shown in Fig. 11.5. The strength of concrete at any maturity can be expressed as the percentage of strength for the maturity of concrete cured at 18°C for 28 days, *i.e.* $(18 + 10) \times (28 \times 24) = 18800$°C hours. The requisite maturity factor recommended for minimum curing for OPC is 4200°C hours (preferably 6000°C hours). In the case of high-early-strength cement, a maturity factor of 2400°C hours is recommended.

226   Concrete Technology

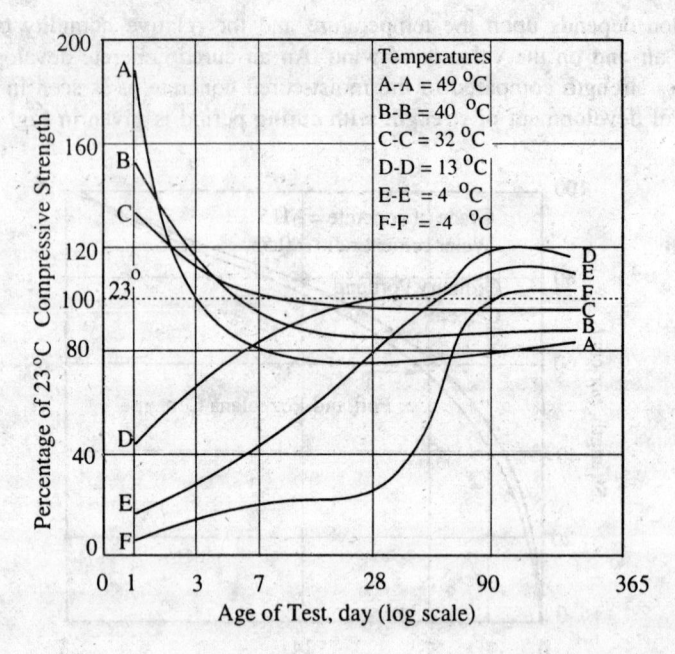

Fig. 11.4 Effect of curing temperature on compressive strength of concrete

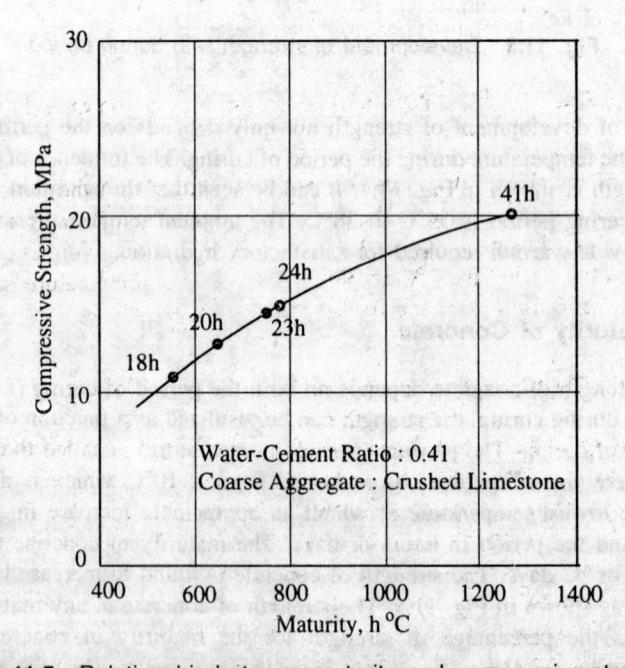

Fig. 11.5 Relationship between maturity and compressive strength

The increase in strength with increased curing temperature is due to the speeding up of the chemical reactions of hydration. This increase affects only the early strengths without affecting the ultimate strengths. Hence, curing of concrete and its gain of strength can be speeded up by raising the temperature of curing, thereby reducing the curing period. This type of curing called *accelerated curing* has many applications in the manufacture of precast concrete products.

11.7.3 Curing Periods

To develop design strength, the concrete has to be cured for up to 28 days. As the rate of hydration, and hence the rate of development of strength, reduces with time, it is not worthwhile to cure for the full period of 28 days.

IS: 456–1978 stipulates a minimum of 7-day moist-curing, while IS: 7861 (Part I)–1975 stipulates a minimum of 10 days under hot weather conditions. High-early-strength cements can be cured for half the periods suggested for OPC. For pozzolana or blast furanace slag cements the curing periods should be increased.

There are many opinions on the length of curing period. Periods varying from 13 to 30 days are specified for highway pavements. There cannot be a definite mandate on this matter as there are too many variables involved, such as the type of cement, ambient temperature, nature of the product, method of curing adopted, etc. Generally, increased curing periods are desirable for high-quality concrete products, concrete floors, roads and airfield pavements. The variation of compressive strength with the curing period is given in Fig. 11.6.

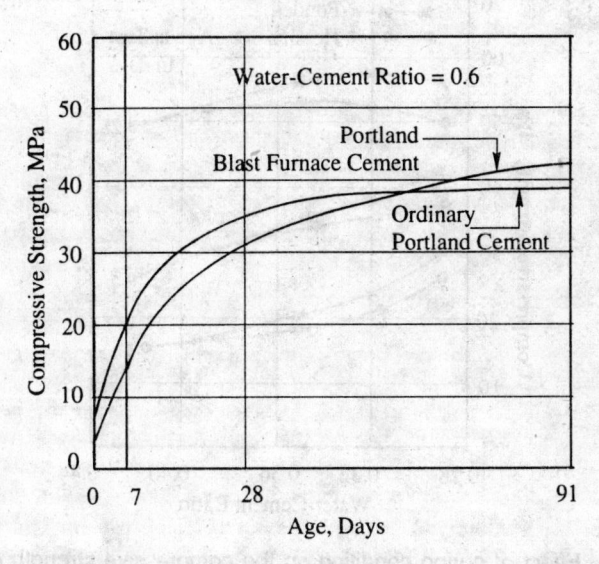

Fig. 11.6 Variation of compressive strength with curing period

11.7.4 Methods of Curing Concrete

There are various methods available for curing. The actual procedures used vary widely depending on the conditions on the site, and on the size, shape and position of the member. The methods can be broadly classified as:

(i) The methods which replenish partly the loss of water by interposing a source of water, or prevent the evaporation, viz.

(a) Ponding of Water over the Concrete Surface after it has set

This is the most common method of curing the concrete slab or pavements and consists of storing the water to a depth of 50 mm on the surface by constructing small puddle clay bunds all around. Ponding may promote efflorescence by leaching.

(b) Covering the Concrete with Wet Straw or Damp Earth

In this method the damp earth or sand in layers of 50 mm height are spread over the surface of concrete pavements. The material is kept moist by periodical sprinkling of water.

(c) Covering the Concrete with Wet Burlap

The concrete is converted with burlap (coarse jute or hemp) as soon as possible after placing, and the material is kept continuously moist for the curing period. The covering material can be used a number of times and, therefore, tends to be economical. The effectiveness of the method as compared with the ponding is shown in Fig. 11.7.

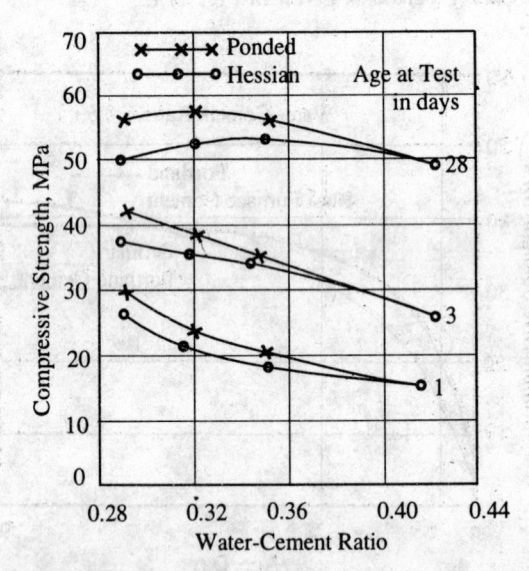

Fig. 11.7 Effect of curing condition on the compressive strength of concrete

(d) Sprinkling of Water

This is a useful method for curing vertical or inclined surfaces of concrete wherein the earlier methods cannot be adopted. The method is not very effective as it is difficult to ensure that all the parts of concrete be moist all the time. The spraying can be done in fine streams through nozzles fixed to a pipe spaced at set intervals.

Flogging is done in the same way except that the flogging nozzles produce a mist-like effect, whereas spraying nozzles shed out fine sprays.

(ii) The methods preventing or minimizing the loss of water by interposing an impermeable medium between the concrete and the surrounding environment are as follows.

(a) Covering the Surface with Waterproof Paper

Waterproof paper prevents loss of water in concrete and protects the surface from damage. The method is satisfactory for concrete slabs and pavements. A good quality paper can be often reused. The paper is usually made of two sheets struck together by rubber latex composition.

Plastic sheeting is a comparatively recent innovation as a protective cover for curing concrete. Being light and flexible, it can be used for all kinds of jobs, effectively covering even the most complex shapes. Several types of sheets, which are guaranteed to give excellent results consistent with economy and can be used over and over again, are available. Most plastic sheetings used in the concrete industry are milky or white in appearance, and this helps keep the concrete temperature at a reasonable level. Plastic sheeting can be welded at the site instead of resorting to large overlaps and made airtight to prevent moisture evaporation from concrete.

(b) Leaving the Shuttering or Formwork on

The thick watertight *formwork* also prevents the loss of moisture in concrete and helps in curing the sides and the base of the concrete.

(c) Membrane Curing of the Concrete

The process of applying a membrane forming compound on concrete surface is termed membrane curing. Often, the term membrane is used not only to refer to liquid membranes but also to a solid sheeting used to cover the concrete surface. The *curing membrane* serves as a physical barrier to prevent loss of moisture from the concrete to be cured. A curing liquid membrane should dry within 3 to 4 hours to form a continuous coherent adhesive film free from pinholes and have no deleterious effect on concrete. Curing with a good membrane for 28 days would give strengths equivalent to two weeks moist-curing. Membrane curing may not assure full hydration as in moist-curing but is adequate and particularly suitable for concrete members in contact with soil.

The different sealing compounds used are:

(i) Bituminous and asphalic emulsion or cutbacks,

(ii) Rubber latex emulsions,

(iii) Emulsions of resins, varnishes, waxes, drying oils and water-repellant substances, and

(iv) Emulsions of paraffin or boiled linseed oil in water with stabilizer.

Sealing compounds are used only after testing for their efficiency. For effective sealing of the surface two uniformly applied coats of the compound may be necessary. These are generally applied to the interior surfaces not directly exposed to the sun.

The quantity of emulsions required per square metre is about 0.1 gallon. Application of membrane should be started immediately after the water sheet has disappeared from the concrete. The solid membranes have been found to be superior to the liquid membranes. Membrane curing should not be adopted if the water-cement ratio is less than 0.5, lest the phenomenon of self desiccation should weaken the concrete by progressively reducing the space available for hydrated products.

(d) Chemical Curing

Chemical curing is accomplished by spraying the sodium silicate (water glass) solution. About 500 g of sodium silicate mixed with water can cover $1m^2$ of surface and forms a hard and insoluble calcium silicate film. It actually acts as a case hardener and curing agent. The application of sodium silicate results in a thin varnish like film which also fill pores and surface voids, thus sealing the surface and preventing the evaporation of water.

(iii) Methods involving the application of artificial heat while the concrete is maintained in a moist condition are used in plant curing where the curing of concrete is accelerated by raising its temperature. The accelerated process of curing has many advantages in the manufacture of precast concrete products since; (a) the moulds can be reused within a shorter time; (b) due to reduced period of curing the production is increased and the cost reduced, and (c) storage space in the factory is reduced.

The temperature can be raised in practice by:
(a) Placing the concrete in steam,
(b) Placing the concrete in the hot water, and
(c) Passing an electric current through the concrete.

11.7.5 Steam Curing

For concrete mixes with *water-cement ratio* ranging from 0.3 to 0.7, the increased rate of strength development can be achieved by resorting to steam curing. The mixes with low water-cement ratio respond more favourably to steam curing than mixes with higher water-cement ratio.

In *steam curing*, the heating of the concrete products is caused by steam either at low pressure or high pressure. The method ensures even heating of products all over, even if the space between the stacked precast concrete products is very small.

A number of considerations govern the choice of steam curing cycle, namely, the precuring period, the rate of increase and decrease of temperature, and the level and time of constant temperature. An early rise in temperature at the time of setting of concrete may be detrimental to concrete because the green concrete may be too weak to resist the air pressure set up in the pores by the increased temperature. Too high a rate of increase or decrease in temperature introduces thermal shocks and the rates should generally not exceed 10 to 20°C per hour. The higher the *water-cement ratio* of concrete, the more adverse is the effect of an early rise in temperature. Therefore, to meet the requirement of compressive strength of concrete, the temperature and/or time required for curing can be reduced by having a lower *water-cement ratio*. While in identical time cycle, the higher the maximum temperature, the greater is the compressive strength. The advantages of curing above 70°C are negated by dilational tendencies due to the expansion of concrete. All the above mentioned factors lead to the conclusion that for a concrete of specified composition and curing period, there is an optimum curing temperature which will result in maximum compressive strength at the end of the curing cycle.

It has been suggested that the steam curing of concrete should be followed by water curing for a period of at least 7 days. This supplementary wet curing is found to increase the later-age strength of steam-cured concrete by 20 to 35 per cent. In the case of concretes with high water-cement ratios, a rapid rate of temperature rise during steam curing may result in lesser 28-day strength than that of normally cured concrete, even though the initial rate of development of strength is higher than for normal curing. The rapid temperature rise may also result in the reduction of bond strength. On the other hand, if a slow rate of temperature rise is adopted, the 28-day strength will almost be equal to that of normally cured concrete and there is no deleterious effect on bond.

In most cases, steam curing is employed only for achieving 50 to 70 per cent of specified strength in a short period instead of full treatment for 2 to 3 days required to obtain specified strength. This would result in the economy in the reuse of moulds and equipment by achieving *stripping-strength* which is normally about 50 per cent of the specified strength. The *stripping strength* will be sufficient to take care of any impacts which may be produced during their demoulding and transportation to the stackyard. The strength which the concrete should develop to enable the despatch and transport of the products from the storeyard should be about 75 per cent of the specified strength. This strength is termed *delivery strength*.

(i) Low Pressure Team Curing

Steam curing at atmospheric pressure can be continuous or intermittent. In the continuous process, the products move on conveyor belts from one end of a long curing chamber to the other end, the length of the chamber and the speed of movement of the conveyors being so designed that the products remain in the curing chamber for the required time. On the other hand, in the intermittent process, the concrete products are stacked in the steam chamber and the steam is allowed into the closed chamber. The steam curing-cycle can be divided into three stages: (i) the heating stage (ii) steam treatment, and (iii) cooling stage.

In the normal steam curing procedure, it is advisable to start the steam curing a few hours after casting. A delay of two to six hours, called the *presetting* or *presteaming period*, depending upon the temperature of curing, is usual. The presteaming period helps to achieve a 15 to 30 per cent higher 24 hour strength than that obtained when steam curing is resorted immediately. The rate of initial temperature rise after presteaming period is of the order of 10 to 20°C per hour and the maximum curing temperature is limited to 85 to 90°C. A temperature higher than this does not produce any increase in the strength of concrete and in fact, as discussed earlier, a temperature of 70°C may be sufficient. For a particular product, the maximum desired temperature is raised at a moderate rate and then the steam is cut off, and the product allowed to soak in the residual heat and moisture of the curing chamber. The product after steam-curing and cooling off to 30°C, should be kept in a warm room at a temperature of about 25°C before being exposed to the outside atmosphere of lower temperature. By adopting a proper steam-cycle, more than 70 per cent of the 28-day compressive strength of concrete can be obtained in about 16 to 24 hours. The steam curing cycle depends upon:

(i) The type of cement, (ii) the aimed stripping and delivery strengths, and (iii) the accelerator.

A typical steam curing cycle is given below:

Presteaming period	3 hours
Temperature rise period	4 hours
Period of maximum temperature	4 hours
Cooling off period	5 hours

(ii) High Pressure Steam Curing

In the case of normal steam-curing at atmospheric pressure, the ultimate strength of concrete may be adversely affected if the temperature is raised rapidly. This difficulty can be overcome by employing the steam at a pressure of 8 atmospheres. The process is termed *high-pressure steam-curing*. High-pressure steam curing is done in the cylindrical steel chambers called autoclaves. The concrete products, after a suitable presteaming period, are wheeled on racks into the autoclaves. The steam is let in gradually until the prescribed pressure or temperature (generally about 1 MPa or 185°C) is reached. This heating stage should be completed and the prescribed pressure reached in about three hours. The increase in temperature allowed is up to 50°C in the first hour, up to 100°C in second hour and up to 185°C in the third hour. The period of treatment under full pressure depends upon the strength requirements. This period is 7 to 10 hours for hollow block products and 8 to 10 hours for slab or beam elements, the period increasing with the thickness of concrete. The steam is cut off and the pressure released after the completion of this stage and the products are left in the autoclaves for two hours for cooling off gradually.

High-pressure steam curing is usually applied to precast products when any of the following characteristics is desired.

(a) *High early strength* With high-pressure steam curing, the compressive strength at 24 hours is at least equal to that of 28-day normally cured products.

(b) *High durability* High-pressure steam curing results in an increased resistance to sulphate action and other forms of chemical attack, and to *freezing* and *thawing*.

During the hydration of cement at higher temperatures, the calcium hydroxide released as the result of primary reaction, reacts with finely divided silica, which is present in the coarse and fine aggregates, forming a strong and fairly unsoluble compound. This results in higher concrete strengths. *Leaching* and *efflorescence* are minimized due to reduction in free calcium hydroxide content. The hydrating dicalcium silicates and tricalcium alluminates react together at high temperatures to form sulphate-resisting compounds. Hence *autoclaved products* show higher resistance to sulphate attack. The *initial drying shrinkage* and *moisture movements* are also considerably reduced. On the debit side, *high-pressure steam-curing* reduces the bond strength to 50 per cent of that obtained with normally cured concrete. Hence steam curing of reinforced concrete members is not recommended.

11.7.6 Curing of Concrete by Infrared Radiation

The curing of concrete by *infrared radiation* has been used in Russia. It is claimed that a much more rapid gain of strength can be obtained than even with steam curing. The rapid initial rise of temperature does not result in a decrease in the ultimate strength as it does in the case of steam-curing. The system is described as particularly applicable to the manufacture of hollow concrete products in which case the heaters are placed in the hollow spaces of the product. The normal operative temperature is 90°C.

11.7.7 Electrical Curing of Concrete

Concrete products can be cured by passing alternating current of low voltage and high amperage through electrodes in the form of plates covering the entire area of two opposite faces of concrete. The potential difference generally adopted is between 30 and 60 V. Evaporation is prevented by using an impermeable rubber membrane on the top surface of the concrete.

Initially up to 3 hours, the resistance of concrete to flow of current decreases due to rise in temperature. There is rise in resistance afterwards, due to decrease in the quantity of free water available in the concrete due to hydration and evaporation. This period of rise is temperature should be about 12 hours. The duration of electrical-curing should be about 48 hours at the temperature of 50°C or 36 hours at the temperature of 70°C. The concrete products are cooled gradually in heat insulated chambers for a minimum period of 24 hours. By electrical-curing concrete can attain the normal 28-day strength in a period of 3 days. The technique is expensive and is not used in India.

11.7.8 Effects of Delayed Curing

The concrete specimens which were placed in laboratory air for varying periods after casting before being moist-cured, have indicated that the strength at 7 to 28 days decreases progressively as the period of air-curing is increased. An exposure for 3 days to air at a temperature of 23°C and having a *relative humidity* of approximately 60 per cent before being moist-cured at 23°C has been found to reduce the 7 day strength by 12 per cent and the 28 day strength by about 10 per cent. The specimens left in air at 23°C for the entire curing period have shown a reduction of 25 per cent in the strength at 7 and 28 days as compared with standard *moist-curing*. The reduction under field conditions would probably have been greater. Similar adverse curing causes greater relative reduction in strength when portland blast furance slag cement and the cements blended with flyash are used.

11.8 FORMWORK

Though formwork generally forms a part of concrete construction practice, but as it influences the performance of hardened concrete appreciably, its salient features are described in brief in the following sections.

The formwork or shuttering may be defined as the moulds of timber or some other material into which the freshly mixed concrete is poured at the site and which hold the concrete till it sets. The formwork includes the total system of support of freshly placed concrete, *i.e.* form lining and sheathing plus all necessary supporting members, bracings, hardware and fasteners.

Concrete construction practices directly affect formwork requirements. It is more than simply making forms of right size. A good formwork should be strong, stiff, smooth and leakproof. As the cost of formwork may be of the order of 20 per cent of the cost of project, it is essential that the forms be properly designed and detailed to effect economy without sacrificing strength and efficiency. It must be realized that smoothness of the external surface is not the main objective.

11.8.1 Requirements of Formwork

(i) *Quality* The formwork is designed and built accurately so that the desired shape, size, position and finish of cast concrete is obtained, and thus: (a) All lines in the formwork should be true and the surface plane so that the cost of finishing the surface of concrete on removal of shuttering is the least, (b) The formwork should be leakproof.

(ii) *Safety* The formwork is built substantially so that it is strong enough to support the dead and live loads during construction without collapse or danger to workmen or the structure. The joints in the formwork should be rigid to minimize the bulging, twisting or sagging due to dead and live loads. Excessive deformations may disfigure the surface of the concrete.

(iii) *Economy* The formwork should be built efficiently to save time and money for the contractor and owner alike. After the concrete has set, the formwork should be easily strippable without damage so that it can be used repeatedly.

11.8.2 Formwork Planning

The formwork is planned in such a way that it becomes an integral part of the total job plan. The above objectives are usually emphasized in the planning, *i.e.* the planning for maximum reuse, economical form construction, efficient setting and stripping practices, and safety from all causes of formwork failure.

Generally, the butted-and-cleated types of joints are preferable in the construction of formwork. All the formwork should be so designed and constructed that it can easily be stripped in the desired order after the setting of concrete, and no piece of formwork gets keyed into the mass of concrete. The shuttering must be chamfered at the junctions to facilitate its free and easy withdrawal. If nails are used, they should be driven until they are in the concrete surface and their heads should be slightly projected outside for easy removal. The nails should preferably be driven at an angle both ways. To prevent the concrete from sticking to the forms, the interior surface of forms should be coated with a thin layer of mineral oil, soft soap or be white-washed. Sometimes linings of oil paper or canvas, etc. are also employed. The formwork should be properly inspected by the engineer-in-charge. Only after making sure that the formwork is properly made should the concreting be allowed. During concreting the formwork should be continuously observed for bulges and other signs of failure. Small cleats, wedges and bolts, etc. should be put in separate boxes and not be thrown indiscriminately.

11.8.3 Types of Formwork

The material used in the formwork largely depends upon the availability and cost. Usually, the timber scantlings consisting of softwood planks and joists are very suitable. However, for big projects where the forms are to be used repeatedly, steel formworks are commonly used.

Timber Formwork

The timber used for formwork should be cheap, easily available and easy to work manually and on machines. A good timber for formwork should be light for easy handling and lifting, stiff for not giving excessive deflections, usually free from knots, knot holes, bad flaws, etc. which may cause failure. Sawn timber is preferable for rough surfaces to be rendered afterwards. Planned timber gives smooth surface. The timber used should not be green, as it would then become dry and shrink, and at the same time not too dry as it would absorb water from concrete. Partially seasoned timber is the most suitable. In case dry or green timber is used, suitable allowances for bulging and shrinkage should be made in preparing the surface. The face of timber that would be in contact with concrete should be properly dressed and its sides should be truly plane for providing water-tight joints with adjoining pieces. To take care of any sags in beams, the forms are given a camber of 1 : 500 along the length.

Timber sheathing can either be square-edged type which is easily strippable and sturdy but liable to leakage, or tongued-and-grooved which does not allow leakage. The latter type, where stripping and cleaning take more time is suited for high class work. For timber formwork, due to its temporary nature, higher stress may be allowed in design than that used in permanent timber works. The soft timber used for the formwork can be assumed to have a linear stress-strain relation with the modulus of elasticity as 9.8×10^3 MPa. The allowable stresses for *Chir* of density of 575 kg/m^3 are as follows:

Bending stress	8.4 MPa
Compressive stress parallel to grains	6.4 MPa
Compressive stress across the grains	2.6 MPa
Shear stress parallel to grains	0.92 MPa

The stresses are based on the assumption that the timber will remain dry. In case it is subjected to alternate wetting and drying, the stresses should be reduced to 0.85 times the value given above and, if continuously wet, to 0.65 times these values.

In practice it would be economical to standardize the size of timber used in formwork so that their repeated use is possible. This would necessarily entail proper planning, and great care is to be exercised by the designer in adjusting the parameters in such a way that the standard scantling can be used. Sometimes it would result in a bit of over-expenditure on concrete but, in the long run, especially in large projects, the saving in formwork will offset this loss. It is better to use clamps and screws, rather than nails, in the formwork to facilitate its stripping and reuse.

Plywood Formwork

Plywood sheets bound with synthetic resin adhesive are being widely used now a days. The thickness of ply varies from 3 to 18 mm. Sizes less than 6 mm thick are used for lining the timber formwork to get neat and smooth surface finish and as a formwork for curved surfaces. The common sizes are 1200 × 1200 mm to 3000 × 3000 mm. The main advantage is that large panel surfaces are available. The fixing of forms is rapid and economical,. It does not warp, swell and shrink during the setting of concrete. Moreover, it has high impact resistance.

Steel Formwork

Steel formworks are commonly employed for big projects where the forms are to be repeatedly used. The steel forms can be easily fabricated and do not require many adjustments as the units are standardized. They give smooth surfaces needing very little finishing. These prove to be economical and are best suited for circular columns and flat slab construction. Joists can be used from wall to wall to support the steel beams used for stiffening the plates. Square steel plates of 500 mm size are generally used. Light steel sheet panels of 500 mm size and stiffened with angles are also available.

11.8.4 Design Loads on the Formwork

The formwork used in the construction of roofs and floors has to carry its own weight, the weight of wet concrete, the live load due to labour, and the impact owing to the pouring of concrete, etc. The surfaces of the formwork should be so dressed that after the deflection due to concrete weight etc. the surface takes the shape desired by the designer. In the design of formwork for columns or walls, the hydrostatic pressure of concrete should be taken into account. This pressure depends upon the water content in the concrete, rate of pouring and the temperature. The hydrostatic pressure of concrete increases with the increase in water content, rate of pouring, and with the reduction in the size of the aggregate and temperature. It is possible to adjust the rate of pouring with the rate of setting of concrete in large and tall structures, so that the formwork can continually move upwards. Therefore, the movable forms need not be very high. A similar procedure can be adopted for columns.

The lateral pressure decreases rapidly after the initial set of concrete, and, therefore, only the height of concrete poured in the preceding half to three-fourths of an hour is considered for calculating the lateral pressure. It may be calculated by considering the concrete as a liquid with a density of 2300 to 1200 kg/m^3 for heights of concrete from 1.5 to 6 m, respectively. Due to large number of variables involved this is only a rough estimate. The experience alone should determine the size of various parts.

11.8.5 Stripping of Form

The removal of forms after the concrete has set is termed *stripping of forms*. The stripping or striking of forms should proceed in a definite order. The formwork should be so designed and constructed as to allow them to be stripped in the desired order. The period up to which the forms must be left in place before they are stripped is called *stripping time*. The factors affecting the stripping time are the position of the forms, the loads coming on the elements immediately after stripping, temperature of the atmosphere, the subsequent loads coming on the element, etc.

Using ordinary portland cement with temperatures above 20°C, the stripping times normally required are given in Table 11.5.

TABLE 11.5 Stripping time for different conditions

Element and supporting conditions	Stripping time, Days
Walls, columns, vertical sides of beams	1 to 2
Slabs with props left in position	3
Beam soffits with props left in position	7
Slabs: removal of props	
(a) Span up to 4.5 m	7
(b) Span over 4.5 m	14
Beam and arches: removal of props (supports)	
(a) Span up to 6 m	14
(b) Span over 6 m	21

For *rapid hardening portland cement*, the *stripping period* can be reduced to three-sevenths of that given in Table 11.5 except for the vertical sides of slabs, beams and columns where the forms are to be retained for 24 hours.

12

Concreting under Extreme Environmental Conditions

12.1 INTRODUCTION

Whenever the concrete is to be placed in extreme weather conditions or underwater, its performance is adversely affected unless appropriate measures are taken to control it. Extreme weather conditions include situations where environmental temperatures during concreting and subsequent curing periods are markedly different from those in normal conditions, *i.e.* either the temperature is too high or too low. The properties and performance of concrete are affected under these situations unless appropriate precautions are taken. In general, an increase in temperature accelerates the rate of *hydration* and therefore, leads to accelerated development of strength. The accelerated growth of hydrates under higher temperature may result in a less uniform microstructure of gel than could be expected were the reactions to proceed at the normal rate. On the other hand, a decrease in temperature retards the rate of hydration and hence of strength development, but the *microstructure* of the *gel* formed is perhaps more orderly and compact.

The situation may become further aggravated by decrease of *humidity* in the atmosphere, increase of wind or a combination of these. This may result in a rapid loss of water due to evaporation which may affect the workability of fresh concrete and cause *plastic shrinkage* and *cracking* that accompany the rapid drying. The subsequent cooling due to evaporation may introduce tensile stresses.

12.2 CONCRETING IN HOT WEATHER

Any operation of concreting done at atmospheric temperature above 40°C or where the temperature of concrete at the time of placement is expected to be beyond 40°C may be categorized as *hot weather concreting*. Concrete is not recommended to be placed at a temperature above 40°C without proper precautions as specified in IS: 7861 (Part I)-1975. The climatic factors affecting concrete in hot weather are a high *ambient temperature* and reduced *relative humidity*, the effects of which may be more pronounced with the increase in wind velocity. The effects of hot weather may be summarized as follows:

(i) Accelerated Setting

A higher temperature results in a more rapid *hydration* leading to accelerated setting, thus reducing the handling time of concrete and also lowering the strength of hardened concrete. The *workability* of concrete decreases and hence the water demand increases with the increase in the temperature of concrete. The addition of water without proper adjustments in mix proportions adversely affects the ultimate quality of concrete. It has been reported that an approximately 25 mm decrease in slump has resulted from 11°C increase in concrete temperature.

(ii) Reduction in Strength

Concrete produced and cured at elevated temperature generally develops higher early strength than normally produced concrete, but the eventual strengths are lower. Regarding the influence of simultaneous reduction in the *relative humidity*, it is seen that specimens moulded and cured in air at 23°C and 60 per cent relative humidity, and at 38°C and 25 per cent relative humidity attained strengths of only 73 and 62 per cent, respectively, in comparison with the specimens which are moist-cured at 23°C for 28 days. High temperature results in greater *evaporation* and hence necessitates increase of mixing water, consequently reducing the strength.

(iii) Increased Tendency to Cracking

Rapid evaporation leads to *plastic shrinkage cracking*, and subsequent cooling of hardened concrete introduces tensile stresses. The rate of evaporation depends on the ambient temperature, relative humidity, wind speed and concrete temperature.

(iv) Rapid Evaporation during Curing

As the hydration of cement can take place only in water-filled capillaries, it is imperative that a loss of water by evaporation from the capillaries be prevented. Furthermore, water lost internally by self-desiccation has to be replaced by water from outside. A rapid initial hydration results in a poor microstructure of gel which is probably more porous, resulting in a large proportion of the pores remaining unfilled. This leads to lower strength.

(v) Difficulty in Controlling the Air Content

At higher temperatures it is more difficult to control the air content in air-entrained concrete. This adds to the difficulty of controlling workability. For a given amount of *air-entraining agent*, hot concrete entrains less air than does concrete at normal temperatures.

12.2.1 Recommended Practices and Precautions

(a) Temperature Control of Concrete Ingredients

The temperature of the concrete can be kept down by controlling the temperature of the ingredients. The *aggregates* may be protected from direct sunrays by erecting temporary sheds or shelters over the aggregate stockpiles. Water can also be sprinkled on to the aggregate before using them in concrete. The mixing water has the greatest effect on lowering the temperature of concrete, because the specific heat of water (1.0) is nearly five times that of common aggregate (0.22). Moreover, the temperature of water is easier to control than that of other ingredients. Under certain circumstances, the temperature of water can most economically be controlled by mechanical refrigeration or mixing with crushed ice. The precooling of aggregates can be achieved at the mixing stage by adding calculated quantities of broken ice pieces as a part of mixing water, provided the ice is completely melted by the time mixing is completed.

(b) Proportioning of Concrete Mix

The mix should be designed to have minimum cement content consistent with other functional requirements. As far as possible, cement with lower *heat of hydration* should be preferred to those having greater *fineness* and heat of hydration. Use of water-reducing or set-retarding admixtures is beneficial. *Accelerators* should not be used under these conditions.

(c) Production and Delivery

The temperature of aggregates, water and cement should be maintained at the lowest practical levels so that the temperature of concrete is below 40°C at the time of placement. The temperature of the concrete at the time of leaving the batching plant should be measured with a suitable metal clad thermometer. The period between mixing and delivery should be kept to an absolute minimum by coordinating the delivery of concrete with its rate of placement.

(d) Placement and Curing of Concrete

The formwork, reinforcement and subgrade should be sprinkled with cool water just before the placement of concrete. The area around the work should be kept wet to the extent possible to cool the surrounding air and increase its humidity. Speed of placement and finishing helps minimize problems in hot weather concreting. Immediately after compaction, the concrete should be protected to prevent the evaporation of *moisture* by means of wet (not dripping) gunny bags, hessian, etc. After the concrete has attained a degree of hardening sufficient to withstand surface damage, moist-curing should begin. Continuous curing is important because the volume changes due to alternate wetting and drying promote the development of surface cracking. On the hardened concrete, the curing shall not be much cooler than the concrete because of the possibilities of thermal stresses and resultant cracking. High velocity winds cause higher rate of evaporation, and hence wind breakers should be provided as far as possible. If possible, the concreting can be done during night shifts.

12.3 COLD WEATHER CONCRETING

Any concreting operation done at a temperature below 5°C is termed *cold weather concreting*. Most codes do not advocate concreting to be done at an atmospheric temperature below 5°C without special precautions. Due to low temperature, the problems are mainly due to the slower development of concrete strength; the concrete in the plastic stage can be damaged if it is exposed to low temperatures which cause ice lenses to form and expansion to occur within the pore structure, and subsequent damage may occur due to alternate freezing and thawing when the concrete has hardened. The effects of cold weather concreting may be summarized as follows.

(i) Delayed Setting

At low temperatures, the development of concrete strength is retarded as compared with the strength development at normal temperatures. The setting period necessary before removal of formwork is thus increased. Although the initial strength of concrete is lower, the ultimate strength will not be severely affected provided the concrete has been prevented from freezing during its early life.

(ii) Early Freezing of Concrete

When *plastic concrete* is exposed to freezing temperature, it may suffer permanent damage. If the concrete is allowed to freeze before a certain *prehardening period*, it may suffer irreparable loss in its properties so much so that even one cycle of *freezing* and *thawing* during the prehardening period may reduce compressive strength to 50 per cent of what would be expected for normal temperature concrete. The prehardening period depends upon the type of cement and environmental conditions. It may be specified in terms of time required to attain a compressive strength of the order of 3.5 to 7.0 MPa; alternatively it can be specified in terms of period varying from 24 hours to even three days depending upon the degree of saturation and *water-cement ratio*.

(iii) Stresses Due to Temperature Differential

A large temperature differential within the concrete member may promote cracking and has a harmful effect on *durability*. Such situations are likely to occur in cold weather at the time of removal of formwork.

12.3.1 Recommended Practice

As per IS: 7861 (Part II)–1981, the following measures should be taken:

(i) Temperature Control of Ingredients

The temperature at the time of setting of concrete can be raised by heating the ingredients of the concrete mix. It would be easier to heat the *mixing water*. The temperature of the water should not exceed 65°C as the *flash set* of cement will occur when the hot water and cement come in contact in the mixers. Therefore, the heated water should come in direct contact with the aggregate, and not the cement, first. The *aggregates* are heated by passing steam through pipes embedded in aggregate storage bins. Another precaution taken along with the heating of ingredients is to construct a temporary shelter around the construction site. The air inside is heated by electric or steam heating or central heating with circulating water. The temperature of ingredients should be so decided that the resulting concrete sets at a temperature of 10° to 20°C.

(ii) Use of Insulating Formwork

A fair amount of heat is generated during *hydration* of cement. Such heat can be gainfully conserved by having insulating *formwork* covers capable of maintaining concrete temperature above the desirable limit up to the first 3 days (or even 7 days) even though the ambient temperatures are lower. The formwork covers can be of timber, clean straw, blankets, tarpaulines, plastic sheeting, etc., and are used in conjunction with an air gap as insulation. The efficiency of the covers depends upon the thermal conductivity of the medium as well as ambient temperature conditions. For moderately cold weather, timber formwork alone is sufficient.

(iii) Proportioning of Concrete Ingredients

The important factor for cold-weather concreating is the attainment of suitable temperature for fresh concrete. Since the quantity of cement in the mix affects the rate of increase in temperature, an additional quantity of cement may be used. It would be preferable to use high alumina cement for concreting during frost conditions, the main advantage being that a higher heat of hydration is generated during the first 24 hours. During this period, sufficient strength (approximately 10 to 15 MPa) is developed to make the concrete safe against frost action. No accelerator should be used if high alumina cement is used. Alternatively, the rapid hardening portland cement or accelerating admixtures used with proper precautions can help in getting the required strength in a shorter period. Air-entraining agents are generally recommended for use in cold weather. Air-entrainment increases the resistance of the hardened concrete to *freezing and thawing* and normally, at the same time, improves the *workability* of fresh concrete. The calcium chloride used as accelerating admixture may cause corrosion of reinforcing steel. In any case, calcium chloride should not be used in prestressed concrete construction.

(iv) Placement and Curing

Before placing the concrete, all ice, snow and frost should be completely removed. Care should be taken to see that the surface on which the concrete is to be placed and eminent parts are sufficiently warm. During the periods of freezing or in near-freezing conditions, water curing is not applicable.

(v) Delayed Removal of Formwork

Because of slower rate of gain of strength during the cold weather, the formwork and props have to be kept in place for a longer time than in usual concreting practice.

The problem of concreting in cold weather can be minimized by adopting precast construction of structures. Precast members are manufactured in the factories where adequate precautions can be taken and concreting can be done in the controlled conditions.

12.4 UNDERWATER CONCRETING

Special precautions need be taken whenever the concrete is to be placed underwater. In regard to the quality of concrete, the recommendations of the Portland Cement Association are as follows:

"The concrete should be plastic and cohesive but should have good flowability. This requires a fairly high slump, usually 150 to 180 mm. A richer mix than generally used for placing under normal conditions is required; usually the cement requirement is not less than eight sacks per cubic metre of concrete. The proportions of fine and coarse aggregates should be adjusted to produce the desired workability with a somewhat higher porportion of fine aggregate than used for normal conditions. The fine aggregate proportion can often be from 45 to 50 per cent of the total aggregate, depending on the grading. It is also important that the aggregate contain sufficient fine material passing the 300-and 150-micron sieves to produce a plastic and cohesive mixture. ASTM standard specifications for concrete aggregate require that not less than 10 per cent of fine aggregate pass the 300 micron sieve and not less than 2 per cent pass the 150 micron sieve. The fine aggregate should meet the minimum requirements and somewhat higher percentage of fines would be better in many cases. For most works coarse aggregate should be graded up to 20 mm or 40 mm."

In addition the coarse aggregate should not contain loam or any other material which may cause *laitance* while being worked.

The demands on the formwork are usually higher than in normal concreting under dry conditions. The formwork not only has to impart the required shape to the structure or its elements, it must also protect the concrete mix during placing until it matures, from the direct action of current and waves. Thus the formwork also serves as a temporary protective casing which during concreting prevents possible washing out of cement and the leakage of cement mortar from the concrete mix. After completion of concreting, it will protect the soft concrete from the impact and abrasive action of the water currents. If necessary, coffer dams are to be constructed to reduce the velocity of flow through the construction zone.

12.4.1 Concreting Methods

The following are the principal techniques which have been used for placing concrete underwater:
  (i) placing in dewatered caissons or coffer dams,
  (ii) tremie method,
  (iii) bucket placing,
  (iv) placing in bags, and
  (v) prepacked concrete.

The placing in dewatered caissons or coffer dams follows the normal *in-the-dry* practice.

Tremie Method

A *tremie* is a watertight pipe, generally 250 mm in diameter, having a funnel-shaped hopper at its upper end and a loose plug at the bottom or discharge end as shown in Fig. 12.1. The valve at the discharge end is used to de-water the tremie

244 Concrete Technology

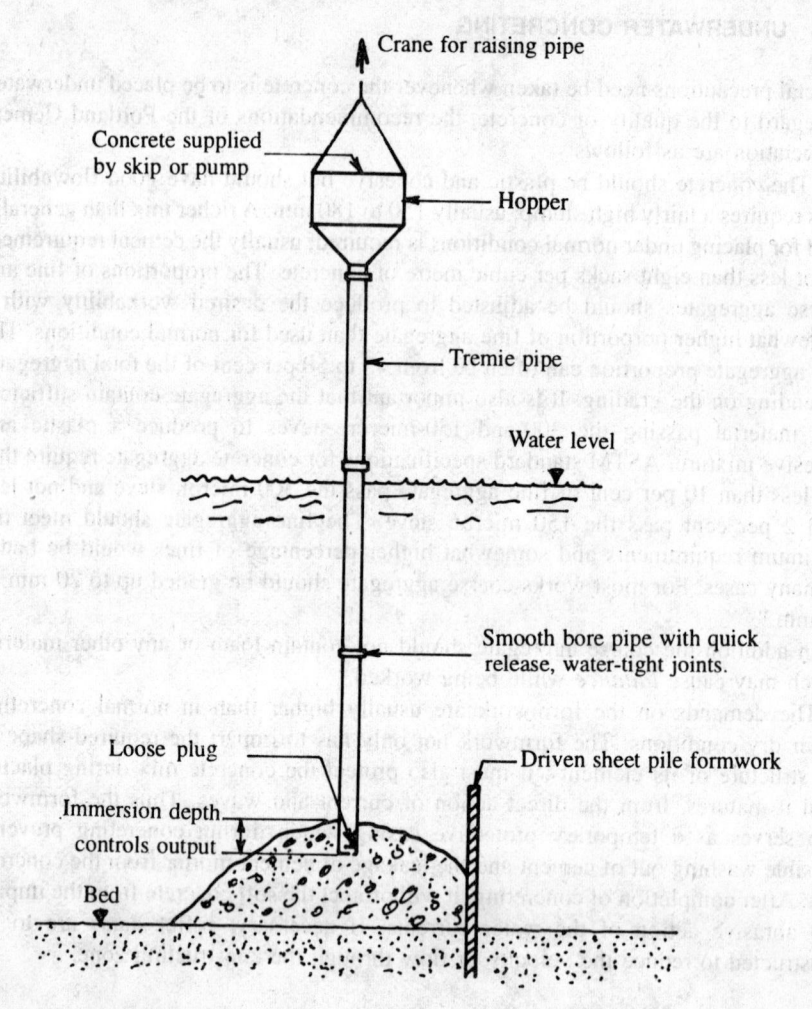

Fig. 12.1 Typical arrangement for a tremie pipe

and control the distribution of the concrete. The tremie is supported on a working platform above water level, and to facilitate the placing it is built up in 1 to 3.5 m sections.

During the concreting, air and water must be excluded from the tremie by keeping the pipe full of concrete all the time; and for this reason the capacity of the hopper should be at least equal to that of the tremie pipe. In charging the tremie a plug formed of paper is first inserted into the top of the pipe. As the hopper is filled, the pressure of fresh concrete forces the plug down the pipe, and the water in the tremie is displaced by concrete.

For concreting, the tremie pipe is lowered into position and the discharge end is kept as deeply submerged beneath the surface of freshly placed concrete as the head of concrete in tremie permits. As concreting proceeds the pipe is raised slightly and the concrete flows outward; care should be taken to maintain continuity of concreting without breaking the seal provided by the concrete cover over the discharge end. Should this seal be broken, the tremie should be lifted and plugged before concreting is recommended. The tremie should never be moved laterally through freshly placed concrete. It should be lifted vertically above the surface of concrete and shifted to its new position.

When large quantities of concrete are to be placed continuously, it is preferable to place concrete simultaneously and uniformly through a battery of tremies, rather than shift a single tremie from point to point. It has been recommended that the spacing of tremies be between 3.5 and 5 m and that the end tremies should be about 2.5 m from the formwork.

The risk of segregation and nonuniform stiffening can be minimized by maintaining the surface of concrete in the forms as level as possible and by providing a continuous and rapid flow of concrete.

Bucket Placing

This method has the advantage that the concreting can be carried out at considerable depths. The buckets are usually fitted with drop-bottom or bottom-roller gates which open freely outward when tripped as shown in Fig. 12.2. The bucket is completely filled with concrete and its top covered with a canvas cloth or a gunny sack to prevent the disturbance of concrete as the bucket is lowered into water. Some buckets are provided with a special base which limits the agitation of the concrete during discharge and also while the empty bucket is hoisted away from the fresh concrete. The bucket is lowered by a crane up to the bottom surface of concrete and then opened either by divers or by a suitable arrangement from the top. It is essential that the concrete be discharged directly against the surface on which it is to be deposited. Early discharge of bucket, which permits the fresh concrete to drop through water, must be avoided. The main disadvantage of the bucket method is the difficulty in keeping the top surface of the placed concrete reasonably level. The method permits the use of slightly stiffer concrete than does tremie method.

Placing in Bags

The method consists in partially (usually about two-third) filling of cloth or gunny sacks with concrete, and tying them in such a way that they can readily be accommodated in a profile of the surface on which they are placed. The properly filled bags are lowered into water and placed carefully in a header-and-stretcher fashion as in brick masonry construction with the help of divers.

The method has advantages in that, in many cases, no formwork is necessary and comparatively lean mixes may be used provided sufficient plasticity is retained. On the other hand, as the accurate positioning of the bags in place can be only accomplished by the divers, the work is consequently slow and laborious. Voids between adjacent bags are difficult to fill, there is little bonding other than that achieved by mechanical interlock between bags. The bags and labour necessary to fill and tie them are relatively expensive; and the method is only suited for placing the concrete in rather shallow water.

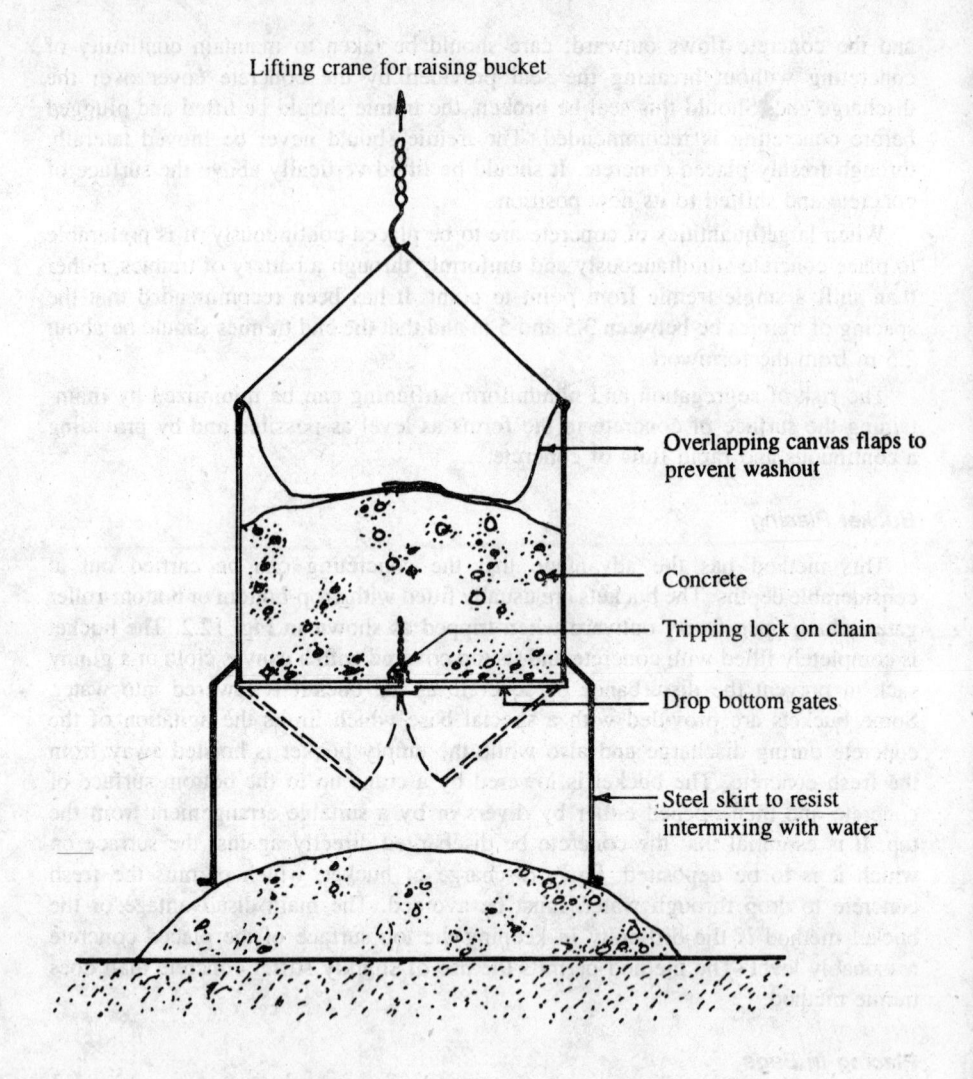

Fig. 12.2 Typical arrangement for a bottom opening bucket

Prepacked Concrete

This technique, also called grouted concrete, consists of placing the coarse aggregate only in the forms and thoroughly compacting it to form a prepacked mass. This mass is then grouted with the cement mortar of the required proportions. The aggregate should be wetted before being placed in position. The mortar that grouts the concrete displaces water and fills the voids.

The aggregate should be well graded to produce a dense and compact concrete. Aggregates up to a maximum size of 80 mm can be conveniently used. Only shutter vibrators can be used for compacting the coarse aggregate. The coarse aggregate may also be allowed to fall from heights of up to 4 meters, without causing any appreciable segregation.

The mortar consists of fine sand, pozzolanic filler material and a chemical agent, which serves (i) to help the penetration, (ii) to inhibit early setting of cement, (iii) to aid the dispersion of the particles, and (iv) to increase the fluidity of mortar. An air-entraining agent is also added to the mortar to entrain about four per cent of air. A small variation of the procedure of preparation of the cement mortar for grouting leads to a process called *colcrete*. In this process the mortar grout is prepared in a special high-speed mixer. No admixtures are used in this process. The high-speed mixing produces a very fluid grout which is immiscible with water. The maximum size of sand used is 5 mm and the sand should be well graded. The mix ratio ranges from 1 : 1.5 to 1 : 4 with a *water-cement ratio* of about 0.45. Rich cement mortar is used for underwater construction and grouting of prestressing cables in post-tensioned bonded construction.

The grouting of prepacked aggregates can be done in any of the following methods:
  (i) The mould can be filled partially with grout, and the coarse aggregate can then the deposited in the grout.
  (ii) The grout can be poured on the top surface of aggregate and allowed to penetrate to the bottom. The method is particulary useful for grouting thin sections.
  (iii) Pumping the grout into the aggregate mass from bottom at carefully designed positions through a network of pipes. The formwork should be constructed at the top of the coarse aggregate in this method.

The quantity of grout in any of these methods should be estimated from the void contents of the coarse aggregates. The grout pressure employed will be of the order of 0.2 to 0.3 MPa.

This technique is very much suited for underwater construction and repair work of mass concrete structures, such as dams, spillways, etc. The prepacked concrete is known to exhibit lower drying shrinkage and higher durability, especially freezing and thawing resistance compared to ordinary concrete for the same proportions. The rate of development of strength is comparatively slow for the first two months and the eventual strengths are the about the same as for normal concrete. In the USA and USSR, the tremie method is most commonly used. In Holland, where large volumes of concrete have been placed under water, the usual method is that of placing by bucket. The bag method is now a days seldom used for important works overseas, but has found some applications in the building up of permanent underwater forms.

13

Inspection and Testing

13.1 INTRODUCTION

The principal aim of conventional *in situ* testing is to ensure that proper materials are used in desired proportions and correct steps of *workmanship* are followed. Recent trends are towards switch over to the *performance-oriented system approach* and *quality control* where a number of items and operations have to be controlled at the right time and in the right measures. From this point of view, the testing of *representative concrete* does not represent the quality of the actual in-place concrete, and quality control cannot be regarded as a mere testing of three concrete cubes at 28 days. In fact, to avoid inferior concrete being placed, the control is to be carried out much before any cubes become available for testing. The cube tests relate to the concrete specimens specially prepared for testing. What is really needed is to carry out tests on concrete in the structure, so that the influence of workmanship in actual *placing, compaction* and *curing* are also reflected. However, a complete switchover to performance-oriented specifications has not been possible, because of difficulties involved in defining what constitutes the satisfactory performance, in setting appropriate performance limits, and in monitoring the performance in the absence of suitable tests.

For *in situ* testing, the aims of investigation should be clearly established at the outset to avoid misleading test results and consequent future disputes over results. By carefully formulating the test programme the uncertainties can often be minimized. Since *in situ* testing of existing structures involve engineering judgement, a complete knowledge of the range of tests available, and their limitations and the accuracies that can be achieved is important.

There are three basic categories of concrete testing, namely,

(i) *Quality control* It is normally carried out by the contractor to indicate adjustments necessary to ensure an acceptable supplied material.

(ii) *Compliance testing* It is performed by, or for, the engineer according to an agreed plan, to check compliance with the specifications.

(iii) *Secondary testing* This test is performed on the hardened concrete in the situations where there is doubt about the reliability of control and compliance results or they are unavailable or inappropriate as in an old structure.

Quality control and compliance tests are normally performed on *standard hardened specimens* from the sample of fresh concrete being used in construction. However, these tests may misrepresent the true quality of concrete actually used in the structure. This is due to differences of compaction, curing and general workmanship. The modern trend is to perform *compliance testing*, using methods which are either *non-destructive* or *cause only limited damage*. Such a test may be used as a backup for conventional testing. The principal usage of *in situ* tests is, nevertheless, as secondary testing for checking compliance with specifications and in assessing the *in situ* quality.

13.2 INSPECTION TESTING OF FRESH CONCRETE

The inspection testing of fresh concrete includes workability test, analysis of fresh concrete, accelerated testing, and nondestructive testing.

13.2.1 Workability Tests

The workability of concrete should be measured at frequent intervals during the progress of work, by means of slump test, compacting factor test or Vee-Bee consistency test as per IS: 1199–1959 specifications or the ball-penetration test (ASTM : C–360). Additional tests should be carried out whenever a change in the materials or mix proportion occurs.

The slump test is of real value as a field control of the mix to maintain the uniformity between different batches of supposedly similar concrete. By control of uniform workability, it is easier to ensure a uniform quality of concrete and hence uniform strength. The advantage of the ball-penetration test lies in its simplicity and the speed at which the test is carried out; there is no necessity of sampling as the test can be performed on the concrete in a wheelbarrow or as placed in the forms.

The compacting factor test is more accurate than the slump test and the results are reproducible. This test may be performed for a wide range of workability *i.e.* for concrete mixes of *high* to *very low workabilities* (CF of 0.92 to 0.68). The Vee-Bee test is suitable for *low* and *very low workabilities*. In the absence of definite correlation between different measures of workability under different conditions, it is recommended that for a given concrete, the appropriate method should be decided beforehand and the workability of concrete should be expressed in terms of such a test, rather than being interpreted from the results of other tests.

If the proportions of the materials are properly maintaind and workability is satisfactory, the results should not differ by more than the tolerance indicated in Table 13.1.

TABLE 13.1 Permissible variations in different workability measurements

Workability measurement	Tolerance or allowable variation
Slump	± 25 mm or ± one-third of required value, whichever is less
Ball penetration	± 12 mm
Compacting factor (CF)	± 0.03 for CF values of 0.90 or more
	± 0.04 for CF values between 0.90 and 0.80
	± 0.05 for CF values of 0.80 or less
Vee-Bee time	± 3 s or ± one-fifth of the required value, whichever is greater

13.2.2 Analysis of Fresh Concrete

The field variations in the actual mix proportions of the concrete can be determined by analyzing the composition of fresh concrete. The quality of concrete can be controlled if a rapid analysis of fresh concrete is carried out allowing the engineer to take the necessary remedial measures, if required. The density and composition of in-place concrete may be determined by analyzing the representative samples taken from the forms. The density values would indicate the degree of compaction achieved. The cement and water contents of the concrete may be determined according to the method given in IS : 1199–1959. The method involves separating the constituents of fresh concrete by wet sieving and determining their proportions after weighing in water. The mass of cement in the sample of concrete is determined from the difference between the mass of concrete in water, and the masses of coarse and fine aggregates in water. The water content is next determined as the difference between the mass of concrete and the combined mass of cement and aggregates. The method requires about 2 hours and a fairly high degree of experience and skill.

There are many other rapid methods for determining the cement content of a sample of fresh concrete. Important among these are: the rapid analysis machine (RAM) method; the EDTA titration method; the HCl heat of solution method, and accelerated strength method. The *rapid analysis machine* method is based on the principle of elutriation. A weighed sample of concrete is fed into the machine which separates the cement and fine particles as slurry, by elutriation. This is followed by subsampling, flocculation and measurement of cement by mass. The sequence of procedure is built into the machine. The determination of cement content takes about 10 minutes.

The EDTA titration method, involves the separation of a sample of mortar from the fresh concrete sample by sieving. The mortar sample is dried at 105°C for one hour. The silica-free acid solution of the sample is titrated against a standard EDTA solution. The percentage of CaO present in the sample is estimated, which gives the cement content of the original sample of concrete.

In the HCl heat of solution method, a sample of concrete is diluted by adding a fixed quantity of water. To the diluted solution is added hydrochloric acid, resulting in an exothermic reaction which decomposes the cement contained in the sample. The heat of reaction reaches a steady temperature quickly. The temperature rise is related to the cement content of sample of concrete. This method has been developed by the Cement Research Institute of India.

A comparison of the four analysis methods has been given in Table 13.2..

TABLE 13.2 Comparison of various analysis methods

Method of analysis	Variation of estimated cement content from the actual values, per cent	Approximate time required for sample preparation	Approximate time required for conducting the test
IS : 1199–1959 method	−11.3 to 2.2	Nil	2 h
RAM method	−10.3 to 6.7	Nil	10 min
EDTA method	−6.6 to −3.1	1.5 h	30 min
HCl heat of solution method	−10.4 to +10	Nil	10 min

The *accelerated strength tests* give a reliable idea about the potential 28-day strength of concrete. The details of accelerated strength tests for the purpose of *quality control of concrete* are available in IS : 9013–1978. Either of the following two methods may be adopted as a standard for the accelerated curing of concrete.

(i) *Warm water method*

The specimens are immersed in water $1\frac{1}{2}$ to $3\frac{1}{2}$ hours after moulding.

Curing water temperature	$55 \pm 1°C$
Curing period	$20\,h \pm 10\,min$

Demould and cool at $27 \pm 2°C$ for 1 h before test.

(ii) *Boiling water method*

Standard moist curing	$23\,h \pm 15\,min$
Water temperature	$100°C$
Curing period	$3\frac{1}{2}\,h \pm 5\,min$
Cooling period	$2\,h$

The actual correlation of accelerated test results to 28-day normally cured specimens depends upon the curing method adopted, the chemical composition of cement and the concrete mix proportions. Typical relationships are shown in Fig. 10.12. It is recommended that the actual relationship under given site conditions should be established using local concrete making materials and such relationships be continuously improved upon as more and more data become available progressively. In the absence of past records with local materials, the relation suggested can be used to predict the *28-day compressive strength* of the normally cured concrete, within ± 15 per cent limits.

13.2.3 Non-destructive Testing of *in situ* Fresh Concrete

Maturity Test

This test is based on the principle that concretes having equal maturities will have equal compressive strengths, regardless of temperature and curing histories. The *maturity of concrete* defined as a product of time and temperature, provides a useful tool. The maturity of the *in situ* concrete at early ages can be determined by a self-contained battery-operated instrument known as the *maturity meter*. The instrument may be used to determine the earliest time when the formwork can be removed during the construction. In factory production of precast elements, the meter can help to obtain an optimum usage of forms. The variation of *compressive strength with maturity* has been shown earlier in Fig. 11.5. The concept of maturity gives valid results, provided that the concretes have initial temperatures between 15°C to 26°C and there is no loss of moisture by drying during the curing period. Maturity is represented by 3 days to 28 days of curing at normal temperatures.

Ultrasonic Pulse Velocity Method

The method consists of producing an ultrasonic longitudinal pulse by an electro-acoustical transducer which is held in contact with one surface of the freshly placed concrete member under test. After traversing a known distance in the concrete, the pulse is converted into an electrical signal by a second electro-acoustical transducer, and an electronic timing circuit enables the transit time of the pulse to be measured from which the pulse velocity is calculated. This procedure is also sometimes called the *ultrasonic method*.

The method can assess the quality of *in situ* fresh concrete soon after it is placed in forms and compacted. The transducers may be brought in contact with concrete through the pre-drilled holes in the formwork, after about 4 to 6 hours of placing the concrete in the formwork. The measurable difference in ultrasonic pulse velocity for different grades of concrete helps in characterizing the *in situ* concrete of different mix proportions. The investigations at CRI have indicated that method is applicable to the concretes which have attained *initial setting*. At the time of *final setting* (i.e., *penetration resistance* to 26.97 MPa as per IS : 8142–1976), the fresh concrete had an ultrasonic pulse velocity of the order of 2 km/s. An inordinate delay in obtaining *ultrasonic pulse velocity* of this magnitude would indicate large variations in the material mix proportions or the quality of workmanship adopted, particularly in compaction.

13.3 ACCEPTANCE TESTING OF HARDENED CONCRETE

The tests on the hardened concrete are carried out to check its compliance with specifications, assess the quality of concrete in the structure, the load carrying capacity, and the strength of concrete at a particular location in the structure.

Checking Compliance with Specification

This type of testing is required to collect additional evidence in contractual disputes following non-compliance to standard specifications. Other instances, involve retrospective checking following deterioration of structure. *Strength* and *minimum cement content* for *durability* form important parts of most specifications. The engineer must select the most appropriate methods of assessing the *in situ* strength on a representative basis, and take into account the expected variations. The compliance to a minimum cement requirement may be confirmed by *chemical* or *petrographic tests*. Tests may also be required to check the adequacy of cover or compaction, and reinforcement quantity and its location.

Assessment of *in situ* Quality

This testing is primarily concerned with the current adequacy of existing structure and its future performance. It is important to distinguish between the need to assess the properties of material, and the performance of a structural member as a whole. The need for testing may be for:

 (i) Assessment of structural integrity or safety following material deterioration, or structural damage caused by fire, blast, fatigue or overload

 (ii) Proposed extension or change of usage of structure

 (iii) Acceptability of structure for purchase or insurance

 (iv) Assessment of cause and extent of deterioration for repair or remedial measures

 (v) Monitoring long-term changes in material properties and structural performance

Assessment of Concrete Strength

Various methods generally used for assessing the quality and strength of concrete are:
1. Surface hardness test method
2. Pulse velocity method
3. Core test method
4. Pullout test method

Core tests provide the most reliable *in situ* strength assessment but cause maximum damage and are slow and expensive. They are often regarded as essential. Though surface hardness and pulse velocity tests are cheap and quick to perform, and enable large areas to be surveyed, their correlation for absolute strength prediction poses many problems. However, these tests are ideal for uniformity assessment.

The above test methods may be classified into two categories.

1. Non-destructive Methods

The standard tests described in chapter 8, conducted on control specimens to determine cube/cylinder strength, modulus of rupure, etc., may be classified as destructive tests in the sense that the specimens are destroyed in the course of tests. In contrast, there is another category of tests known as non-destructive tests which may be performed directly on the *in situ* concrete without the removal of a sample. These tests do not impair the intended performance of the element or member being tested. They also include methods which cause localized surface zone damage and may be called partially destructive.

2. Methods Requiring Sample Extraction

These tests are performed in the laboratory on the samples commonly taken in the form of cores drilled from the concrete. Some chemical tests may be performed on smaller drilled powdered samples taken directly from the structure. The tests have large practical limitations, and their reliability and accuracy vary widely.

Visual inspection may provide valuable information related to workmanship, structural serviceability and material deterioration, and it is particularly important that the investigator be able to differentiate between the various types of cracks. Some typical crack types are shown in Fig. 16.1. *Segregation* or excessive *bleeding* at shutter joints reflect a problem with concrete mix whereas *honeycombing* may be an indication of poor *workmanship*. *Excessive deflection* or *flexural cracking* point towards structural inadequacy, whereas *distortion* of door frames, cracking of windows or cracking of a structure or its finishes may indicate long-term *creep deflections* and *thermal movements*. *Material deterioration* is often indicated by *surface cracking* and *spalling* of concrete, and an examination of crack patterns may provide a preliminary indication of the cause. The most common cause is *reinforcement corrosion* due to an inadequate *cover* or a *high chloride concentration*, and is indicated by *splitting* and *spalling* along the line of bars possibly with *rust staining*. Binoculars, telescopes and borescopes may be useful where access is difficult.

The commonly used method of *surface tapping* is useful in locating delamination near the surface and has been developed with more complex *pulse-echo techniques*.

Engineering judgement is required to determine the number and location of tests, and the relevance of the results to the element as a whole. The appropriate number of tests is a compromise between accuracy, effort, cost and damage. The recommended number of tests which may be considered equivalent to a single result are given in the Table 13.3.

TABLE 13.3 Number of test results for various test methods

Test Method	Number of results at a location
Standard core test	3
Non-standard small core test	9
Schmidt rebound hammer test	12
Ultrasonic Pulse velocity test	1
Pull-out test	6

Where *specification compliance* is being investigated, it is recommended that no fewer than four cores be taken from the suspect patch of concrete. Weaker top zones of members should be avoided. Testing at around mid-height is recommended for beams, columns and walls, and surface zone tests on slabs must be restricted to soffits unless the top layer is first removed. Care must similarly be taken to discard material from the top 20 per cent (or atleast 50 mm) of slab when extracting cores. Columns may be expected to be reasonably uniform except for a weaker zone in the top 300 mm or 20 per cent of their depth.

Like dry cubes the cores generally yield strengths which are approximately 10–15 per cent higher than those tested in saturated condition.

13.3.1 Surface Hardness Method

The surface hardness is commonly determined by a rebound test. Rebound test hammer method developed by a Swiss engineer, Ernst Schmidt is a practical rebound test. The Schmidt test hammer weighs less than 2 kg and has an impact energy of 2.2 Nm. The spring controlled hammer mass slides on a plunger within a tubular casing as shown in Fig. 13.1. The plunger retracts against a spring when pressed against the concrete surface, and this spring is automatically released when fully tensioned, causing the hammer mass to impact against the concrete through the plunger. When the spring-controlled mass rebounds, it takes with it a rider which slides along a graduated scale and is visible through a small window in the side of the casing. The rider can be held in position on the scale by depressing the locking button. The equipment can be operated horizontally or vertically (up or down). The plunger is pressed strongly and steadily against the concrete surface to be tested at right angles, until the spring loaded mass is triggered from its locked position. After impact the *scale index* is read while the hammer is still in the test position. The measurement taken is an arbitrary quantity referred to as *rebound number*. A *calibration curve* relating the compressive strength of the concrete with the rebound number is shown in Fig. 13.2. The test is suitable for the concretes having strength in the range of 20–60 MPa. The reading is very sensitive to the local variations in the concrete, especially to the aggregate particles near to the surface. The hammer may strike an aggregate particle, thereby giving a misleading result. It is therefore

Inspection and Testing 255

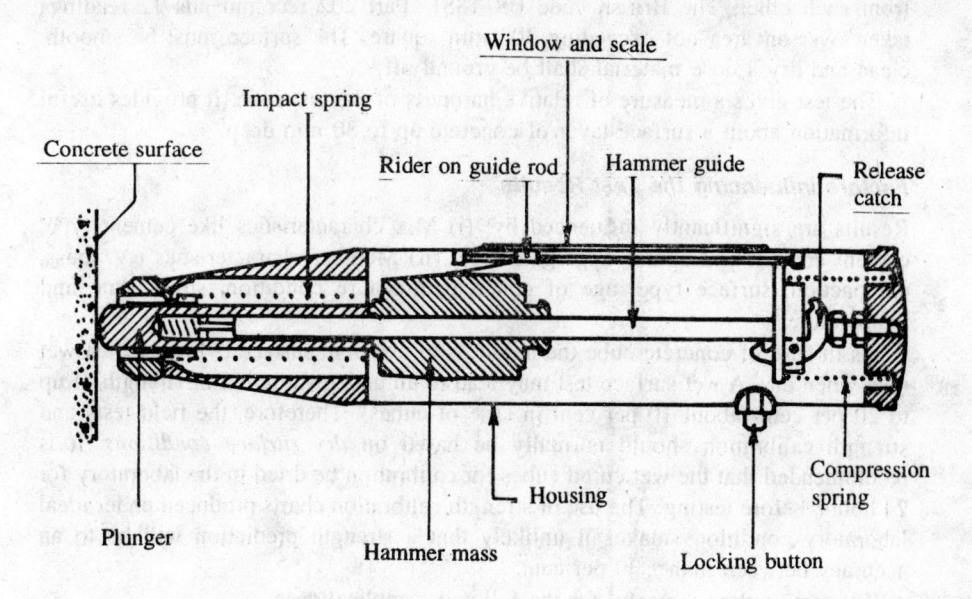

Fig. 13.1 Typical rebound test hammer

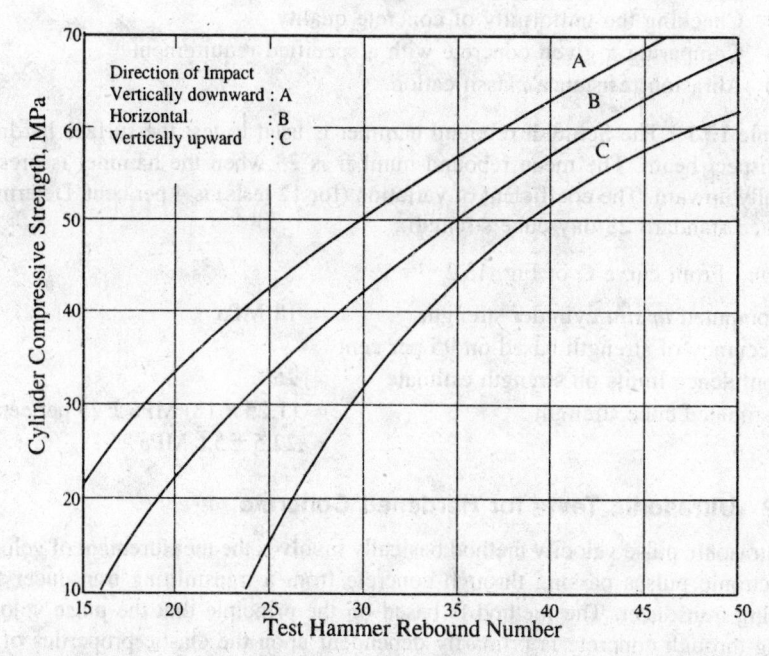

Fig. 13.2 Relationship between rebound number of test hammer and compressive strength of concrete

necessary to take several readings at each test location, and to determine their average. Normally *a grid* is used to locate impact points not less than 20 mm apart from each other. The British code BS–1881 : Part 202 recommends 12 readings taken over an area not exceeding 300 mm square. The surface must be smooth, clean and dry. Loose material shall be ground off.

The test gives a measure of relative hardness of the test zone. It provides useful information about a surface layer of concrete up to 30 mm deep.

Factors Influencing the Test Results

Results are significantly influenced by: (i) Mix characteristics like cement type, cement content and coarse aggregate type, (ii) Member characteristics e.g. mass, compaction, surface type, age of concrete, moisture condition, stress state and temperature.

As in case of concrete cube the hardness of concrete surface is lower when wet than when dry. A wet surface test may lead to an underestimation of strength of up to 20 per cent (about 10 per cent in case of cubes). Therefore, the field tests and strength calibration should normally be based on *dry surface conditions*. It is recommended that the wet cured cubes for calibration be dried in the laboratory for 24 hours, before testing. The use of strength calibration charts produced under ideal laboratory conditions makes it unlikely that a strength prediction will be to an accuracy between than ± 30 per cent.

The test method is useful for the following applications:
   (i) Approximate estimation of strength
   (ii) Checking the uniformity of concrete quality
   (iii) Comparing a given concrete with a specified requirement
   (iv) Abrasion resistance classification.

Example 13.1 The Schmidt rebound hammer is used to test the surface hardness of a suspect beam. The mean rebound number is 25 when the hammer is pressed vertically upward. The coefficient of variation (for 12 tests) is 4 per cent. Determine estimated standard 28-day cube strength.

Solution From curve C of Fig. 13.2

Computed *in situ* cylinder strength	= 18 MPa
Accuracy of strength based on 95 per cent confidence limits on strength estimate	= 25%
Estimated cube strength	= (1.25 × 18) MPa ± 25 per cent
	= 22.5 ± 5.6 MPa

13.3.2 Ultrasonic Tests for Hardened Concrete

The ultrasonic pulse velocity method basically involves the measurement of velocity of electronic pulses passing through concrete from a transmitting transducer to a receiving transducer. The method is based on the principle that the pulse velocity passing through concrete is primarily dependent upon the elastic properties of the materials and is independent of geometry. The pulse velocities range from about 3 to 5 km/s. The methods employ pulses in the frequency range of 15–175 kHz, generated and recorded by electronic circuits.

The test equipment consists of an arrangement for generating electronic pulse at regular intervals of time (usually between 10 to 50 s) transmitting these to concrete, receiving and amplifying the pulse, and measuring the time of transit between the transducers. A typical test circuit in shown in Fig. 13.3. Electronic timing device

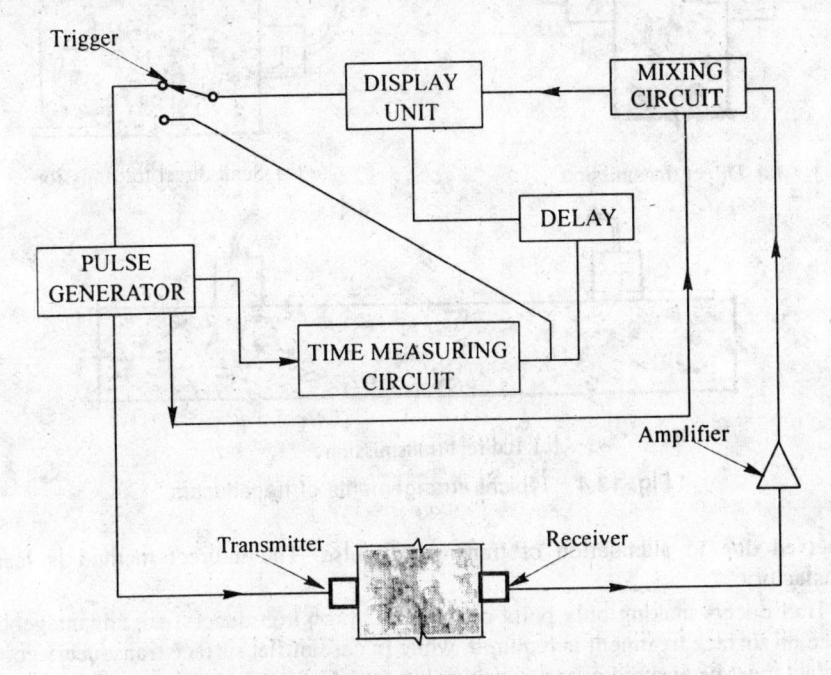

Fig. 13.3 Typical ultrasonic pulse velocity test circuit

accurately measures the interval between the onset and reception of the pulse to an accuracy better than ±1 per cent. The transit time is displayed either on an oscilloscope or as a digital read out. Transducers with natural frequencies between 20 kHz–150 kHz are most suitable for use with concrete, and these may be of any type, although the piezo-electric crystal is most popular.

Transducer Arrangement

The transducers may be arranged in three ways as shown in Fig. 13.4.

1. Opposite faces (direct transmission)
2. Adjacent faces (semi-direct transmission)
3. Same-face (indirect transmission)

Since the maximum pulse energy is transmitted at right angles to the face of the transmitter, the direct method is most reliable from the point of view of transit time measurement; also the path is clearly defined and can be measured accurately. This approach is recommended wherever possible for assessing the quality of concrete. The semi-direct method can sometimes be used satisfactorily if the angle between the transducers, and the path length is not large, otherwise clear signals will not be

258 Concrete Technology

(a) Direct transmission

(b) Semi-direct transmission

(c) Indirect transmission

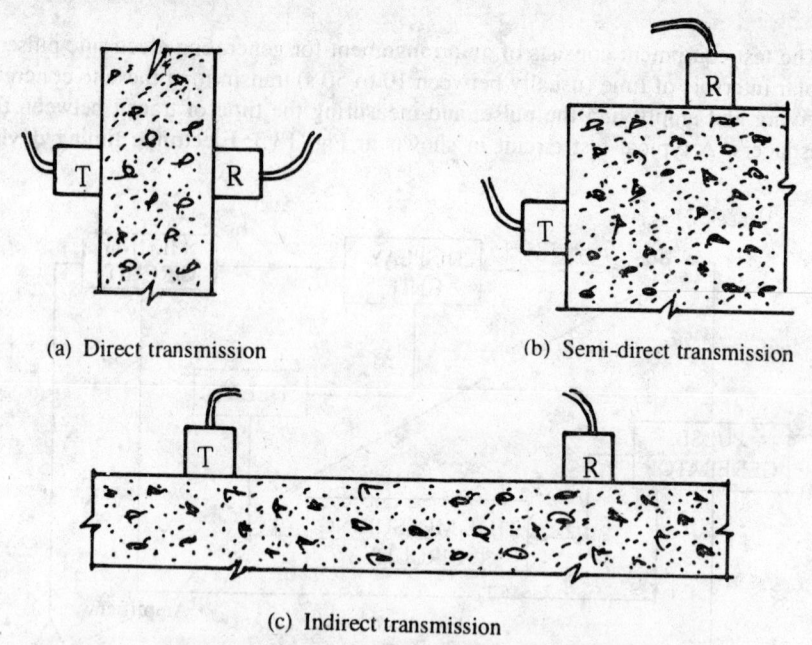

Fig. 13.4 Typical arrangements of transducers

received due to attenuation of transmitted pulse. The indirect method is least satisfactory.

Transducers making only point contact (i.e., prob transducers) are advantageous since no surface treatment is required, while in case of flat surface transducers good contact must be ensured over a considerable area. However, prob transducer is more sensitive to operator pressure.

The testing can also detect internal flaws like inadequate compaction, voids or cracks and segregation in concrete of the structure. In the region of imperfections, the ultrasonic pulse is diffracted around the periphery of the defect and takes more time to reach the receiving transducer. Thus the *transit time* is more and the *ultrasonic pulse velocity* is reduced. The magnitude of reduction in the pulse velocity indicates the extent of imperfections and, therefore, the level of workmanship employed. The Cement Research Institute of India has evolved guidelines, based on an extensive data on concrete constructions in the country for assessing the quality of concrete. The guidelines characterize the quality of concrete in terms of *good, medium* and *poor gradings* as given in Table 13.3.

TABLE 13.3 Quality gradings for concrete

Ultrasonic pulse velocity, km/s	*Quality of concrete*
Above 3.5	Good
3.0 to 3.5	Medium
Below 3.0	Poor

The above gradings are based on the principle that comparatively higher pulse velocities are obtained when the quality of concrete in terms of density, homogeneity and uniformity, and lack of imperfections, is good. In the case of concretes of poorer quality, lower velocities are obtained. The actual pulse velocity chiefly depends upon the material and mix proportions of the concrete. The aggregates through their density and modulus of elasticity make a significant contribution. The other parameters affecting the ultrasonic pulse velocity are age, surface conditions, moisture content and temperature of concrete; the amount and orientation of reinforcement bars in relation to the direction of propagation of pulses, and stress level in the concrete. Due to the difference in the extents to which these parameters influence the pulse velocity and compressive strength, the assessment of compressive strength from ultrasonic pulse velocity is not accurate.

The pulse velocity through saturated concrete may be higher up to 5 per cent than through the same concrete in a dry condition. The relationship between pulse velocity and dynamic elastic modulus of concrete (a composite material consisting of cement mortar matrix and aggregate having different elastic and strength properties) measured by *resonance tests* on prism specimen as fairly reliable.

Application of the Ultrasonic Testing

Laboratory Applications

Ultrasonic testing may be used in the detection of the onset of microcracking during loading test on structural members.

In Situ Applications

The ultrasonic testing may be used for the
  (i) Measurement of concrete uniformity
  (ii) Detection of cracks and honeycombing
  (iii) Estimation of strength
  (iv) Assessment of concrete deterioration due to fire, mechanical, frost or chemical actions
  (v) Measurement of layer thickness
  (vi) Measurement of elastic modulus, and
  (vii) Strength development monitoring.

13.3.3 Partially Destructive Strength Tests

The test methods used to assess *in situ concrete strength* which cause some *localized damage* sufficiently small to cause no loss in structural performance are called *partially destructive strength tests*. These methods which involve penetration, pull-out and pull-off techniques and use correlation charts for the estimation of strength, are not sensitive to as many variables as are rebound hammer or pulse-velocity testing. A major advantage of this type of testing is that strength can be estimated immediately as compared to core testing but the estimate may not be as accurate.

Penetration Resistance Test

In this test, also called Windsor probe test specially designed hardened steel alloy bolts are fired into concrete surface. The test is very popular in North America for monitoring strength development on the site subject to ASTM : C803. The method has also been included in BS : 1881 (Part 207). It is a form of hardness testing and the measurements relate only to the quality of concrete near the surface (normally up to a depth of about 50 mm).

Pull-out Testing

As the name suggests, this method involves the measurement of the force required to pull-out a specially shaped steel rod or some similar device from a concrete surface. The test can be performed either on an insert cast into concrete in the formwork or on an insert fixed into a hole drilled into hardened concrete. The former is called the *cast-in method* which is preplanned, and will thus be of value only in the testing for *specification compliance*, whereas the latter called *drilled-hole method* offers a greater flexibility and is more appropriate for field surveys of hardened concrete. In both cases, the value of the test depends upon the ability to relate pull-out force to concrete strength. An important feature of method is that the relation between pull-out force and concrete strength is relatively unaffected by mix characteristics and curing history. The approach offers the advantage of providing a more direct measure of strength, and at a greater depth than surface hardness testing by rebound methods, but still require only one exposed surface.

Cast-In Method

The method consists of incorporating the pull-out assembly in the formwork at the time of concreting the actual structural member. Alternatively, the pull-out assemblies may be incorporated in large concrete blocks cast at the time of concreting the member. The blocks shall receive the compaction and curing identical to that of actual member. The pull-out test could then be performed on the blocks during construction. The details of the pull-out assembly are shown in Fig. 13.5.

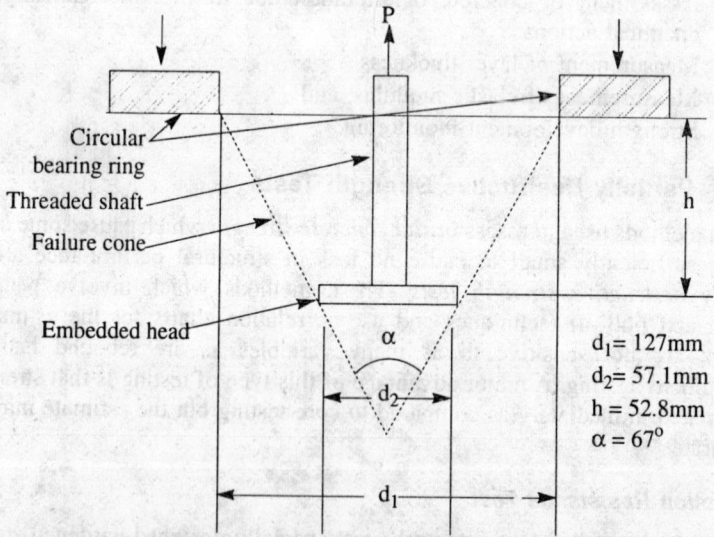

$d_1 = 127$ mm
$d_2 = 57.1$ mm
$h = 52.8$ mm
$\alpha = 67°$

(a) Relative position and dimension of the circulation bearing ring and the embedded assembly.

Fig. 13.5 (a) Relative position of the ring bearing plate and embedded head

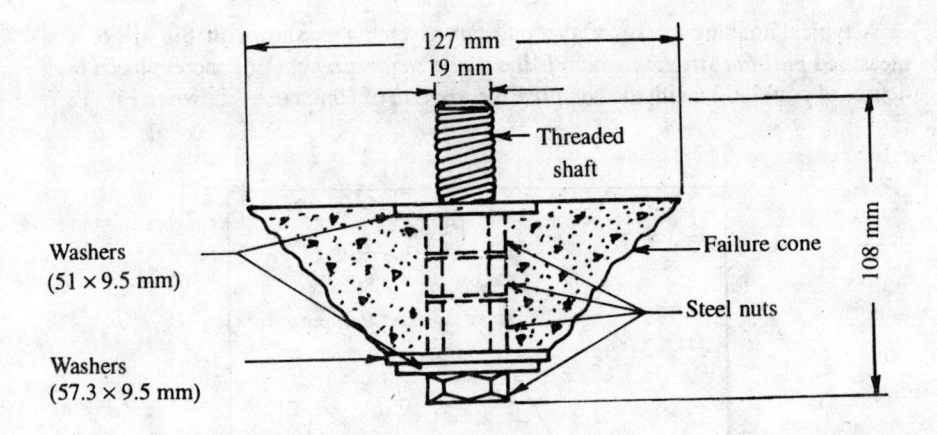

(b) The pullout assembly together with the cone of concrete

Fig. 13.5 (b) Typical pull-out assembly

The assembly is pulled out hydraulically against a circular bearing ring. A cone of concrete is pulled out with the assembly and the force required to achieve this is translated to compressive strength by the use of an empirical calibration.

$$\text{Pull-out strength, } f_p = \frac{P}{A} \quad (13.1)$$

where P and A are the pulling force and failure surface area, respectively. The area A may be calculated from:

$$A = \frac{\pi}{4}(d_1 + d_2)[4h^2 + (d_1 - d_2)^2]^{1/2} \quad (13.2)$$

where d_1 = larger diameter of pulled out concrete cone frustum, i.e. internal diameter of bearing ring
d_2 = diameter of pull-out insert head, and
h = distance from insert head to the surface.

Drilled-hole Method

The main advantage of this method is that its use need not be pre-planned. The method is extremely valuable for *in situ* concrete strength assessment, especially when the concrete mix details are unknown. To illustrate the principles an *expanding sleave method* is describe below. The method basically consists in drilling a hole 30–35 mm deep into the concrete using a rotohammer drill with a nominal 6-mm bit. The hole is then cleared of dust with an air blower and a 6-mm wedge anchor bolt with expanding sleave is tapped lightly into the hole until the sleave is 20 mm below the surface. The bolt is loaded at a standardized rate against a tripod reaction or bearing ring of 80 mm diameter with three feet, each 5 mm wide and 25 mm long.

The method has a *high test variability* due to localized nature of test, the imprecise load transfer mechanism and variations due to drilling.

262 Concrete Technology

A typical loading curve of the pull-out assembly is shown in Fig. 13.6. The measured *pull-out strength* which is the *direct shear strength* of concrete has a high degree of correlation with the *compressive strength* of concrete as shown in Fig. 13.7.

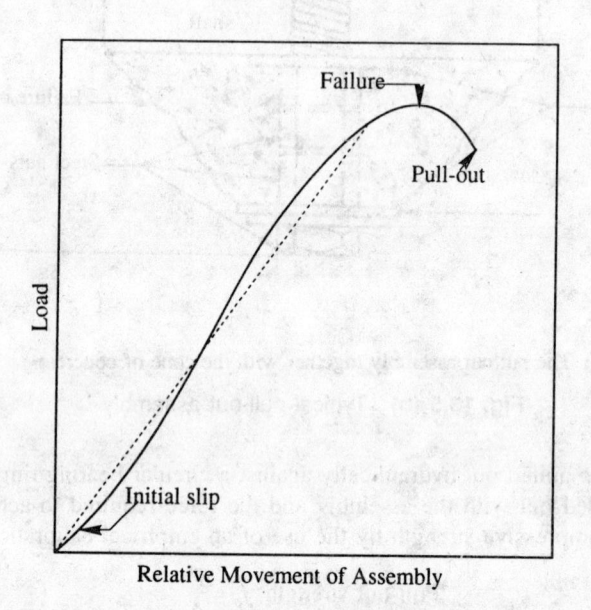

Fig. 13.6 Typical loading curve of pull-out assembly

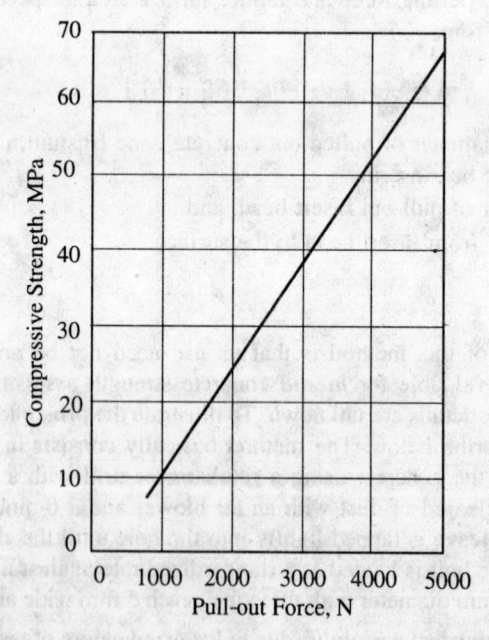

Fig. 13.7 Relationship between pull-out force and compressive strength

The test is reproducible with an accepted degree of accuracy. Moreover, in order not to damage the concrete, the pull-out test need not necessarily be carried out to its completion. It is just enough to apply a predetermined force to the embedded pull-out assembly, and if it is not pulled out, the *in situ* concrete can be assumed to have attained the given strength. In the western countries, the test has become very popular.

13.3.4 Concrete Core Tests

The examination and compressive strength testing of cores drilled from hardened concrete enable the visual inspection of interior regions of the structural member along with estimation of actual concrete strength. The cores are also frequently used to measure *density, water absorption, indirect tensile strength,* and finally as samples for *chemical analysis*.

Location

The choice of core location is governed primarily by the basic purpose of the testing, e.g., for serviceability assessment the cores should normally be taken at the points where minimum strength and maximum stress coincide. However, in a slender member core cutting may impair future performance and hence cores should be taken at the nearest *non-critical locations*. Aesthetic consideration concerning appearance may also sometimes influence the choice of location. For checking specification compliance, the cores should be located to avoid *unrepresentative concrete* (normally occupying top 25 per cent in case of columns, walls or deep beams, and top thickness of slab). The reinforcement bars should be avoided wherever possible.

Size

For compression testing the standards require that the core diameter be atleast three times the nominal maximum size of aggregate. The accuracy of results, generally, decreases as the ratio of aggregate size to core diameter increases. Codes further require that a minimum diameter of 100 mm be used (75 mm in Australia). In case of small-sized members where large holes would be unacceptable smaller diameters are used. The choice of core diameter is also influenced by the length of core specimen which can be obtained. It is generally accepted that the cores for compression testing should have a length/diameter ratio of between 1.0 and 2.0. For reducing drilling costs, damages, variability along length, and geometric influence on testing, the cores should be kept as small as possible (l/d = 1.0–1.2). However, the uncertainties in relating the core strength to cylinder or cube strength are minimized if the core length/diameter ratio is close to 2.0.

The cores must be in a sufficient number to represent concrete under examination as well as provide a strength estimate of acceptable accuracy. Code requires that atleast three standard cores be used in testing.

Drilling

The cores are usually cut by means of portable equipment having rotary cutting tool with diamond bits, and water supply to lubricate the cutter. Care must to taken to prevent lateral movement which will result in a distorted or broken core and to ensure uniformity of pressure. Once a sufficient depth has been drilled, the core is removed by using the drill or tongs and the hole is filled by ramming a dry, low shrinkage concrete into the hole or by wedging a cast cylinder of suitable size into the hole with cement grout or epoxy resin. Each core must be clearly labelled for identification.

Testing

Each core must be trimmed and ends either ground or capped before visual examination, assessment of void percentage, and density. To obtain a core of suitable length having parallel ends which are normal to the axis of the core, trimming with a diamond saw is used. The ends of the core are capped with high alumina cement mortar to provide parallel end surfaces perpendicular to the axis of the core. The caps should be kept as thin as possible.

(i) *Visual examination* The aggregate distribution, honeycombing, cracks, defects and drilling damage are easily seen on a dry surface. However, for the assessment of aggregate type, size and characteristics a wet surface is preferable. Precise details of location and size of reinforcement passing through the core must also be recorded. Voids are normally classified as small voids (0.5 to 3 mm), medium voids (3 to 6 mm), and large voids (> 6 mm), and honeycombing when the voids are interconnected.

(ii) *Strength tests* Following strength tests may be performed on the cores:

(a) *Compressive strength* The cores are tested in *saturated surface dry condition* after capping and immersion in water for atleast two days. The core length and mean diameter (average of diameter measured at quarter and mid-points along the length) to the nearest 1 mm. Compression testing is carried out at a rate within the range 12–24 MPa/min on a suitable testing machine and mode of failure is noted. In case the cap cracks or gets separated from the core, the results should be considered as being of doubtful accuracy.

(b) *Tensile strength* The tensile strength may be measured by the *splitting test* on the core as explained in the section 8.2.3.

Factor Influencing Core Compressive Strength

The significant factors are outlined below:

(i) *Moisture and voids* The moisture condition of the core influences the measured strength; a saturated specimen has a value of 10 to 15 per cent lower than comparable dry specimen. It is therefore important that the relative moisture conditions of core and *in situ* concrete are taken into account while estimating the actual *in situ* concrete strengths. Voids in core will reduce the measured strength.

(ii) *Length/diameter ratio of core* As l/d ratio increases, the measured strength will decrease due to the effect of specimen shape, and stress distribution during test. For establishment of relation between core strength and standard cube strength, a ratio $l/d = 2.0$ is regarded as the datum of computation.

(iii) *Diameter of core* The diameter of the core may influence the measured strength and variability. Measured concrete strength decreases with the increase in the size of specimen; for sizes above 100 mm this effect will be small, but for smaller sizes this effect may become significant.

(iv) *Direction of drilling* As a result of layering effect the measured strength of specimen drilled vertically relative to the direction of casting is likely to be greater than that for a horizontally drilled specimen from the same concrete; an average difference of 8 per cent has been reported in the literature.

(v) *Reinforcement* Due to presence of reinforcement, the measured strength of concrete is underestimated up to 10 per cent. Reinforcement must therefore be avoided wherever possible, but in the case where it is present, the measured core strength may be corrected as follows:

$$\text{Corrected strength} = \text{measured strength} \times \left[1.0 + 1.5\left(\frac{\phi_r h}{\phi_c l}\right)\right] \qquad (13.3)$$

where ϕ_r = reinforcement bar diameter
ϕ_c = core diameter
h = distance of bar axis from the nearer end of core, and
l = core length (uncapped).

For the case of core having multiple bars

$$\text{Corrected strength} = \text{measured strength} \times \left[1.0 + 1.5\left(\frac{\Sigma\phi_r \cdot h}{\phi_c \cdot l}\right)\right] \qquad (13.4)$$

If the spacing of two bars is less than the diameter of the larger bar, only the bar with the higher value of $(\phi_r \cdot h)$ should be considered.

Estimation of Cube Strength

The *equivalent cube strength* can be estimated in two steps. In the first step a correction for the effect of length/diameter ratio is applied to convert the core strength to an *equivalent standard cylinder strength*. In the second step appropriate relationship between strength of cylinders and cubes is used to convert the equivalent standard cylinder strength obtained in step 1 to equivalent cube strength. This conversion to a cube strength may be based on the generally accepted average relationship:

$$\text{Cube strength} = 1.25 \times \text{cylinder strength}$$

$$\left(\text{for } \frac{l}{d} = \lambda = 2.0\right) \qquad (13.5)$$

The corresponding relation taking into account the strength differential of 6 per cent between a core with cut surface relative to cast cylinder, and strength reduction of 15 per cent for weaker top surface zone of a corresponding cast cylinder, as adopted by BS : 1881 are given below.

For vertically drilled core:

$$\text{Estimated } in\ situ \text{ cube strength} = \frac{2.3 f_c}{1.5 + (1/\lambda)} \qquad (13.6)$$

where f_c is the measured strength of a core with length/diameter = λ. An 8 per cent difference between vertical and horizontally drilled cores is incorporated resulting into following expression:

For horizontally drilled core:

$$\text{Estimated } in\ situ \text{ cube strength} = \frac{1.08 \times 2.3 f_c}{1.5 + (1/\lambda)} \simeq \frac{2.5 f_e}{1.5 + 1/\lambda} \qquad (13.7)$$

However, the target (potential) strength of a standard specimen made from a particular mix is about 30 per cent higher than the actual fully compacted *in situ* strength (41). The expression for cube strength is given as:

For vertically drilled core:

$$\text{Estimated potential cube strength} = \frac{3.00 f_c}{1.5 + 1/\lambda} \qquad (13.8)$$

For horizontally drilled core:

$$\text{Estimated potential cube strength} = \frac{3.25 f_c}{1.5 + 1/\lambda} \qquad (13.9)$$

If smaller diameter cores are drilled, then to obtain comparable accuracy, at least thrice the number of standard cores should be used in testing.

The method of preparation of cores after drilling and the procedure of test are described in IS : 516–1959. As per IS : 456–1978 the concrete in the member represented by a core test shall be considered aceptable if the *average equivalent cube strength* of the cores is equal to atleast 85 per cent of the cube strength of the grade of concrete specified for the corresponding age, and no individual core has a strength less than 75 per cent. The main drawback of this test is the difference in the intrinsic quality of concrete in structure and control specimens in the laboratory. In case the core test results do not satisfy the above requirements, the load test is resorted to. The core tests are performed when:

(i) The standard 28-day cube strength test gives lower results than acceptable and the primary aim of the core test is to ascertain whether the structural element is of adequate strength.

(ii) It is essential to estimate the load-carrying capacity of the structure for its safety under change of loading or usage contemplated for the structure.

The typical values of coefficient of variation and maximum accuracies of expected *in situ* strength prediction for a single site-made unit constructed from a number of batches as reported in literature (38,41,190) are given in Table 13.4. The

TABLE 13.4 Typical values of coefficient of variation of test results and maximum accuracies of prediction

Test method	Coefficient of variation for individual member with good degree of control, per cent	Best 95 per cent confidence limits on strength estimate, per cent
Windsor prob	4	±20 (3 tests)
Rebound hammer	4	±25 (12 tests)
Ultrasonic pulse velocity	2.5	±20 (1 test)
Pull-out method	8	±20 (6 tests)
Concrete cores		
(i) standard	10	±10 (3 specimens)
(ii) small (non-standard)	15	±15 (9 specimens)

values offer only an approximate guide and are applicable under ideal conditions with specific calibrations for the particular concrete mix. If any factor varies from this ideal condition the accuracies of prediction will be reduced.

Example 13.2 A typical 50 mm non-standard diameter core drilled vertically from a transmission tower foundation contained one 18 mm ϕ reinforcement bar normal to the core axis which was located 27 mm from one end. The crushing load on the core of length (after capping) of 80 mm was 35 kN. Determine the *in situ* cube strength.

Solution   (a)   *In situ* concrete strength as per BS : 1881-Part 120(34):

$$\text{Measured concrete core strength}, f_c = \frac{(35 \times 10^3)}{\pi \times 50^2/4} = 17.8 \text{ MPa}$$

$$\frac{\text{Length}}{\text{diameter}} \text{ ratio of the core} = 80/50 = 1.6$$

$$\text{Estimated in-situ concrete strength} = \frac{2.3 \times 17.8}{[1.5 + (1/1.6)]} = 19.2 \text{ MPa}$$

$$\text{Correction factor for the reinforcement} = 1 + 1.5 \left(\frac{18}{50} \times \frac{27}{80}\right) = 1.18$$

Corrected in situ concrete strength = $19.2 \times 1.18 = 22.66$ MPa
± 12 per cent for an individual result
$= 22.66 \pm 2.7$ MPa

(b)   *Potential strength (41)*:

Estimated potential cube strength (for vertically drilled core)

$$= \frac{3.0 \times 17.8}{[1.5 + (1/1.6)]} = 25.13 \text{ MPa}$$

Corrected potential cube strength = $25.13 \times 1.18 = 29.65$ MPa

13.3.5 Load Tests

In the cases where member strength can not be adequately determined from the results of *in situ* concrete strength tests, load testing may be necessary. These are aimed at checking the structural capacity, and are hence concentrated on suspect or critical locations. Except in the cases where variable loading dominates, the *static load tests* are conducted. Loading tests may be divided into two main categories.

1. *In situ load testing*   The principal aim is to demonstrate satisfactory performance under an *overload* above the *service load conditions*. The performance is usually judged by measurement of deflection under this load sustained for a specified period. The need for the test may arise from the doubts about the quality of construction or design, or where some damage has occurred, and the approach is particularly valuable where public confidence is involved. In some international codes the static load tests are an established component of *acceptance criteria*. These tests are normally accompanied by some sort of monitoring of structural behaviour under *incremental loads*.

The *in situ* tests should not be performed before the *characteristic strength of concrete* has been reached (i.e. concrete is 28 days old). Preliminary work is always required to ensure safety in the event of a collapse under test conditions, and that the members under test are actually subjected to the calculated *test load*. Scaffolding must be provided to support at least twice the total load from any member liable to collapse together with the test load. This should be set to catch falling members after a minimum drop, but at the same time should not interfere with expected *deflections*.

Test Loads

IS : 456–1978 requires that the structure should be subjected to a total *design dead load* of structure plus 1.25 times the *imposed design load* for a period of 24 hours, and then the imposed load should be removed.

Thus   Test load = design dead load + 1.25 × (design imposed load)

The performance is based initially on the acceptability of the *measured deflection* and *cracking* in terms of the design requirements coupled with the examination of unexpected defects. The deflection due to imposed load only shall be recorded. If within 24 hours of removal of imposed load, the structure does not recover at least 75 per cent of the deflection under superimposed load, the test may be repeated after a lapse of 72 hours (second load cycle). If the *recovery* under second load cycle is less than 80 per cent, the structure shall be deemed to be unacceptable. If significant deflections occur, the deflection recovery rates after removal of load should also be examined. On the other hand if the maximum deflection in mm is less than 40 l^2/D (where l is the effective span in metres and D is the overall depth in mm), it is not necessary for the recovery to be measured.

Load Application

The load should be provided as cheaply as possible. The rate of application and distribution of load must be controlled and the magnitude must be easily assessable. Bricks, bags of cement, sand bags, steel weight, and water are amongst the materials which may be used and the choice will depend upon the nature and magnitude of the load required as well as the availability of materials and ease of access. Care must be taken to avoid arching of the load as deflection increases. If loading is to be spread over a larger area, ponding of water is the most appropriate method providing load. Slabs may be pounded by providing suitable containing walls and water proofing. Water is particularly useful in the locations with limited space or difficult access. However, leakage should be minimized to avoid damage to the finishes. Loads should always be applied in predetermined increments and in a way which will cause minimum lack of symmetry or uniformity. Similar precautions should be taken during unloading.

Deflection dial gauges must be carefully observed throughout the loading cycle, and if there are signs of deflections increasing with time under constant load, further loading should be stopped and the load reduced as quickly as possible. The potential speed of load removal is thus an important safety consideration. *Non-gravity loading* offers advantages of greater control, which can also be affected from a distance from the immediate test area. However, it is restricted to specialized or complex loading arrangements.

The deflection measurements are made by mechanical dial gauges which must be clamped to an independent rigid support. Gauges are normally located at mid-span and quarter points to chek symmetry of behaviour. The gauges must be set so that they can be easily read with a minimum risk to personnel and chance of disturbance during test is small. Telescopes may often be convenient for this purpose. Readings should be taken at all incremental stages throughout the test cycle. Measurement accuracy of ± 0.1 mm is generally possible with dial guages. A plot of *load deflection curve* is recommended. An examination of the plot can yield valuable information about the behaviour of the test member.

In situ load tests may be used in the following circumstances:

(i) Where the structure is suspected to be substandard due to the quality of design and construction and it is required to check the adequacy of the structure.

(ii) Where non-standard design concept has been used and it is required to demostrate the validity of the concept.

(iii) Where structural modification has been carried out for change in occupancy which may require increased loading.

(iv) Where a proof of improved performance is required following major repairs.

Ultimate Load Testing

It is frequently used as a quality control check on standard precast elements. *Ultimate load testing* is an imporant approach where *in situ* overload tests are inadequate. These are generally carried out in a laboratory where carefully controlled hydraulic load application and recording system are available. The results of a carefully monitored test provide conclusive evidence regarding the behaviour of the component examined.

13.3.6 Chemical Testing

Chemical analysis of hardened concrete may be used to check the *specification compliance* involving cement content, aggregate-cement ratio or alkali content determination. Water-cement ratio, and hence strength, are difficult to assess to any worthwhile degree of accuracy. Thus the analysis may be used only in cases of uncertainty, or in resolving disputes, rather than as a means of quality control of concrete. Specialised laboratory facilities are required for most forms of chemical testing. One of the major problems of basic chemical testing is the lack of a suitable solvent which will dissolve *hardened cement* without affecting the aggregates, and if possible samples of aggregates and cement should also be available for testing. The basic steps involved are outlined as follows:

1. Sampling

Sufficient samples are taken to represent the body of concrete under examination at a particular location in the structure. The basic requirements for a sample for chemical analysis are that:

(i) The concrete sample should preferably be in a single piece with the minimum linear dimension being atleast five times the maximum aggregate size (weighing about 2 kg). Several samples are taken from different points.

(ii) The sample should be free from reinforcement and foreign matter.

The sample shall be clearly labelled giving all relevant details and sealed in a heavy-duty polythene bag which should also be labelled.

2. Chemical Analysis

The method is based on the fact that the lime compounds and the silicates in portland cement are readily decomposed by, and soluble in dilute hydrochloric acid (HCl), than the corresponding compounds in aggregate. The quantity of soluble silica or calcium oxide is determined by a simple analytical procedure, and if the composition of cement is known, the cement content of the original volume of the sample can be calculated. Allowance must be made for any material which might have been dissolved from the aggregate. The representative sample of aggregate should, therefore, be analyzed by identical procedures to determine the *correction* to be made.

(i) *Preparation of sample* The sample is initially broken into lumps not larger than 40 mm, taking care as far as possible to prevent aggregate fracture. These lumps are dried in an oven at 105°C for 15–24 hours, allowed to cool to room temperature, and divided into subsamples. A portion of the dried sample is crushed to pass a 4.75-mm sieve and a subsample of 500–1000 g is obtained which is then crushed to pass a 2.36-mm sieve and quartered to give a sample which is pulverized in a ball mill to pass a 600-µm sieve. This is also quartered and further ground to a powder to pass 150-µm sieve. This final sample is freed from particles of metallic iron abraded from pulverizer ball mill, by means of a strong magnet.

(ii) *Determination of calcium oxide content* A portion of the prepared sample weighing $5 + 0.005$ g is treated with boiling dilute hydrochloric acid. Triethanolamine, sodium hydroxide and calcein indicator are added to the filtered solution which is then titrated against a standard EDTA solution. The CaO content may be calculated to the nearest 0.1 per cent. If the CaO content of the aggregate is less than 0.5 per cent, this analysis may be considered adequate. However, additional determination of *soluble silica content* is recommended.

(iii) *Determination of soluble silica content* Soluble silica is extracted from a 5 ± 0.005 g portion of the prepared sample by treatment with hydrochloric acid and its insoluble residue collected by filtration. The filtrate is reduced by evaporating and treating with hydrochloric acid and polyethylene oxide, before being filtered again and diluted to provide a stock solution.

The filter paper containing the precipitate produced at the last stage, is ignited in a weighed platinum crucible at $1200 \pm 50°C$ until constant mass is achieved, before cooling and weighing. The soluble silica content can be calculated to the nearest 0.1 per cent from the ratio of the mass of the ignited residue to that of prepared sample. The calcium oxide content is determined from the stock solution using the procedure given in (i) above. The insoluble residue is determined from the material retained during the initial filtration process by repeated treatment with hot ammonium chloride solution, hydrochloric acid and hot water followed by ignition in a weighed crucible to $925 \pm 25°C$.

3. Calculation of Cement and Aggregate Contents

The cement content should be calculated separately from both the measured calcium oxide and soluble silica contents, unless the calcium oxide content of the aggregate is less than 0.50 per cent or greater than 35 per cent in which case results based on CaO are not recommended. In the latter case, if the soluble silica content of the aggregate is greater than 10 per cent the analysis should be undertaken to determine some other constituent present in a larger quantity in the cement.

The calculation of cement content is based on the assumption that the combined water of hydration is 0.23 times the cement content, and that 100 g of oven dried concrete consists of C g of cement + A g of aggregate + 0.23C g of combined water of hydration.

$$100 = C + A + 0.23C = 1.23C + A \qquad (13.10)$$

Thus if,

a = calcium oxide or soluble silica content of cement, per cent,
b = calcium oxide or soluble silica content of aggregate, per cent, and
c = measured calcium oxide or soluble silica content of the analytical sample, per cent;

Then
$$\frac{c(1.23C + A)}{100} = \frac{aC}{100} + \frac{bA}{100}$$

or
$$(c - b)A = (a - 1.23c)C \qquad (13.11)$$

Then from Eqs. (13.10) and (13.11) the cement content

$$C = \frac{c - b}{a - 1.23b} \times 100 \text{ per cent (to nearest 0.10 per cent)}$$

and

Aggregate content, $A = \dfrac{a - 1.23c}{a - 1.23b} \times 100$ per cent (to nearest 0.10 per cent)

Thus the aggregate-cement ratio A/C may be obtained to the nearest 0.10. The cement content by mass is given by

$$\frac{C \times \text{oven dry density of concrete}}{100} \text{ kg/m}^3 \text{ to nearest 0.10 kg/m}^3$$

The above calculations require that an analysis of both the cement and aggregate to be availabe. If an analysis of cement is not available, OPC or RPC complying with the relevant Indian Standard Code may be assumed. If the two estimated cement contents are within 25 kg/m^3 or 1 per cent by mass, the value is adopted. Thus, method of analysis suffers from the drawback that it cannot be used for concretes which contain aggregates or admixtures or additives such as flyash or pozzolanas, which liberate soluble silica under the conditions of test.

14

Special Concretes and Concreting Techniques

14.1 INTRODUCTION

Notwithstanding its versatility, cement concrete suffers from several drawbacks, such as *low tensile strength*, *permeability to liquids* and consequent *corrosion of reinforcement*, susceptibility to *chemical attack*, and *low durability*. Modifications have been made from time to time to overcome the deficiencies of cement concrete yet retaining the other desirable characteristics. Recent developments in the material and construction technology have led to significant changes resulting in improved performance, wider and more economical use. The improvements in performance can be grouped as:

(i) Better mechanical properties than that of conventional concrete, such as compressive strength, tensile strength, impact toughness, etc.,

(ii) Better durability attained by means of increased chemical and freeze-thaw resistances,

(iii) Improvements in selected properties of interest, such as impermeability, adhesion, thermal insulation, lightness, abrasion and skid resistance, etc.

The mechanical properties can be improved by using one or more of the following approaches:

(i) modifications in microstructure of the cement paste,
(ii) reduction in overall porosity,
(iii) improvements in the strength of aggregate-matrix interface, and
(iv) control of extent and propagation of cracks.

14.1.1 Modification in the Microstructure

Considerable improvements in inter-particle cohesive forces can be realized by reducing the inter-particle spacing of the hydrate phase. Perhaps the most notable attempt to modify the microstructure is the application of the *hot pressing technique*. By application of pressures of up to 350 MPa during moulding at temperatures up to 150°C, compressive strengths of the order of 520 MPa have been obtained. The electron micrographs of such *hot pressed cement plates* have revealed a marked improvement in the microstructure in comparison to those cured at ordinary temperatures in that they show dense and relatively homogeneous structures. Though not yet used in construction, this method reveals the potential of future concrete.

14.1.2 Reduction in Porosity

The *mechanical properties* and *durability* of concrete can be improved by filling the pores, voids and cracks by incorporating or impregnating the concrete with *polymers*. In *polymer-impregnated concrete*, the pores in conventional concrete after normal curing are emptied under vacuum and then a *monomer* is sucked in, which is later polymerized by the application of heat or radiation. Considerable increase in tensile and compressive strengths, and modulus of elasticity and hardness results. Compressive strengths of the order of about 280 MPa have been obtained. Commercial applications of polymer-impregnated concrete (PIC) include piles, tunnel liners, precast prestressed bridge deck panels and in wide ranging repairs.

Sulphur-impregnated concretes, in a similar manner, have resulted in high strength concretes from lean conventional concrete mixes. A typical value of compressive strength of sulphur-impregnated concrete has been reported to be 55 MPa from a reference moist-cured ordinary concrete having a strength of 5.5 MPa i.e. a ten-fold increase. In India the applications of sulphur-impregnated concrete are limited due to high cost of sulphur.

14.1.3 Stronger Aggregate-matrix Interface

The *mechanical properties* of cement concrete which consists of a relatively inert aggregate bounded by hydrated cement binder of *matrix*, depend upon the strength of aggregate and the stability of concrete through the matrix. In particular, the interface between the aggregate and the matrix must be capable of transferring the stresses due to loads to aggregate. This is generally achieved in cement concrete through the strong Van der-waals bonds between the micro-crystalline components of hydrated cement paste and the aggregate. However, the bonds are not so strong as to transfer tensile or shear stresses, and hence the composite, i.e., cement concrete is relatively weak in tension and shear. Only the compressive stresses are effectively transmitted. As the aggregate is usually very strong, the aggregate strength can be fully exploited by achieving greater force transfer capability. Beyond a level, the conventional cement matrix is unable to accomplish this.

It is possible to supplement the cement matrix in the composite with another matrix or, if the cement matrix is replaced by a more efficient matrix, it should be possible to obtain concrete of much higher strength. If the binder or the matrix exhibits ionic or covalent bonds with aggregate at the interface, the resulting composite will also be sufficiently strong to transmit large tensile forces. Efforts in this direction have resulted in the use of polymers, either as sole matrix or supplement to the cement matrix.

With the addition of polymers, the failure of concrete specimens does not occur through the aggregate-mortar interface, but through the aggregates themselves, thereby showing improvement in bond strength at the interface. With an improvement in bond strength at the interface, the aggregate strength can be fully exploited, i.e. the concrete strength is limited by the mechanical strength of the aggregate.

14.1.4 Control of Extent and Propagation of Cracks

The most notable development in this direction is the use of *ferrocement* and *fibre-reinforced concretes*. In ferrocements, meshes of thin steel wires of various configurations and sizes are incorporated as reinforcement in cement-mortars. However, in fibre-reinforced concrete (FRC), steel, glass or polymeric fibres of suitable mechanical and chemical properties and having optimum *aspect ratios* are incorporated with other concrete materials at the mixing stage. In a way both can be viewed as reinforced concretes. The wire-mesh or fibres hold the matrix together after *localized cracking*, and provide improved *ductility* and *post-cracking load-carrying capacity*. The compressive strength improves slightly (say by 25%), but the *tensile strength, first-crack tensile strength, impact strength* and *toughness* or *shock absorption capacities* show a two-to-four fold improvement.

Ferrocement has found wide applications in boat-hull building, construction of shells, and similar structural components of thin sections. Applications of fibre-reinforced concrete include pavements and runways, industrial floors, hydraulic structures, breakwater, armour units, pile foundations, etc.

The combinations of fibre-reinforced concrete and polymer impregnation technique are seen as the potential method of utilizing the advantages of both, i.e. a ductile material of high toughness equal to 228 times that of normal mortars. Similarly, a *fibrous ferrocement* composite can be regarded as a future composite of high potential.

14.2 LIGHT-WEIGHT CONCRETE

The conventional cement concrete is a heavy material having a density of 2400 kg/m^3, and high *thermal conductivity*. The dead weight of the structure made up of this concrete is large compared to the imposed load to be carried, and a relatively small reduction in dead weight, particularly for members in flexure e.g., in highrise buildings, can save money and manpower considerably. The improvement in thermal insulation is of great significance to the conservation of energy. The reduction in dead weight is normally achieved by cellular construction, by entraining large quantities of air, by using *no-fines* concrete and lightweight aggregates which are made lighter by introducing internal voids during the manufacturing process.

The term no-fines indicates that the concrete is composed of cement and coarse aggregate (commonly 10 or 20 mm grading) only, the product has uniformly distributed voids. Suitable aggregates used are natural aggregates, blast-furnace slag, clinker, foamed slag, sintered flyash, expanded-clay, etc.

Lightweight aggregate is a relatively new material. For the same crushing strength, the density of concrete made with such an aggregate can be as much as 35 per cent lower than the normal weight concrete. In addition to the reduced dead weight, the lower modulus of elasticity and adequate ductility of lightweight concrete may be advantageous in the seismic design of structures. Other inherent advantages of the material are its greater fire resistance, low thermal conductivity, low coefficient of thermal expansion, and lower erection and transport costs for pre-fabricated members. For pre-fabricated structures a smaller crane is required or the same crane can handle larger units due to reduction in dead weight. For cast *in situ* structures, its smaller dead weight makes foundations less expensive.

Moreover, continued extraction of conventional dense natural aggregate from the ground is bound to be accompained by severe environmental problems leading to deterioration of the countryside and its ecology. On the other hand, use of manufactured aggregates made of industrial wastes (slags etc.), preferably those containing sufficient combustible materials (pulverized fuel ash) to provide all or most of the energy for their production, may help in alleviating the problem of disposal of industrial waste.

14.2.1 Light-weight Aggregates

Light-weight aggregates may be grouped in the following categories:

(i) Naturally occurring materials which require further processing, such as expanded-clay, shale and slate, etc.,

(ii) Industrial by-products, such as sintered pulverized fuel ash (flyash), foamed or expanded-blast-furnace slag, and

(iii) Naturally occurring materials, such as pumice, foamed lava, volcanic tuff and porous limestone.

Aggregates Manufactured from Natural Raw Material

The artificial lightweight aggregates are mainly made from clay, shale, slate or pulverized-fuel ash, subject to a process of either *expansion (bloating)* or *agglomeration*. During the process of expansion the material is heated to fusion temperature at which point pyroplasticity of material occurs simultaneously with the formation of gas. Agglomeration on the other hand occurs when some of the material fuses (melts) and various particles are bonded together. Thus to achieve proper expansion a raw material should contain sufficient gas-producing constituents, and pyroplasticity should occur simultaneously with the formation of gas. The gas may form due to decomposition and combustion of sulphide and carbon compounds; removal of CO_2 from carbonates or reduction of Fe_2O_3 causing liberation of oxygen. The common examples of *natural minerals* suitable for expansion are clay, shale, slate and perlite and exfoliated vermiculite.

(i) *Expanded or bloated-clay* Bloated-clay aggregates are made from a special grade of clay suitable for expansion. The ground clay mixed with additive which encourages bloating, is passed through a rotary or vertical shaft kiln fired by a mixture of pulverized coal and oil with temperature reaching to about 1200°C. The material produced consists of hard rounded particles with a smooth dense surface texture and honeycomb interior.

(ii) *Expanded-shale* The crushed raw material such as colliery waste, blended with ground coal is passed over a sinter strand reaching a temperature of about 1200°C. At this temperature the particles expand and fuse together trapping gas and air within the structure of the material with a porous surface texture.

(iii) *Expanded-slate* The crushed raw material is fed into a rotary kiln with temperature reaching 1200°C. The material produced is chemically inert and has a highly vitrified internal pore structure. This material is then crushed and graded.

(iv) *Exfoliated-vermiculite* The raw material resembles mica in appearance and consists of thin flat flakes containing microscopic particles of water. On being suddenly heated to a high temperature of about 700–1000°C, the flakes expand (exfoliate) due to steam forcing the laminates apart. The material produced consists of accordion granules containing many minute air layers.

Industrial by-product Light-weight Aggregate

These include sintered-pulverized fuel ash, foamed blast furance slag and pelletized slag.

(i) *Sintered-pulverized fuel ash* The fly-ash collected from modern power stations burning pulverized fuel, is mixed with water and coal slurry in screw mixers and then fed onto rotating pans, known as pelletizers, to form spherical pellets. The *green* pellets are then fed onto a sinter strand reaching a temperature of 1400°C. At this temperature the fly ash particles coagulate to form hard brick-like spherical particles. The produced material is screened and graded.

(ii) *Foamed-blast furnace slag* It is a by-product of iron production formed by introducing water or steam into molten material. The material produced after annealing and cooling is angular in shape with a rough and irregular glassy texture, and an internal round void system.

Naturally Occuring Light-weight Aggregates

The common examples are pumice and diatomite. *Pumice* is light and strong enough to be used in its natural state, but has variable qualities depending upon its source. It is chemically inert and usually has a relatively high silica content of approximately 75 per cent. Diatomite on the other hand is a semiconsolidated sedimentary deposit formed in cold water environment.

Production In India, raw light-weight aggregates are produced by using any of the following:
  (i) bloated-clay aggregates by bloating suitable clays with or without additives,
  (ii) sintered-flyash aggregates by sintering the flyash, and
  (iii) light-weight aggregate from blast furnace slag.

In one of the processes for manufacturing light-weight concrete, the cement and pulverized sand are first mixed in a certain proportion (1 : 1 for insulation and 1 : 2 for partitioning purposes). The mixture so formed is then made into slurry with the addition of a pre-determined quantity of water. The sand-cement slurry is next foamed to the extent of pre-determined volume with the help of a foaming compound. The foam product is thereafter poured into moulds. The moulded blocks are finally cured under elevated *hydrothermal conditions* in autoclaves which imparts strength, reduces drying shrinkage and gives the block a creamy colour.

In another product, lime and sand are used as raw materials. Both are first ground to fine powder in huge ball mills. The mixture is then made into slurry with the addition of water. Adding aluminium powder and gypsum to the slurry triggers a chemical reaction, and the hydrogen gas evolved gives the cellular concrete its lightness. After initial hardening it is cut into convenient sizes and the moulded blocks are finally cured under elevated *hydrothermal conditions* (under a pressure of 12 atmospheres and temperature of 196°C).

The suitability of a particular light-weight aggregate is determined by the specified compressive strength and the density of concrete.

14.2.2 Properties of Light-weight Aggregates

The properties of the manufactured light-weight aggregates depend mainly on the raw material, and the process of manufacture. The properties of aggregates manufactured from materials which occur as industrial by-products can be altered to a limited extent only by the processes of bloating, foaming, sintering, agglomerating and crushing. Since the aggregate make up approximately 75 per cent of the total volume of the concrete, it influences the workability, strength, modulus of elasticity, density, durability, thermal conductivity, shrinkage, and creep properties of concrete. The structural concrete should have a high strength with low density, high modulus of elasticity, and low rate of shrinkage and creep. On the other hand, a light-weight aggregate concrete should possess low thermal conductivity. The thermal conductivity decreases with decreasing density, therefore the density of the concrete must be as low as possible. The most suitable aggregates for structural light-weight concrete are expanded-clay, shale and slate, fly ash and colliery waste. Adequate strength for structural light-weight aggregate concrete can be obtained with foamed and expanded-blast furnace slag. For light-weight aggregate concrete for thermal insulation the suitable aggregates are pumice, perlite, vermiculite, diatomite and expanded-polystyrene.

A surface texture with tiny and uniformly distributed pores is preferred. Particle size and shape as well as surface condition of aggregates influence properties of fresh concrete. Crushed and angular light-weight aggregate requires high mortar content resulting in a higher density than that with rounded aggregate. The strength of the light-weight aggregate particles decreases with decreasing density. The density, bulk density and water absorption capacity of some of the commonly used light-weight aggregates are given in Table 14.1. The compressive strengths and unit weights of typical concretes produced by these aggregates are also given in the table.

14.2.3 Mix Proportions

Due to large variations in the characteristics of light-weight aggregates, it is difficult to seek a single approach to mix design for structural light-weight aggregate concrete. However, following points should be considered.

As in case of normal weight concrete, the light- weight aggregate concrete can attain the strength of mortar matrix only if the strength and stiffness of the aggregate are at least as high as those of the mortar. Below this limit, internal stress transfer takes place in the same way as in normal weight concrete. In this case concrete strength is approximately equal to the strength of mortar. The *water-cement ratio* and mix proportions applicable to ordinary concrete can be adopted, however instead of total water only the effective water must be taken into account.

TABLE 14.1 Physical properties of light-weight aggregate

Aggregate Type	Particle shape, and surface texture	Density, (kg/m³)	Bulk density (kg/m³)	24-hour water absorption capacity, per cent	Typical concrete Compressive strength (MPa)	Typical concrete Unit weight (kg/m³)
(A) Aggregate for high strength concrete ($f_{ck} > 15$ MPa)						
Expanded-clay	Rounded and slightly rough particles	*Coarse* 600 to 1600 *Fine* 1300 to 1800	300 to 900	5 to 30	10 to 60	1000 to 1700
Expanded-shale and slate	Often angular and slightly rounded, smooth surface	*Coarse* 800 to 1400 *Fine* 1600 to 1900	400 to 1200	5 to 15	20 to 50	1300 to 1600
Fly ash	Similar to expanded-clay	1300 to 2100	600 to 1100	20	30 to 60	1500 to 1600
Foamed-blast furnace slag	Irregular angular particles with rough and openpored surface	1000 to 2200	400 to 1100	10 to 15	10 to 45	1800 to 2000
Sintered-colliery waste	Angular with open-pored surface	1000 to 1900	500 to 1000	15	10 to 40	1400 to 1600
(B) Aggregate for medium strength concrete (3.5 to 15 MPa)						
Pumice	Rounded particles with open-textured but rather smooth surface	550 to 1650	350 to 650	50	5 to 15	1200 to 1600
(C) Aggregate for low strength concrete (0.5 to 3.5 MPa)						
Perlite	Rounded and of angular shape and rough surface	100 to 400	40 to 200	—	1.2 to 3.0	400 to 500
Vermiculite	Cubical	100 to 400	60 to 200		1.2 to 3.0	300 to 700

It is, of course, also possible to manufacture light-weight concretes with higher strength than the limit strength mentioned above by using a stronger mortar (having greater stiffness) with a higher density. In this concrete the mortar matrix will transmit higher stress at the same deformation. For economic reasons, it is preferable to select a stronger aggregate such that the required concrete strength can be attained with the mortar of lower strength. For a concrete of given compressive strength, a strong aggregate requires a low mortar strength, a weak aggregate a high mortar strength.

Since aggregate strength and its modulus of deformation is not usually available, the suitability of a light-weight aggregate for a specific application is generally assessed by means of the particle density or bulk density. For structural light-weight concrete, the maximum nominal size of the aggregate is limited to 20 mm since the modulus of deformation, strength and density of aggregate particles decrease as particle size increases. On the other hand a lower maximum size and a large proportion of fines may lead to higher strength but the concrete density will increase.

Natural sand is often used to improve the *workability* and reduce the *shrinkage* of fresh concrete and increases its strength, but it will increase the density of concrete.

The conventional *water-cement ratio* rule is not suitable for light-weight concrete. In light-weight concrete the water content to be taken into account for calculation of water-cement ratio is not the total quantity of water present but only the effective or free water. The relationship between strength and water-cement ratio varies from aggregate to aggregate. The effect of cement strength on the strength of concrete is not linear.

The main problem of light-weight concrete mix design lies in the advance determination of effective water and air contents of cement matrix at the moment of completion of compaction of concrete. The prediction is difficult to make since throughout the mixing process the effective (free) water content is progressively reduced through absorption by aggregate, except when completely saturated aggregate is used. The combined free-water and air contents can be approximately estimated as the *residual absolute volume*, when the density of fresh concrete, the mix proportions and particle density are known. The residual absolute volume, V_{res}, is obtained by subtracing the volume of the solid cement and aggregates from the total volume of concrete.

$$V_{res} = V_w + V_{air} = 1000\left(1 - \frac{C}{S_c} - \frac{A}{S_a}\right)$$

where, V_w and V_{air} are the free-water and air contents in litre/m^3, respectively. C and A denote cement and aggregate contents in kg/m^3, respectively. S_c and S_a are the density of cement and mean particle density of aggregates in kg/m^3, respectively. In contrast to normal concrete the relationship between *residual absolute volume-cement ratio* and the strength of light-weight concrete vary from aggregate to aggregate. As the residual absolute volume-cement ratio decreases the concrete strength increases, however the increase is less than for normal concrete. For every type of light-weight aggregate the compressive strength of the concrete bears a definite relationship to the residual absolute volume-cement ratio and to the cement strength. This characteristic can facilitate the design of light-weight concrete mixes.

280 Concrete Technology

The *optimum cement content* may be determined by *trial mixes*. In general for first trial, the cement content required for ordinary sand and gravel concrete may be used, but more cement is normally required for most light-weight aggregate concretes.

There are several methods to determine the aggregate content. In this section a method using *effective water-cement ratio* for the calculation of aggregate content is described.

This method of light-weight aggregate concrete mix design which is based on water-cement rule and is an adaptation of the well-known British mix design method has been suggested by F.I.P. The steps involved to obtain the mix proportions for the stipulated 28-day strength of concrete are:

(i) The target mean strength of the concrete is determined from the characteristic strength.

(ii) The water-cement ratio for the required target strength is read off from Fig. 14.1.

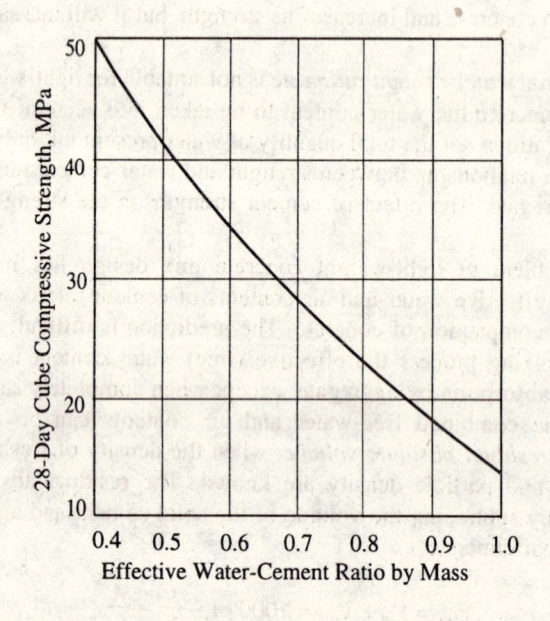

Fig. 14.1 Typical relationship between water-cement ratio and compressive strength of an light-weight aggregate concrete

(iii) For the water-cement ratio determined in step (ii) aggregate-cement ratio (by volume), cement content in kg/m^3 and optimum percentage of fine aggregate for the desired workability are selected from Table 14.2.

(iv) The cement and aggregate volumes are converted to mass contents by multiplying with their dry bulk densities.

(v) The water content is adjusted for the absorption and moisture content of aggregates to obtain effective free-water content for the mix.

(vi) A trial mix is prepared and water content is adjusted to maintain the desired workability. The density of fresh wet compacted concrete is calculated and cement content checked. If it is not correct, minor corrections are made by adding or subtracting cement and subtracting or adding the same volume of fine aggregate as follows:

Volume of fresh concrete = Absolute volume of ingredients

$$1000 \text{ litres} = \frac{\text{weight of cement}}{\text{density of cement}} + \frac{\text{weight of fine aggregate}}{\text{saturated surface dry* density of fine aggregate}} + \frac{\text{weight of coarse aggregate}}{\text{saturated surface dry* density of coarse aggregate}} + \text{air} + \text{free water}$$

TABLE 14.2 Variation of cement content, aggregate-cement ratio, and fine aggregate content with water-cement ratio for the light-weight aggregate concrete

Water-cement ratio	Medium: Compacting factor = 0.90, slump = 15–75 mm			High: Compacting factor = 0.95, slump = 75–150 mm		
	Cement, (kg/m^3)	aggregate-cement ratio	Fine aggregate, per cent	Cement (kg/m^3)	aggregate-cement ratio	Fine aggregate, (per cent)
0.40	525	3.00	28	591	2.65	27
0.45	468	3.50	30	533	3.03	28
0.50	419	4.04	32	474	3.50	30
0.55	372	4.63	35	425	4.01	32
0.60	332	5.35	40	383	4.55	34
0.65	293	6.05	49	342	5.23	38
0.70	260	6.83	58	306	5.99	44
0.75	230	8.02	69	273	6.75	53

For other light-weight aggregates, necessary reference curves of effective water-cement ratio against 28-day cube crushing strength, and tables should be prepared with the help of experimental results on laboratory mixes using varying fine- coarse aggregates ratios.

Design example Suppose a light-weight concrete mix having a 28-day characteristic compressive strength of 30 MPa is required. The degree of workability envisaged is 0.95 CF and the degree of quality control available at the site may be termed as good. The unit weights of cement, fine and coarse aggregates available are 1425 kg/m^3, 1100 kg/m^3 and 850 kg/m^3, respectively. The moisture contents of fine and coarse aggregates (by mass) are 10 and 6 per cent, respectively, and short term absorptions (by mass) are 5 and 11 per cent, respectively.

*mean particle density after 1/2 hour soaking.

282 Concrete Technology

Procedure:

$$f_t = 30 + 1.65 \times 6 \approx 40 \text{ MPa}$$

(i) From Fig. 14.1

    Water-cement ratio $\qquad = 0.53$

(ii) From Table 14.2

    Aggregate-cement ratio $\qquad = 3.794 \approx 3.8$

    Fine aggregate, per cent $\qquad = 31.2 \approx 31.0$

    Cement-content $\qquad = 444.6 \approx 445 \text{ kg/m}^3$

(iii) *Dry volumes per cubic metre of concrete*

Cement	=	445/1425	= 0.312 m³
Aggregate	=	3.8 × 0.312	= 1.186 m³
Fine aggregate	=	31% of 1.186	= 0.368 m³
Coarse aggregate	=	1.186 − 0.368	= 0.818 m³

(iv) *Batch masses of ingredients for 1 m³ of concrete, kg*

Ingredient	water	: cement	: fine aggregate	: coarse aggregate
	0.53 × 445	445	0.368 × 1100	0.818 × 850
Apparent masses	236	445	405	695
Adjustment for moisture content	(−)40.5 (−)41.5		(+)40.5	(+)41.5
Absorption of aggregate				
fine 5% × 405	+20.0			
coarse 11% × 695	+76.0			
Corrected dry batch masses	250 0.56	445 1.0	445.5 1.0	736.5 1.65

14.2.4 Properties of Structural Light-weight Concrete

With suitable light-weight aggregates, a concrete can be produced with densities which are 25 to 40 per cent lower but with strengths equal to the maximum normally achieved by ordinary concrete.

    The fracture mechanism of light-weight aggregate concrete under compression differs from that of a normal weight concrete. In normal weight concrete the tensile stress is generated on the interface between aggregate and mortar matrix directed transversely to the direction of external loading. With light-weight aggregate concrete, the tensile stress also acts transversely to the direction of external loading, but immediately above and below the aggregate due to diversion of greater part of force around the aggregate because of its lower modulus of deformation. The internal cracks which lead to fracture are initiated at all the places in concrete where tensile stress exceeds the actual tensile strength.

The important characteristics of lightweight concrete are:

(i) *Low density* The density of the concrete varies from 300 to 1200 kg/m^3. The lightest grade is suited for insulation purposes while the heavier grades with adequate strength are suited for structural applications. The low density of cellular concrete makes it suitable for precast floor and roofing units which are easy to handle and transport from the factory to the sites.

(ii) *High strength* Cellular concrete has high compressive strength in relation to its density. The compressive strength of such concrete is found to increase with increasing density. The tensile strength of aerated cellular concrete is about 15 to 20 per cent of its compressive strength.

Due to a much higher strength-to-mass ratio, the cellular concrete floor and roof slabs are approximately one quarter the weight of normal reinforced concrete slabs.

(iii) *Thermal insulation* The insulation value of light-weight concrete is about three to six times that of bricks and about ten times that of concrete. A 200 mm thick wall of aerated concrete of density 800 kg/m^3 has the same degree of insulation as a 400 mm thick brick wall of density 1600 kg/m^3.

(iv) *Fire resistance* Light-weight concrete has excellent fire resisting properties. Its low thermal conductivity makes it suitable for protecting other structures from the effects of fire.

(v) *Sound insulation* Sound insulation in cellular concrete is normally not as good as in dense concrete.

(vi) *Shrinkage* Light-weight concrete is subjected to shrinkage but to a limited extent. The autoclaving of cellular concrete reduces drying shrinkage to one-fifth of that occurring during air-curing.

(vii) *Repairability* Light-weight products can be easily sawn, cut, drilled or nailed. This makes construction easier. Local repairs to the structure can also be attended to as and when required without affecting the rest of the structure.

(viii) *Durability* Aerated concrete is only slightly alkaline. Due to its porosity and low alkalinity it does not give rust protection to steel which is provided by dense compacted concrete. The reinforcement used, therefore, requires special treatment for protection against corrosion.

(ix) *Speed of construction* With the adoption of prefabrication, it is possible to design the structure on the concept of modular coordination which ensures a faster rate of construction.

(x) *Economy* Due to light weight and high strength-to-mass ratio of cellular concrete products, their use results in lesser consumption of steel. Composite floor construction using precast unreinforced cellular concrete blocks and reinforced concrete grid beams (ribs) results in appreciable saving in the consumption of cement and steel, and thereby reduces the cost of construction of floors and roofs considerably. A saving of as much as 15 to 20 per cent in the cost of construction of floors and roofs may be achieved by using this type of construction compared to conventional construction.

(xi) *Quality control* A better quality control is exercised in the construction of structure with light-weight concrete products owing to the use of factory made units.

14.2.5 Applications

Different uses of light-weight (aerated) concrete can be summarized as follows:

(i) as load bearing masonry walls using cellular concrete blocks,
(ii) as precast floor and roof panels in all types of buildings,
(iii) as a filler wall in the form of precast reinforced wall panels in multistoreyed buildings,
(iv) as partition walls in residential, institutional and industrial buildings,
(v) as *in situ* composite roof and floor slabs with reinforced concrete grid beams,
(vi) as precast composite wall or floor panels, and
(vii) as insulation cladding to exterior walls of all types of buildings, particularly in office and industrial buildings.

The Bureau of Indian Standards has published several codes for controlling the production of autoclaved cellular concrete in India. The relevant code of practice for ordinary concrete, viz. *IS* : 456–1978, is applicable to the design of reinforced light-weight concrete structural elements.

14.3 ULTRA-LIGHT-WEIGHT CONCRETE

The ultra-light-weight concrete, with unit weights or densities ranging from 600 to 1000 kg/m^3, is made from a mixture of cement, sand (omitted for concrete having unit weight or density less than 600 kg/m^3) and expanded-polystyrene beads one to six millimeters in diameter. This concrete has a high *thermal insulation efficiency* and is mainly used for pre-fabricated non-load bearing panels, hollow and solid blocks, light-weight sandwich panels and in highway construction as a part of sub-base where frost could endanger the stability of the subgrade.

Polystyrene beads or foam manufacture is essentially a polymerization process which utilizes liquid styrene monomer dispersed in an aqueous medium containing foaming or expanding agent and polymerization catalyst. The foam or expanded bead products may be treated with bromine solutions for improving the *fire retardance* and *self-extinguishing characteristics*. The expanded beads normally have a density between 12 to 25 kg/m^3, however sheet form produces densities up to 30 to 75 kg/m^3. The commonly used size range of expanded beads is 1 to 3 mm. When exposed to sunlight, the foam or beads deteriorate producing a characteristic yellowing.

The expanded-polystyrene concrete mixes can be designed to have compressive strengths up to 15 to 20 MPa at densities around 1600 kg/m^3. The mix design aims at achieving an economic and optimum balance between density, thermal insulation and strength. The expanded-polystyrene beads become electrostatically charged during processing which makes them difficult to wet (hydrophobic) during mixing. *Proneness to segregation* can be overcome by using a *bonding agent* (normally water-emulsified epoxies, and aqueous dispersions of polyvinyl propionate) and by controlling the fluidity of the paste or mortar.

Compressive strength and thermal insulation properties of expanded-polystyrene concrete increase with its density. Due to reduced *specific thermal capacity* the *heat of hydration* of the cement causes greater and more rapid increase in temperature in this concrete than in concrete made with conventional aggregate resulting in accelerated *rate of setting* and *hardening*. The setting and hardening rates can be controlled by selecting the appropriate cement and using *water-reducing admixtures*.

The conventional workability tests namely slump test, compacting factor test, Vee-Bee test, and flow table test are unsuitable in case of expanded-polystyrene concrete. Conventional techniques can be used for casting and placing expanded-polystyrene concrete. The mechanical properties of expanded-polystyrene concrete are a function of density as in the case of other light-weight aggregate concretes, but have comparatively lower values. The elastic and shrinkage deformations are considerably greater than for normal-weight concrete.

14.4 VACUUM CONCRETE

In concreting thin sections like slab and walls a fluid mix with *water-cement ratio* of 0.50 to 0.65 is required to facilitate the placing and compaction. Such a mix will lead to relatively low strength and poor abrasion resistance. In such situations, the vacuum treatment of concrete, involving the removal of excess water and air by using suction can be helpful. An arrangement for vacuum treatment of concrete using suction through a surface mat connected to a vacuum pump is shown in Fig. 14.2.

The duration of treatment depends upon the water-cement ratio and the quantity of water to be removed. It generally ranges from 1 to 15 minutes for slabs varying in thickness from 25 mm to 125 mm. The effect of treatment is more pronounced

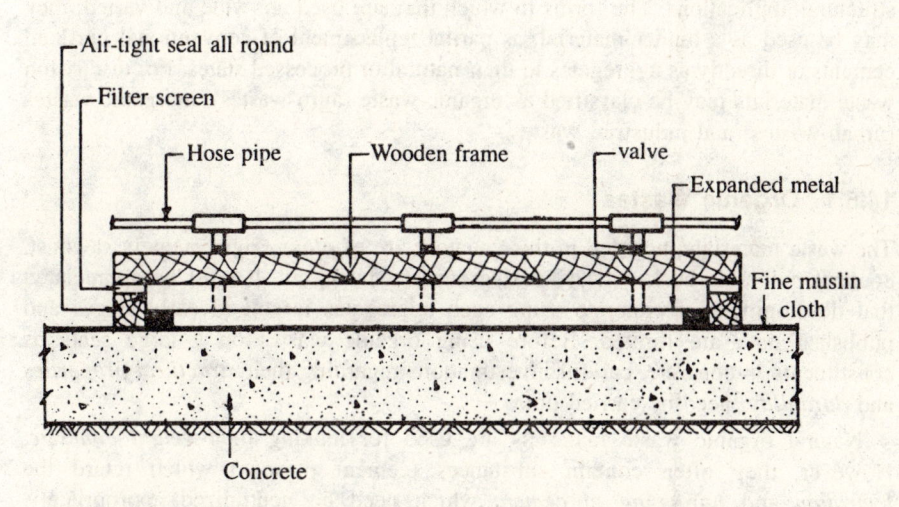

Fig. 14.2 An arrangement for vacuum treatment of concrete

in the beginning and falls off rapidly. Hence, it is of no advantage to prolong the periods of treatment beyond these values.

The *vacuum treatment* is not very effective for water-cement ratios below 0.4. The suction pressure on the concrete is about one-third the atmospheric pressure. The vibration of concrete before vacuum treatment can assist the process. The application of vibration simultaneously with vacuum treatment after initial vibration is very effective. Continued vibration beyond 90 s may damage the structure of concrete, and hence the vibrations should be stopped beyond this period and only vacuum needs to be applied for the remaining duration of the treatment.

The vacuum treatment has been found to considerably reduce the time of final finishing of floor and stripping of wall forms. The strength of concrete and its *resistance to wear* and *abrasion* increases and total shrinkage is reduced. Vacuum-treated concrete provides a good bond with the underlying concrete.

The vacuum-processed concrete has been extensively used for factory production of precast plain and reinforced concrete units. The other important application is in the construction of horizontal and sloping concretes slabs, such as floor slabs, road and airport pavements, thin load-bearing and partition walls. Vacuum treatment can also be effectively used in the resurfacing and repair of road pavements.

14.5 WASTE MATERIAL BASED CONCRETE

Recent investigations have made it possible to make concrete using agro-, urban- and industrial waste materials. Successful utilization of a waste material depends on its use being economically competitive with the alternate natural materials. These costs are primarily made up of handling, processing and transportation. The *stability* and *durability* of products made of concrete using waste materials over the expected life span is of utmost importance, particularly in relation to building and structural applications. The forms in which they are used are wide and varied: they may be used as a binder material, as partial replacement of conventional portland cements or directly as aggregates in their natural or processed states. For discussion waste materials may be classified as organic waste (agro-wastes), inorganic wastes (urban-wastes) and industrial wastes.

14.5.1 Organic Wastes

The waste materials included in this category are of *plant origin*, namely sawdust, coconut pith, rice husk, wheat husk, groundnut husk, etc. It must be appreciated that development of concrete using such aggregates is still in early stages and published data are limited. Before using organic wastes on a large scale as constituents in concrete, careful investigations regarding their *structural properties* and *durability*, need be carried out.

Natural organic waste materials are used for making *light-weight concrete*. However, they often contain substances (cement poisons) which retard the *hydration* and *hardening of cement* which need be neutralized appropriately. Moreover, it is difficult to obtain a waste material which is not a mixture of several species. Consequently, there is a considerable variation in results from batch to batch. The light-weight concrete produced using organic wastes have comparatively high *moisture movement* and show relatively higher percentage of *volume changes*. A very high *shrinkage* limits its use to designs where freedom of movement is possible.

Rice husk Huge quantities of rice husk are generated in rice mills. Each tonne of paddy produces about 200 kg of husk. Because of its very *low density*, rice husk requires large space for storage and hauling. In India it is disposed off by burning, thus reducing the bulky waste to manageable volumes of ash of less than 50 per cent of its initial volume but open burning creates sever pollution problems.

Rice husk contains only very small quantities of water-soluble cement poisons as compared to saw dust. It has a low bulk density of only 100 to 150 kg/m^3. The light-weight concrete (bulk density 600 kg/m^3) produced using rice husk as an aggregate is suitable mainly for making precast blocks and slabs for walls and partitions.

The modern trend is to incinerate the rice husk under the conditions which results in ash containing a highly *reactive form of silica*. The ash produced can be easily pulverized to the required fineness. Typically rice-husk ash contains 80–90 per cent of amorphous SiO_2, 1 to 2 per cent of K_2O and the rest being unburnt carbon. The ground reactive rice-husk ash can be blended with ordinary portland cement to produce satisfactory *hydraulic acid resistance cements*. The compressive strengths of some of the blends are given in Table 14.3.

Hydration of portland cement produces $Ca(OH)_2$ which quickly combines with highly reactive silica of rice-husk ash to form additional calcium silicates. Thus the rice-husk ash does not fall in the category of ordinary reactive silica materials commonly known as *pozzolanas* which add to strength only at later stages. If controlled burning of rice husk cannot be carried out, it is still possible to obtain pozzolanic material from rice husk.

TABLE 14.3 Strengths of blended cement made from rice-husk ash and portland cements

\multicolumn{3}{c}{Proportions of blends (by mass), per cent}	\multicolumn{3}{c}{Compressive strength, MPa}
Portland cement (IS : 8112–1976)	Rice-husk ash	Quicklime	3 days	7 days	28 days
100	—	—	23	33	43
80	20	—	28	46	61
70	30	—	32	46	60
50	50	—	26	40	59
30	70	—	24	36	43
0	80	20	10	24	35

Central Building Research Institute, Roorkee (India) has developed a cheap *cementitious material* from rice husk and waste lime sludge available from the sugar and paper industries. The dried cakes of mixture of sludge and rice husk are burnt, and the burnt material on grinding yields a fast-setting grey-coloured cementitious material which can be used in place of cement or lime in mortars for brick masonry work, plastering and foundation concreting.

Ordinary portland cement contains approximately 60 to 65 per cent of CaO, part of which is released upon hydration as free $Ca(OH)_2$ and it is this product which makes portland cement concrete prone to deterioration in acidic environments. On the other hand the rice-husk ash cement though has similar strength characteristics contains as little as 20 per cent CaO. Upon hydration none of free lime would be present as $Ca(OH)_2$, the products of hydration being *calcium silicate hydrates* and silica gel, thus the rice-husk ash concrete is more resistant to acid environments.

Rice-husk ash cement concrete The compressive strengths of concretes made with rice-husk ash cement using siliceous gravel and crushed limestone are given in

Table 14.4. The 28-day compressive strength using crushed limestone aggregate is about 23 per cent higher than that obtained using gravel aggregate, probably due to the formation of a stronger interfacial bond between the cement paste and aggregate. Because of black colour of rice-husk cements, these cements can be used to make permanent *black concrete* for *glare-free pavements* and architectural applications. These concretes show better *long-term colour stability* than that obtained by using *colouring pigments*.

TABLE 14.4 Compressive strength of rice-husk cement concrete

Aggregate type	Compressive strength, MPa		
	3 days	7 days	28 days
Crushed aggregate	29	39	47
Gravel aggregate	24	33	38

Rice-husk ash when mixed with sand and lime in suitable proportions with an appropriate quantity of water can be used to cast bricks. These will require some air-curing followed by wet-curing before being used. The *rice-husk ash bricks* have a *density* of 1400–1600 kg/m^3 and *compressive strength* of 5–6 MPa. The *water absorption* is 15–20 per cent.

14.5.2 Inorganic Waste (Urban Wastes)

The inorganic wastes which are hard, particularly, the demolition waste such as broken concrete, broken bricks and crushed glass can be used to produce concretes of requisite strength and durability.

Broken Concrete

Huge quantities of building rubble becomes available each year during the demolition of old structures to make way for new and modern ones due to rapid urbanization. *Disposal* of such materials is difficult in view of the scarcity of suitable *dumping grounds*, and meeting the *environmental requirements*. Hence, the broken concrete is increasingly being recycled. *Recycled concrete* is simply the old concrete that has been removed from pavements, foundations or buildings and crushed to the specified size. The reinforcing bars and other embedded items, if any, must be removed and care must be taken to prevent excessive contamination of the concrete by dirt, plaster or gypsum from the building. The other undesirable materials are wood and asphalt.

Investigations have indicated that the recycled concrete waste can at best be used as a substitute for coarse aggregate only from the point of view of compressive strength, volume changes, and freeze-thaw resistance. A reduction in compressive strength by 10–30 per cent and an increase in water absorption of about 5 per cent have been reported when broken concrete aggregates replace natural aggregates. The concrete using recycled concrete aggregate loses its workability more rapidly than the conventional concrete, because concrete aggregate is more porous than natural aggregate. Thus making concrete with recycled concrete aggregate may require 5 to 10 per cent more mixing water to achieve the same workability as gravel concrete. Workability and mix proportions being the same, the compressive strength of recycled aggregate concrete is in the range of about 75 per cent, and the modulus of elasticity about 65 per cent of conventional concrete. It has been reported that the properties of aggregates from demolished concrete are affected more by the method of crushing than the properties of the original concrete. The relative densities of crushed concrete fine and coarse aggregates are 2.1 and 2.3, respectively. Before using the crushed concrete as an aggregate in concrete for roads and buildings, investigations should be carried out for establishing its suitability and feasibility especially in helping the solid-waste disposal problem.

Applications Recycled concrete has been used as an aggregate in lean-concretes *(econo-crete)*. This concrete is suitable for base or sub-base for pavements, and may reduce the cost of construction by 30 per cent.

Broken-brick Aggregate Concrete

Broken brick is a waste product obtained as rejected overburnt or damaged bricks in brick works and at construction sites. Broken-brick aggregate is obtained by crushing waste bricks and have a density varying between 1600–2000 kg/m^3. It is used in concrete for foundation in light buildings, flooring and walkways. Broken-brick aggregate may also be used in *light-weight reinforced concrete* floors. Brick aggregate gives a very low slump even when reasonably workable and may thus require use of plasticizers. Broken-brick concrete can be designed to obtain the compressive strength (between 15 MPa to 35 MPa) generally required in practice and such a concrete possesses satisfactory structural properties.

14.5.3 Industrial Wastes

Some of the industrial *by-product wastes* can be profitably used in the concrete construction industry which requires large quantities of low cost raw materials. This utilization offers triple benefits, namely, conservation of fast-declining natural resources, planned gainful exploitation of waste materials and release of valuable land for more profitable use. The most influential factor that dictates the utilization of industrial by-products is the economic cost in comparison to the conventional materials that would have been otherwise used. A brief description of waste by-products is given here.

1. Blast-furnace Slag

Large quantities of slag are generated during the production of iron and steel. Granulated or foamed or dense blast-furance slag can be produced depending on the rate and manner of cooling the molten slag. The granulated slag can be used in the manufacture of slag cements. *Blast-furnace slag cements* contain slag up to 60 per cent, hence there is considerable reduction in the rate of *heat evolution* and a significant increase in the *resistance to chemical attack*. The low rate of heat evolution and the fact that the early strength is less affected in hot weather make blast-furnace slag cements attractive for use in tropics where *thermal contraction cracking* often poses problems. For the same reasons, these cements can be advantageously used in *mass concrete*, and for high chemical resistance in marine structures. The dense air-cooled slag aggregate may be used as a replacement of natural aggregate in concrete. On the other hand, foamed blast-furnace slag, a light-weight aggregate, is mainly used for blockmaking and insulating roofs and floor screeds, and is suitable for structural reinforced concrete.

2. Coal Ash from Power Stations

The main by-product is fly ash or pulverized fuel ash which is fine dust carried upward by combustion gases and collected in cyclones or wet scrubbers, and electrostatic precipitators. The bulk ash which is greyish in colour becomes darker with increasing proportions of unburnt carbon. It is used as a *cement replacement*. The contribution of *fly ash* to the strength of concrete has been attributed to: (i) direct water reduction, (ii) increase in the effective volume of paste in the mix, and (iii) pozzolanic reaction. However, fly ash reduces the rate of development of strength and increases *drying shrinkage* and *creep strains*. Since the early strength of fly ash concrete is less than that of portland cement, its proportion is generally limited to 30 per cent in the situations where early strength is important. The low rate of heat evolution makes fly ash useful in *mass concrete*. The fly ash concrete has high *resistance to sulphate attack*.

3. Red Mud Aggregate

The red-coloured waste by-product resulting from the production of alumina from bauxite by Bayer's process is known red mud. The compressive, tensile and bending strengths of concrete made with red mud aggregate have been reported to be considerably higher than those of concrete made with gravel. *Light-weight aggregates* may be manufactured from mixtures of red mud and fly ash, and blast-furnace slag and some types of pumice. The addition of 2–5 per cent red mud to portland and slag cements increases strength at an early age but gives lower strength at a later age as compared to that obtained with pure cements.

4. Silica-fume Concrete

Silica fume is a by-product of the reduction of high purity quartz with coal in electric arc furnaces in the production of ferro-silicon metal. Because of its extreme fineness (about 20000000 mm^2/g) and high glass content, silica-fume is a very effecient pozzollanic material, i.e., it is able to react efficiently with the hydration products of portland cement in concrete.

Silica-fume is generally more efficient in concretes having high water-cement ratios. Investigations indicate that in concretes with a water-cement ratio of about 0.55 and higher, the silica-fume has an efficiency factor of 3 to 4. This means that (within the usual 0–10 per cent range of replacement) 1 kg of silica-fume can replace 3 to 4 kg of cement in concrete without changing the compressive strength of concrete.

Ultra high strength concrete (of the order of 70 to 120 N/mm^2) is now possible for field placeable concrete with silica-fume admixture. Such high strength concrete has increased modulus of elasticity, low creep and drying shrinkage, excellent freeze thaw resistance, low permeability and increased chemical resistance.

Silica-fume in concrete can be used:
 (i) To conserve cement
 (ii) To produce ultra high strength concrete
 (iii) To control alkali-silica reaction
 (iv) To reduce chloride associated corrosion and sulphate attack
 (v) To increase early age strength of flyash/slag concrete.

Recently the attributes of silica-fume found their use in shotcrete applications in the pulp, mining, and chemical industries. The high strength concrete made with silica-fume and local aggregates provided greatest abrasion-erosion resistance for the eroded stilling basin of Kinzua Dam in Pennsylvania.

14.6 MASS CONCRETE

The concrete placed in massive structures like dams, canal locks, bridge piers, etc., can be termed mass concrete. A large-size aggregate (up to 150 mm maximum size) and a low slump (stiff consistency) are adopted to reduce the quantity of cement in the mix to about 5 bags per cubic metre of mass concrete. The mix, being relatively harsh and dry, requires power vibrators of immersion type for compaction. The concrete is generally placed in open forms. Because of the large mass of the concrete, the *heat of hydration* may lead to a considerable rise of temperature. Placing the concrete in shorter lifts and allowing several days before the placement of the next lift of concrete can help in the dissipation of heat. Circulation of cold water through the pipes buried in the concrete mass may prove useful. Alternatively, where possible concreting can be done in the winter season such that the peak temperature in concrete can be lowered, or the aggregates may be cooled before use.

The high temperature of mass concrete due to the heat of hydration may lead to extensive and serious shrinkage cracks. The shrinkage cracks can be prevented by using *low heat cement* and by continuous curing of concrete. The mass concrete develops high early age strength but the later age strength is lower than that of continuously cured concrete at normal temperatures. The volume changes of mass concrete during setting and hardening are small, but the concrete is susceptible to *large creep* at later ages.

14.7 SHOTCRETE OR GUNITING

Shotcrete is mortar or very fine concrete deposited by jetting it with high velocity (pneumatically projected or sprayed) on to a prepared surface. The system has different proprietary names in different countries such as Blastcrete, Blowcrete, Guncrete, Jet-crete, Nucrete, Pneukrete, Spraycrete, Torkrete, etc., though the principle is essentially the same. Shotcrete offers advantages over conventional concrete in a variety of new construction and repair works. Shotcrete is frequently more economical than conventional concrete because of less *formwork* requirements, requiring only a small portable plant for manufacture and placement. It is capable of excellent bonding with a number of materials and this may be an important consideration.

Shotcrete has wide applications in different constructions, such as thin overhead vertical or horizontal surfaces, particularly the curved or folded sections; canal, reservoir and tunnel lining; swimming pools and other water-retaining structures and prestressed tanks. Shotcrete is very useful for the *restoration and repair of concrete* structures, fire damaged structures and *waterproofing* of walls.

Shotcrete has been sucessfully used in the stabilization of rock slopes and temporary protection of freshly excavated rock surfaces. Its utility has been proved for protection against long-term corrosion of piling, coal bunkers, oil tanks, steel building frames and other structures, as well as in encasing structural steel for *fireproofing*.

Special shotcrete has been developed for high temperture applications, such as refractory lining of kilns, chimneys, furnaces, ladles, etc.

14.7.1 Types of Shotcreting

There are two basic types of shotcreting processes, as described in the following sections.

(i) The Dry Mix Process

In the dry mix process the mixture of cement and damp sand is conveyed through a delivery hose pipe to a special mechanical feeder or gun called delivery equipment. The mixture is metered into the delivery hose by a feed wheel or distributor. This material is carried by compressed air through the delivery hose to a special nozzle. The nozzle is fitted with a perforated mainfold through which water is introduced under pressure and intimately mixed with other ingredients. The mortar is jetted from the nozzle at high velocity on to the surface to be shotcreted. In this process any alteration in the quantity of water can be easily accomplished by the nozzleman (i.e. the worker in charge of the nozzle). If the water content is more, then concrete tends to slump when jetted onto the vertical surface. On the other hand in the case of deficiency of water the material which will rebound from the surface will be excessive. Alteration of water content can be made accordingly. The amount of water should be so adjusted that wastage of material by rebounding is minimum The *water-cement ratio* should be between 0.33 and 0.50.

Several forms of equipment are available for shotcreting by this technique. A common layout includes an air-compressor, a material hose, air and water hoses, a nozzle gun and a pressure tank or pump for water supply, and transporting equipment. The equipment ensuring continuous supply of the mortar can convey the material to a distance of 300 to 500 m horizontally and 45 to 100 m vertically.

(ii) The Wet Mix Process

In this process, all the ingredients, i.e. cement, sand, small-sized coarse aggregate and water, are mixed before entering the chamber of delivery equipment. The *ready-mixed concrete* is metered into the feeding chamber and conveyed by compressed air at a pressure of 5.5 to 7 atmosphere to a nozzle. Additional air is injected at the nozzle to increase the velocity and improve the gunning pattern. Equipment capable of placing concrete at the rate of 3 to 9 cubic meters per hour is available.

The phenomenon of falling back of a part of mortar or concrete jetted on to surface to be treated, due to high velocity of jet, is called *rebound* and depends upon the *water-cement ratio*, and the nature and position of the surface treated. The rebound decreases with higher water cement ratio, and has higher values for vertical and overhead surfaces. The approximate range of rebound is 5 to 15 per cent for horizontal slabs, 15 to 30 per cent for sloping and vertical surfaces and 20 to 50 per cent for the treatment of overhead surfaces and corners. The rebounding material mostly consists of sand or/and coarse particles and very little quantity of cement falls back. The rebound material falling on surface is cleaned before being treated by shotcreting. The vertical surfaces should be treated from bottom to top. The rebound also depends upon the angle of jetting with respect to the surface being treated; it is minimum when the nozzle is held at right angles to the surface. The nozzle should be kept at a distance of 900 mm from the surface. Addition of pozzolanic material to the mix reduces the rebound by improving its plasticity. Since rebounded material is a wastage, it is economical to reduce the rebound by adding as much water at the nozzle as conditions permit.

The *dry-mix process* is preferred in case the *light-weight concrete* is used. The lower water-cement ratio used results in higher *strengths*, less *creep* and *drying shrinkage*, and higher *durability*. Whereas in the wet process the higher durability can easily be achieved by using *air-entraining agents*. The water-cement ratio can be very accurately controlled in the wet process. Further, the wet process does not cause dust problems. The larger capacity availabe in wet mix process results in higher rates of placing of concrete. The procedure of shotcreting a surface involves the following steps:

(a) *Preparation of surface to receive shotcrete* Where the shotcrete is to be placed against earth surfaces as in canal linings, the surfaces should first be thoroughly compacted and trimed to line and grade. Shotcrete should not be placed on any surface which is frozen, spongy, or where there is free water. The surface should be kept damp for several hours before applying shotcreting.

For repairing *deteriorated concrete* it is essential to remove all unsound material. Chipping should be done to remove all the offsets in the cavity which may cause abrupt change in thickness of the repair work. No square shoulders should be left at the perimeter of the cavity, and all edges should be tapered. After ensuring that the surface to which the shotcrete is to be bounded is sound, it should be sand-blasted. The nozzleman usually scours clean the area before applying the shotcrete with an air-water jet, and then the water is shut off and all free water blown away by compressed air.

(b) *Construction of forms* The forms are usually of plywood sheeting, true to line and dimension. They are adequately braced to ensure protection against excessive vibration. The forms should be constructed to permit the escape of air and rebound during the gunning operation. They should also be oiled or dampened. Adequate and safe scaffolding is necessary so that the nozzleman can hold the nozzle at a distance of 1 to 1.5 m from the surface.

(c) *Placement of reinforcement* Sufficient clearance should be provided around the reinforcement to permit complete encasement with the shotcrete. The minimum clearance between the reinforcement and the form may vary between 12 mm for the case of a mortar mix and wiremesh reinforcement to 50 mm for the case of concrete and reinforcing bars.

(d) *Preparation for succeeding layers* The receiving layer should be allowed to take its *initial set* before applying a fresh layer of shotcrete. All laitance, *loose material* and *rebound* should be removed by brooming. Any laitance which has been allowed to take *final set* should be removed by sand blasting and the surface cleaned with an air-water jet.

(e) *Finishing of the surface* Natural gun finish is prefered from both structural and durability standpoints. There is a possibility that further finishing may disturb the section, harming the bond between shotcrete and underlying material, and creating cracks in the shotcrete. Where greater smoothness or better apperance is required special finishes must be applied. Sometimes, for finer finish, a *flash coat* consisting of finer sand than normal, and with the nozzle held well back from the surface, is applied to the shotcrete surface as soon as possible after the screeding.

14.7.2 Properties of Shotcrete

The properties of the shotcrete are essentially the same as for conventional concrete of same materials, proportions and void system. However, the following points should be borne in mind.

  (i) In shotcrete, generally, a small-maximum-size aggregate is used and cement content is high. These should enhance durability in most cases.
  (ii) Whereas conventional concrete is consolidated by vibration, shotcrete is consolidated by the impact of a high-velocity jet impinging on the surface. This process not only increases the cement content due to rebound but also brings about different *air-void systems* affecting the durability of shotcrete.
  (iii) The application procedures have a greater effect on the in-place properties of shotcrete than the mix proportions.
  (iv) Shotcrete specimens are usually sawed from test panels of about one metre square and 75 mm thick made by gunning out a plywood form.

14.7.3 Durability of Shotcrete

The *low water-cement ratio* enhances the durability of shotcrete for most types of exposures. The preferred range for most shotcretes is 0.30 to 0.45. The wetter mixes give a poorer quality of shotcrete, and tend to sag or fall out during application. On the other hand, the drier mixes have hardly enough water for hydration and cause excessive rebound which makes it difficult to get sound shotcrete under field conditions and increases the waste and aggravates the problems of disposing it.

The properties of cement and aggregate affect the durability of shotcrete to the same extent as that of the ordinary concrete. Exposure conditions may make it advisable to specify a particular type of cement or restrict its alkali content. Shotcrete is somewhat vulnerable to sand pockets and laminations which adversely affect its durability. There are doubts regarding its ability to withstand severe freezing and thawing.

14.7.4 Air-entrainment in Shotcrete

It is generally believed that it is impossible to entrain air with dry-mix shotcrete, because of the absence of the usual concrete mixing action and the high velocity of impingement of the material on to the application surface. However, in wet process, the addition of air-entraining admixtures has been found to reduce the size of air-voids, thereby increasing the durability marginally. Some investigators believe that air-entrainment is helpful from the application standpoint; the mix is a little stickier and the rebound is reduced. The simplest procedure, however, is to use air-entraining cements.

14.7.5 Nature of Failures in Shotcrete

Most shotcrete failures involve the peeling off of sound shotcrete because of *bond failure*. This occurs in spite of the fact that one of shotcrete's greatest attributes is its excellent ability to form a bond with concrete or another shotcrete layer. The other type of failure is the delamination between shotcrete layers where the surface preparation has not been good. In repair technology applications, shotcrete has been found to *spall* because of the *corrosion of reinforcement*.

The *drying shrinkage* is somewhat higher than for most of the low slump conventional concretes, but generally falls within the range 0.06 to 0.10 per cent.

14.7.6 Special Shotcretes

(i) Steel Fibrous Shotcrete

The plain unreinforced shotcrete like unreinforced concrete is a *brittle* material, with little capacity to resist pronounced tensile stresses or strains without cracking and disruption. If reinforced with steel fibres, its *strength increases* considerably. This reinforcement of shotcrete also improves its *ductility, energy absorption* and *impact resistance*. Steel fibres control the cracking and hold the material together even after excessive cracking. A fibrous shotcrete containing up to 2 per cent of steel fibres (by volume), has shown an increase in *flexural strength* of an order of 50 to 100 per cent and in *compressive strength* by 50 per cent. The *toughness* and *impact resistance* are found to increase by ten times or more.

An important improvement is evident in the mode of failure which requires large deformations to cause failure and the material continues to carry a significant load even after cracking. This large increase in strain capacity provides *post-crack resistance* which is advantageous in applications such as tunnel and mine linings, where there may be large deformations.

It has been noticed that a greater percentage of fibres makes the aggregate rebound from the surface. High-speed photography has shown that many steel fibres were in the outer portion of air stream and that many of them were blown away radially from near the point of intended impact shortly before or after they hit the surface. If less air is used, the amount and velocity of remnant air currents is reduced and the rebound of fibres is less. Other measures include jetting at the wettest stable consistency.

It has been noticed that *hooked steel fibres* can be used in the field with a conventional shotcrete machine. They have a substantial higher load carrying capacity after the development of cracks. These shotcretes have a very high *toughness index* indicating excellent *energy-absorbing capacity*. Due to the excellent *anchorage* established between the hooked fibres and the matrix, the composite has *high ductility* and *flexural strength*. Shotcretes with hooked fibres show a tremendous ability to absorb *impact loading*. For higher fibre contents, the impact resistance increases dramatically. The reduction in *drying shrinkage* is proportional to the quantity of fibres added.

(ii) Refractory Shotcrete

Shotcrete containing hydraulic cement (i.e., calcium aluminate cement) as binding agent which is suitable for use at high temperatures, is termed refractory shotcrete. This shotcrete utilizes aggregates and binders suitable for use up to a temperature of 1900°C. It exhibits variable properties throughout its thickness after firing due to the temperature gradient from the hot to the cold surface. In addition to this property, other factors, such as *thermal cycling*, *thermal shock*, *chemical attack*, *abrasion* and *erosion* should be considered in the design of refractory shotcrete. Another characteristic is that its 24-hour strength is similar to the 28-day strength of portland cement shotcrete.

Ingredients Hydraulic binders are available in three types: (i) low purity (ii) intermediate purity, and (iii) high purity. The higher the purity, higher is the aluminium oxide content and lower is the iron oxide content. The higher the service temperature, the higher is the aluminium content required. Among the aggregates used in increasing order of service temperature are slag, limestone, trap rock, expanded-shale, perlite, calcined fire clay, calcined bauxite, kaolin, etc. The water used should be potable and free from acids. Portland cement and calcium aluminate cement combined in a mix will accelerate the hardening process in the shotcrete.

Curing, drying and firing Refractory shotcrete achieves its ultimate strength in 24 hours. Therefore, it is important that the curing procedure be instituted immediately after placing and continued for 24 hours to achieve complete *hydration* and control *drying shrinkage*. The usual methods of curing like covering with wet burlaps, fine spraying or resin type *curing membrane* are effective.

After curing and before placing the refractory shotcretes in service, it is essential that the lining be dried to eliminate both free and combined water. Thorough drying minimizes any chances of explosive spalling resulting from the internal formation of steam. A well-executed heating procedure will assure the integrity of a lining, thereby assuring a longer service life.

Refractory shotcreting is particularly effective where forms are impractical, access is difficult, thin layer and one of variable thickness are required, normal casting techniques cannot be employed from considerations of economy.

14.7.7 Disadvantages with Shotcretes

The method in which raw materials, aggregate and cement are handled may be objectionable to environmentalists. Dust from either the fine aggregate and/or cement can settle on the ground around the application area. This potential hazard must be considered when designing or applying a shotcrete coating. Sometimes special type of enclosures may be necessary to confine the area designed for batching, mixing or charging the gun. In addition to the dusting problem, rebound also has to be cleaned up and hauled to an approved waste area.

The high cost of shotcrete combined with wastage due to rebound has to be weighted with the difficulties involved with other techniques before shotcreting is adopted for any particular situation. Inspite of all these problems, shotcrete has its own merits, such as the need of formwork only on one side of the work, its suitability for concreting thin sections and in sites where access is difficult. Moreover, shotcrete bonds perfectly with the existing old concrete masonry, exposed rock or suitably prepared steel surface, and hence is very effective in the repair of the structures concerned.

14.7.8 Guniting

The technique of depositing very thin layers of mortar in each pass of the nozzle than that available with the shotcrete, is termed *guniting*. A typical arrangement in the gunite system is shown in Fig. 14.3. In addition to the general requirements for *quality control* of normal concrete, guniting requires careful and skilful handling of nozzle for high quality finished work. The surface to be gunited should be cleaned

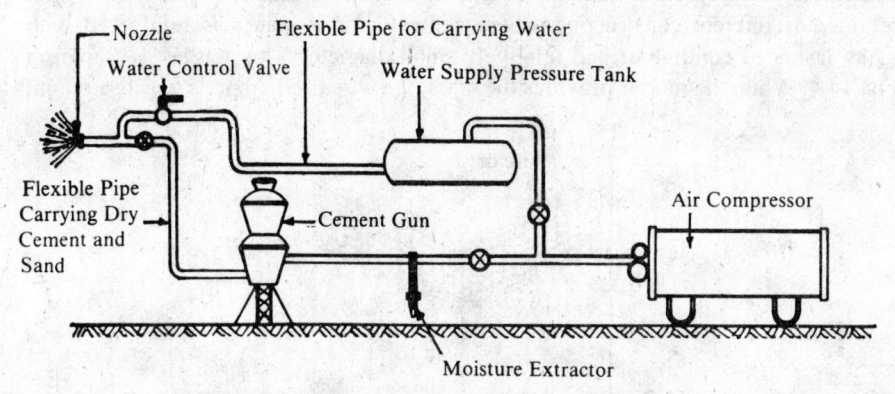

Fig. 14.3 Typical arrangement for gunite-system

thoroughly of grease or oil or any other loose or defective material by applying either air blast or high pressure water jet. If necessary the surface can even be sand-blasted by omitting the cement using the same gun and reducing the velocity of jet. The sand blasting can also help in removing the loose rust on the reinforcement. The surfaces likely to absorb water should be kept wet up to six hours before guniting. The mix generally used for guniting is 1 : 3 to 1 : 4.5 with a *water-cement ratio* of about 0.30. The maximum size of sand is limited to 10 mm. A 7-day strength of the order of 70 MPa can be achieved with a 1 : 3 mix. This high-quality mortar with low water-cement ratio results in low *permeability*, good *resistance to weathering* and *chemical attack*, etc. The resistance to the chemical attack can be increased by using sulphate-resisting portland cement or high-alumina cement. Addition of pozzolana up to 5 per cent by a mass of cement can improve plasticity and reduce the *rebound*. Due to good bonding between reinforcement and gunite, the gunite acts as a part of the structure and not merely as an added cover. Each layer of the gunite is usually provided with a spot-welded steel-wire-mesh fabric of 5 mm diameter wires, to reduce *initial shrinkage* and to prevent *cracks* in the freshly placed material. The meshes should overlap each other by one mesh to maintain continuity whenever the meshes are joined. The reinforcing fabric should be maintained at a distance of 6 to 10 mm from the original concrete surface during guniting. The guniting can effectively be used for the repair of dams, spillways, bridge piers, sewerage pipes and water mains, and for protection of canal banks. It has been extensively used for the protection of steel girders from corrosion etc., and waterproofing of reservoirs and tunnels. Another important application is in the repair of all types of concrete structures where the concrete has spalled off due to corrosion of reinforcement.

14.8 FERROCEMENT

The concept of use of fibres to reinforce brittle materials dates back to ancient constructions built in India using mud walls reinforced with woven bamboo mats and reeds. In the present form, ferrocement may be considered as a type of thin reinforced concrete construction where cement mortar matrix is reinforced with many layers of continuous and relatively small diameter wire meshes as shown in Fig. 14.4. While the mortar provides the mass, the wire mesh imparts tensile strength

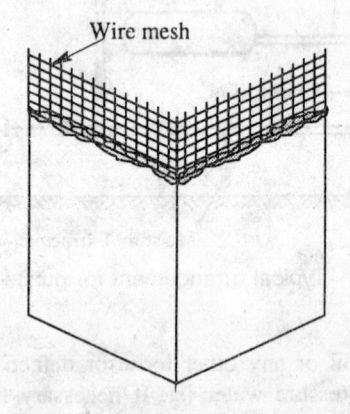

Fig. 14.4 Typical section of ferrocement

and ductility to the material. In terms of structural behaviour ferrocement exhibits very high tensile strength-to-weight ratio and superior cracking performance. The distribution of a small diameter wire mesh reinforcement over the entire surface, and sometimes over the entire volume of the matrix, provides a very high *resistance against cracking*. Moreover, many other engineering properties, such as *toughness*, *fatigue resistance*, *impermeability*, etc. are considerably improved. Sometimes, conventional reinforcing bars in a skeleton form are added to thin wire meshes in order achieve a stiff reinforcing cage. The commonly used composition and properties of ferrocement made with steel wire mesh reinforcement are summraized in Table 14.5.

TABLE 14.5 Normal ranges of composition and properties of ferrocement

Perimeter	Range
Wire-mesh	
Wire diameter	$0.5 \leq \phi \leq 1.5$ mm
Type of mesh	Chicken wire or square woven-or welded-wire galvanized mesh, expanded metal
Size of mesh openings	$5 \leq s \leq 25$ mm
Distance between mesh layers	Distance between 2 layers ≥ 2 mm
Volume fraction of reinforcement	Up to 8 per cent in both directions corresponding to 650 kg/m^3 of concrete
Specific surface of reinforcement	Up to 4 cm^2/cm^3 in both directions
Skeletal reinforcement (if used)	
Type	Wires; wire fabric; rods; strands
Diameter	$3 \leq d \leq 10$ mm
Grid size	$50 \leq g \leq 100$ mm
Typical mortar composition	
Portland cement	Any type depending on application
Sand-cement ratio	$1.0 \leq S/C \leq 2.5$ (by mass)
Water-cement ratio	$0.35 \leq W/C \leq 0.6$ (by mass)
Fine aggregate (sand)	Fine sand all passing IS : 4.75 mm sieve and having 5 per cent by mass passing IS : 1.18 mm sieve, with a continuous grading curve in between
Composite properties	
Thickness	$10 \leq t \leq 60$ mm
Steel cover	$1.5 \leq c \leq 5$ mm
Ultimate tensile strength	34.5 MPa
Allowable tensile stress	10.0 MPa
Modulus of rupture	55.0 MPa
Compressive strength	27.5 to 60.0 MPa

14.8.1 Materials

Cement Mortar Matrix

As described above, the ferrocement composite is a rich cement-mortar matrix of 10 to 60 mm thickness with a reinforcement volume of five to eight per cent in the form of one or more layers of very thin wire mesh and a skeleton reinforcement consisting of either welded-mesh or mild steel bars.

Normally portland cement and fine aggregate matrix is used in ferrocement. The matrix constitutes about 95 per cent of the ferrocement and governs the behaviour of the final product. This emphasises the need for proper selection of constituent materials, their mixing and placing.

The choice of cement depends on the service conditions. To maintain the quality of cement, it should be fresh, of uniform consistency and free of lumps and foreign matter. Cement should be stored under dry conditions and for as short duration as possible.

The fine aggregate (sand) which is the inert material occuping 60 to 75 per cent of the volume of mortar must be hard, strong, non-porous and chemically inert. The aggregate should be free from silt, clay and other organic impurities. The particle sizes of 2.36 mm and above, if present in substantial quantities, may cause the mortar to be porous. On the other hand very fine particles, if present in a substantial amount, will require more water to achieve the required *workability*, thereby adversely affecting the strength and *impermeability*. The fine aggregates conforming to grading zones II and III with particles greater than 2.36 mm and smaller than 150 μm removed are suitable for ferrocement. Therefore, sands with maximum sizes of 2.36 mm and 1.18 mm with optimum grading zones II and III are recommended for ferrocement mixes. Use of fine sand in ferrocement is not recommended.

The *water content* which governs the strength and workability of mortar primarily depends upon the *maximum grain size*, the *fineness modulus*, and the *grading of the sand*. The water used for making mortar should be free from impurities such as clay, loam, acids, salts, vegetable matter, etc.

Plasticizers and other *admixtures* may also be added for achieving: (i) an improved *workability*, (iii) water reduction for increase in strength and reduction in *permeability*, (iii) *water-proofing*, (iv) increase in *durability*. In addition admixtures (containing chromium trioxide) may be used to prevent galvanic-corrosion of galvanized steel reinforcement. Pozzolanas such as flyash may be added as cement replacement materials (up to 30 per cent) to increase the durability.

Mix Proportions

The mix proportions in terms of sand-cement ratio (by mass) normally recommended are 1.5 to 2.5. The water-cement ratio (by mass) may vary between 0.35 and 0.6. In order to reduce permeability, the water-cement ratio must be kept below 0.4. The moisture content of the aggregate should be taken into account in the calculation of required water. The amount of water can be reduced by the use of appropriate admixtures.

The slump of fresh mortar should not normally exceed 50 mm, and 28-day compression strength of moist cured cubes should be around 350 MPa for most applications. Sand being the principal constituent of ferrocement, its properties have a major influence on the amount of water and hence on the mix design. Improvements in the grading composition of sand may allow considerable reduction of water requirement. Sand with maximum nominal size less than 2.36 mm or 1.18 mm should be avoided in ferrocement mixes.

The mixes should have compositions such that the total absolute volume of cement and fines is about 300 cm^3 per litre fo mortar. A change in the amount of cement must be accompanied by a corresponding change in that of fines.

Reinforcement

As explained earlier, the reinforcement used in ferrocement is of two types, viz. *skeleton steel* and *wire mesh*. The skeleton steel frame is made conforming exactly to the geometry and shape of structure, and is used for holding the wire meshes in position and shape of the structure.

Skeleton Steel

The skeleton steel comprises of relatively large-diameter (about 3 to 8 mm) steel rods typically spaced at 70 to 100 mm. It may be tied-reinforcement or welded wire fabric. The welded-wire fabrics normally contain larger diameter wires spaced at 25 mm or more. Welded-wire fabrics of 3 to 4 mm diameter wires welded at 80 to 100 mm centre-to-centre have been successfully used for making skeleton frames for the cylindrical or other ferrocement surfaces where these meshses can be bent easily. They provide better and uniform distribution of steel and save time in fabrication but may cost a little more when compared to mild steel bar frames. In the case of structures where higher stresses may occur, as in case of boats, barges, etc. the mild steel bars provided to act as skeletal steel are also counted as reinforcement imparting structural strength, stiffness and durability. However, a minimum possible size of bars should be used in order to obtain the effect of wire meshes and hence the composite effect. The spacing of the skeletal transverse and longitudinal steel bars of diameter of 5 to 7 mm depends upon the type and shape of structure. In the case of boathulls, a spacing of 75 to 100 mm is adequate whereas in water tanks, bins, etc. the spacing may vary between 200 and 300 mm. The bars are mostly tied with binding wires but can also be welded.

The reinforcement should be free from dust, loose rust, coatings of paint, oil or similar undesirable substances.

Wire Mesh

The wire mesh consisting of galvanized wire of diameter 0.5 to 1.5 mm spaced at 6 to 20 mm centre-to-centre, is formed by welding, twisting or weaving. Specific mesh types include woven or interlocking mesh, woven cloth, and welded-mesh. The welded-wire mesh may have either hexagonal or square openings as shown in Fig. 14.5. Meshes with hexagonal openings are sometimes referred to as *chicken*

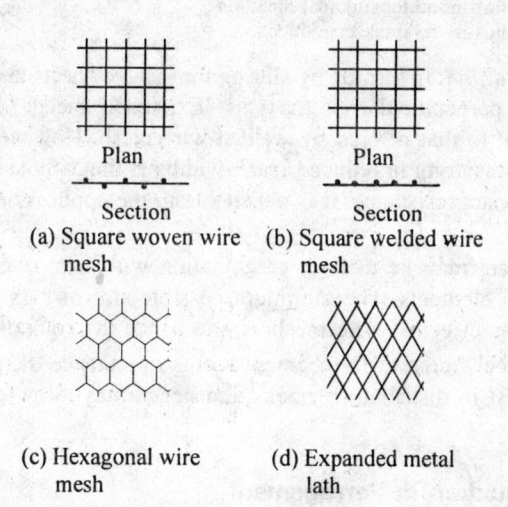

Fig. 14.5 Different types of welded wire meshes

wire meshes. The hexagonal wire mesh is cheaper but structurally less efficient than the mesh with square openings because the wires are not oriented in principal (maximum) stress directions. Moreover, the rectangular meshes have better rigidity when placed or tied over the skeleton frame, and do not sag during placing the mortar. Meshes with square openings are available either in the form of welded-wire mesh or in the woven form. The welded wire meshes have a higher Young's modulus and hence provide a higher stiffness and less cracking in the early stages of loading. On the other hand, woven-wire meshes are a little more flexible and easy to work with than the welded meshes. In addition, welding anneals the wires and limits the tensile strength. Generally, the square woven meshes consisting of 1.0 or 1.5 mm diameter wires spaced at about 12 mm are preferable. Wire meshes are also available in the galvanized form. Galvanizing, like welding, reduces the tensile strength. However, to control cracking the welded wire fabric should be used in combination with wire meshes. The minimum yield strength of wire used in fabric should be 415 MPa for plain wires and 500 MPa for deformed wires. The wire diameter should be less than 12 mm except in case of very thick plates. Mechanical properties of steel wire meshes and reinforcing bars are given in Table 14.6.

TABLE 14.6 Mechanical properties of steel wire meshes and reinforcing bars

Property	Woven square mesh	Welded square mesh	Hexagonal mesh	Expanded metal lath	Logitudinal bars
Yield strength, f_y, MPa	450	450	310	380	410
Effective modulus, E_{RL}, GPa	140	200	100	140	200
Effective modulus, E_{RT}, GPa	160	200	70	70	—

E_{RL} = value of modulus in the longitudinal direction
E_{RT} = value of modulus in the transverse direction

Expanded-metal lath is formed by slitting thin gauge sheets and expanding them in the direction perpendicular to the slits. Expanded metal offers strength approximately equal to that offered by welded-wire mesh. However, they result in a stiffer composite resulting in reduced crack-widths at the early stage of loading and provide better impact resistance. It is unsuitable in the applications involving sharp curves.

Reinforcing bars may be used in combination with wire meshes for relatively thick ferrocement elements. The minimum possible size of bars should be used in order to obtain the effect of wire meshses and hence the composite effect.

Addition of steel fibres to ferrocement seems to enhance the properties considerably. They assist in distributing cracks and hence may allow the use of heavier meshes.

14.8.2 Construction in Ferrocement

The construction in ferrocement can be divided into four phases: (i) fabrication of skeleton framing system, (ii) fixing of bars and mesh, (iii) application of mortar, and (iv) curing. The quality of mortar and its application is the most critical phase. Mortar can be applied by hand or by *shotcreting*. Since no formwork is required as in conventional reinforced concrete construction, ferrocement is suitable especially for structures with curved surfaces such as shells and other free-form shapes.

The required number of layers of wire mesh are fixed on both sides of the skeleton frame. First, the external mesh layers are fixed and tied to the frame bars. The meshes should be fixed by staggering the hold-positions in such a manner that the effective hold size is reduced. A spacing of at least 1 to 3 mm is left between two mesh layers. Wherever two pieces of the mesh are joined, a minimum *overlap* of 80 mm should be provided and tied at a close interval of 80 to 100 mm centre-to-centre.

The weighed quantities of the ingredients, namely, cement, good quality graded-sand, waterproofing and antishrinkage compounds, are dry mixed. The liquid additives are added to the mixing water taken in the required quantity. About half of the mixing water is put in the mixer before charging the mixer with dry mixed mortar ingredients. The mixing is carried out and the remaining water is then added gradually. The cement-aggregate ratio is generally kept between 1 : 1.75 to 1 : 2.5 (by mass) and water-cement ratio may be 0.35 to 0.40 depending upon the required workability. Generally a 3 minute mixing time is enough. The mortar should be mixed in batches of such a quantity as can be utilized in one hour of working, so that mortar can be placed before its setting starts.

The placing of the mortar is termed the *impregnation of meshes with matrix*. This is the most critical operation in ferrocement casting. If the mortar impregnation is not proper the structure is bound to fail in its performance. A sufficient quantity of mortar is impregnated through mesh layers so that the mortar reaches the other side and there are no voids left in the surface. A wooden hammer of about 100 mm diameter with a 150 mm long wooden handle can be used for mild hammering over the temporarily held form. This will give sufficient vibrations for compacting the mortar. As soon as it is ensured that the mortar penetration through the mesh is satisfactory, the form is shifted to the next position.

For structures like boat hulls and shells, the mortar is placed using a technique called the two-operation *mortar impregnation*. In this system the outside of the mesh is plastered first and the inner layer left exposed. The excess mortar is scrapped using trowels and wire brushes. The mortar is left for setting till it attains enough strength for carrying the load from inside during the application of a second layer of mortar. Cement slurry is sprayed or brushed over the entire inner surface and the second layer of mortar is applied from inside.

In structures where many layers are used as reinforcement and the thickness is more than 20 mm, it is advisable to do the casting in three layers. The *core* or *middle* layer is applied first covering the skeleton steel and one layer of wire mesh. This core provides a firm surface for mortar application on top and bottom. The core is cured for at least 3 days before the other two layers of mortar are applied. Cement slurry may be brushed over the middle layer for getting a good bond between old and new mortars.

For normal applications, the mortar provides adequate protection against *corrosion of reinforcement*, but where the structure is subjected to chemical attack by the environment as in sea water, it is necessary to apply suitable *protective coatings* on the exposed surface. These coatings should be such that they do not react with either the mortar or the reinforcement, and at the same time not be susceptible to the environmental attack. Vinyl and epoxy coatings have been found to be satisfactory especially on structures exposed to sea water and also in most other *corrosive environments*. For protection against a less severe environment, cheaper asphaltic and bituminous coatings are generally satisfactory.

14.8.3 Properties of Ferrocement

Though ferrocement is often considered to be just a variation of conventional reinforced concrete which may be true for the ferrocement with small quantities of reinforcement, however, it is not true for the quantity of reinforcement provided in most of the applications. Moreover, a system of construction using layers of closely spaced wire mesh separated by skeleton bars and filled with cement mortar presents all the mechanical characteristics of a *homogeneous material*.

Tensile strength of ferrocement depends mainly on the volume of reinforcement in the direction of force and the tensile strength of the mesh. The tension behaviour may be divided into three regions, namely, pre-cracking stage, post-cracking stage and post-yielding stage. A ferrocement element (member) subjected to increasing tensile stresses behaves like a linear elastic material till the development of *first crack* in the matrix. Once the cracks have developed the material enters the stage of *multiple cracking* and this stage continues up to the point where wire meshes start to yield. In this stage number of cracks keep on increasing with an increase in tensile stress without any significant increase in crack width. With the yield of reinforcement, the composite enters the stage of *crack-widening*. The number of cracks remains essentially constant and the crack widths keep increasing. The behaviour is primarily controlled by the reinforcement bars.

In the elastic pre-cracking stage the modulus of ferrocement composite E_c can be expressed in terms of modulii of mortar and reinforcement E_m and E_r, respectively, and volume fraction of reinforcement in longitudinal direction, V_{rl}:

$$E_c = (1 - V_{rl})E_m + V_{rl}E_r \approx E_m + V_{rl}E_r = E_m(1 + \eta V_{rl})$$

where $\eta = E_r / E_m$.

During the multiple cracking stage, the contribution of mortar to the stiffness of composite is negligible. Hence the stiffness of composite is approximately represented by:

$$E_c = V_{rl}E_r$$

The value of E_r may be substantially different for woven-mesh from that for a welded-mesh. It has been noticed that higher the volume of reinforcement and smaller the diameter of wires, longer is multiple cracking stage with a larger number of cracks developed in the same *gauge length*.

An inverse relationship between the first crack strength and average wire spacing based on linear elastic fracture mechanics has been established. The load carrying capacity of ferrocement is correlated with the *specific surface area of reinforcement*, S, which is defined as the total surface area of the wires in contact with cement mortar divided by the volume of composite. However, it should be noted that some investigators have used the surface area of the wires in the load resisting direction S_L only. The specific surface area has been found to influence the *first crack load* in tension, as well as the *width and spacing of cracks*. For example, a 12 mm thick ferrocement section with five layers of a 12 mm square welded or woven 1.0 mm diameter wire mesh reinforcement has about ten times as much specific surface areas as the conventional reinforcement. This results in a considerably increased load carrying capacity. Consequently, ferrocement has tensile strength as high as its compressive strength i.e. 27 MPa, and the widths of cracks are very small even at failure (about 0.05 mm). Ferrocement structures can be designed to be watertight at service loads.

The maximum composite stress at first crack increases in direct proportion to the *specific surface*. A specific surface area S_L equal to 1 cm^2/cm^3 has been suggested as the lower limit for a composite to be the ferrocement. The other parameter which is a direct measure of the ultimate strength of ferrocement is the percentage of reinforcement, p, defined as either the volume of wires per unit volume of composite in the loaded direction or the area of wires per unit cross-sectional area of composite in the loaded direction. There is a unique geometric relationship between S and p, but their relationships to the physical properties are quite different. S is mostly associated with the cracking behaviour whereas p is a direct measure of the ultimate strength of ferrocement because the *ultimate load* is resisted entirely by the wire mesh.

Thus depending upon the cracking stage a typical *tension stress-strain curve* for ferrocement exhibits three distinct regions namely, *elastic*, *quasi-elastic* or *elastoplastic*, and *plastic regions*. In region I, the material is linearly elastic because both the reinforcement and matrix deform elastically. The cracking of cement mortar is the beginning of region II and the slope of the stress-strain curve decreases. The point of decrease of the slope of the stress-strain curve indicates the first crack visible to the naked eye or with special lighting arrangement. In region III, the wire reinforcement supports the total load and the ultimate capacity can be estimated from the maximum load capacity of the wire reinforcement alone. The boundaries of the elastoplastic region are found to shift with the specific surface of the mesh, mesh size, geometry and orientation of the mesh, yield and ultimate strengths of wire.

The behaviour of thin ferrocement element under compression is primarily controlled by the properties of the cement-mortar matrix, i.e., thin ferrocement plate elements can be treated as plain mortar plates for most practical applications. Like in the reinforced and prestressed concrete beams, the *fatigue behaviour* of ferrocement flexural element is governed by the tensile fatigue properties of the mesh. The ferrocement beams show poor resistance to fatigue under cyclic loading. Impact tests on ferrocement slabs show that the *impact resistance* increases almost linearly with the increase in specific surface (volume fraction) and ultimate strength of mesh reinforcement. For the same reinforcement fraction, ferrocement using welded-wire meshes offers highest impact resistance and the one reinforced by chicken-wire meshes offers the lowest. Woven mesh reinforcement provides an impact strength higher than that obtained by chicken-wire meshes but lower than that by welded-wire meshes.

14.8.4 Applications

Due to the very high percentage of well distributed and continuously running steel reinforcement, the ferrocement behaves as steel plates. As discussed earlier, its *cracking resistance*, *ductility*, *impact* and *fatigue resistance* are higher than those of concrete. In addition, the *impermeability* of ferrocement products is far superior to that of ordinary reinforced concrete products. Ferrocement combines easy mouldability of concrete to any desired shape, lightness, tanacity and toughness of steel plates. Due to very high tensile strength-to-weight ratio and superior cracking behaviour, the ferrocement is an attractive material for light and water-tight structure and other portable structures such as mobile homes. The other specialized applications include water tanks as detailed in Fig. 14.6, silos and bins, boat hulls, biogas

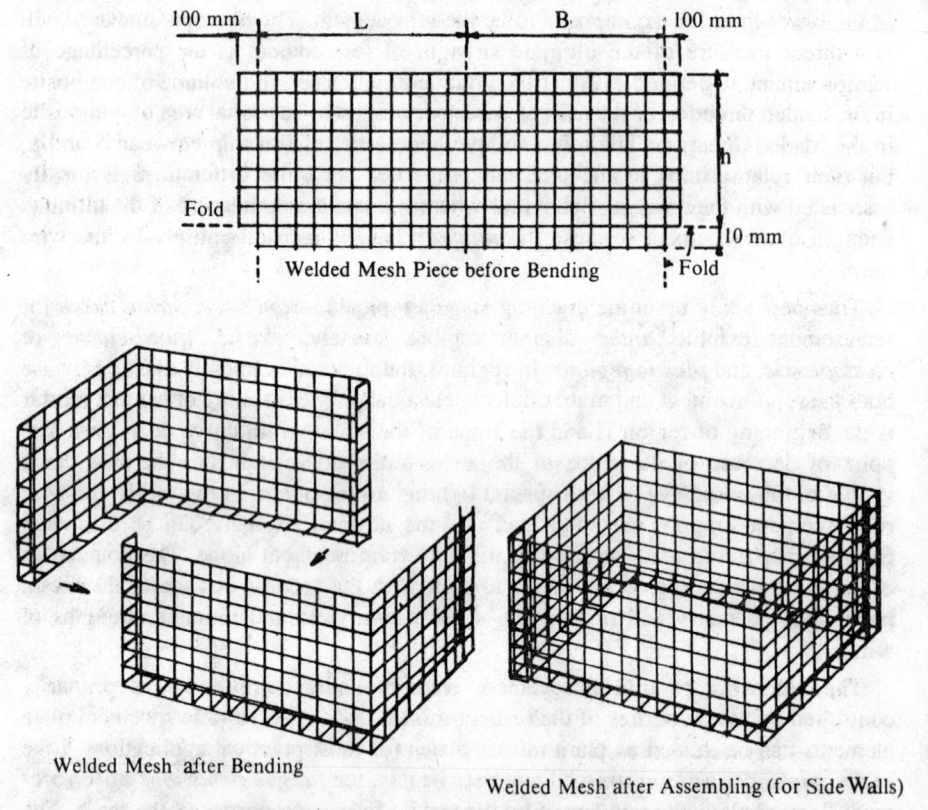

Fig. 14.6 Details of welded mesh for rectangular ferrocement tank

holders, pipes, folded plates and shell roofs, floor units, kiosks, service core units, wind tunnels, modular housing, swimming pools, and permanent forms of concrete columns.

The major advantages of the ferrocement can be summarized as follows:
 (i) Ferrocement structures are thin and light. Therefore, a considerable reduction in the self-weight of structure and hence in foundation cost can be

achieved. A 30 per cent reduction in dead weight on supporting structure, 15 per cent saving in steel consumption and 10 per cent in roof cost has been estimated in USSR.
(ii) Ferrocement is suitable for manufacturing the precast units which can be easily transported.
(iii) The construction technique is simple and hence does not require highly skilled labour.
(iv) Partial or complete elimination of formwork is possible.
(v) Ferrocement construction is easily amenable to repairs in case of local damage due to abnormal loads (such as impact).

14.8.5 Fibrous Ferrocement

Fibre-reinforced concrete possesses higher *compressive strength*, *toughness*, *increased resistance to wear and tear*, and *higher post-cracking strength*. Its permeability is very low, and the tensile and flexural strengths are lower than those of ferrocement. Another major practical difference is that fibre-reinforced concrete must be cast in forms whereas ferrocement can be shaped into surfaces of desired configuration. In the case of ferrocement, a very fine wire mesh is required to control the cracking and skeleton steel to support the wire mesh. Use of fine meshes with thin wires at closer spacings for effective crack control is the limitation of ferrocement. The strength of conventional thin ferrocement mortar panels is also limited, and therefore prone to localized damage resulting from impact etc. On the other hand, fibre-reinforced concrete possesses better *impact resistance*.

To improve some of the mechanical properties of ferrocement, such as *toughness* and *impact resistance*, a new composite material known as *fibrous ferrocement* has been developed using fibres in plain mortar and meshes with larger diameter wires at longer pitches. Atcheson and Alexander fabricated fibrous ferrocement panels using 25 mm long steel fibres of 0.4 mm square section in plain mortar with high-strength tensile wire meshes of thrice the usually accepted maximum diameter at twice the normal spacing which withstood stresses up to 57 MPa. This is very high compared to the stresses of 10 to 20 MPa in conventional ferrocement panels. Further intensive research is required in exploiting the potential of the new product obtained by combining two modern economical composites.

14.9 FIBRE-REINFORCED CONCRETE

14.9.1 Introduction

The presence of microcracks at the mortar-aggregate interface is responsible for the inherent weakness of plain concrete. The weakness can be removed by inclusion of fibres in the mix. The fibres help to transfer loads at the internal microcracks. Such a concrete is called fibre-reinforced concrete. Thus the fibre reinforced concrete is a composite material essentially consisting of conventional concrete or mortar reinforced by the random dispersal of short, discontinuous, and discrete fine fibres of specific geometry. The fibres can be imagined as an aggregate with an extreme deviation in shape from the rounded smooth aggregate. The fibres interlock and entangle around aggregate particles and considerably reduce the *workability*, while the mix becomes more cohesive and less prone to *segregation*. The fibres suitable for reinforcing the concrete have been produced from steel, glass and organic polymers. Naturally occurring asbestos fibres and vegetable fibres, such as jute, are also used for reinforcement. Fibres are available in different sizes and shapes. They can be classified into two basic categories, namely, those having a higher elastic modulus than concrete matrix (called *hard intrusion*) and those with lower elastic modulus (called *soft intrusion*). Steel, carbon and glass have higher elastic modulii than cement mortar matrix, and polypropylene and vegetable fibres are the low modulus fibres. *High modulus fibres* improve both *flexural* and *impact resistances* simultaneously whereas *low modulus fibres* improve the impact resistance of concrete but do not contribute much to flexural strength.

The major factors affecting the characteristics of fibre-reinforced concrete are: water-cement ratio, percentage (volume fraction) of fibres, diameter and length of fibres. The location and extent of cracking under load will depend upon the orientation and number of fibres in the cross-section. The fibres restrain the shrinkage and creep movements of unreinforced matrix. However, fibres have been found to be more effective in controlling compression creep than tensile creep of unreinforced matrix.

In contrast to reinforcing bars in reinforced concrete which are continuous and carefully placed in the structure to optimize their performance, the fibres are discontinuous and are generally randomly distributed throughout the concrete matrix. As a result, the reinforcing performance of steel fibres, for example, is inferior to that of reinforcing bars. In addition, the fibres are likely to be considerably more expensive than the conventional steel rods. Thus, fibre-reinforced concrete is not likely to replace conventional reinforced concrete. However, the addition of fibres in the *brittle cement* and *concrete matrices* can offer a convenient, practical and economical method of overcoming their inherent deficiencies of poor tensile and impact strengths, and enhances many of the structural properties of the basic materials such as fracture toughness, flexural strength and resistance to fatigue, impact, thermal shock or spalling. The provision of small-size reinforcement as an integral part (or ingredient) of fresh concrete mass enhances its potential in the manufacture of thin sheet products and fabrication of structural components.

Essentially, fibres act as crack arrestor restricting the development of cracks and thus transforming an inherently brittle matrix, i.e., portland cement with its *low tensile* and *impact resistances*, into a strong composite with superior *crack resistance*, improved *ductility* and distinctive *post-cracking behaviour* prior to failure. Steel fibres are probably the best suited for structural applications. Due to superior properties like increased tensile and bending strengths, improved ductility, resistance to cracking, high impact strength and toughness, spalling resistance, and high energy absorption capacity, fibre-reinforced concrete (FRC) has found special application in hydraulic structures, airfield and highways pavements, bridge decks, heavy duty floors and tunnel linings.

14.9.2 Mechanism of Fibre-matrix Interaction

In contrast to fibre composites in resin and metal matrices where the fibres are aligned and constitute about to 60 to 80 per cent of composite volume, FRC contains much less fibres which are randomly oriented. The tensile *cracking-strain* of cement matrix (less than 0.02) is very much lower than the *yield* or *ultimate strain of steel* fibres. As a result, when a fibre reinforced composite is loaded, the matrix will crack long before the fibres can be fractured. Once the matrix is cracked the composite continues to carry the increasing tensile stress; the peak stress and the peak strain of the composite are greater than those of the matrix alone. During the inelastic range between *first cracking* and the peak, *multiple cracking* of matrix occurs as indicated in Fig. 14.7. Until the initial cracking of the matrix, it is

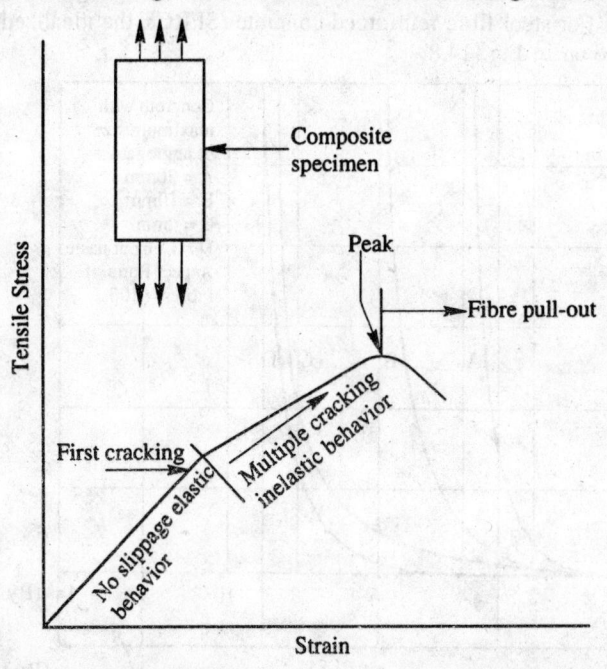

Fig. 14.7 Behaviour of fibre reinforced concrete under tensile load

reasonable to assume that both the fibres and the matrix behave elastically and there is no slippage between the fibres and the matrix. After initial cracking has occurred, the composite will carry increasing load only if the *pull-out resistance of fibres* is greater than the load at the initial cracking, since in the post-cracking stage, the failure of composite is generally due to fibre-pullout rather than fibre-yielding or fracture.

In FRC, the fracture is a continuous process wherein the cracking occurs over a wide range of loading and the debonding of fibres occurs over several stages. The *bond* or the *pull-out resistance of* fibres depends on the average bond strength between the fibres and the matrix, the number of fibres crossing the crack, and the length and diameter of the fibres. The ratio l/d is called the *aspect ratio* where l is the length and d the diameter of the fibres.

Improvement in the structural performance of FRC depends on the strength characteristics of the fibres themselves, volume of fibre reinforcement, dispersion and orientation of fibres, and their shape and aspect ratio. Higher strength, larger volume, larger length, and smaller diameter of fibres have been found independently to improve strength of the composite. The orientation and dispersion effects may depend, among other things, on loading conditions. Unidirectional fibres uniformly distributed throughout the volume are most efficient in uniaxial tension. While flexural strength may depend on a unidirectional alignment of fibres dispersed away from the neutral plane, flexural shear strengths may call for random orientation. A proper shape and higher aspect ratio are also needed to develop adequate bond between the concrete and the fibres so that the fracture strength of fibres may be fully utilized. For steel fibre reinforced concrete (SFRC), the idealized stress-strain relation is shown in Fig. 14.8.

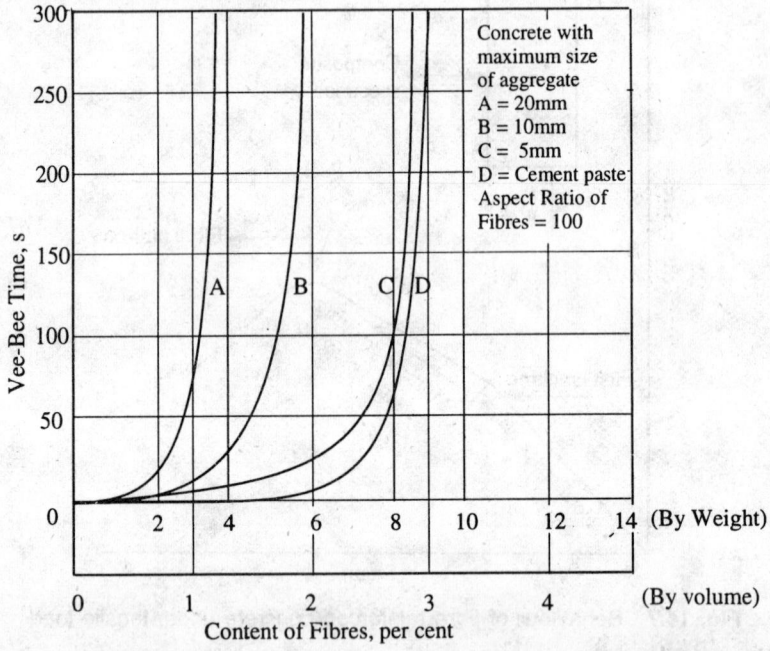

Fig. 14.8 Effect of aggregate size on workability of fibre reinforced concrete

In a FRC member subjected to flexure, the load at the first crack will increase due to the crack arresting mechanism of the closely spaced fibres. After the concrete cracks in tension, the fibres continue to take the load, provided the bond is good. When the fibre strain reaches its breaking strain, the fibres fracture resulting in load transfer to the fibres of adjacent layers which on reaching their breaking strain fracture and result in the shifting of the neutral axis. Failure occurs when the concrete in compression reaches the ultimate strain. The most important factors affecting the ultimate load are the volume of fibres and their aspect ratio.

14.9.3 Concrete Matrix

The cement required is OPC or PPC conforming to IS : 269–1976 or IS : 1489–1976, respectively. The aggregates are usually crushed quartz conforming to IS : 383–1970. A fibre-reinforced concrete requires a considerably greater amount of fine material than plain concrete so that it may be conveniently handled and placed. To be fully effective, each fibre needs to be completely embedded in the matrix and this determines the proportion of fine to coarse aggregate. Fibre concrete, therefore, generally requires a greater proportion of cement paste than conventional concrete for handling and placing by using the equipment meant for ordinary concrete. Normal concrete contains 25 to 35 per cent of cement paste of the total volume of concrete, and fibre-reinforced concrete requires paste content of the order of 35 to 45 per cent of the total volume of concrete, depending upon the fibre geometry and fibre volume.

14.9.4 Different Types of Fibres

The most commonly used man-made fibres have been *steel* and *polypropylene*, principally in concrete, and glass, principally in cement mortar for thin section applications. Properties of some of the commonly used fibres are given in Table 14.7.

TABLE 14.7. Physical properties of various types of fibres and matrices

Material	Specific gravity	Effective Modulus, GPa	Tensile Strength, MPa	Elongation at breaking point, per cent
Acrylic	1.10	2.1	210–420	25.0–45.0
Asbestos (Chrysotile)	2.55	8.4–14	200–1800	2.0–3.0
Carbon				
(i) high modulus	1.9	380	1800	0.5
(ii) high strength	1.9	230	2600	1.0
Cellulose	1.5	10–40	500	—
Cotton	1.5	5	420–700	3–10
Glass (Cem-FIL filament)	2.7	80	1050–3870	1.5–3.5
Nylon	1.1	4.2	780–850	16.0–20.0
Polyester	1.4	8.5	750–880	11.0–13.0
Polyethelene (high modulus)	0.96	15–40	300–700	3.0–10.0
Polypropylene	0.91	3–15	560–780	8.0
Rayon	1.50	7.3	420–630	10–25
Steel	7.86	200	280–420	3.5
OPC paste	2.0–2.2	10–20	2–6	0.01–0.05
OPC concrete	2.30	20–35	1–4	0.005–0.015

1. Steel-fibre Reinforced Concrete

A number of steel-fibre types are available as reinforcement. Round steel fibres, the commonly used type, are produced by cutting round wires into short lengths. The typical diameters lie in the range of 0.25 to 0.75 mm. Steel fibres having a rectangular cross-section are produced by slitting the sheets about 0.25 mm thick. For improving the mechanical bond between the fibre and matrix, indented, crimped, machined and hook-ended fibres are normally produced. The aspect ratio (=length/diameter) of fibres which have been employed vary from about 30 to 250.

Fibres made from mild steel drawn wire conforming to IS : 280–1976 with the diameter of wire varying from 0.3 to 0.5 mm have been practically used in India. *Round steel fibres* are produced by cutting or chopping the wire, *flat sheet fibres* having a typical cross-section ranging from 0.15 to 0.41 mm in thickness and 0.25

to 0.90 mm in width are produced by slitting (shearing) flat sheets. *Deformed fibres* which are loosely bonded with water soluble glue in the form of a bundle are also available. Since individual fibres tend to cluster together, their uniform distribution in the matrix is often difficult. This may be avoided by adding fibre bundles which separate during the mixing process. The properties of various types of fibres are compared in Table 15.7.

Properties of Fresh Steel-fibre Reinforced Concrete For satisfactory performance in the hardened state, fibre reinforcement should be uniformly distributed and fresh concrete be well compacted. Before adding fibres during mixing, it is essential that the clumps of tightly-bound fibres be broken up. For bulk steel-fibre mixes, a mixing sequence is recommended which is to blend fibre and aggregate before charging the mixer, e.g. by combining fibre and aggregate on a conveyor belt or chute.

The ease with which the fibre concrete can be compacted during construction depends on the nature and amount of the fibre used and, most importantly for short fibres, on their *aspects ratio*. The *slump test* has been judged to be a poor indicator of relative workability of steel-fibre concretes, since the addition of fibres to the mix changes the slump out of proportion to the workability change. The Vee-Bee test which incorporates the effects of vibration has been found to give a realistic assessment of workability of fibre concretes. The unsuitability of conventional workability tests for fibre concrete is essentially because of the fact that internal structure and flow characteristics of fibre-reinforced concrete are distinctly different from those of conventional concrete due to the presence of fibres. The composite forms a relatively stable system due to the interlocking of fibres which resists the flow of fresh concrete. This makes the tests like slump and compacting factor ineffective for fibre concrete because the mobilizing force in these tests (self weight) is inadequate to overcome the effective cohesion in the presence of fibres.

Typical relationships between Vee-Bee time, fibre content and aspect ratio for fibre-reinforced mortars are shown in Fig. 14.9 which indicate that the workability

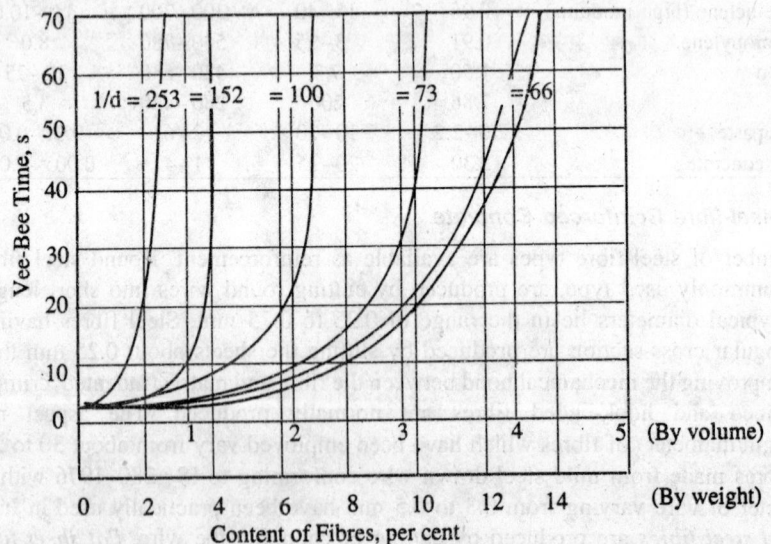

Fig. 14.9 Effect of fibre aspect ratio on workability of fibre reinforced concrete

of mix decreases with an increase in fibre concentration and aspect ratio. There is a critical fibre content for each aspect ratio beyond which the response to vibration decreases rapidly. Figure 14.8 shows that a reduction of maximum aggregate size facilitates the inclusion of fibres, although little is gained by using aggregate size smaller than 4.75 mm. Use of pulverized fuel-ash as a partial replacement of cement (30% by mass of cement) and a water-reducing admixture may be recommended to facilitate compaction.

Measurement of workability ACI Committee: 544 (1978) has recommended the use of inverted slump-cone test for the measurement of *workability*. The test measures the time to empty the steel-fibre concrete mix from an inverted slump-cone resting 75 mm above the bottom of a nine-litre (yield) bucket, after a 25–30 mm diameter vibrator prob has been inserted. The prob is allowed to fall and touch the bottom of the bucket. The time recorded in the range of 11 to 28 seconds indicates a steel-fibre concrete of good workability. This test has not been fully evaluated and is somewhat cumbersome.

In the workability measurement by conventional tests it is basically the cohesion of the mix which is indirectly measured. This cohesion of mix results in shear strength of the mix in the fresh state. It has been observed that the resistance to penetration by a cone of plastic material is dependent on the shear strength of the fresh concrete. Based on this observation, a cone penetration test has been suggested to measure the workability of fibre-reinforced concrete wherein a standard cone penetrates by its own weight through a mass of fresh mix. The depth of penetration in millimetres may be taken as a measure of workability. The penetration depth of a metallic cone with an apex angle of 30° and having a weight of 40 N has been reported to give the representative workability. The choice of cone with 30° apex angle and 40 N weight is based on the observation that the penetration depths obtained with this cone are neither too large nor too small, and are suitable for the normal range of mixes. For normal mixes the depth of penetration has been found to vary from 200 mm to 50 mm.

The cone penetration test is easy to conduct and can be conveniently adopted in the field conditions. The test has the comparative simplicity of a slump test while being suitable even for low workability mixes for which conventional tests fail. The test data have a consistent relationship with the other measures of workability given by slump, Vee-Bee time, compacting factor and ACI inverted cone method. The relationships between workabilities measured by different methods are given in Figs 14.10 to 14.12.

Factors affecting workability The factors having a predominant effect on the workability are: aspect ratio (l/d) and fibre volume concentration. Long thin fibres ($l/d > 100$) tend to mat together while short stubby fibres ($l/d < 50$) cannot interlock and can be dispersed by vibration. A *minimum fibre volume concentration* called *critical concentration* is needed to increase the strength. The critical concentration is generally inversely proportional to the aspect ratio. For $l/d = 100$, a volume concentration of 0.5 per cent for flexural strengthening and 1.7 per cent for tensile strengthening is required. However, for a 1.7 per cent concentration, an adequate workability can be obtained only with cement paste, and cement-sand mortar; whereas a 0.5 per cent concentration can perfectly be provided in the concrete. Thus there is a restricted range of practical fibre reinforced concrete with improved

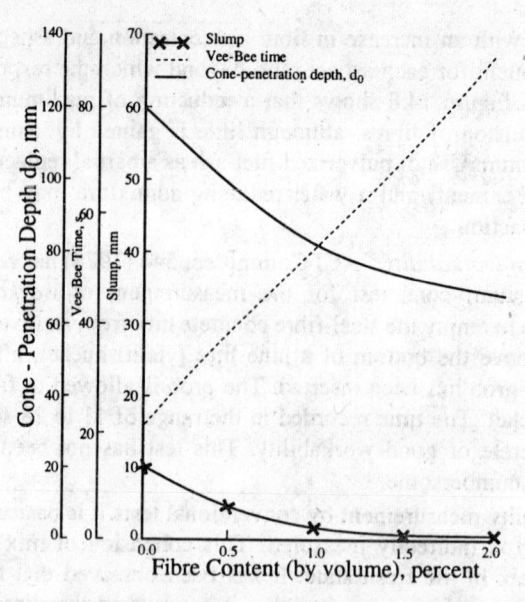

Fig. 14.10 Variation of workability values with fibre content

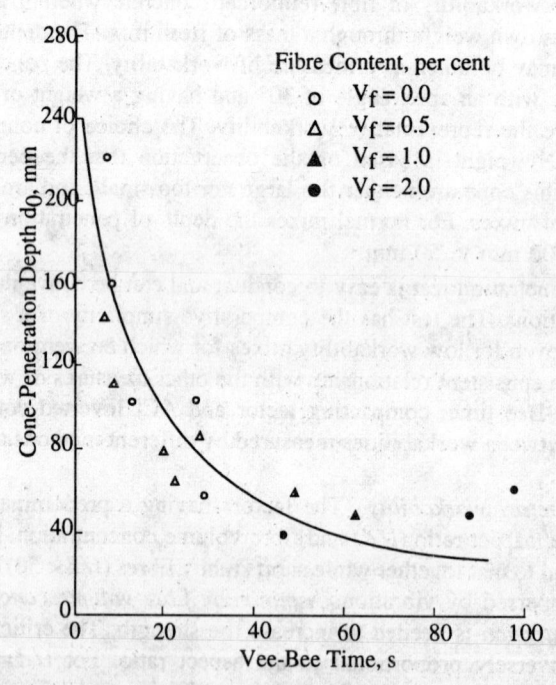

Fig. 14.11 Relation between Vee-Bee time and cone penetration depth

strengths. The performance of hardened concrete depends upon the *specific fibre surface (SFS)* which is defined as the total surface area of all the fibres present within the unit volume of the composite. The specific fibre surface depends upon

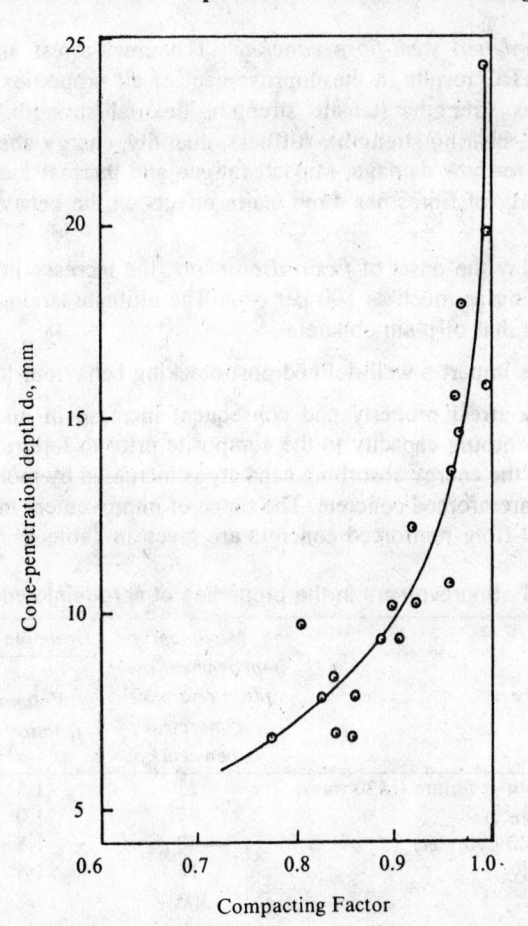

Fig. 14.12 Relation between compaction factor and cone penetration depth

the fibre volume concentration, fibre size and aspect ratio. For a fibre volume concentration of V_f per cent, the specific fibre surface in a unit volume of composite is given by

$$SFS = n(\pi dl)$$

where, n, l and d are the number, length and diameter of the fibres, respectively, and $\pi\, dl$ is the surface area of each fibre. The number of fibres is given by

$$n = \frac{V_f \times 100}{\pi d^2 l/4}$$

Thus,
$$SFS = \frac{400 V_f}{d} = \frac{400 V_f A}{l}$$

where, $A = l/d$ is the aspect ratio. The above expression indicates that for the given fibre volume concentration and aspect ratio, the specific fibre surface is inversely proportional to the fibre length.

Properties of hardened steel-fibre concrete The crack-arrest and crack-control mechanism of SFRC results in the improvement of all properties associated with cracking, such as strengths (tensile strength, flexural strength, shear strength, torsional strength, bearing strength), stiffness, ductility, energy absorption, and the resistance to freeze-thaw damage, impact, fatigue and thermal loading. The crack controlling property of fibres has three major effects on the behaviour of concrete composite.

(i) Fibres delay the onset of *flexural cracking*, the increase in tensile strain at the first crack being as much as 100 per cent. The ultimate strain may be as large as 20 to 50 times that of plain concrete.

(ii) The fibres impart a well-defined post-cracking behaviour to the composite.

(iii) The crack-arrest property and consequent increase in ductility imparts a greater energy absorbing capacity to the composite prior to failure. With a 2.5 per cent fibre content the energy absorbing capacity is increased by more than ten times as compared to unreinforced concrete. The range of improvement in the mechanical properties of steel-fibre-reinforced concrete are given in Table 14.8.

TABLE 14.8 Improvement in the properties of fibre-reinforced concrete

Property	Maximum improvement over plain reference concrete, per cent	Optimum fibre parameters — Volume fraction V_f	Optimum fibre parameters — Aspect ratio l/d
Compressive strength at failure (M20 mix)	25	1.5	—
Tensile strength (direct)	45	1.0	80
Tensile strength (split cylinder)	40	1.5	80
Modulus of elasticity	15	1.5	80
Ultimate strain	300	—	—
Flexure strength			
(i) at first crack	40	1.5	80
tensile strain	100	—	—
(ii) at failure	60	1.5	80
tensile strain	20 to 50 times		
Modulus of rupture	10	—	—
Energy absorption	500	1.5	80
	1000	2.5	100
Impact resistance (due to explosive charges and dropped weight)	400–900	—	—
Flexural fatigue			
Static load	125	—	—
Indurance to 2×10^6 cycles at a strain rate equal to that in refe- rence specimens subjected to static load			
(i) Non reversal	90 of static	—	—
(ii) Full reversal	70 of static	—	—
Post flexural fatigue, flexural strength	10–30 of similar beams of non fatigue histories	—	—

The fibre concretes reinforced by conventional steel bars have substantially improved *serviceability conditions* obtained by crack and deflection control, besides increasing flexural strength marginally.

(i) *Tensile strength* Within the practical limitations imposed on volume and aspect ratio (l/d) of fibres for the ease of mixing, there is a modest increase in tensile strength due to fibre reinforcement, but more substantial is the increase in *toughness* as measured by the area under the *stress-strain curve*.

(ii) *Compressive strength* The presence of fibres in concrete produced no or only modest increase in compressive strength, although the increased ductility resulting from the addition may be advantageous, particularly in *over-reinforced concrete beams* where a brittle failure can be changed into a ductile one.

(iii) *Flexural strength* The increase in the flexural strength (characterized by modulus of rupture) for mortars and concrete is of the order of 25 per cent for the values of $V_f l/d$ from 40 to 120 (a practical limit from workability consideration). The *fatigue flexural strength* after 10^5 cycles is increased by a similar proportion.

The *ultimate load carrying capacity* of fibre-reinforced concrete beam depends mainly on the adequacy of bond. In the absence of excellent *interfacial* bond the fibres are debonded as soon as the load is transferred to them immediately after cracking of the matrix and the ultimate load will not be greater than the ultimate load of beams without fibre reinforcement. If the bond is excellent, the fibres can withstand loads even after the cracking of the matrix, and this results in an increase in the *ultimate strength*. An improvement in bond can be achieved by the introduction of *indented*, *crimped*, or *bent fibres*.

The flexural strength depends on the volume fraction and aspect ratio of fibres. Steel-fibres up to 4 per cent by volume have been found to increase the first crack flexural strength of concrete up to 2.5 times the strength of unreinforced composite.

(iv) *Toughness and impact strength* The area under the complete load-deflection curve (or under a prescribed part of the curve) can be described as a measure of *toughness* or *energy absorption* capability of the material. Improvements in impact strength for fibre-reinforced concretes are highly dependent on the type of fibre and the method of test. It is estimated by using falling weight method or pendulum-type impact machine. Crimped fibres yield a greater improvement in impact strength than straight fibres.

The impact strength against dynamic tensile and compressive loads due to dropped weights or explosives is 8 to 10 times that of plain concrete. The test results of the specimens containing 0.50 mm diameter high tensile strength crimped steel fibres indicate an improvement in the toughness under impact loading, by more than 400 per cent. An increase in fatigue strength with increasing percentage of steel fibres has been noticed.

(v) *Durability* As in case of conventional reinforced concrete, steel fibres will be protected from *corrosion* provided the alkalinity of the matrix is maintained in the vicinity of the fibres. *Carbonation* of concrete matrix may lead to corrosion of the fibres, and any deterioration may be accelerated if the concrete is cracked. Since fibre-concrete normally fails due to *fibre pull-out* rather than fibre fracture the uncorroded fibre strength is not fully utilized, a considerable reduction in fibre diameter due to corrosion could be tolerated provided that corrosion does not reduce the interfacial bond strength.

The studies have indicated a greater rate and extent of chloride penetration for fibre-reinforced concrete than for conventional plain concrete. This suggest that the fibres extending from the surface may create an entry for the chlorides in addition to normal capillary system and make fibre reinforced concrete more vulnerable to *corrosion* damage than conventional steel reinforcement.

Application of steel-fibre reinforced concrete Steel-fibre reinforced concrete (SFRC) provides additional strength in flexure, fatigue, impact and spalling. These properties lead to smaller concrete sections, improved surface quality and reduced maintenance. SFRC can be applied in the following areas.

The main applications of SFRC are in highway and airfield pavements, hydraulic structures, tunnel linings, industrial floors, bridge decks, repair works, etc.

(i) *Highway and airfield pavements* The steel-fibre concrete can be used in new pavement constructions or in the repair of existing pavements by the use of *bonded* or *unbonded overlays* to the slab beneath. The major advatages are : a higher flexural strength results in the reduction of required pavement thickness; the resistance to impact and repeated loading is increased. The transverse and longitudinal joint spacings may be increased. Under conditions of restrained shrinkage, the greater tensile strain capacity of steel-fibre concrete results in lower maximum crack widths than in plain concrete.

SFRC gives a smooth riding surface without irregular depressions. The overlays for the rehabilitation of runways, taxiways, bridge decks, and the strengthening of existing runways and taxiways to comply with the rigid requirements of the newer generation heavy-duty jet aircrafts, are extensively used. SFRC can be advantageously used in the repair of damaged patches in existing runways, and highway pavement slabs.

The thickness of pavements constructed with concrete having a cement content of 410 kg/m^3, water-cement ratio of 0.6, maximum size of aggregates as 20 mm using 1.4 per cent (by volume, i.e., 106 kg/m^3) trough type steel-fibres could be 25 per cent less than normal concrete pavements.

(ii) *Hydraulic structures* The major advantage of using steel-fibre concrete in hydraulic structures is its resistance to *cavitation* or *erosion damage* by high velocity water flow. The steel-fibre concrete has been successfully used in the repair of spilling basin at Tarbela Dam in Pakistan. The fibre concrete contained about 1 per cent (by volume) of $25 \times 0.25 \times 0.55$ mm slit steel fibres.

(iii) *Fibre shotcrete* Fibre shotcrete has been used in rock slope stabilization, tunnel lining and bridge repair. A thin coating of plain shotcrete applied monolithically on top of the fibre shotcrete, may be used to prevent *surface staining* due to *rusting*. The conventional sprayed concrete techniques can be used by including fibre mixing with the pneumatic conveying of fibres from a rotary fibre feeder to the nozzle via a 75 mm diameter flexible hose. In addition to usual shotcrete advantages, the fibres are aligned in two dimensions (in a plane) by the mode of application of relatively thin coating. The fibre shotcrete can be used in the protection of structural steel work particularly in the support structure.

(iv) *Refractory concrete* Steel-fibre reinforced refractory concretes have been reported to be more durable than their unreinforced counterpart when exposed to high thermal stress, thermal cycling, thermal shock or mechanical abuse. The increased service span is probably due to combination of crack control, enhanced toughness, and the spall and abrasion resistance imparted by the steel fibres. Through the use of shotcrete technique, the material can be used for lining ash hoppers and lining flame exhaust ducts.

(v) *Precast applications* They include manhole covers, concrete pipes, machine bases and frames. Improved flexural and impact strengths may allow the use of steel-fibre concrete components in rough handling situations.

(vi) *Structural applications* Structural applications of steel-fibre concrete are rare. However, the following possibilities may be considered.

(a) Fibre reinforcement can provide an increased impact resistance to conventionally reinforced beams, and thus an enhanced resistance to local damage and spalling.

(b) Fibre reinforcement can inhibit crack growth and crack widening, this may allow the use of high strength steel without excessive crack widths or deformations at service loads.

(c) Fibre reinforcement provides ductility to conventionally reinforced concrete structures, and hence enhances their stability and integrity under earthquake and blast loading.

(d) Fibre reinforcement increases the shear strength of concrete. As a consequence punching shear strength of slabs in increased and sudden punching failure may be transformed into gradual ductile one.

Mix design for steel-fibre reinforced concrete The mix should contain minimum fibre content and maximum aggregate for the specified strength and workability. The *cement paste content* depends upon three factors

(i) Volume fraction of fibres
(ii) Shape and surface characteristics of fibres, i.e., specific fibre surface
(iii) Water-cement ratio

For the commonly encountered SFRC mixes, the following range of parameters is associated:

Cement content	300 to 500 kg/m^3
Water-cement ratio	0.45 to 0.60
Ratio of sand to total aggregate, per cent	50 to 100
Maximum size of aggregate	10 and 20 mm
Fibre content	1.0 to 2.5 per cent
Fibre aspect ratio	50 to 1000

Mix design procedure Following are the steps involved in the mix design of fibre reinforced concrete.

(i) Corresponding to the required 28-day field design flexural strength of steel fibre-reinforced concrete, the design strength for laboratory mix is determined.

(ii) For fibres of known geometry and for stipulated volume fraction, the water-cement ratio is selected between 0.45 and 0.6.

(iii) Depending on the maximum size of aggregate and fibre concentration, the paste content is determined by mass.

(iv) The fine-to-coarse aggregate ratio varies from 1 : 1 to 1 : 3, a ratio of 1 : 1.5 is a good start for a volume percentage of fibre up to 1.5 and length of fibre up to 40 mm.

(v) For the water-cement ratio and paste content determined in the steps (ii) and (iii), respectively, the cement and water contents may be worked out.

(vi) The fibre content (by mass) is calculated by taking the density of fibres as 7850 kg/m^3.

(vii) The total quantity of the aggregate is determined from:

$$W_A = W_{FRC} - (W_W + W_C + W_F)$$

where W_A, W_{FRC}, W_W, W_C and W_F are the masses of total aggregate, fibre reinforced concrete, water, cement and fibres, respectively.

(viii) The quantities of fine and coarse aggregates are worked out by using the step (iv).

(ix) The trial mix is prepared and the paste content adjusted if the mix shows any tendency to segregate.

(x) The workability of the mix is checked using appropriate test.

Design illustration For the illustration of the above procedure, select a fibre concentration of 1.5 per cent (by volume) of trough-shaped fibres with aspect ratio of 80 with 0.45 mm diameter.

For the maximum nominal size of aggregate of 20 mm, assume the paste content to be 40 per cent, and fine-to-coarse aggregate ratio to be 1 : 1.5. Adopt,

$$\text{water-cement ratio} = 0.55$$

mass of fibres per cubic metre of SFRC = $7850 \times 0.015 = 117.7$ kg (say 118 kg)

For 1 kg (= 1/3.15 = 0.317 litre) of cement, the water content required is 0.55 litre giving a total paste content of 0.867 litre. Therefore, for a cement paste content of 40 per cent, i.e., 400 litres per cubic metre of concrete, the cement and water contents are 461(=400/0.867) and 254(=0.55 × 461) kg, respectively. The total aggregate content = 2400 − (254 + 461 + 118) = 1567 kg. For the assumed fine-to-coarse aggregate ratio of 1 : 1.5, the coarse and fine aggregates are 940 and 627 kg, respectively.

Practical Mix Proportions Though the high fibre content brings about large improvements in mechanical properties, it makes the concrete unworkable. On the other hand, a low fibre content in workable concretes show no significant improvements in the desirable properties. Thus a practical concrete is a compromise between these situations. Typical mixes using fibre volume concentrations of 0.8 to 1.5 per cent with water-reducing admixtures and/or flyash have been extensively used. With steel-fibres, the typical mix proportions by mass are:

Cement	:	Water-cement ratio	:	Sand	:	10 mm aggregate
1	:	(0.4 to 0.6)	:	(2 to 3)	:	(0.8 to 3)

Similar mixes have also been used for polypropylene fibres.

2. Polypropylene Fibre-reinforced (PFR) Cement-mortar and Concrete

Polypropylene is one of the cheapest and abundantly available polymers. Polypropylene-fibres are resistant to most chemicals and it would be the cementitious matrix which would deteriorate first under aggressive chemical attack. Its melting point is high (about 165° C), so that a working temperature as high as 100°C may be sustained for short periods without detriment to the fibre properties.

Polypropylene short fibres in small volume fractions between 0.5 to 1.0 per cent have been commercially used in concrete to achieve considerable improvement in impact strength of the hardened concrete. They have low modulus of elasticity. Polypropylene fibres are available in two forms: monofilaments produced from spinnarets, and film fibres produced by extrusions. The film fibres are commonly used and are obtained from fibrillated film twisted into twine and chopped, usually into 25–50 mm lengths for use in concrete. The fibrillated film may also be opened to produce continuous networks for use in thin sheet manufacture. Fibrillated film may also be woven to produce flat meshes which may be used as thin cement sheet reinforcement.

Polypropylene fibres being hydrophobic can be easily mixed as they do not need lengthy contact during mixing and only need to be evenly dispersed in the mix. These are therefore added shortly before the end of mixing the normal constituents. Prolonged mixing may lead to undesirable *shredding of fibres*. There is no physico-chemical bond between fibre and the matrix, only a mechanical bond is formed as cement paste penetrates the mesh structure between individual fabrics of chopped length or continuous network.

Properties of fresh PFR concrete The compacting factor test has been reported to be most suitable. The inclusion of polypropylene fibres reduces the workability considerably, e.g., a normal concrete mix of medium workability (C.F about 0.88) may reduce to a low workability mix (C.F about 0.75) following the addition of 1 per cent of chopped 35 mm polypropylene fibres. Polypropylene monofilaments can be used in small volume fractions of about 0.1 to 0.2 per cent to alter *rheological properties* of the material, e.g., highly *air-entrained concretes* can be stabilized by fibres.

Properties of hardened PFR concrete The tensile strength of concrete is essentially unaltered by the presence of a small volume of short polypropylene fibres. Although the change in flexural strength of polypropylene reinforced-concrete is marginal, the *postcracking behaviour* has shown its ability to continue to absorb *energy* as fibres-pullout. The energy absorbing capacity has been found to increase with the length of fibres, e.g. the 75 mm polypropylene fibres may result in an energy absorption comparable to that of the less efficient of steel fibres, and at a considerably lower cost.

Durability Polypropylene may deteriorate under attack from ultra violet radiation or by thermal oxidation process. The cement matrix appears to prevent the former. To combat thermal oxidation, sophisticated stabilizers have been developed to delay degradation, and enhance durability.

Applications of PFR mortar and concrete

(i) *Clading panels* Inclusion of polypropylene fibres instead of steel mesh reinforcement may allow reduction in panel thickness.

(ii) *Shotcreting* Surface coatings of polypropylene reinforced-mortar may be provided by shotcreting using normal equipment. The fibres of about 20 mm length enable smooth transport of the dry mix through air hoses and nozzles. Water is then added at the gun orifice. Shotcreting can be advantageously used in wet environments where polypropylene fibres can eliminate the need for steel (corrodable) mesh on which spray of mortar is required.

(iii) Polypropylene concrete can be advantageously used in the energy dissipating blocks.

The potential market for polypropylene reinforced-cement is, principally, as a substitute for absestos-cement roofing and cladding panels.

3. Glass-fibre Reinforced Concrete (GFR)

Glass fibres are made up from 200 to 400 individual filaments which are lightly bonded to make up a strand. These strands can be chopped into various lengths or combined to make cloth, mat or tape. Using the conventional mixing technique for normal concrete, it is not possible to mix more than about two per cent (by volume) of fibres of up to a length of 25 mm.

The major application of glass fibre has been in reinforcing the cement or mortar matrices used in production of thin-sheet products. The commonly used varieties of glass-fibres are E-glass used in the reinforcement of plastics, and AR-glass. E-glass have inadequate resistance to alkalies present in portland cements whereas AR-glass have improved alkali-resistant characteristics. Sometimes polymers are also added in the mixes to improve some physical properties such as *moisture movement*.

The process of manufacture of glass-fibre cement products may involve spraying, premixing or incorporation of continuous rovings. In the *spray-suction* process, the glass-fibre strand is chopped into lengths between 10 to 50 mm and blown in spray simultaneously with the mortar slurry on to a mould or flat bed followed by suction to remove excess water. On the other hand in the technique involving premixing, short strands (about 25 mm in length) are mixed into mortar paste or slurry before further processing by casting into open moulds, pumping into closed moulds, etc. Care must be taken to avoid fibre tangling and matting together, and to minimize the fibre damage during mixing.

In the process incorporating continuous rovings, the rovings are impregnated with cement slurry by passing them through a cement bath before they are wound on to an appropriate mandrel. Additional slurry and chopped fibres can be sprayed on to the mandrel and compaction can be achieved by the application of roller pressure combined with suction.

Properties of hardened GFR concrete The behaviour of glass-fibre cement sheets under tensile force is typified by *multiple cracking* of the matrix. Longer fibres improve the ultimate failure stress. In wet environments, significant reduction in strength takes place. The material may become brittle on ageing.

One of the most important improvements in the property achieved by glass fibre is the spectacular improvement in *impact strength*. With the addition of just 5 per cent glass fibres, an improvement in the impact strength of up to 1500 per cent can be registered as compared to plain concrete. With a two per cent fibre content (up to 25 mm in length), the flexural strength is almost doubled. The second important improvement is in the resistance to *thermal shock*. Ductility also improves with an increase in strength and modulus of rupture.

The flexural strength of water stored and weathered specimens reduces with time and nearly equals that of the matrix alone. The reduction in *energy absorption* is similar to that in flexural strength. The long-term *durability* of glass fibre-reinforced cement can be improved by the addition of 15 per cent polymer to the mortar matrix. The increase in matrix cost is balanced by the use of cheaper E-glass fibres.

Applications The glass fibre-reinforced cement finds its use in formwork systems, ducting, roofing elements, sewer lining, swimming pools, fire-stop partitioning, tanks and drainage elements, etc. Sometimes it is used in combination with polymer impregnated *in situ* concrete.

4. Asbestos Fibres

The naturally available inexpensive mineral fibre, asbestos, has been successfully combined with portland cement paste to form a widely used product called *asbestos cement*. Asbestos fibres have thermal, mechanical and chemical resistance making them suitable for sheet products, pipes, tiles and corrugated roofing elements. Asbestos-cement products contain about 8 to 16 per cent (by volume) of asbestos-fibres. The flexural strength of asbestos cement board is approximately two to four times that of unreinforced matrix. However, due to relatively short length (10 mm), the fibres have low impact strength. There are some health hazards associated with the use of asbestos cement. In the near future, it is likely that *glass fibre-reinforced concrete* will replace asbestos completely.

5. Carbon Fibres

Carbon-fibres form the most recent and probably the most spectacular addition to the range of fibres available for commercial use. Carbon fibres come under the high E-type fibres. These are expensive. Their strength and stiffness characteristics have been found to be superior even to those of steel. But they are more vulnerable to damage than even glass fibres, and hence are generally treated with resin coating.

6. Organic Fibres

Organic fibres, such as polypropylene or natural fibres may be chemically more inert than either steel or glass fibres. They are also cheaper, especially if natural. The polypropylene-fibre concrete has been described earlier. A large volume of vegetable fibres (7 per cent, 50 mm length) may be used to obtain a multiple cracking composite. The problem of mixing and uniform dispersion may be solved by adding a superplasticizer.

Polypropylene, nylon and other organic fibres due to their low modulus of elasticity are not effective in crack control, and also the organic fibres may decay. However, these fibres improve *impact resistance*.

7. Vegetable Fibres

The commonly used fibres are jute, coir and bamboo. They possess good tensile strength in their natural dry state. Their tensile strengths do not suffer significantly even after being immersed in 10 per cent normal solution of sodium hydroxide for up to 28 days. However, longterm *durability* is doubtful.

In contrast to glass fibres, steel, asbestos and polypropylene fibres are chemically stable in a cement paste matrix. The high alkalinity of cement paste protects steel from being corroded. The corrosion of steel fibres can however become a problem when the matrix has cracked.

Irrespective of the type, size and shape of fibres to be used in a mix, the fundamental requirement of fibre-reinforced concrete is that all the individual fibres should be uniformly distributed throughout the matrix. The mix should have sufficient paste content to coat the fibres and aggregate, so that the ingredients can be placed and compacted in the final position without any *segregation*.

The mix proportions generally depend on the intended applications of the composite. The *prime considerations* are uniform *dispersion of fibres, adequate workability* for placing and compaction with the available equipment. The *workability of fibre-reinforced concrete* is influenced by *maximum size of aggregate* (Fig. 14.8), *volume fraction, geometry* and *aspect ratio of fibres* as shown in Fig. 14.9. As the size of aggregate increases, it becomes more difficult ot achieve uniform fibre dispersion, since the fibres are bunched into mortar fraction which can move freely past the aggregate during compaction. To obtain a better dispersion the coarse aggregate content is kept lower than in a normal mix and the maximum size of aggregate is preferably limited to 10 mm. The mortar matrix (consisting of particles less than 4.75 mm) should be around 70 per cent, and aggregate-cement ratio as low as 3 : 1. A fine-to-coarse aggregate ratio of 1 : 1 is often a good strarting point for a mix trial. *Water-cement* ratio between 0.4 and 0.6, *cement-content* of 250 to 430 kg/m^3 are recommended for providing adequate paste content to coat large surface of fibres. Beyond a certain optimum content of fibres the workability of the composite decreases rapidly.

14.9.5 Batching, Mixing, Placing, Compaction and Finishing

The fibres are usually added to the aggregates before the introduction of cement and water into the mixer. For laboratory testing, fibres can be added in small amounts to the rotating drum charged with cement, aggregate and water. For large batches, the fibres are blown into the previously charged rotating drum.

A fibre mix generally requires more time and vibration to move the mix and to compact it into the forms. Surface vibration of forms and exposed surface is preferable to prevent segregation. The properties of fibre-reinforced concrete depend upon fibre alignment. More energy is required to compact fibre concrete than conventional concrete. Some of the precautions taken while mixing, placing and compacting fibre-reinforced concrete are as follows.

(i) While mixing small quantities of fibre-reinforced concrete by hand, there is a possibility of steel fibres shooting up and hitting the eyes of the worker or even pricking the hand. To avoid these hazards, the hands should be protected by gloves and the eyes with safety glasses.

(ii) A pan mixer of the counter-flow type should be used for mixing fibre-reinforced concrete.

(iii) For uniform distribution of steel fibres, a dispenser should be used. While dispensing the fibres into concrete, the rate at which the fibres are fed to the mixer should be synchronized with rate of mixing.

(iv) Forks and rakes can prove helpful for handling low slump mixes.

(v) Standard screeding methods and trowels can be used for finishing fibre concrete. A textured surface can be obtained by using a stiff brush.

Standard workability tests, such as the slump, compacting factor and Vee-Bee tests are suitable for conventional concrete but not for mixes containing fibres. For instance, the slump of a mix, even with a low fibre content, can be zero though the mix responds well when vibrated. A workability test should provide the condition of flow on vibration, because FRC responds well to conventional vibrating table as it does not easily segregate from the mix due to its low specific gravity.

14.10 POLYMER CONCRETE COMPOSITES (PCCs)

Polymer concrete composites are obtained by the combined processing of polymeric materials with some or all of the ingredients of the cement concrete composites. Depending on the process by which the polymeric materials are incorporated, polymer concrete can be classified as follows.

14.10.1 Polymer-impregnated Concrete (PIC)

In *polymer-impregnated concrete*, low viscosity liquid *monomers* or *prepolymers* are partially or completely impregnated into the *pore systems* of hardened cement composites and are then polymerized. The partial or surface impregnation improves durability and chemical resistance. Overall improvements in the structural properties are modest. On the other hand, total or in depth impregnation improves structural properties considerably.

The hardened concrete, after a period of moist curing, contains a considerable amount of free water in its *voids*. The water-filled voids form a significant component of the total volume of concrete ranging from 5 per cent in dense concrete to 15 per cent in *gap-graded concretes*. In polymer-impregnated concrete, it is these water-filled pores that are sought to be filled with polymers, i.e. the major parameters affecting monomer loading are the moisture and the air in the voids in concrete. The total or in-depth polymer impregnation of concrete, therefore, involves the following states:

(i) Availability of well-designed cement concrete, which is adequately moist cured with optimum strength.

(ii) Removal of moisture by drying the concrete by heating to develop surface temperatures of the order of 120 to 150°C. The small specimens can be heated in an air-oven. For large cast-*in situ* surfaces a thick blanket of sand (usually 10 mm thick) can be used to prevent a steep *thermal gradient*. Infra-red heaters may be used. About 6 to 8 hours of heating is required to expel a large part of the free water in the concrete.

(iii) Cooling of concrete surfaces to safe levels (about 35°C) to avoid flammability.

(iv) Removal of air by subjecting the dry concrete specimen to vacuum in a process vessel. The degree of vacuum applied and the duration have significant influence on the quantity of monomer that can be impregnated and therefore, on the depth of impregnation.

(v) Application of monomer by soaking the concrete specimen in it for a sufficiently long time to achieve the desired depth of penetration. The soaking time depends on the viscosity of monomer, preparation of the specimen prior to soaking and the characteristics of the concrete. To reduce the time required to achieve a desired depth of monomer penetration, external pressure using nitrogen gas or air is generally employed.

(vi) Covering the surface with a *plastic sheet* to prevent evaporation of monomer.

(vii) Polymerization by heating the catalyzed monomer to the required temperature levels (usually between 60 and 150°C depending upon the type of monomer) also called thermal catalytic technique. The heating can be done under water, or by low pressure steam injection, or by infra-red heaters or in an air-oven. Depending on the polymer, 2 to 6 hours are required for this stage. The heating decomposes the catalyst and initiates the polymerization reaction. This reaction is called a *thermal catalytic reaction*. When monomer has penetrated into concrete, polymerization can also be initiated using ionizing radiation such as gamma rays. The polymers, when fully polymerized or cross-linked, are solids occupying the volume in which they have been impregnated. As such, at the impregnation stage, the polymer has to be in a prepolymer liquid form, generally called monomer. The state of polymerization of monomers, or of prepolymer resins, is brought about also by adding initiators, and cross-linking agents.

Polymers can be broadly categorized as *thermoplastics* and *thermosetting resins*. Thermoplastics soften at an elevated temperature (usually between 100 and 150°C and called glass transition temperature), and as such the advantage of using thermoplastic impregnated concrete is lost at such temperatures. Thermoplastic monomers have a low viscosity and are able to penetrate hardened concrete well and fill a large part of the pores. Their polymerization is accomplished by addition reactions not leading to low molecular weight by-products. Thermosetting resins, on the other hand, are more viscous and difficult to impregnate into concrete. However, they can withstand higher temperatures without softening. But the condensation reactions which occur may lead to the formation of low molecular weight by-products which would occupy some of the space.

It is necessary that a monomer or its polymer is chemically compatible with the compounds of cement and the constituents of hydrated cement paste to prevent their adverse effects.

Monomer/resin systems used for polymer impregnated concrete are styrene, polyester, methylmethacrylate, butylacrylate, acrilonitrile, epoxies and their copolymer combinations. The types and strength properties of some of the commonly used systems are given in the Table 14.9.

TABLE 14.9 Types of polymers and strength properties of polymer impregnated concrete

Polymer type	Technique employed	Polymer loading, per cent (by mass)	Compresive strength, MPa	Strength improvement ratio	
Styrene	Specimens vacuum-treated and pressure-impregnated	4 to 6	60–90	2.6–3.0	Thermal catalytic
Styrene	Predried specimens just soaked in monomer	1 to 2	25–30	1.5–2.0	Thermal catalytic
60% styrene + 40% trimethol-propane trimethacrylate (TMPTMA)	Vacuum-treated and pressure-impregnated	6 to 7	50–60	1.5–2.0	Thermal catalytic
Methyl-methacrylate (MMA)	Vacuum-treated and pressure-impregnated	5 to 7	100–125	3.5–4.0	Thermal catalytic
MMA	Vacuum-treated and pressure-impregnated	5 to 7	120–140	4.0–4.5	Radiation
MMA + 10% TMPTMA	Vacuum-treated and pressure-impregnated	5.5 to 7.5	150	5	Radiation
MMA + 10% TMPTMA	Predried specimens just soaked with monomer from one face only	2	70	2.3	Thermal catalytic
MMA + 10% TMPTMA	High pressure steam cured concrete, dried, vacuum-treated and impregnated under pressure	6 to 8	170–190	5.7–6.3	Radiation
Acrylonitrile	Vacuum-treated and pressure-impregnated	3.5 to 5.5	80	2.7	Thermal catalytic
10% polyester +90% styrene	Vacuum-treated and pressure-impregnated	5 to 6.5	130	4.3	Thermal catalytic
Vinyl chloride	Vacuum-treated and pressure-impregnated	3 to 5	70	2.3	Thermal catalytic
Epoxy	Vacuum-treated and pressure-impregnated	—	105	3.5	Thermal catalytic

The applications of polymer impregnated concrete are as follows:

(i) Surface Impregnation of Bridge Decks
The aim of impregnating the bridge decks is to render them impervious to the intrusion of moisture, deicing chemicals and chloride ions.

(ii) Applications in Irrigation Structures
The effect of caviation and erosion in dams and other hydraulic structures can be catastrophic. Conventional repairs of the damage are expensive and huge losses may be caused due to loss of benefits from irrigation, power generation, flood control, etc. In such cases the polymer-impregnated treatement may be cost-effective. The concrete may be removed from the place of severe damage and the damaged area patched, dried and treated by impregnation.

(iii) Structural Members
Polymer-impregnated concrete has potential as a structural material. Polymer-impregnated prestressed concrete beams have shown remarkable improvements over conventional concrete. The maximum tendon force could be up to four times that in unimpregnated concrete. The *creep deflection* is of the order of 1/19 to 1/16 that of *static deflection*. Shear strength improves by the same factor as compressive strength. The stress and strain curves of polymer concretes are shown in Fig. 14.13.

(iv) Marine and Underwater Applications
Greatly improved structural properties and negligible water absorption and permeability make polymer-impregnated concrete an excellent material for marine and underwater applications, such as in desalination plants and sea floor structures. Even a partial impregnation of concrete piles in sea water reduces the corrosion of reinforcing bars by 24 times.

(v) Repair of Structures
Polymer impregnation has a very good potential for the repair of damaged structures. Restoration and preservation of stone monuments is an interesting application.

14.10.2 Polymer Concrete (Resin Concrete)
Polymer concrete is a composite wherein the polymer replaces the cement-water matrix in the cement concrete. It is manufactured in a manner similar to that of cement concrete. Monomers or pre-polymers are added to the graded aggregate and the mixture is thoroughly mixed by hand or machine. The thoroughly mixed polymer concrete material is cast in moulds of wood, steel or aluminium, etc. to the required shape or form. Mould releasing agents can be added for easy demoulding. This is then polymerized either at room temperature or at an elevated temperature. The *polymer phase* binds the aggregate to give a strong composite. Polymerization can be achieved by any of the following methods.
  (i) Thermal-catalytic reaction
  (ii) Catalyst-promoter reaction
  (iii) Radiation

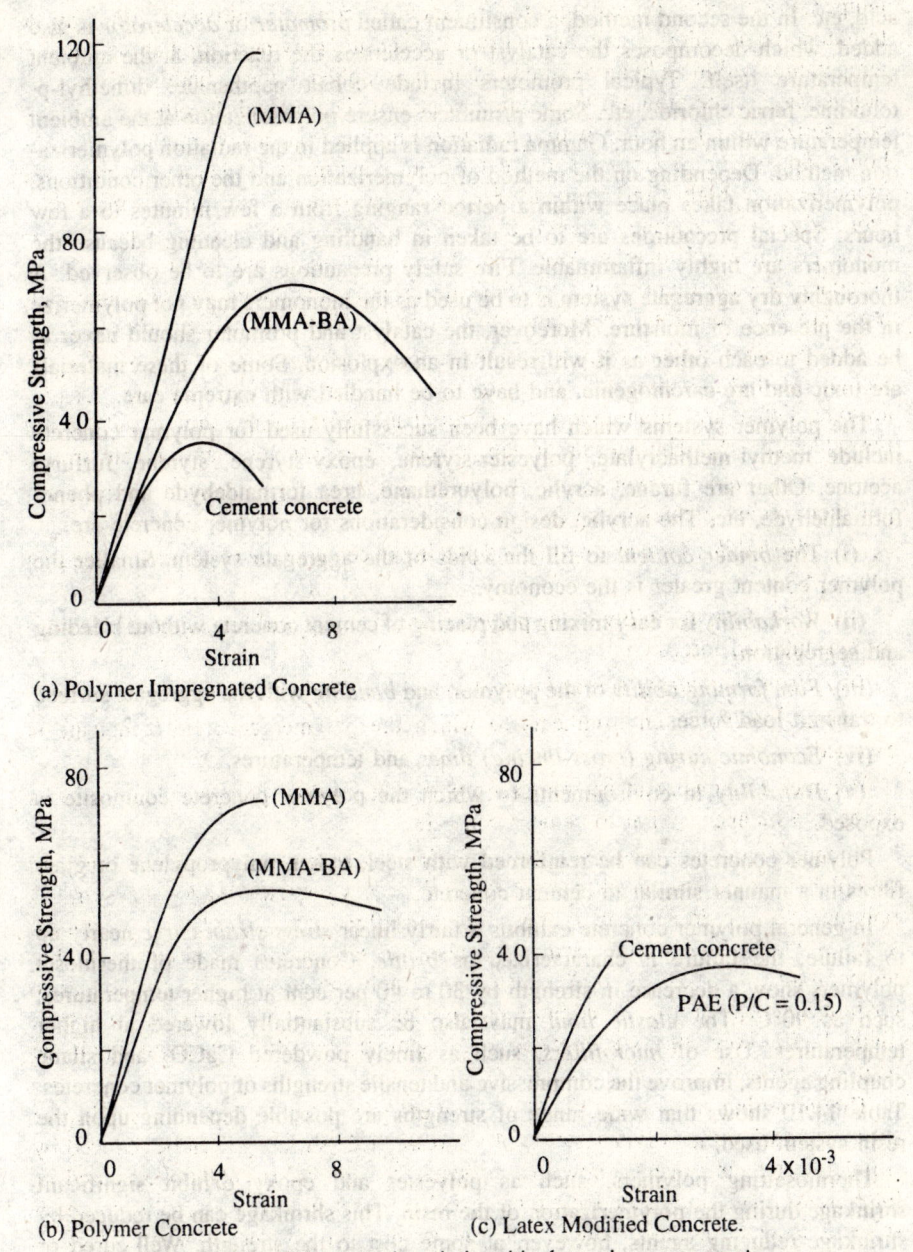

Fig. 14.13 Stress-strain relationship for polymer concrete

(a) Polymer Impregnated Concrete
(b) Polymer Concrete
(c) Latex Modified Concrete.

In the first method, only the catalyst is added to the monomer (thermoplastic) and polymerization is initiated by decomposing the catalyst by the application of elevated temperatures up to 90°C. Typical catalysts used for different monomer systems include, benzoyl peroxide, methyl-ethyl-ketone peroxide, benzenesulphonic acid, etc. In the second method, a constituent called *promoter* or *accelerator* is also added, which decomposes the catalyst or accelerates the reaction, at the ambient temperature itself. Typical promoters include cobalt napthanate, dimethyl-p-toluidine, ferric chloride, etc. Some promoters ensure polymerization at the ambient temperature within an hour. Gamma radiation is applied in the radiation polymerization method. Depending on the method of polymerization and the other conditions, polymerization takes place within a period ranging from a few minutes to a few hours. Special precautions are to be taken in handling and cleaning because the monomers are highly inflammable. Fire safety precautions are to be observed. A thoroughly dry aggregate system is to be used as the monomers may not polymerize in the presence of moisture. Moreover, the catalyst and promoter should never to be added to each other as it will result in an explosion. Some of these materials are toxic and are carcinogenic, and have to be handled with extreme care.

The polymer systems which have been sucessfully used for polymer concrete include methyl-methacrylate, polyester-styrene, epoxy-styrene, styrene, furfuryl acetone. Other are furane, acrylic, polyurethane, urea formaldehyde and phenol formaldehyde, etc. The acrylic, design considerations for polymer concrete are:

(i) The *binder content* to fill the voids of the aggregate system. Smaller the polymer content greater is the economy.

(ii) *Workability* for easy mixing and placing of cement concrete without bleeding and segregation.

(iii) *Film forming ability* of the polymer, and *bonding* with the aggregate surface to transmit load forces.

(iv) *Economic curing (cross-linking) times* and temperatures.

(v) *Durability* in environments to which the polymer concrete composite is exposed.

Polymer concretes can be reinforced with steel, nylon, polypropylene or glass fibres in a manner similar to cement concrete.

In general polymer concrete exhibits a fairly linear *stress-strain curve* nearly up to failure; the failure is characterized as *brittle*. Concretes made of thermoset polymers show a decrease in strength by 30 to 40 per cent at higher temperatures, such as 90°C. The *elastic limit* may also be substantially lowered at higher temperatures. Use of *microfillers*, such as finely powdered $CaCO_3$ and silane coupling agents, improve the compressive and tensile strengths of polymer concretes. Table 14.10 shows that wide range of strengths are possible depending upon the resin system used.

Thermosetting polymers, such as polyester and epoxy exhibit significant shrinkage during the polymerization of the resin. This shrinkage can be reduced by shrinkage reducing agents, however, at some cost to the strength. Well-cured or fully cross-linked polymer concrete has excellent resistance to acids, salts, common solvents and petroleum products. *Fatigue strength* of polymer concrete, with or without fibres, is excellent. Polymer concrete having up to 5 per cent steel fibres has better *ductile* and *impact-resistant properties*.

TABLE 14.10 Compressive and tensile strengths of polymer concrete

Type of polymer system	Compressive strength, MPa	Tensile strength, MPa	Modulus of elasticity, ($\times 10^2$) MPa
Isophthalate or Orthophthalate polyester	50–140	7–10	9–30
Vinylester	114	7–9	—
Epoxy	45–130	6–16	7–31
Methyl-metha-acrylate + trimethol-propane trimetha-acrylate	60–80	8–9	36
Furane	70–80	5–8	20–32
Methyl-metha-acrylate (MMA)	60–120	8–9	15–18
Acrylic	130	30	—
Carbamide	40–60	25–50	10–12

While the general purpose polyester-styrene systems require an elevated temperature of about 60–70°C for complete polymerization; resins or monomer systems are available which can cross-link at low temperatures such as 0°C within one or two hours.

Condensation polymers which are relatively inexpensive, like phenol formaldehyde have been used successfully to develop polymer concrete.

Polymer concretes have good potential as repair material and for overlays. Thin sand-filled overlays (12 to 30 mm thick) reduce water permeability and chloride penetration. Polymer concrete can be used for rapid *repair of damaged airfield pavements* and *industrial structures*. Polymer concrete can be used for treating the sluiceways and stilling basin of the dam.

Polymer concrete pipes have been used for the transporting a variety of chemicals, for carrying effluents and wastewater, etc.

Polymer concrete can be used in rock bolts. It provides necessary corrosion protection to ground anchors. Polymer concretes possess good electrical properties and can be used for high voltage insulator application. Electrical structures such as poles for electrical transmission lines have been manufactured from polymer concrete.

14.10.3 Polymer Modified Concrete

Polymer modified concrete (PMC), more specifically called *polymer cement concrete*, is a composite obtained by incorporating a polymeric material into concrete during the mixing stage. However, the polymer so added should not interfere with the hydration process. Since many polymers are insoluble in water, their addition can only be in the form of emulsion or dispersion or latex. The composite is then cast into the required shape in the conventional manner and cured in a manner similar to the curing of cement concrete. The hydrated cement and the polymer film formed due to the curing of the polymeric material constitute an interpenetrating matrix that binds the aggregate.

The polymeric materials in the form of latices and prepolymers may be added to modify cement concretes. Depending upon the type of modifier, polymer modified cement concretes can be subdivided as:

(i) Latex-modified cement concrete
(ii) Prepolymer-modified cement concrete

In general, the quantities of polymers required for polymer-modified cement concretes are relatively small, being in the range of one to four per cent by mass of the composite. In constrast polymer-impregnated concretes require 5 to 8 per cent and polymer concretes 8 to 15 per cent of polymer. Polymer modified cement concretes, are therefore, the least expensive. The processing of PMCC is also simplest. Conventional plant and equipment could be adopted. However, the improvements in mechanical properties have not been as high as observed in PIC or PC.

14.10.4 Latex-modified Cement Concrete

Latices are white milk like suspension consisting of very small-sized polymer particles suspended in water with the help of emulsifiers and stabilizing agents. It contains about 50 per cent of polymer solid by mass.

Both *elastomeric* and *glassy polymers* have been employed in latices for modifying cement concrete. The elastomeric polymers are characterized by their rubber-like elongation and by their relatively low modulus of elasticity at ambient temperatures. Some of the commonly used elastomeric latices are: natural rubber latex, styrene-butadiene rubber latex, acrilonitrile-butadiene rubber latex and neoprene.

Glassy polymers are characterized by high modulus of elasticity, higher strength, and relatively brittle type of failure. Common examples are polyvinyl acetate, polyvinylidene chloride, styrene-butadiene copolymer latex, and acrylic polymers. The use of polyvinyl acetate latex due to its sensitivity to the moisture is discontinued. Polyvinylidene copolymer latex, due to its residual chloride and possible corrosion of reinforcement, is used only in unreinforced concrete applications. The latex systems for modifying cement concrete are not available in India. The optimum curing procedure involves the moist curing of composites for 1 to 7 days, followed by dry curing at room temperature. At 28 days, the latex modified composites reach about 80 per cent their final strength.

14.10.5 Prepolymer-modified Cement Concrete

Some of the prepolymer systems used are polyester-styrene based system, epoxy systems and furane systems. With exception of epoxies, prepolymers (unlike latices) do not improve the workability of cement concrete.

The strength improvement of PMC over conventional concrete is of the order of 50–100 per cent. Its adhesion to plain concrete is good. The ductility is significantly improved and early microcracking is avoided. Consequently, the tensile strength and modulus of rupture are more than twice those of control concrete. There is considerable improvement in durability over conventional concrete due to lower *water-cement ratio* and filling of pores with polymer. Further research is required since the high cost of polymer addition has not been commensurately reflected in improved strength.

The excellent bond of latex concrete to existing concrete, superior shear bond strength, good freeze-thaw resistance, resistance to the penetration of chloride ions, improved ductility, and superior tensile and flexural strengths makes latex modified concrete an eminent material for overlays and resurfacing applications for bridge decks, industrial flooring, food processing factories, fertilizer stores, damp resistant floors, for railway platforms, and nuclear processing areas.

Surface deterioration is a major problem in marine and irrigation structures. Excellent resistance to salt water makes LMCC very effective repair material. LMCC are used for fixing ceramic tiles, lining effluent ducts, reservoirs, and sewerage and industrial waste handling structures.

Latex and fibre-reinforced composites have a great potential in cement composites due to their synergestic behaviour and improvement in matrix fibre bond.

14.10.6 Prepolymer Cement Concrete (PCC)

PCC is used for flooring in food processing and chemical industries, in wear-resistant floors, and in decks over steel bridges. Due to the early development of strengths, it is suited for repair of sea defence structures.

The development of *polymer-concrete composites* has opened up the possibility of extending the very range of applicability of concrete like composites. It has become possible to tailor a polymer-concrete composite to meet the requirements of any given application. Polymer-concrete composites are far superior to cement concretes in their resistance to chemicals, such as acids, and salt solutions. Polymer-impregnated ferrocement, a thin, lightweight and highly durable composite has a great potential for applications in coastal, off-shore and chemical industrial structures.

Polymer-concrete composites are very cost effective in applications requiring high degrees of durability and chemical resistance and where so far, costlier materials and composites have been employed. In such situations developing nations could ill-afford either the use of inefficient material of construction or the employment of costlier conventional alternatives. The improved durability of polymer-impregnated concrete is shown in Fig. 14.14.

14.11 SULPHUR CONCRETE AND SULPHUR-INFILTRATED CONCRETE

Sulphur concrete is a composite material consisting of elemental sulphur, coarse aggregate and fine aggregate. It contains neither cement nor water. Its high compressive strength at early age (of the order of 35 MPa at 8 h), makes it ideal for small precast units for outdoor use; its good chemical durability makes it an excellent material for use in industrial plants. However, its low melting point (119°C), its vulnerability to combustion and the production of toxic gases, its corrosive effect on reinforcing steel under wet and humid conditions, and its brittleness make it unfit for most structural applications.

334 Concrete Technology

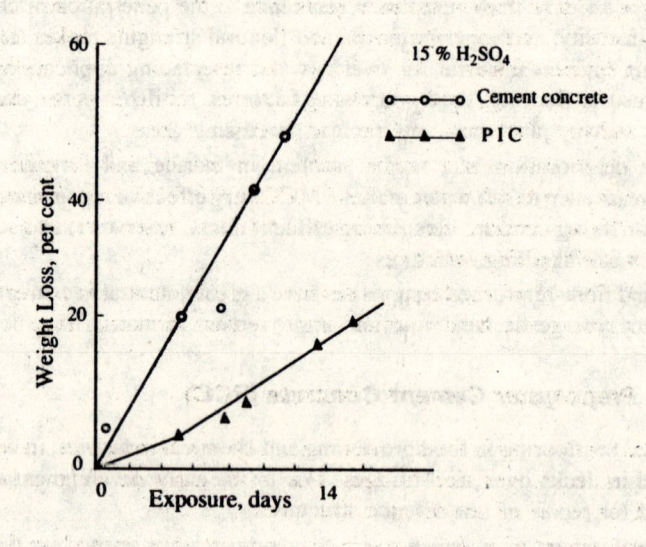

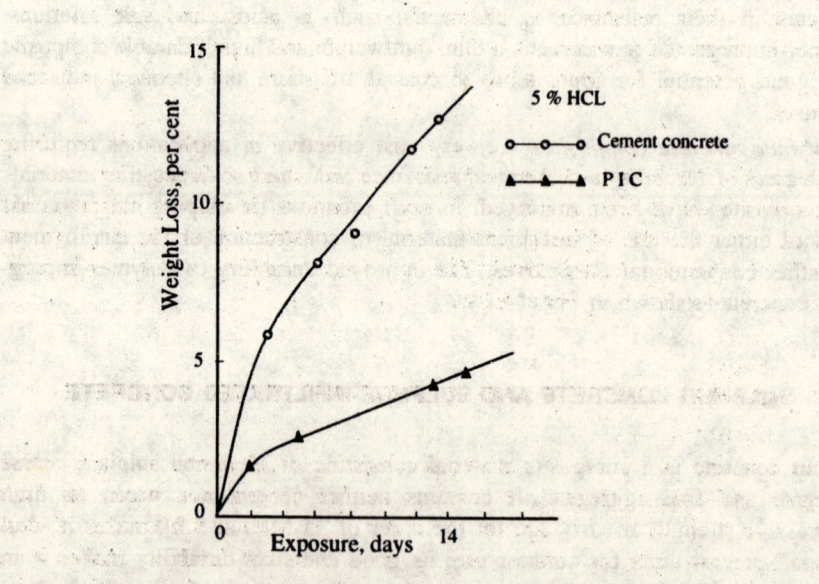

Fig. 14.14 Durability of polymer-impregnated concrete

Sulphur-infiltrated concrete, on the other hand, is produced by infiltrating conventional portland cement concrete having *water-cement ratios* of the order of 0.70 with molten sulphur. The infiltrated composite has improved mechanical properties, its compressive and flexural strengths being of the order of 100 and 10 MPa, respectively. Sulphur-infiltrated concrete is more durable than conventional concretes in acidic environments but is unstable in alkaline solutions and when left submerged in water over long periods.

Both sulphur concrete and sulphur-infiltrated concrete are specialized products which are useful as construction materials in the situations where high life expectancy of concrete is desired but have serious limitations.

14.11.1 Sulphur Concrete (SC)

Production

The sulphur concrete is produced by mixing powder sulphur and aggregates preheated to about 180°C in a conventional mixer. The preheated coarse aggregates are placed in a tilting mixer which is then started and a sufficient amount of sulphur is added so as to finely coat the aggregates. This is followed by the addition of sand, the remaining sulphur and silica flour (added as workability agent). Mixing is continued for one more minute during which the sulphur and aggregates combine to form a flowable homogeneous mixture which can be cast in moulds.

Mix Proportions

The mix proportions are primarily fixed for achieving a workable mixture giving the highest strength properties with minimum amount of sulphur. A sulphur content of 20 per cent (by mass) of total mix for concrete using 20 mm nominal size aggregate with fine-to-coarse aggregate ratio of 1 : 1.5 will be found to be suitable. Typically consider a sulphur concrete mix of proportions 1 : 1.6 : 2.4 (by mass) of sulphur, fine aggregate and coarse aggregate, respectively, with silica content of 5 per cent. A comparison of this mix of sulphur concrete with conventional portland cement concrete mix of equal strength having proportions 1 : 1.8 : 2.6 with water-cement ratio of 0.41, indicates that sulphur concrete gains strength very rapidly and reaches its ultimate strength in 6 to 8 hours under ambient temperature and humidity conditions whereas normal portland cement concrete attains its ultimate strength in about an year.

Mechanical Properties of Sulphur Concrete

A summary of structural properties of sulphur concrete is given in Table 14.11.

Table 14.11 Mechanical Properties of Sulphur Concretes

Property	Range
Tensile strength, MPa	10–20
Compressive strength, MPa	25–70
Modulus of rupture, MPa	3–10
Modulus of elasticity, GPa	20–40

As for normal portland cement concrete, strength properties of sulphur concrete are affected by the type of aggregate used. Sulphur concrete made with crushed aggregate gives much higher strength values than that for the corresponding concrete obtained by using natural gravel. The strength values obtained with larger specimens have been found to be lower. The decrease in the strength of large specimens is probably due to the combined effect of specimen size and slower rate of cooling while they were still in the moulds. This aspect of sulphur concrete could pose serious problems in thick structural members.

Sulphur concrete is relatively a brittle material and failure of specimens coincides with the ultimate stress at a strain of approximately 0.0015. The *brittleness* of the material can be reduced by adding *fibre reinforcement*. Sulphur concretes exhibit significantly better structural properties in terms of *fatigue resistance* than normal portland cement concrete. The sulphur concrete with fly ash as an additive has higher fatigue life.

Sulphur concrete has been reported to exhibit considerably more *creep* than portland cement concrete which can be very disadvantageous for structural concrete members.

Durability of Sulphur Concrete

Sulphur concrete when subjected to a temperature higher than 119°C, which is the melting point of elemental sulphur, melts and loses all its structural strength. Since sulphur has a low melting point, the use of sulphur concrete is limited to applications where temperature does not exceed 80°C. Sulphur is combustible and burns to form sulphur dioxide, a toxic gas, in the presence of oxygen. Since sulphur oxidizes in the presence of moisture, it is possible that sulphur concrete, when exposed to a moist environment in the presence of ultraviolet rays, may produce sulphuric acid.

Sulphur mortars are susceptible to attack by bacteria. The sulphur concretes prepared from silica or limestone aggregates and modified-sulphur have been found to be unaffected on exposure to acids and alkalies. However, unmodified sulphur is attacked by strong oxidizing agents (acids and alkalies) and certain organic chemicals like carbon disulphide, phenols, etc.

As sulphur concrete contains no water or cement and has low water absorption (0–1.5 per cent), its performance in *freeze-thaw cycling* is principally a problem of resistance to *thermal cycling*. Due to a very high *thermal coefficient of expansion* ($8-35 \times 10^{-6}$ per °C), the sulphur concrete has *low durability* against thermal cycling.

One of the attractive features of sulphur concrete is that it can be melted to recover sulphur and aggregates, and the recycled materials can be reused for concrete. The properties of concrete made with recycled sulphur are comparable to the original strength values. Due to an abundant supply of sulphur as an industrial byproduct, sulphur concrete may prove to be competitive with conventional concrete. Sulphur is mainly produced as a byproduct during the processing of natural gas. It is also a waste product of the textile industry.

14.11.2 Sulphur-infiltrated Concrete (SIC)

Sulphur-infiltrated concrete was developed as an economical alternative to *polymer-impregnated concrete* (PIC) to be used for higher strength and durable precast concrete elements. Sulphur is a material which is considerably cheaper than the monomers used in the production of PIC. Sulphur-infiltrated concrete is obtained by infiltrating the lean concrete with molten sulphur.

Production of SIC

The concrete to be infiltrated should be produced using normal aggregates with aggregate-cement ratios between 3 : 1 to 5 : 1 and having water-cement ratio preferably in the range 0.60 to 0.80. The infiltration procedure normally used consists in moist-curing of concrete elements for 24 hours at about 23°C followed by drying (at 121°C) for a period of 24 hours, immersing in molten sulphur at 121°C under vacuum for two hours, releasing the vacuum and soaking for an additional half an hour, and then removing the elements from molten sulphur to cool. In case of low water-cement ratio concretes which are relatively dense external pressure may be applied following the release of vacuum to force sulphur into concrete. The foregoing procedure may be modified to suit individual job conditions. However, the following points should be kept in mind.

(i) For concretes with a water-cement ratio of the order of 0.65, the one-day old elements must be handled with care to avoid damage.

(ii) The drying temperature should be kept as high as possible but not exceeding 150°C since a higher temperature may damage the gel-structure of the young hydrated cement paste. The period of drying will depend on the type and size of element.

(iii) The period of vacuum (evacuation time) appears to be less critical than the immersion time in molten sulphur after evacuation. For concretes with water-cement ratio of about 0.55 increased immersion time is essential to achieve full infiltration.

Mechanical Properties of SIC

Filling of *capillary voids* in the *hydrated cement paste* and *larger voids* present at the interface between aggregate and cement paste with infiltrated sulphur modifies the physical and mechanical properties of concrete considerably. The final *porosity* determines the mechanical properties of SIC regardless of mix proportions. This is in aggreement with the behaviour of PIC. The improvements in the properties of concrete due to infiltration of sulphur are given in Table 14.12.

TABLE 14.12 Improvements in structural properties due to sulphur-infiltration

Property	Improvements over reference unfiltrated concrete, per cent
Compressive strength	
(i) non air-entrained concrete	150
(ii) air-entrained concrete	200
Splitting tensile strength	100
Elastic modulus	100

The stress-strain relationship of SIC is linear and the strain at failure is not substantially increased because of brittleness of sulphur component.

Durability

Generally the performance of sulphur-infiltrated concrete is satisfactory against freezing and thawing, seawater attack, and wetting and drying. The sulphur-infiltrated concrete is more durable than conventional concrete in higher concentrations of H_2SO_4 and HCl. When left submerged in stagnant water over extended periods of time slight *leaching of sulphur* may take place and concrete may eventually show undesirable expansion followed by some cracking. The instability of SIC in aqueous media is apparently related to the presence of polysulphide anions formed during infiltration and found to be highly soluble in alkaline pore solutions of wet concrete. The polysulphide and calcium ions (dissolved from concrete) form concentrated calcium polysulphide a yellow orange leachate. Under moist aerated conditions it reacts with oxygen to form *sulphur efflorescene*.

The strength properties of sulphur-infiltrated concrete are not significantly affected when it is exposed to short term temperatures up to 100°C. At these temperatures SIC exhibits certain amount of *ductile behaviour* before failure.

The magnitude of increase in *abrasion resistance* of SIC depends on the *sulphur loading* of the test specimens. However, the sulphur filling of the pores in concrete provides an uninterrupted path for heat flow resulting in increased values of *thermal conductivity* over that of normal dry concrete.

The *ductility* of sand-cement mortars infiltrated with sulphur increases significantly by inclusion of *fibres*. The addition of fibres increases both *ultimate flexural strength* and *peak deflection*. The addition of fibres may increase the peak deflection by as many as six times.

The sulphur-infiltrated concrete provides a *corrosive protection* to embedded steel. The sulphur loading required for a given corrosive protection depends upon water-cement ratio used in concrete. Higher the water-cement ratio, higher the sulphur loading required. The minimum sulphur loading varies from 10 per cent for 0.70 water-cement ratio to 5 per cent for 0.40 water-cement ratio.

Applications

The sulphur-infiltrated concrete is ideally suited for precast units such as patio slabs, sidewalks, kerbs, sewer pipes, and precast units for tunnel linings. *High impermeability* makes SIC suitable for industrial applications requiring *high corrosion resistance*. SIC can also be used for the repair of deteriorated structures and bridge decks.

The conventional concrete sewer pipes are prone to corrosion due to attack by sulphuric acid. The sewerage gases, principally H_2S, are oxidized by thiobacillus bacteria present in sewer to sulphuric acid which in turn attacks normal portland cement concrete. The problem may be further compounded by the presence of sulphate in water which penetrates concrete pores causing it to swell and disintegrate. The sulphur infiltration can enhance the life span of the sewer pipe due to extremely low impermeability of infiltrated product and good chemical resistance. It is estimated that the life span of SIC pipes could be doubled at a cost of approximately 25 per cent higher than for unfiltrated concrete pipes.

14.12 JET (ULTRA-RAPID HARDENING) CEMENT CONCRETE

The jet cement which has come into the market in the early 1970's, has many characteristics superior to those of ordinary portland cement. Due to very *short setting time*, it develops *super high initial strength* making it suitable for use in a wide range of placing and curing temperatures. The development of strength under low temperatures is excellent. Jet cement is also called *one hour cement* as it is easily possible to obtain high early strength within the hour. It contains about 20 per cent of reactive calcium fluoroaluminate which is the source of high early strength. The setting time of mortars and concretes made with jet cement can be freely controlled by adding the required amount of *retarder*. Jet cement shows stable strength development extending over a long time and has *high ultimate strength*. In contrast to aluminous cement, there is no loss of strength with age. It has *low drying shrinkage* and *low permeability*.

The jet cement is manufactured by mixing mainly specially selected anhydrite (II-$CaSO_4$) and cement clinker powder interground with sodium sulphate and calcium carbonate (about 1 per cent), and boric acid (about 0.2 per cent). The jet cement clinker is usually made from a ground homogenized mixture of limestone, clay, bauxite and fluorite by burning at a fairly low temperature of 1250–1350°C in order to prevent the formation of a tricalcium aluminate phase. Clinker and anhydrite are ground to 400–450 m^2/kg and 600–800 m^2/kg, respectively. The specific gravity of jet cement is about 3.03–3.05 and specific surface area is about 500–550 m^2/kg. Thus the specific gravity of this cement is lower than that of ordinary portland cement and the specific surface area is considerably higher. The setting time of jet cement is extremely short, the final setting being from 10 to 15 minutes. The *initial setting* time can be prolonged in proportion to the amount of retarder added. The anhydrite is usually manufactured by burning by-product gypsum and desulphurization waste from power plants.

The one day compressive and flexural strengths of jet cement mortar with cement : sand ratio of 1 : 2 and water-cement ratio of 0.65 are approximately equal to 7-day and 3-day strengths of ordinary portland cement mortar having the same mix proportions.

With the use of jet cement improved *workability* of freshly mixed concrete is obtained due to *enhanced cohesiveness* and *resistance to segregation*. However, it is necessary to increase the water content by 1.25 to 1.75 per cent in order to increase the concrete slump by approximately 10 mm. The Vee-Bee time of jet cement concrete is higher than that of ordinary portland cement concrete of same water content. There is an optimum fine aggregate percentage for each type of cement, at which the Vee-Bee time reaches a minimum value. The jet cement generally reduces the optimum value by 4 to 5 per cent because of higher *fineness* of the cement.

The *setting time* of concrete can be regulated by controlling the amount of *retarder* added. It is necessary to adopt an optimum amount of retarder based on the temperature and working conditions in order to retain sufficient handling time for the fresh concrete. At the job site having high temperature, the mixing of concrete materials is preferable to the use of *ready-mixed concrete*. The *bleeding* of fresh concrete made with jet cement is insignificant in mixes having slump values lower than 150 mm (used for normal concrete work). Consequently, concrete surface must be finished as soon as possible after placing the concrete.

The concrete made with jet cement shows good strength development at low temperature, and hence is suitable for *winter concreting*. The rate of strength development of jet cement it quite different from that of ordinary portland cement. The moist curing of concrete at early ages is important, since the concrete cured in dry state immediately after *stripping* yields a lower strength development. The 28-day strength of jet cement concrete is about 20 per cent higher than that of ordinary portland cement concrete at the same water-cement ratio, and a curing temperature of 20°C. The ratio of *tensile strength* to *compressive strength* varies from 1/10 to 1/14, and the value is almost the same as that of regular concrete using ordinary portland cement. *Bond strength* between reinforcing bars and concrete using jet cement is considerably higher than that of ordinary portland cement concrete. The adhesive strength of concrete construction joints of jet cement concrete is 1.5 to 1.8 times higher than that of regular concrete when the concrete surface is treated carefully.

The jet cement concrete yields high *modulus of elasticity* at early ages. The relationship between the dynamic *modulus of elasticity*, E_d and *compressive strength* of concrete f_{ck} (MPa) is given by the following equations:

Jet cement concrete : $E_d = 8920 f_{ck}^{0.376}$ MPa

Ordinary portland cement : $E_d = 11980 f_{ck}^{0.320}$ MPa

When the concrete strengths are same the modulus of elasticity of concrete using jet cement is slightly lower than that of regular concrete because of the *lower specific gravity* of cement.

The jet cement concrete gives lower values of *drying shrinkage* than that of concrete made with ordinary portland cement. However, the *creep* values are higher at early ages and lesser after 2 months. The *watertightness* at early ages is considerably higher than that of ordinary portland cement concrete. This can further be improved by extending *curing period* and by increasing the *cement content*. The rise in the temperature of concrete, caused by hydration of cement, is considerable.

14.12.1 Application of Jet Cement Concrete

Jet cement has been found to be most suitable for urgent repairwork, and winter concreting. The cost of this cement is about five times that of ordinary portland cement.

(i) *Building construction* The jet cement concrete can be used for the purpose of urgent building construction at low temperature. The surface should be finished immediately after placing, and concrete slab may be cured under canvas sheets. The column and wall forms may be removed after one day, and slab forms after two days. Since the handling time is short and the slump loss tends to be higher, quick handling is required in construction work.

(ii) *Concrete pavements* The jet cement concrete may enable the road to be used within hours after placing with little or without curing.

(iii) *Repair work* The cracks in reinforced concrete piers, and damage in expansion joints in railway or highway bridges may be repaired with jet cement concrete during the period when no train or traffic passes over the bridge. The jet concrete has been used satisfactorily in renewing the concrete pavings on an earth sub-base, repair of machine bases and concrete sleepers in Japan.

(iv) *Winter concreting* The jet cement has found major applications in winter concreting. At very low temperatures concrete may be cured with heaters to obtain the required strength.

(v) *Concrete products* To increase production efficiency by allowing early removal of form or stripping and early transportation, the jet cement can be used for the manufacture of concrete blocks, precast concrete panels, concrete curtain walls, reinforced concrete pipes etc.

(vi) *Grouting* Grouting mix consisting of jet cement, water, sand and an admixture may yield a strength of 1.5 to 2.5 MPa at one hour. The grouting may be used in stiffening the construction, consolidation of earth, etc.

14.13 GAP-GRADED CONCRETE

This type of concrete is obtained when a *gap-graded aggregate* is used in the production of concrete. In case of gap-grading certain undesirable sizes of aggregates are omitted from the conventional *continuous gradings*. The undesirable sizes are those which prevent the efficient packing of the other sizes. Sometimes available *single-sized aggregate* only is used.

The gap-grading is normally aimed at achieving strength from the efficient packing of the aggregate. A well-packed aggregate will require minimum cement paste to fill the *minor voids*. For discussion consider the coarse aggregate to be mathematically modelled as spheres of diameter D called *major spheres*. A multitude of these spheres will have a rhombohedral form of packing. The voids between the major spheres can be fitted with spheres of diameter $0.414\,D$, known as *major occupational spheres*. The fine aggregate would then mathematically consist of *minor occupational spheres* of diameter $0.225\,D$ which would fit into the remaining voids. The remaining minor voids can now be fitted by *admittance spheres* of diameter $0.155\,D$, and these could also be provided by the fine aggregate. Cement paste would then occupy the remaining voids and a mathematically perfect compact mix would result. Such a mix, however, cannot be cast in practice and consequently only the major, and admittance spheres are considered to be of practical value in a mix design.

Mixes, therefore, are often designed with *single-sized* aggregate and a sand, all the particles of which can pass through the voids in the compacted coarse aggregate. However, the particles of sand must not be smaller than necessary to restrict the surface area to be coated with cement paste. Irrespective of the calculation suggested above, the sand content should be sufficient to distribute itself uniformly throughout the mix under practical conditions.

The workability can be increased be reducing the surface area of all ingredients in a unit volume. This can be achieved by using largest size aggregate consistent with other constraints.

Gap-grading enables leaner and drier mixes than conventional concrete of equivalent strength to be used resulting in lesser *shrinkage*. However, a leaner mix makes the vibration almost essential. Compressive forces on gap-graded concrete are transmitted from particle to particle of the coarse aggregate and not through cement-sand matrix. Consequently the *creep* associated with such concrete is low. Due to the use of single-sized aggregate the *segregation tendency* is checked.

A number of investigators have recommended the use of two single-sized coarse aggregates with sand and cement in a gap-graded mix. Because of efficient packing of aggregates in gap-graded concrete, *vertical shuttering* can often be removed shortly after casting.

The gap-grading is very sensitive to undesirable particles and the mix obtained will be of reduced efficiency.

14.14 NO-FINES CONCRETE

As the name suggests, this concrete does not contain fine aggregate. The coarse aggregate particles have been found to possess a cement paste coating of up to 1.3 mm around them. Hence no fine concrete contains a multitude of voids which is responsible for its low strength. However, large voids give good *thermal insulation*, and these voids being large enough prevent the movement of water through the concrete by *capillary action*.

The compressive strength of no-fines concrete is considerably lower than that of conventional concrete and depends on the cement content and grading of aggregate. The strength generally varies from 1.5 MPa to 15 MPa. In lean mixes, cement content may be as little as 70 to 130 kg per cubic metre of concrete, this is due to the absence of large surface area of fine aggregate particles which would have otherwise to be coated with cement paste. Thus the cost of no-fines concrete is lower than that of conventional concrete. It does not segregate, hence can be dropped from a considerable height. However, it should be vibrated for a very short period otherwise cement paste would run off.

The water-cement ratio does not seem to be the controlling factor in this case. It varies from 0.38 to 0.52. The density of no-fines concrete depends on grading of aggregate, and with normal aggregate it varies from 1600 to 2000 kg/m^3. *Shrinkage* is generally lower than in the ordinary concrete.

Normally no-fines concrete is not suitable for reinforced concrete work. However, due to good thermal insulation no-fines concrete walls have been used in cold countries for housing. It has been found that rain beating on a wall penetrates only a short horizontal distance before falling down to the bottom of the wall, there being no capillary paths to conduct the water completely through it. It is, however, often desirable to paint exposed no-fines concrete walls. High absorption of water makes no-fines concrete unsuitable for use in foundation and in situations where it may be in contact with water.

15

Deterioration of Concrete and its Prevention

15.1 INTRODUCTION

Though concrete is quite strong mechanically, it is highly susceptible to chemical attack, and thus concrete structures get damaged and even fail unless some measures are adopted to counteract deterioration of concrete and thereby increasing the durability of the concrete structure.

The durability of concrete can be defined as its resistance to the deteriorating influences of both external and internal agencies. The external or environmental agencies causing the loss of durability include weathering, attack by natural or industrial liquids and gases, etc. Whereas the internal agencies responsible for the lowering of durability are harmful alkali-aggregate reactions, volume changes due to noncompatible thermal and mechanical properties of aggregates and cement paste, presence of sulphates and chlorides from the ingredients of concrete, etc. In the case of reinforced concrete the ingress of moisture or air may lead to corrosion of steel, and cracking and spalling of concrete cover.

A durable concrete is dense, workable having as low a permeability as possible under the given situation. The recommendations for making durable concrete usually envisage limits for maximum *water-cement ratio*, thickness of cover, type of cement and the amounts of chlorides and sulphates in the concrete.

15.2 CORROSION OF CONCRETE

The gradual deterioration of concrete by chemically aggressive agents is called *concrete corrosion*. Basically corrosion of concrete is a physico-chemical process and the extent of deterioration caused to it by the aggressive agents is dependent upon the properties of the constituents of concrete and the corrosive media. Any factor which may help in the development of cracks in concrete will promote the penetration of the aggressive solution and gases and will lead to the faster deterioration of concrete structure.

15.2.1 Cracking-corrosion Interaction

The two main factors responsible for the loss of durability of concrete structures are:
  (i) deterioration of concrete, and
  (ii) corrosion of steel reinforcement.

These two phenomena cannot be separated out because the deterioration of concrete cover over the steel reinforcement leads to the corrosion of steel and, on the other hand, corrosion of steel reinforcement promotes destruction of concrete due to the development of internal stresses on the formation of voluminous corrosion products of iron. Thus the durability of concrete is greatly dependent on the interaction of cracking and corrosion.

Microcracks, which are always present at the aggregate-cement interface and at the reinforcement-cement interface, do not affect the durability of concrete as long as they are limited in number and size, and are discontinuous. But when these microcracks become continuous and enlarged under the influence of stresses or due to the leaching of cement paste, they facilitate the transport of aggressive ions and gases, thereby affecting the durability of concrete to a great extent.

15.2.2 Types of Deterioration

Deterioration of concrete is caused not only by acids in the form of water solutions or acidic gases which form acids on dissolving in water, but by salt solutions and even by alkalies. A large number of other substances, such as fertilizers, insecticides and certain organic compounds are harmful to concrete.

Deterioration of concrete due to corrosion caused by the various aggressive chemicals can be classified into three categories as shown in Fig. 15.1. This classification will help in developing the methods to increase durability of concrete and protection to reinforcement. The principal forms of destruction due to corrosion are as follows:

(i) *Decomposition of Concrete* In this form the decomposition of concrete is caused by action of liquids (aqueous solutions) which are able to dissolve the ingredients of hardened cement. Water percolating through the mass of concrete can greatly speed up decomposition by increasing the ionic strength of solution. A common example of this type of destruction is the leaching action.

(ii) *Chemical Reaction* In this form of destruction a chemical interaction between hardened cement constituents and a solution takes place. The easily soluble reaction products are removed from the internal structure of concrete by diffusion or percolation. This happens when concrete is attacked by a solution of acids and certain salts.

(iii) *Crystallization* This form of destruction involves accumulation, crystallization, and polymerization of reaction products which increase the volume of solid phase within the pore structure of the concrete.

In addition to the principal forms of destruction cited above there are other specific influences, such as destruction due to the interaction between the hardened

Deterioration of Concrete and its Prevention 345

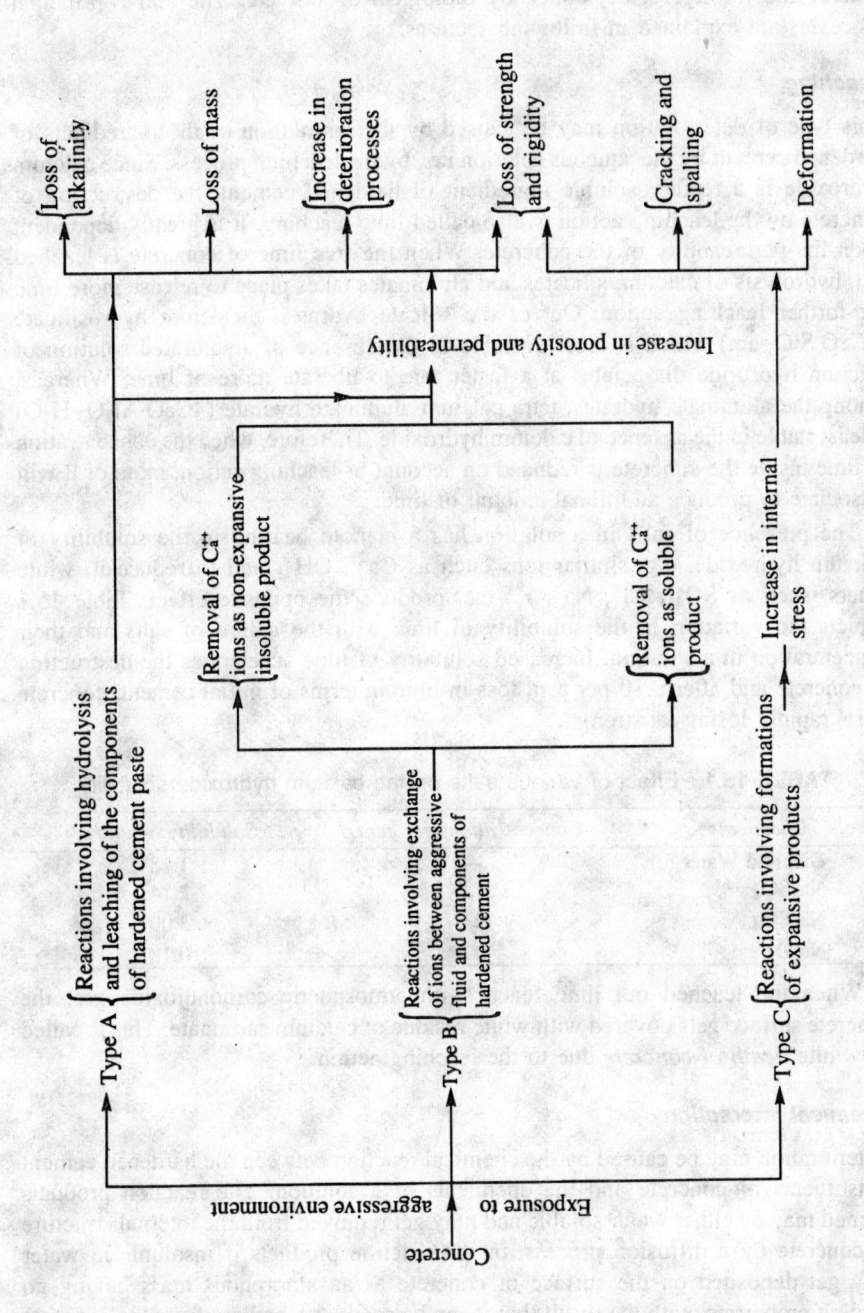

```
                    Loss of      Loss    Increase in    Loss of      Cracking     Deformation
                    alkalinity   of mass deterioration  strength     and
                                         processes      and rigidity spalling
                       ↑           ↑        ↑              ↑           ↑            ↑
                       └───────────┴────────┼──────────────┴───────────┘            │
                                            │                                       │
                              Increase in porosity and permeability                 │
                                            ↑                                       │
                       ┌────────────────────┼──────────────────┐                    │
                Removal of Ca²⁺      Removal of Ca²⁺                                │
                ions as non-expansive ions as soluble                      Increase in internal
                insoluble product    product                               stress
                       ↑                    ↑                                       ↑
        Reactions involving    Reactions involving exchange        Reactions involving formations
        hydrolysis and         of ions between aggressive          of expansive products
        leaching of the        fluid and components of
        components of          hardened cement
        hardened cement paste
             Type A                     Type B                              Type C
                ↑                          ↑                                   ↑
                └──────────────────────────┼───────────────────────────────────┘
                                 Exposure to aggressive environment
                                            ↑
                                         Concrete
```

Fig. 15.1 Deterioration of concrete due to aggressive agencies

cement and the aggregate; attack by biological agents, etc. The above principal processes are explained in following sections.

Leaching

This type of deterioration may be caused by the dissolution of the ingredients of hardened cement by the aqueous solution i.e., by the leaching process. Since calcium hydroxide is a readily soluble ingredient of hardened cement, the destruction of concrete by the leaching action is also called lime leaching. It is greatly dependent upon the permeability of the concrete. When the free lime of concrete is leached out, hydrolysis of calcium silicates and aluminates takes place to release more lime for further leaching action. Out of the silicate hydrates, dicalcium hydrosilicate ($2CaO \cdot SiO_2 \cdot aq.$) which is most unstable in the absence of a saturated solution of calcium hydroride dissociates at a faster rate to liberate more of lime. Whereas, among the aluminate hydrates, tetra calcium aluminate hydrate ($4CaO \cdot Al_2O_3 \cdot H_2O$) is least stable in the absence of calcium hydroxide. Therefore, when the concentration of lime inside the concrete is reduced on account of leaching action, more of it will dissociate to produce additional amount of lime.

The presence of salts in a solution has a marked bearing in the solubility of calcium hydroxide, e.g. similar ions cuch as Ca^{2+}, OH^- tend to reduce it, while others, such as SO_4^{2-}, Cl^-, Na^+, K^+ etc. produce the opposite effect. Table 15.1 depicts the variation in the solubility of lime with the nature of salts and their concentration in a solution. Increased solubility of lime accelerates the destruction of concrete and after a 10 per cent loss in lime in terms of initial cement, concrete starts rapidly losing its strength.

TABLE 15.1 Effect of various salts on the calcium hydroxide solubility

Chemicals	Concentration, per cent	Solubility, gm/litre
Distilled Water	—	1.18
Na_2SO_4	1.0	2.14
Na_2SO_4	2.0	3.00
NaOH	0.5	0.18

When the leached out lime reacts with atmospheric carbondioxide gas, the concrete surface gets covered with white residue of calcium carbonate. This is called the white *death of concrete* due to the leaching action.

Chemical Interaction

Deterioration may be caused by the chemical reaction between the hardened cement constituents of concrete and the chemicals of a solution. The reaction products formed may be either water soluble and may get removed from the internal structure of concrete by a diffusion process, or the reaction products if insoluble in water may get deposited on the surface of concrete as an amorphous mass having no binding properties, with the result that it can be easily washed out from the concrete surface.

Acids first react with free lime of concrete forming calcium salts and later on attack the hydrosilicates and hydroaluminates forming the corresponding calcium salts, whose solubility will govern the extent of deterioration caused to the concrete.

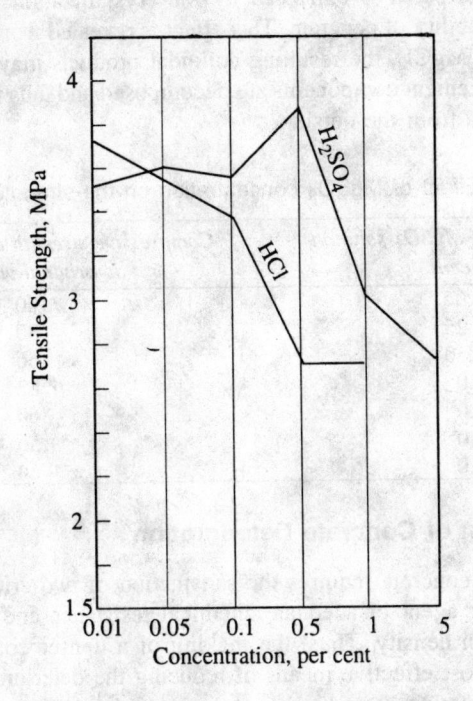

Fig. 15.2 Deterioration of concrete due to acid attack

As can be observed from Fig. 15.2, the hydrochloric acid corrodes the concrete to a greater extent in comparison to the sulphuric acid at low concentrations because H_2SO_4 forms a less soluble $CaSO_4$ on the reacting with lime of concrete, which seals the pores of concrete for further permeation and offers resistance to acid corrosion. But at higher concentrations of H_2SO_4, the concrete strength is reduced due to the accumulation of $CaSO_4$ in the pores and the development of internal stresses.

Crystallization

Concrete may get deteriorated by the accumulation or crystallization of salts in its pores, which leads to the development of internal stresses and formation of cracks. These salts in the pores of concrete may be either formed as a result of chemical reaction between the corrosive media and the constituents of hardened cement or may be brought from outside by the penetration of salt solution and released there on the evaporation of water.

Alkaline solutions of low concentration are less harmful to concrete; however, the concrete gets deteriorated on exposure to concentrated solutions of alkalies, as they combine with atmospheric carbondioxide producing crystallizable carbonates.

Alternate wetting and drying of structural members with salt solutions increase the deleterious effect of salt due to the phenomena of accumulation or crystallization of salts. The same holds good for partially immersed structural members.

Magnesium ions are particularly damaging in combination with sulphate ions, since magnesium corrosion is enhanced by the crystallization of gypsum which increases the permeability of concrete. This effect is revealed from the data of Table 15.2. In the case of $MgCl_2$, the resulting colloidal products may seal the concrete before the hydrated cement components are decomposed and thus prevent the inward diffusion of Mg ions from the outside.

TABLE 15.2 Effect of $MgSO_4$ concentration on the strength of concrete

Concentration of $MgSO_4$ solution, per cent	Compressive strength as percentage of original value
0.0	100
0.25	80
0.50	66
1.0	62
2.0	60
3.0	56
4.0	54

15.2.3 Prevention of Concrete Deterioration

A durable structural concrete requires the satisfaction of two criteria, namely that of a suitable binding agent of adequate chemical resistance and that of thorough compaction to a high density. Thus, the making of a denser concrete having the least porosity is a most effective means of reducing the deterioration of concrete. A quantitative information regarding the effects of the range of parameters like water-cement ratio, cement content, curing conditions together with effects of cement admixtures and replacements on the corrosion of concrete helps determine the durability of concrete empirically. The effects of these parameters are described below.

An increase in the water-cement ratio above 0.45 or 0.50 is found to increase the permeability of cement paste exponentially. Thus from the considerations of permeability the water-cement ratio is usually limited to 0.45 to 0.55 except in mild environment. For a given water-cement ratio, the cement content is governed by the required workability. In addition, the cement content should be such that to ensure sufficient alkalinity (pH value of concrete) to provide passive environment against corrosion of reinforcement. In the concrete for marine environment or in sea water applications, a minimum cement content of 350 kg/m^3 or more is required. Moreover, the water-cement ratio and cement content must provide enough paste to overfill the *voids* in compacted aggregate. The void content of aggregates depends upon the type and *nominal maximum size of aggregate* used, e.g. crushed rock and rounded river gravels of 20 mm nominal size have approximately 27 and 22 per cent of aggregate voids. A cement content of 400 kg/m^3 and water-cement ratio of 0.45 will produce paste volume of 30 per cent which is sufficient to overfill the voids of crushed rock. On the other hand a cement content of 300 kg/m^3 and water-cement ratio of 0.50 will result in 25 per cent paste volume which may be suitable for rounded gravels aggregate. A further increase in cement-content will result in higher workability.

The concrete in sea water or exposed directly should be at least of M15 grade in case of plain concrete and M20 in case of reinforced concrete. The use of portland slag or pozzolana cement is advantageous under such conditions. The ordinary portland cement having C_3A content less than 5 per cent has got the maximum resistance against sulphatic environment. The supersulphated cement is supposed to provide an acceptable durability against acidic environment, when concrete is dense with a water-cement ratio of 0.40 or less.

As the addition of hydraulic additives reduces the rate of leaching considerably, their addition will also be helpful in the prevention of deterioration of concrete. Since the carbonated layer on the surface of concrete increases the resistance of concrete to deterioration by leaching, it is possible to attain a marked improvement in the quality of concrete by encouraging natural or artificial carbonation of surface layer. Deterioration of concrete can also be prevented by treating concrete with solutions of salts or even acids in minor concentration to attain on the surface of hardened cement a layer of calcium salts less soluble than calcium hydroxide. This can be accomplished by treating the surface with solutions of 3 per cent fluosilicic acid, 5 per cent oxalic acid and saturated solution of mono calcium phosphate.

Durability of concrete can also be increased by impreganating the pores of concrete with a suitable polymer.

As the destructive processes in concrete are complex, a clear understanding of the destructive mechanism may help the selection of an appropriate technique to protect or improve the resistance of structural concrete to the aggressive agents.

15.3 CORROSION OF REINFORCEMENT

Concrete normally provides a high degree of protection against corrosion to embedded steel reinforcement. This is because concrete inherently provides a highly alkaline environment for the steel which protects and passivates the steel against corrosion. In addition concrete of low *water-cement ratio* and well cured has a low permeability which minimizes penetration of corrosion inducing agents like oxygen, chloride ion, carbon dioxide and water.

With concrete of suitable quality, corrosion of steel can be prevented provided the structure is properly designed for the intended environmental exposure. In instance of very severe exposure, the use of other protective measures such as corrosion inhibitors, coatings on steel or concrete or cathodic protection may be utilized. However, corrosion of steel and accompanying distress can result if the concrete is not of suitable quality, the structure is not properly designed for the anticipated environment or the actual environment or other factors were not as anticipated or changed during the life of the structure.

Sometimes, the first evidence of distress is brown staining of the concrete around the embedded steel. This brown staining resulting from rusting or corrosion of the steel may permeate to the concrete surface without cracking of the concrete but usually is accompanied by cracking. Sometimes, cracking of concrete occurs shortly thereafter. Concrete cracking occurs because the corrosion products of steel, an iron oxide or rust, has a volume twice as much as that of metallic iron from which it is formed. The forces generated by this expansive process can far exceed the tensile strength of the concrete with resulting cracking. Steel corrosion not only causes distress because of staining, cracking and ravelling of the concrete but may also cause structural failure resulting from the reduced cross-section and hence reduced tensile force capacity of the steel, this normally being more critical with thin prestressing steel tendons than with larger reinforcing bars. To understand the corrosion phenomenon of embedded reinforcement, it is imperative to study the corrosion of steel itself.

Steel corrosion can take place by several mechanisms, but corrosion of steel by direct oxidation or due to stress corrosion is of little concern in concrete structures. Indirect oxidation of steel in concrete resulting because of the existence of difference in metals or nonuniformities of the steel or nonuniformities in the chemical or physical environment provided by the surrounding concrete, is believed to be the main cause of corrosion distress in concrete. This type of corrosion is termed *electrochemical corrosion.*

15.3.1 Mechanism of Electro-chemical Corrosion

The metals have a tendency to oxidize to a metal ion in an aqueous solution of normal ionic activity at standard temperature. This ionization of metal, i.e., oxidation of metal at the anode is often referred to as the primary stage of the corrosion reaction called anodic reaction and can be represented by:

$$Fe \rightarrow Fe^{++} + 2e$$

This reaction results in the anodic region of the metal to have an excess of electrons. Therefore, to maintain equilibrium of electric charges an equivalent quantity of hydrogen is plated out at adjacent surfaces of the metal. This thin invisible protective film of hydrogen around the cathode inhibits further progress of corrosion reaction, unless the hydrogen film is removed in some manner.

The destruction of hydrogen film may take place in one of the two ways: (i) oxygen depolarization at the cathode, and (ii) evolution of hydrogen as a gas. These processes called cathodic reactions are usually represented by:

$$\tfrac{1}{2}O_2 + H_2O + 2e \rightarrow 2OH^-$$

$$2H^+ + 2e^- \rightarrow H_2$$

These cathodic reactions which are often called the secondary reactions control the rate of corrosion of the structural steel. The chemical reactions are depicted in Fig. 15.3. Therefore, any environmental condition which influences these reactions will influence the rate of corrosion. Since cathodic depolarization is dependent on the concentration of dissolved oxygen next to the metal, it is influenced by the degree of aeration, temperature, salt concentration, etc.

The secondary reactions permit the primary reaction to proceed with the accumulation of ferrous ion in the solution which in the presence of water and oxygen are oxidized and precipitated as rust. However, two stages of oxidation may exist depending upon the availability of oxygen. The products of the first stage namely ferrous hydroxide is more soluble than the second stage product i.e. hydrated ferric hydroxide. The first is usually formed directly at the metal surface and is converted to the latter at a little distance away from the surface where it is in contact with more oxygen as shown in Fig. 15.3.

(a) Chemical reaction at anodic and cathodic areas

(b) Formation of rust at anodic region

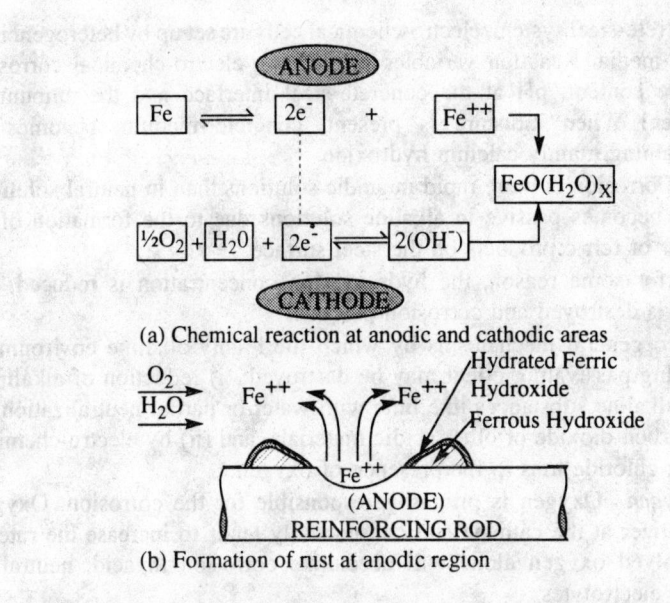

Fig. 15.3 Chemical reaction resulting into the formation of rust

The structure and composition of the rust varies considerably with the conditions prevailing during its formation and the structure of rust plays an important role in the subsequent corrosion process, e.g. if the rust layer is hard, dry and fairly adherent to the metal surface, it may retard corrosion by forming a protective coating. On the other hand, if the rust layer is spongy and readily detachable, it will absorb oxygen and moisture from the surrounding media and consequently add to further corrosion.

15.3.2 Corrosion of Steel Reinforcement Embedded in Concrete

The corrosion of the reinforcement embedded in concrete is extremely complicated and is influenced by numerous factors, both external and internal. The effect of these factors can be studied under two heads.

Factors Associated with Steel

The nonhomogeneities in the metal surface due to difference of chemical composition over the surface, discontinuous surface layers or differences in texture tend to increase the probability of corrosion due to the development of potential difference. But from the total corrosion standpoint these factors are not as important as the external conditions that may exist. The difference in potential may also be due to stresses in the reinforcement. A crystalline structure in the strained area has a somewhat different configuration from that in the unstrained areas and is anodic to it, thereby setting up an electro-chemical cell.

Factors Associated with Concrete

In the basic concrete-steel system, electro-chemical cells are set up by heterogencities of the concrete media. Reaction variables influencing electro-chemical corrosion are the moisture content, pH at the concrete-steel interface and the amount of available oxygen. When moisture is present, concrete medium becomes an electrolyte containing mainly calcium hydroxide.

Effects of pH   Corrosion is more rapid in acidic solutions than in neutral solutions (pH = 7). Steel becomes passive in alkaline solutions due to the formation of an impervious layer of ferric products on the steel surface.

However, if for some reason, the hydroxyl ion concentration is reduced, the protective layer is destroyed and corrosion proceeds.

There are two general mechanisms by which the highly alkaline environment and accompanying passivating effect may be destroyed: (i) reduction of alkalinity by leaching of alkaline substances like lime with water or partial neutralization by reaction with carbon-dioxide or other acidic materials, and (ii) by electro-chemical action involving chloride ions in the presence of oxygen.

Influence of oxygen   Oxygen is primarily responsible for the corrosion. Oxygen acts as a depolarizer at the cathode and consequently tends to increase the rate of corrosion. Dissolved oxygen alone will accelerate corrosion in acid, neutral or slightly alkaline electrolytes.

Though the main action of oxygen is as a depolarizer at the cathode, but at the anode it may lead to the formation of protective layer of insoluble ferric hydrates, which influences the rate and probability of corrosion. Thus oxygen may play a dual role: as a depolarizing agent it tends to enhance corrosion, on the other hand it may produce protective layers on the anodic regions prohibiting further corrosion.

The cracking of concrete and its permeability allow the penetration of oxygen to local areas of the reinforcement and in the unequal distribution of oxygen over the steel surface, which results in the setting up of anodic and cathodic regions. Thus the presence of oxygen in varying concentrations along the reinforcement will tend to increase the probability of corrosion.

Influence of moisture   As the corrosion reactions occur only if moisture is present, all corrosive factors become ineffective in its absence. In addition, the moisture penetration is the means whereby any exterior substance, such as chloride salts, carbon dioxide and dissolved oxygen may gain access to the reinforcement.

Influence of chloride ion concentration   The presence of salts provides two opposing effects: (i) it increases the conductivity of the electrolyte, thus raising the corrosion rate, and (ii) at high concentrations it diminishes the solubilty of oxygen, thereby lowering the corrosion rate. However, there is a *critical* or threshold *concentration of chloride* ions which must be exceeded before the initiation of corrosion. Any increase in chloride ions concentration beyond this critical value results in an increased rate of corrosion up to some limit at which the availability of oxygen necessary for corrosion to occur may be significantly reduced. The increased concentration of chloride ions (destructive to the protective film) can be tolerated without any resulting corrosion provided the alkalinity is increased, thereby promoting the formation of a protective oxide film.

Chlorides may be present in the concrete from several different sources, e.g. soluble chlorides, or using saline water as a concrete mix water. Calcium chloride may be introduced in fresh concrete by the use of aggregates containing chlorides, or using saline water as a concrete mix water. Calcium chloride may be used as an accelerator or a concrete admixture containing chlorides may be used. Some cements may also contain small amounts of chlorides. Chlorides may also enter the concrete from the environment, deicing salts, etc.

Influence of carbonation The carbon dioxide absorbed into the concrete may convert the calcium hydroxide into calcium carbonate thereby reducing the pH value and, consequently, the protective value of the concrete. Carbonation also tends to increase the shrinkage of concrete and thus promotes the development of cracks. The increase in the permeability of concrete due to shrinkage cracks allows the penetration of moisture and other external chemicals which may promote corrosion.

Influence of the quality of concrete *Permeability* of concrete is probably the most important single factor affecting the corrosion of reinforcement. Concrete of high permeability will have a high electrical conductivity and allow the penetration of deleterious substances to the reinforcement. Concrete permeability depends on numerous factors: some of which are *water-cement ratio, cement content, nominal size of aggregate* and its *grading,* methods of compaction, curing, etc.

Low-quality concrete is characterized by voids adjacent to the reinforcement, which may retain high moisture content, leading to rapid corrosion attack.

Effect of thickness of concrete cover over steel The thickness of concrete cover over steel is of great importance as this cover protects the steel from the factors that promote corrosion. The diffusion of chloride ions into the cement paste results in the formation of partially insoluble calcium chloroaluminate. This reaction reduces the concentration of chloride ions at any particular location and hence the tendency for inward diffusion is further reduced.

Influence of humidity In the regions having a relative humidity of 50 per cent or less, corrosive actions may be negligible. Likewise structures permanently immersed in water exhibit little or no corrosion of the reinforcement. The concrete of submerged structures generally maintains a high pH value and uniform salt concentration, thus reducing the formation of corrosion cells. Probably the main reason for this protection is absence of air at the concrete surface and thus oxygen cannot penetrate to the reinforcement.

Influence of inhibitors A number of admixtures, both organic and inorganic have been suggested as specific inhibitors of iron corrosion. Inorganic inhibitors are potassium dichromate, stannous chloride, zinc and lead chromates, calcium hypophosphite and sodium nitrite, while organic inhibitors are sodium benzoate, ethyl aniline and mercaptobenzothiazole. Some of the admixtures may retard setting of concrete or be detrimental to later-age strength. With some inhibitors, inhibition may occur at an optimum percentage of the inhibitor, whereas at lower or higher ratios the inhibitor may actually promote corrosion.

15.3.3 Effects of Corrosion

In most cases, the corrosion rate is extremely slow and the normal life span of a structure is not largely affected. However, if the external and internal conditions are such that a corrosive environment exists, a destructive action may take place at an increased rate and create serious problems.

The distress due to corrosive action may be in the form of deep pitting and a severe loss of cross-section of the reinforcement. This is particularly serious if the reinforcement is subjected to high stresses as in the case of structures carrying heavy loads. A combination of high stress and intense corrosion will produce stress concentrations that may result in rupture of the reinforcement. The corrosion of embedded steel can be minimized by using the following recommendations.

(i) For the reinforced-concrete members totally immersed in sea-water, the cover should be increased by 40 mm beyond that specified for normal conditions. However, for the members periodically immersed in the sea-water, this increase in cover should be restricted to 50 mm. In the case of high-strength concretes of grade M25 or above, the additional thickness of cover specified above may be reduced to half.

(ii) The additional cover thickness ranging from 15 to 50 mm beyond the values for normal conditions may be provided when the surfaces of the concrete members are exposed to the action of harmful chemicals (e.g. concrete in contact with soil contaminated with such chemicals), saline atmosphere, acid vapours, sulphurous smoke, etc. However, the total cover is limited to 75 mm.

(iii) To reduce the corrosion of reinforcement, the chloride ions in the concrete should be limited to its *threshold* or *critical value*. IS: 456–1978 has prescribed the limit of total amount of chloride in concrete at 0.15 per cent by mass of cement. However, for prestressed concrete, the total amount of chloride ions in concrete should be limited to 0.06 per cent.

(iv) In the case of an excessively aggressive environment, or where for practical reasons it is not possible to meet the requirements of cover and quality of concrete recommended above, special protection systems should be considered. Corrosion inhibitors may be added to concrete to prevent the corrosion of embedded steel. They may be either anodic or cathodic or mixed type. Anodic inhibitors form an extremely insoluble film adhering firmly to the surface of the bar at the anodic areas. Alkalies, phosphates and chromates are typical examples. In the same way, cathodic inhibitors stifle the reaction at the cathode by depositing a nonconductive film on the steel bar. Calcium carbonate, aluminium oxide, calcium nitrite are common examples of cathodic inhibitors. The use of special steels to overcome the problem of corrosion of reinforcement is a costly solution. However, the sacrificial protection provided by coating the reinforcement with either a metallic or nonmetallic material has been found to be satisfactory. Irrespective of the type, it is essential that the coating should completely envelope the bar and should remain unbroken. In addition, it should remain passive under all conditions. Hot-dip galvanizing providing zinc coating is effective as it is metallurgically bonded to the reinforcing bar and it does not affect the bond with concrete or yield and ultimate

Deterioration of Concrete and its Prevention 355

strength or ductility of the bar. Organic coatings have also been used as an alternative to metallic coatings. The epoxy coated steel bars have shown significant increase in protection and reduction in cracking of cover.

The concrete surface coatings providing a barrier at the surface which penetrate and seal the pores of concrete are popular for this type of protection. Waterproof membranes are also being extensively used.

16

Repair Technology for Concrete Structures

16.1 INTRODUCTION

Though concrete is a relatively durable building material, it may suffer *damage* or *distress* during its life period due to a number of reasons. Because of the varying conditions under which it is produced at various locations, the quality of concrete suffers occasionally either during production or during service conditions resulting in distress. The structural causes of distress of concrete may include externally applied and environmental loads exceeding the design stipulations, accidents and subsidences. Sometimes distress in a structure is brought about by *poor construction practices, error in design and detailing*, and *construction overloads*. The other causes may be *drying shrinkage, thermal stress, weathering, chemical reactions* and *corrosion of reinforcement*. In addition to the distress in hardened concrete, the plastic concrete may also suffer damage due to *plastic shrinkage* and *settlement cracking*. Sometimes on stripping off the forms a number of surface defects such as bulges, ridges, *honeycombing, bolt-holes* etc. are noticed on the fresh concrete members. Such defects can be avoided to a large extent by providing a watertight and rigid formwork in such a way that stripping can be done without the use of crowbars or other tools. In addition to these defects, *blow-holes* develop during concreting operations due to improper design of *formwork*. These are formed in the surface of concrete by trapped air and water bubbles against the face of formwork. These can be reduced if the form face is slightly absorbent and adequate compaction is provided. In case the blow-holes are exceptionally large, or if a smooth surface is required, they must be filled with 1 : 1 or 1 : 2 cement-sand mortar. The sand should be sieved through a 300-μm or 600-μm sieve, depending upon the smoothness of the finish required. Crushed limestone dust is preferable. The mortar should be rubbed over the affected area with a rubber-faced float, and finally rubbed down with a smooth stone or mortar block for a smooth finish.

The *honeycombing* consisting of groups of interconnected deep voids normally indicate inadequate compaction or loss of grout through joints in formwork or between formwork and previously cast-concrete. The affected area is delineated with a saw cut to a depth of 5 mm. The unsound material is chipped out to the solid concrete. After the surface has been prepared a bonding coat should be applied to all exposed surfaces, and new concrete should be placed against the prepared surface. The bonding coat may consist of a slurry of cement and water, but it is desirable to incorporate a polymer admixture.

The repair of bulges and projections can be carried out by chipping off the concrete from the surface and then rubbing the surface with a grinding stone. Scouring of a vertical surface of hardened concrete making it resemble a map of delta of a river is caused by water moving upward against the face of formwork. This is a sign of excessive wet or harsh concrete. It is a superficial defect so, unless it is unusually deep and the cover to reinforcement is unusually small, the remedial measures consist of early facing up in the same way suggested for filling blow-holes.

This type of repair (more appropriately called *finishing*) should be done as soon as possible after the forms are stripped and before the concrete becomes too hard. The repair should preferably be completed within 24 hours of the removal of the forms. This is done to develop a good bond and make the repaired portion as durable and permanent as the original work.

If the repairs are not properly done, the newly placed concrete becomes *loose* and *drummy* with the passage of time and finally gets detached from the main concrete. The darker colour of the repaired patches can be corrected by adding 10 to 20 per cent white portland cement to the patching mortar to obtain uniform surface colour. The various defects occurring during construction are outlined in the Table 16.1.

TABLE 16.1 Summary of defects occurring during construction

Symptom	Cause	Prevention	Remedy
Cracks in horizontal surface, as concrete stiffens or very soon thereafter	Plastic shrinkage: rapid drying of surface	Shelter during placing. Cover as early as possible. Use air-entrainment	Seal by brushing in cement or low-viscosity polymer
Cracks form above ties, reinforcement etc., or at arrisses, especially in deep lifts	Plastic settlement: concrete continues to settle after starting to stiffen	Change mix design. Use air-entrainment	Re-compact upper part of concrete while still plastic. Seal cracks after concrete has hardened
Cracks in thick sections, occurring as concrete cools	Restrained thermal contraction	Minimize restraint to contraction. Delay cooling until concrete has gained strength	Seal cracks
Blowholes in form faces of concrete	Air or water trapped against formwork. Inadequate compaction. Unsuitable mix design. Unsuitable release agent	Improve vibration. Change mix design. Use appropriate release agent. Use absorbent formwork	Fill with polymer-modified fine mortar

(contd.)

TABLE 16.1 (Cont.)

Symptom	Cause	Prevention	Remedy
Voids in concrete	Honeycombing; Inadequate compaction. Grout loss	Improve compaction. Reduce maximum size of aggregate. Prevent leakage of grout	Cut out and make good. Inject resin
Erosion of vertical surfaces, in vertical streaky pattern	Scouring; Water moving upwards against form face	Change mix design, to make more cohesive or reduce water content	Rub in polymer-modified fine mortar
Colour variations	Variations in mix proportions, curing conditions, materials, characteristics of form face, vibration, release agent. Leakage of water from formwork	Ensure uniformity of all relevant factors. Prevent leakage from formwork	Apply surface coating
Powdery formed surface	Surface retardation, caused by sugars in certain timbers	Change form material. Seal surface of formwork. Apply lime-wash to form face before first few uses	Generally none required
Rust-strains	Pyrites in aggregates. Rain streaking from unprotected steel. Rubbish in formwork. Ends of wire ties turned out	Avoid contaminated aggregates. Protect exposed steel. Clean forms thoroughly. Turn ends of ties inwards	Clean with dilute acid or sodium citrate /sodium dithionite. Apply surface coating
Plucked surface	Insufficient release agent. Careless removal of formwork	More care in applications of release agent and removal of formwork	Rub in fine mortar, or patch as for spalled concrete
Lack of cover to reinforcement	Reinforcement moved during placing of concrete, or badly fixed. Inadequate tolerances in detailing	Provide better support for reinforcement. More accurate steel fixing. Greater tolerances in detailing	Apply polmermodified cement and sand rendering Apply protective coating

16.2 SYMPTOMS AND DIAGNOSIS OF DISTRESS

In addition to minor structural defects outlined above, the other distresses can be observed in the form of cracks, spalling and scaling of concrete. Cracking is the most common indication of the distress in a concrete structure. It may affect appearance only, or indicate significant structural distress or lack of durability. Cracks may represent the total extent of the damage, or they may point to problems of greater magnitude. These, in turn, may cause corrosion of reinforcement due to the entry of moisture and oxygen.

All the concrete structures crack in some form or the other. Most buildings develop cracks in their fabric which are superficial and occur soon after the construction. Cracks, even if harmless, may have an adverse psychological effect. However, cracking in concrete structures is not necessarily a cause for accusing the designer, builder or supplier. What really matters is the type of structure and the nature of *cracking*. Cracks that are acceptable for building structures may not be acceptable for water-retaining structures.

Cracking of concrete structures can never be totally eliminated, but the practitioner should be aware of the causes, evaluation techniques, and the methods of repair. The approach to diagnosis of the problem of cracking should be identical to that of a doctor to the patient. An engineer should have a sound knowledge of all the facets of concrete technology, i.e. of the behaviour of construction materials, construction techniques, types of cracks likely to occur, their causes and respective remedial measures. In short, treatment of cracks involves detection, diagnosis and remedy. Before remedies are sought, correct diagnosis will decide whether satisfactory repair is possible. The development of cracks and their repair is a perpetual problem involving considerable cost and inconvenience to the occupants. The problem should be tackled on two fronts, i.e. by adopting preventive measures and repairing them. However, prevention is better than repair. The designer and builders should attempt to reduce the formation of cracks by using appropriate construction materials, and by adopting appropriate design and construction techniques.

The cracks in a structure are broadly classified in two categories: *superficial cracks* and *structural cracks*. The structural cracks may be *active* and *dormant*. A crack where a movement is observed to continue is termed active, whereas the crack where no movement occurs is termed dormant or static. The following information may help in diagnosing the cracks:

(i) whether the crack is new or old,
(ii) type of crack, i.e. whether it is active or dormant,
(iii) whether it appears on the opposite face of the member also,
(iv) pattern of the cracks,
(v) soil condition, type of foundation used, sign of movement of ground, if any,
(vi) observations on the similar structures in the same locality,
(vii) study of specifications, method of construction used and the test results at the site, if any,
(viii) views of the designer, builder, occupants of the building, if any, and
(ix) weather during which the structure has been constructed.

From the above discussion it is evident that the cracking is a complex phenomenon. The various aspects of the problem are discussed as follows.

The latent defects in a concrete structure may be caused by inadequacy of design, materials or construction practices which may not become evident until sometime after its completion. The immediate cause of deterioration may be a chemical action or corrosion of reinforcement, but in majority of the cases the basic cause may be traced back to something such as unrealistic detailing or poor workmanship.

The incompatible dimensional changes caused by drying shrinkage and thermal movements during and after the hardening period may also cause cracks in concrete members. Before any repair work is taken in hand, the cause of damage must be clearly identified, for which careful investigation is required. Some of common causes are discussed in the following sub-sections.

16.2.1 Cracking of Plastic Concrete

When the exposed surfaces of freshly placed concrete are subjected to a very rapid loss of moisture caused by low humidity, wind, and/or high temperature, the surface concrete shrinks. Due to restraint provided by the concrete below the drying surface layer, tensile stresses develop in the weak, stiffening plastic concrete, resulting in shallow cracks that are usually short, discontinuous running in all directions and very seldom extend to the free edge. In an unreinforced slab they are typically diagonal as shown in Fig. 16.1 (a). In the presence of reinforcement their pattern may be modified. Plastic shrinkage usually occurs prior to final finishing before curing starts. The cracks are often fairly wide at the surface. They range from few centimeters to many metres in floors or slabs or other elements with large surface areas. *Plastic shrinkage cracks* may extend the full depth of elevated thin structural elements.

Plastic shrinkage cracks can be controlled by reducing the relative volume change between the surface and the interior concrete by preventing a rapid moisture loss due to hot weather and dry winds. This can be accomplished by using fog nozzles to saturate the air above the surface and use of plastic sheeting to cover the surface between the final finishing operations. Wind breakers reducing the wind velocity, and sunshades reducing the surface temperatures are also helpful, and it is good practice to schedule flatwork after the walls have been erected. The remedial measures after the cracks have formed usually consists of sealing them against entry of water by brushing in cement or low viscosity polymer.

After initial placement, vibration and finishing the concrete has a tendency to continue to consolidate or settle, especially in deep sections after it has started to stiffen. During this period the plastic concrete may be locally restrained by reinforcing bars, previously placed concrete, or formwork tie-bolts. These local restraints may cause cracks and/or voids adjacent to the restraining element as shown in Fig. 16.1 (e). When associated with the reinforcing bars, this *settlement cracking* increases with increasing bar size, increasing slump, and decreasing cover. The degree of settlement cracking may be magnified by insufficient compaction or the use of leaking or highly flexible forms. The use of lowest possible slump, an increase in concrete cover, provision of a sufficient time interval between placement of concrete in various elements, adequate vibration and proper form design will reduce settlement cracking. Air entrainment may help in avoiding these cracks. The remedial measures, after concrete has hardened, consists in sealing the cracks in order to protect the reinforcement.

16.2.2 Cracking of Hardened Concrete

The moisture-induced volume changes are characteristic of concrete. A loss of moisture from cement paste results in a volume shrinkage by as much as 1 per cent, whereas the internal restraint provided by the aggregate reduces the magnitude of this volume change to about 0.05 per cent. On the other hand, an increase in moisture in the concrete tends to increase its volume. It these volume changes are restrained (usually by another part of structure or by the subgrade), the tensile stresses develop. When the tensile strength of concrete is exceeded, it will crack. The cracks may propagate at much lower stresses than are required to cause *crack initiation*. In massive concrete elements, tensile stresses are caused by differential shrinkage between the surface and the interior concrete. The larger shrinkage at the surface causes the cracks to develop that may with time penetrate deeper into the concrete. *Surface cracking* on walls and slabs usually occurs due to drying shrinkage when the surface layer of concrete has a higher water content than the interior concrete. The *surface crazing* appears in the form of a series of shallow, closely spaced fine cracks.

The extent of shrinkage cracking depends upon the amount of shrinkage, degree of restraint, modulus of elasticity, and amount of creep. The amount of drying shrinkage is influenced mainly by the amount and type of aggregate and the water content of the mix. The shrinkage decreases with the increase in the amount of aggregate, and the reduction in *water content*. The higher the stiffness of the aggregate, the more effective it is in reducing the shrinkage of concrete, e.g. the shrinkage of concrete containing sandstone aggregate may be more than twice that of concrete with basalt or granite. Therefore, the drying shrinkage can be reduced by using the maximum practical amount of aggregate and lowest usable water content in the mix. Shrinkage cracking can be controlled by using properly spaced *contraction joints* and proper *steel detailing*. The shrinkage cracking can also be controlled by using *shrinkage compensating cement*.

Thermal Cracking

The temperature difference within a concrete structure result in differential volume change. When the tensile strain due to differential volume change exceeds the tensile strain capacity of concrete, it will crack. The temperature differentials associated with the *hydration of cement* affect the *mass concrete* such as in large columns, piers, footings, dams, etc. whereas the temperature differentials due to changes in the ambient temperature can affect any structure.

The liberation of the heat of hydration of cement causes the internal temperature of concrete to rise during the initial curing period, so that it is usually slightly warmer than its surroundings. In thick sections and with rich mixes the temperature differential may be considerable. As the concrete cools it will try to contract. Any restraint on the free contraction during cooling will result in tensile stresses which are proportional to the temperature change, coefficient of thermal expansion, effective modulus of elasticity (which is reduced by creep) and degree of restraint. The more massive the structure, the greater is the potential for temperature differential and degree of restraint. Thermally induced cracking can be reduced by controlling the maximum internal temperature, delaying the onset of cooling by insulating the formwork and exposed surfaces, controlling the rate of cooling, and increasing the tensile strain capacity of the concrete. Special precautions need to be taken in the design of structures in which some portions are exposed to temperature changes while the other portions of structures are either partially or completely protected. A drop in temperature may result in the cracking of the exposed element while increased in temperature may cause cracking in the protected portion of structure. *Temperature gradients* cause deflection and rotation in structural members; if these are restrained serious stresses can result. Allowing for movement by using properly designed contraction joints and correct detailing will help alleviate these problems. If the cracks do form, remedial measures are similar to those for cracks that form after a structure in service.

Cracking Due to Chemical Reactions

The most important constituent of concrete namely cement is alkaline; so it will react with acids or acidic compounds in presence of moisture, and in consequence the matrix becomes weakened and its constituents may be leached out. The concrete may crack, as a result of expansive reactions between aggregate containing active silica and alkalies derived from cement hydration, admixture or external sources (e.g. curing water, ground water, alkaline solutions stored). The *alkali-silica reaction* results in the formation of a swelling gel, which tends to draw water from other portions of concrete. This causes local expansion and accompanying tensile stresses which if large may eventually result in the complete deterioration of the structure. Control measures include proper selection of aggregate, use of low-alkali cement, and use of pozzolana. Typical symptoms in unreinforced and highly reinforced concrete are *map cracking*, usually in a roughly hexagonal mesh pattern as shown in Fig. 16.1 (b), and gel excluding from cracks.

The *alkali-carbonate reaction* occurs with certain limestone aggregates and usually results in the formation of alkali and silica between aggregate particles and the surrounding cement paste. Here also the affected concrete is characterized by a *network pattern of cracks* as shown in Fig. 16.1 (b). The problem may be minimized by avoiding *reactive aggregate*, use of a smaller-size aggregate and use of low-alkali cement.

When the sulphate bearing waters come in contact with the concrete, the sulphate penetrates the hydrated paste and reacts with hydrated calcium aluminate to form calcium sulpholuminate with a subsequent large increase in volume, resulting in high local tensile stresses causing the deterioration of concrete as illustrated in Fig. 16.1 (c). The blended or pozzolana cements impart additional resistance to sulphate attacks.

The calcium hydroxide in hydrated cement paste will combine with carbon dioxide in the air to form calcium carbonate which occupies smaller volume than the calcium hydroxide resulting in the so-called *carbonation shrinkage*. This situation may result in significant *surface grazing* and may be especially serious on freshly placed concrete surface kept warm during winter by improperly vented combustion heaters.

Cracking Due to Weathering

The *environmental factors* that can cause cracking include (i) freezing and thawing, (ii) wetting and drying, and (iii) heating and cooling. Except in tropical regions, the damage from freezing and thawing is the most common weather-related physical deterioration. In the aggregate particles saturated above the *critical degree of saturation*, the expansion of absorbed water during freezing may crack the surrounding cement paste and/or damage the aggregate itself. The control measures include the use of the lowest practical *water-cement ratio* and total *water content*, *durable aggregate* and *adequate air-entrainment*. Adequate curing prior to exposure to freezing conditions is also important. Other weathering processes that may cause cracking in concrete are alternate wetting and drying, and heating and cooling. If the volume changes due to these processes are excessive, cracks may develop and give the impression that the concrete is on the verge of disintegration. The fire and frost actions may also damage the structure. The damage due to these factors may appear in the form of general flacking and spalling of concrete from the surface. Concrete gradually loses strength with increase in temperature above about 300°C, damage being greater with aggregate having high coefficient of thermal expansion.

Cracking Due to Corrosion of Reinforcement

It is the most frequent cause of damage to reinforced concrete structures. This aspect of the cracking problem has been discussed in detail in Chapter 15. However, the salient features are outlined for ready reference. The *corrosion of steel* produces iron oxides and hydroxides, which have a volume much greater than the volume of the original metallic iron. This increase in volume causes high radial bursting stresses around reinforcing bars and results in local radial cracks. These *splitting cracks* may propagate along the bar, resulting in the formation of *longitudinal cracks* parallel to the bar or *spalling* of concrete as illustrated in Fig. 16.1 (d). Cracks provide easy access to oxygen, moisture, and chloride, and thus even a minor split can create a condition in which corrosion continues and causes further cracking.

Reinforcing steel usually does not corrode in concrete because a tightly adhering protective oxide coating forms in a highly alkaline environment. This is known as *passive protection*. However, if the alkalinity of the concrete is reduced through *carbonation* or the passivity of this steel is destroyed by aggressive chloride ions, the reinforcing bars may corrode. Cracks transverse to reinforcement usually do not cause continuing corrosion of reinforcement if the concrete has low *permeability*. If the combination of density and cover thickness is sufficient to restrict the flow of oxygen and moisture, corrosion slows down or ceases.

For general concrete construction, the best control measure against corrosion-induced *splitting* is the use of concrete with low permeability. Increased concrete cover over the reinforcing bar is effective in delaying the corrosion process and also in resisting the splitting and spalling caused by corrosion or transverse tension. In very *severe exposure conditions*, additional protective measures, such as coated reinforcement, sealers or overlays on concrete, and corrosion-inhibiting admixtures can be adopted.

Cracking Due to Poor Construction Practices

Poor construction practices, such as adding water to concrete to improve workability, lack of curing, inadequate form support, inadequate compaction, and arbitrary placement of construction joints, can result in cracking in concrete structures. Adding water to improve workability has the effect of reducing strength, increasing settlement and ultimate drying shrinkage. The early termination of curing will allow for increased shrinkage at the time when the concrete has low strength. Incomplete hydration due to drying will reduce not only the long-term strength but also the *durability* of the structure. Lack of support for forms or inadequate compaction can result in the *settlement cracking* of concrete before it has developed sufficient strength to support its own weight, while improper location of construction joints can result in cracking at the planes of weakness. Some of the defects occuring during construction are summarized in the Table 17.1.

Cracking Due to Construction Overloads

The loads induced during construction can be far more severe than those experienced in service. Unfortunately, these conditions may occur at the early ages when the concrete is most susceptible to damage and often result in permanent cracks.

A common error occurs when the precast members are not properly supported during transportation and erection. The use of arbitrary or convenient lifting points may cause severe damage. A big element lowered too fast, and stopped suddenly carries significant momentum which is translated into an *impact load* that may be several times the dead weight of the element.

Storage of materials and equipment can easily result in loading conditions during construction far more severe than any load for which the structure is designed. Damage from unintentional construction overloads can be prevented only if the designers provide information on load limitations for the structure and if the construction personnel heed to these limitations.

Cracking Due to Errors in Design and Detailing

The design and detailing errors that may result in unacceptable cracking include use of poorly detailed re-entrant corners in walls, precast members and slabs; improper selection and/or detailing of reinforcement; restraint of members subjected to volume changes caused by variations in temperature and moisture; lack of adequate contraction joints, and improper design of foundations resulting in *differential settlement* within the structure. Re-entrant corners provide a location for *stress concentration* and, therefore, are prime locations for initial cracks, as in the case of window and door openings in concrete walls and dapped beams. Additional properly anchored diagonal reinforcement is required to keep inevitable cracks narrow and prevent them from propagating further.

An inadequate amount of reinforcement may result in excessive cracking. A common mistake is to lightly reinforce an element because it is a *non-structural element* and tying it to the rest of the structure in such a manner that it is required to carry a major portion of the load once the structure begins to deform. The *non-structural element* will carry a load in proportion to its stiffness. Since this element is not detailed to act structurally, unsighty cracking may result even though the safety of the structure is not threatened. The restrained members subjected to volume changes frequently develop cracks. A slab, wall or a beam restrained against shortening, even if prestressed, can easily develop tensile stress sufficient to cause cracking. Beams should be allowed to move.

Improper foundation design may result in excessive *differential movement* within a structure. If the differential movement is relatively small, the cracking problem may be only visual in nature. However if there is a major differential settlement, the structure may not be able to redistribute the loads rapidly enough, and a failure may occur. One of the advantages of the reinforced concrete is that, if the movement takes place over a sufficiently long period of time, *creep* will allow some redistribution of load.

Special care need to be taken in the design and detailing of structures in which cracking may cause a major serviceability problem. These structures also require continuous inspection during all phases of construction to supplement the careful design and detailing.

Cracks Due to Externally Applied Loads

Load-induced tensile stresses may result in cracks in concrete elements. A design procedure specifying the use of reinforcing steel, not only to carry tensile forces, but also be obtain both an adequate crack distribution and a reasonable limit on crack width is recommended. Flexural and tensile crack widths can be expected to increase with time for members subjected to either sustained or repetitive loading. A well-distributed reinforcing arrangement offers the best protection against undesirable cracking.

16.3 EVALUATION OF CRACKS

As in the case of a medical practitioner prescribing medicine without thoroughly examining the patient, it is difficult for a repair engineer to advocate any repair technology without making a thorough investigation. Before proceeding with repair, the investigations should be made to determine the location and extent of cracking, the causes of damage, and the objectives of repair. Calculation can be made to determine stresses due to applied loads. For detailed information, the history of the structure, structural drawings and specifications, and construction and maintenance records should be reviewed.

The objectives of repair include restoration and enhancement of durability, structural strength, functional requirements and aesthetics. The evaluation of cracks is necessary for the following purposes:
  (i)   To identify the cause of cracking
  (ii) To assess the structure for its safety and serviceability
  (iii) To establish the extent of the cracking
  (iv) To establish the likely extent of further deterioration
  (v)   To study the suitability of various remedial measures
  (vi) To make a final assessment for serviceability after repairs

Apart from visual inspection, tapping the surface and listening to the sound for hollow areas may be one of the simplest methods of identifying the weak spots. The suspected areas are then opened up by chipping the weak concrete for further assessment.

The comparative strength of concrete in the structure may be assessed to a reasonable accuracy by *non-destructive testing* and by the tests on the *cores* extracted from the concrete. The commonly used non-destructive tests are the *rebound-hammer* test and *ultrasonic-pulse-velocity test*

16.3.1. Visual Examination

The appearance of concrete surface may suggest the possibility of chemical attack by a general softening and leaching of matrix, or in case of sulphate attack by whitening of concrete. Rust stains often indicate corrosion of reinforcement but they may also be caused due to the contamination of aggregate with iron pyrites. If the cracked concrete is broken out, the appearance of the crack surface gives useful information; dirt or discolouration show that the crack has been there for some time. General flaking of an exposed concrete surface suggests frost damage. In fire damaged structure, the colour of concrete gives an indication of the maximum temperature reached.

The crack pattern may be informative, a mesh pattern suggests drying shrinkage, and surface crazing may indicate frost attack or in rare cases *alkali-aggregate reaction*. Typical crack patterns are illustrated in Fig. 16.1. The cracks caused by unidirectional bending will be the widest in the zone of maximum tensile stress and will taper along their length, while cracks caused by direct tension will be of roughly uniform width. *Pop-outs* in concrete are usually associated with particles of coarse aggregate just below the surface.

The location and width of cracks should be noted on a sketch of the structure. A grid marked on the surface of the structure can be used to accurately locate cracks on the sketch. *Crack widths* can be measured to an accuracy of about 0.025 mm using a *crack comparator*, which is a small hand-held microscope with a scale on the lens closest to the surface being viewed. Location of observed spalling, exposed reinforcement, surface deterioration, and rust staining should be noted on the sketch.

The use of brittle liquid coatings on the suspected structure can also help detect the crack or growth of cracks over a period of time. The movement of the cracks can be monitored with the help of mechnical movement indicators or crack monitors using electrical resistance thin filaments, which amplify crack movement and indicate the maximum range of movement occurring during the measurement period, i.e. the extent of progressive growth of crack. *Linear variable differential transformers* (LVDTs) and *data acquisition systems* (ranging from strip chart recorder to a computer based system) are available.

The cracks in concrete may be evaluated at macro, micro, submicro, and atomic levels (Angstroms $°A$). In the present discussion the macrostructure cracks having a size (i.e. width/depth) in the range of 0.1 to 0.3 mm are of interest.

16.3.2 Non-destructive Testing

Non-destructive tests may be performed to determine the presence of internal cracks and voids, and the depth of penetration of cracks that are visible at the surface. Useful information can often be obtained by tapping the surface of concrete with an ordinary hammer. The difference in sound when concrete is struck may identify areas of delamination of concrete that has been damaged, say by fire. A hollow sound indicates a separation or crack below the surface.

(a) Plastic shrinkage
(unreinforced slab)

(b) Alkali-aggregate reaction
(unstressed)

(c) Sulphate attack

(d) Reinforcement corrosion

Settlement of concrete surface

Void under reinforcing bar

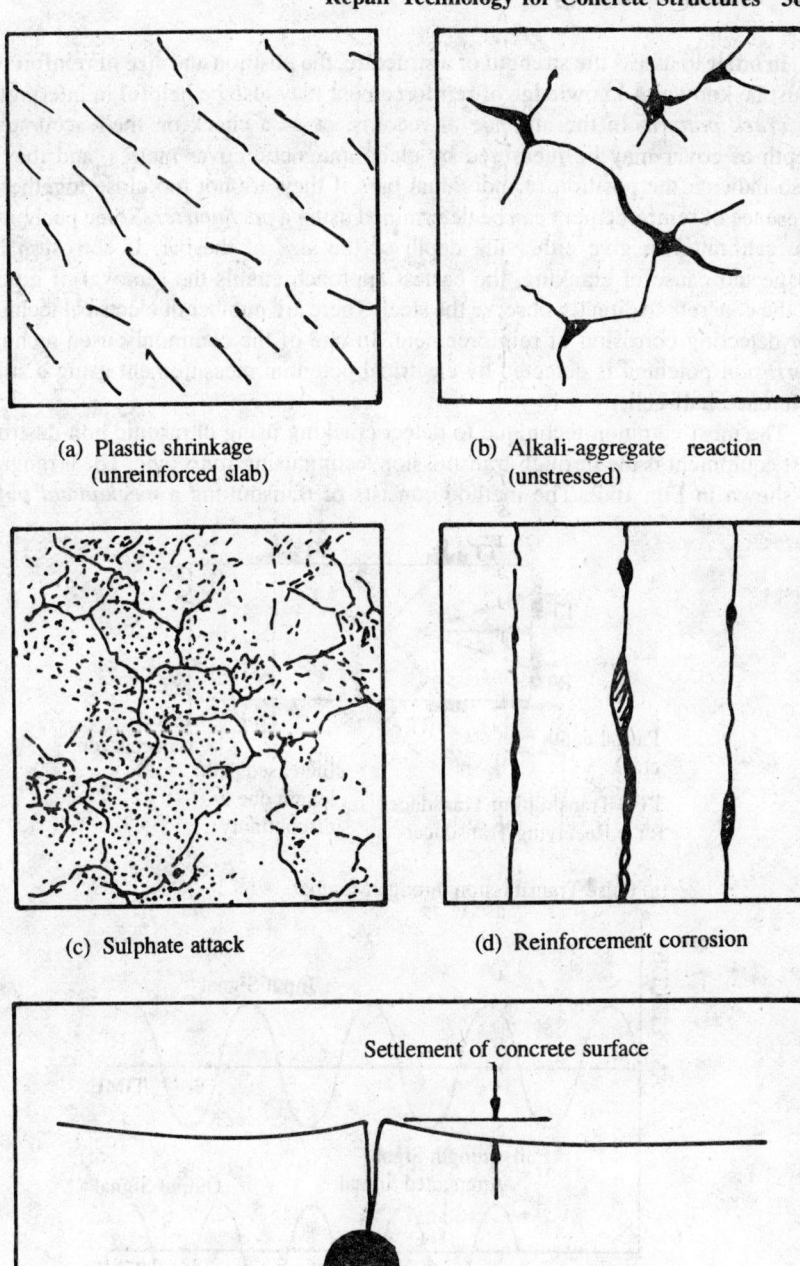

(e) Typical plastic settlement crack over reinforcing bar

Fig. 16.1 Some typical crack patterns

368 Concrete Technology

In order to assess the strength of a structure, the position and size of reinforcement must be known. A knowledge of reinforcement may also be helpful in interpretation of *crack pattern*. In the absence of records, or as a check on their accuracy, the depth of cover may be measured by electromagnetic cover metres, and they may also indicate the position of individual bars if they are not too close together. The presence of reinforcement can be determined using a *pachometer*. Some pachometers are calibrated to give either the depth or the size of the bar. If corrosion is the suspected cause of cracking, the easiest approach entails the removal of a portion of the concrete to directly observe the steel. There are number of electrical techniques for detecting corrosion of reinforcement. In one of the commonly used techniques *corrosion* potential is detected by electrical potential measurement using a suitable reference half-cell.

The most common technique to detect cracking using ultrasonic non-destructive test equipment is the through-transmission testing using *soniscope*. The arrangement is shown in Fig. 16.2. The method consists of transmitting a *mechanical pulse* to

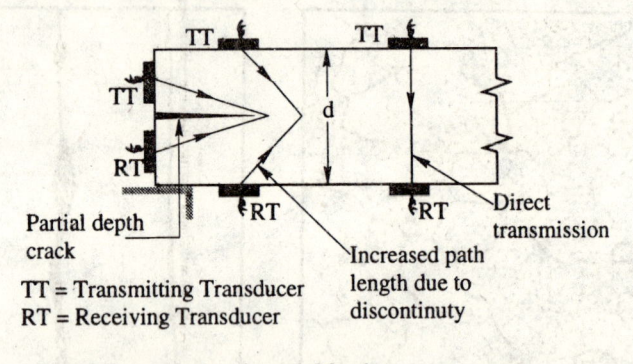

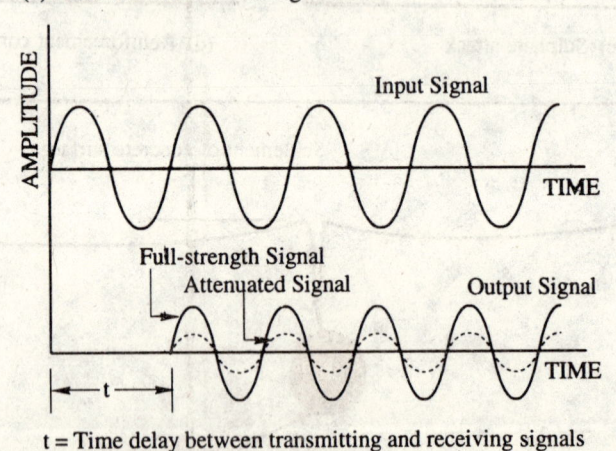

(a) Pulse Transmission through Member

TT = Transmitting Transducer
RT = Receiving Transducer

t = Time delay between transmitting and receiving signals
Ultrasonic Pulse Velocity (UPV) = d/t

(b) Oscilloscope Signal

Fig. 16.2 Through-transmission ultrasonic testing

one face of the concrete member and receiving it at the opposite face, and calculating the *pulse velocity* from the time taken by the pulse to pass through the member, and the distance between the transmitting and receiving transducers. A significant change in measured pulse velocity may occur if an internal discontinuity results in an increase in the path length of the signal as it passes around the end of crack. Generally, the higher the pulse velocity, the better the quality and durability of the concrete. Internal discontinuities can also be detected by attenuation of signal strength if the signal is displayed on an oscilloscope. If no signal arrives at the receiving transducer, a significant internal discontinuity, such as a crack or void, is indicated. An indication of the extent of the discontinuity can be obtained by taking readings at a series of positions on the member. The results of ultrasonic testing should be interpreted cautiously, e.g. with fully saturated cracks the ultrasonic testing will generally be ineffective. Another method called the *pulse echo method* for location of cracks is shown in Fig. 16.3.

(a) Pulse Echo

(1) Solid member   (2) Internal crack

(b) Oscilloscope Signals

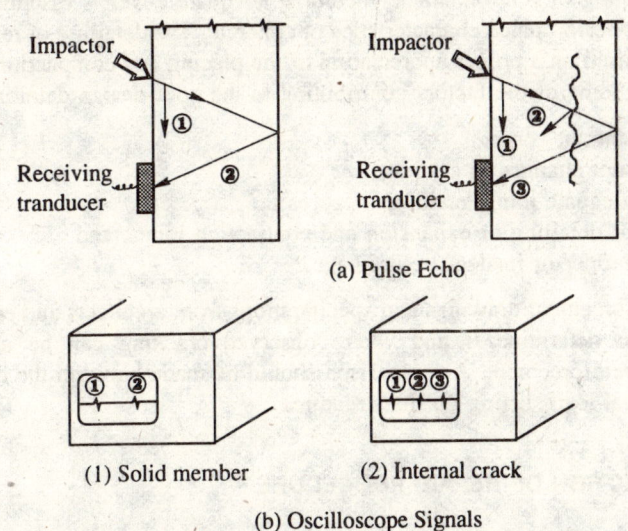

Fig. 16.3 Pulse echo method for crack location

16.3.3 Tests on Concrete Cores

The basic objective of testing the hardened *concrete cores* is to check its compliance with specifications. In order to resolve the disputes arising out of the testing of concrete by destructive tests using conventional procedures, concrete cores are drilled and tested to estimate the strength of concrete in the actual structures. However, significant information can be obtained from cores taken from selected locations within the structure. Cores and core holes afford the opportunity to accurately measure the width and depth of cracks. Core material and crack surface can be examined petrographically to determine the presence of *alkali-silica reaction products* or other deleterious substances. The cores can also be used to detect *segregation* or *honeycombing* or to check the bond at construction joints. Usually a core is cut by means of rotary cutting tool with diamond bits. Thus a cylindrical specimen is obtained, sometimes containing embedded fragments of reinforcement.

The cores are soaked in water, capped and tested in compressions in a moist condition. The height/diameter ratio of the core is kept nearly two to compare its strength with standard cylindrical specimens. In some cases, beam (prism) specimens can be sawn from road or air-field pavement slabs using a diamond or carborundum saw. Such specimens are tested in flexure. The cutting of beams is, of course, a very cumbersome and expensive process and not much used. While comparing the strength of cores with that of standard cylinder specimens, the factors regarding site, curing and the position of the cut-out concrete in the structure should be considered. Cores usually have the lowest strength near the top surface of the structure, be it a column, wall, beam or even a slab. With an increase in depth below the top surface, the strength of cores increases up to about a depth of 300 mm.

16.3.4 Review of Drawings for Reinforcement Details

Proper detailing of reinforcement, including adequate cover is essential to ensure the required performance characteristics of concrete. The detailing of reinforcement should be based on a proper appreciation of the placing and compaction techniques to be used. Some of the factors contributing to the poor design detailing are:

  (i) re-entrant corners,
 (ii) abrupt changes in section,
(iii) inadequate joint detailing,
 (iv) poor detailing of expansion and contraction joints, and
  (v) improper or inadequate drainage.

The reinforcement drawings and specifications from architects and engineers are examined to determine if and where observed cracking can be attributed to inadequate reinforcement. A comparison should be made between the design loads and the actual loads acting on the structure.

16.4 SELECTION OF REPAIR PROCEDURE

The repair of concrete structures may vary between just giving a cosmetic treatment and a total replacement. By a proper investigation and by using well-designed equipment, tools and materials, a number of structures which may appear to have been damaged beyond repair can be reinstated economically. An appropriate repair method can be selected depending upon the cause and extent of damage, importance of the structural element, and its location. The choice of the method will determine its success. A procedure may be selected to accomplish one or more of the following objectives:

   (i) To increase strength or restore load carrying capacity
  (ii) To restore or increase stiffness
 (iii) To improve functional performance
  (iv) To provide water tightness
   (v) To improve appearance of concrete surface
  (vi) To improve durability
 (vii) To prevent access of corrosive materials to reinforcement

Depending upon the nature and extent of the damage, one or more repair methods may be selected, e.g. tensile strength can be restored across a crack by injecting it with an epoxy. However, it may be necessary to provide additional strength by adding reinforcement. Epoxy injection alone can be used to restore flexural stiffness if further cracking is not anticipated. Cracks causing leaks in water retaining structures should be repaired unless the leakage is considered minor. They can be repaired when cracks result in an unacceptable appearance. However, if the crack location is still visible, some form of coating over the entire surface may be required. To minimize further deterioration due to corrosion of reinforcement, cracks exposed to a moist environment should be sealed.

Success of the long-term repair procedures chiefly depends on the nature of cracks as well as their cause. For example, if the cracking is primarily due to drying shrinkage, it is likely that after a period of time the cracks will stabilize. On the other hand, if the cracks are due to continuing foundation settlement, repair will not be effective until the settlement problem is corrected. The repair procedure also depends on the capabilities and facilities available with the builder, and on the availability of the repair materials.

16.5 REPAIR OF CRACKS

Once the cracked structure has been evaluated and the causes of cracking established, a suitable repair procedure may be selected which takes these causes into account. The methods of crack repair including the characteristics of cracks that may be repaired with each procedure and the types of structure that have been repaired are described in the following sections. The repair of concrete structures is carried out in the following stages:

    (i) Pre-treatment of surface and reinforcement, i.e. removal of damaged concrete. This process is termed as the preparation of surface for repairs.

    (ii) Application of repair material.

16.5.1 Preparation of Surface

Prior to the execution of any repair, one of the essential requirements common to all repair techniques is that all deteriorated or damaged concrete should be removed. This can be accomplished by using tools and equipment, the type of which depends to a large degree on the size, depth, and extent of repair. For smaller jobs, the removal of concrete can be accomplished by hand tools whereas for larger repairs, the surface can be prepared by using light and medium-weight air hammers fitted with spade-shaped bits. Care should be taken to avoid any damage to the unaffected portions. For cracks and other narrow defects, sawtoothed bits can be used to obtain sharp edges and suitable under cuts as shown in Fig. 16.4 (c). The preparation of a surface for repair involves the following steps:

    (i) Complete removal of unsound material
    (ii) Undercutting along with the formation of smooth edges
    (iii) Removal of the cracks from the surface
    (iv) Formation of a well-defined cavity geometry with rounded inside corners.

For the damaged area shown in Fig. 16.4 (a), the correct cavity geometry is illustrated in Fig. 16.4 (c)

    (v) Providing rough but uniform surfaces for repair

372 Concrete Technology

(a) Plan of damaged area

(b) Incorrect method of cutting out

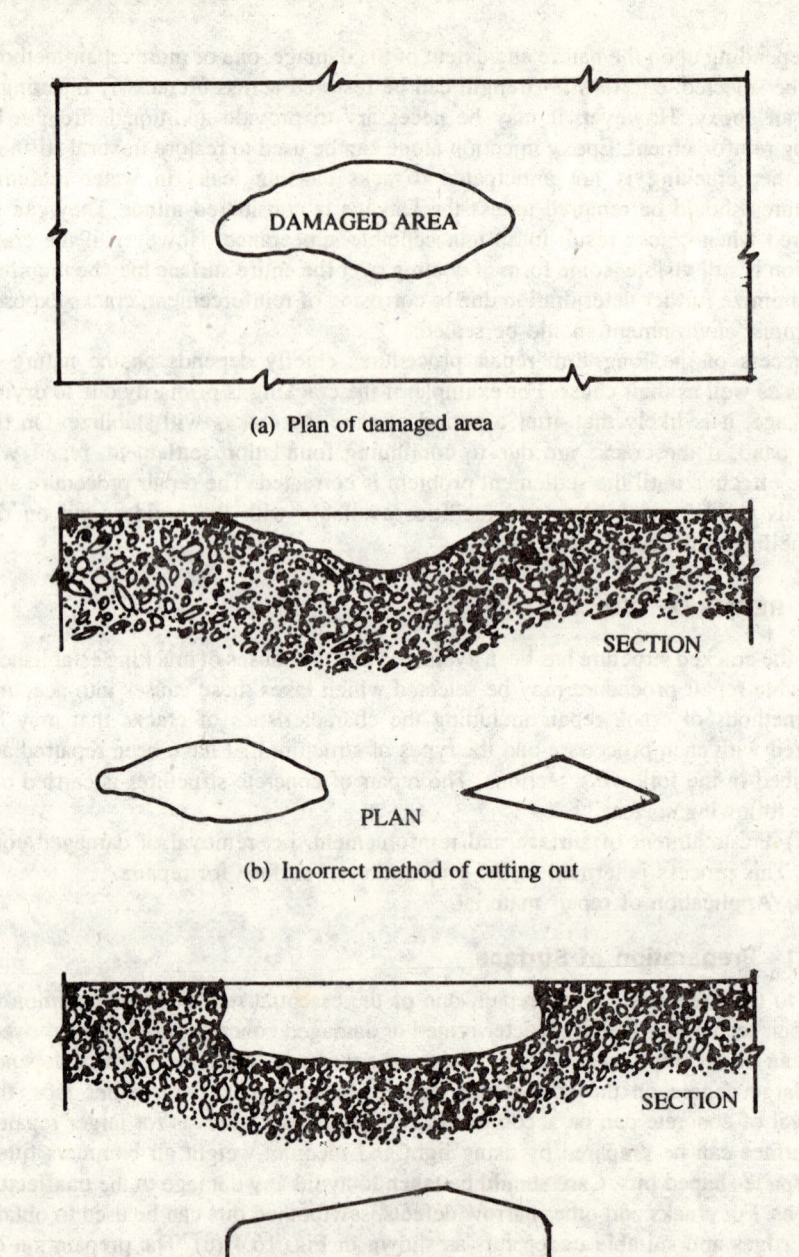

(c) Correct method of cutting out

Fig. 16.4 Preparation of surface by cutting out the damaged area

The surface so prepared should be clean, dry, free of laitance and strong. A *clean surface* means that there should be no foreign matter (contaminants) such as dirt, loose particles, grease or oil, paints, resins, etc. on the surface. *Dry surface* in most cases means that no free water shall be present. *Free of laitance* means that the skin of high water-cement gel, which appears on the surface during concrete placement, should be removed. This skin has poor integrity and adhesion to the parent concrete. *Strong concrete* below the surface refers to the ability of concrete to resist fractures due to the stresses exerted on it by the repair material and techniques.

Dust is any fine visible foreign matter present on the concrete surface and can be detected by wiping across the surface of concrete with a dark cloth, the presence of white powder on the cloth indicates the presence of dust, and such a surface is not suitable for the application of most of the repair materials.

Dust and other loose foreign materials present on the surface of the concrete can be removed by *wire brushing*. If this method is not adequate, and other method may be used which will actually remove some concrete along with contaminants. *Grinding, scarifying* and *sand-blasting* are the commonly used methods. Grinding or paring is slow but suitable for smaller areas. Scarifying chips away a thin layer of concrete. Scabblers can be used to scarify up to 6 mm from the concrete surface. These are air-operated units having tungsten carbide bits that pound the concrete surface. The subsequent scarifying may be performed using an abrasive blasting machine.

Sand or steel shot-blasting is perhaps the most effective method of cleaning a concrete surface. The water blasting with low pressure jets may be effective in some cases. Oil can be detected by sprinkling water on the surface. If the water stands in droplets without spreading out immediately, it indicates that the surface is contaminated with an oily substance which will interfere with the bonding of most repair materials.

Oil, grease and animal fat may be removed by chemical cleaning, i.e. scrubbing the surface with detergents, caustic soda solution or trisodium phosphate. A vigorous scrubbing action with a stiff broom should be carried out during the washing proceduce. The surface should be washed off thoroughly with a pressure hose to remove all traces of loosened oil and as well as the cleaning solution.

The *laitance* which can be detected by the presence of fine powder on surface, when it is scrapped with the knife blade, may be removed by *acid etching*. However, if the acid etching is to be used the surface should be precleaned by removing any build-up of dirt, oil, grease or any other foreign matter. In case the presence of chlorides in the concrete is not objectionable, a 10 per cent solution of hydrochloric acid in water is applied in the ratio of 1.15 litre/m^2 of the surface. A stiff bristle broom or brush should be used to spread and vigorously scrub the acid solution uniformly on the surface. After the foaming action has subsided, the surface is thoroughly washed off with water still scrubbing it with stiff brooms. This is necessary to remove the salts which may have been formed by the reaction of the acid with the cement. The acid at the surface after washing should be checked with a litmus paper. When the presence of chlorides is objectionable, a 15 per cent solution of phosphoric acid can be substituted. The presence of acids from other sources may also be ascertained by placing a litmus paper in a thin film of water on the surface of concrete. A pH value below 4 indicates that the acidity of concrete is too high for the successful application of repair or barrier systems.

Form release agents and curing compounds should be avoided when it is known that a barrier system is to be applied on the concrete surface. If used, they should be such that they can be completely removed before the application of barrier system. It is recommended to avoid chemical release agents altogether. Proprietary paint system which can be applied to the forms to prevent contamination of concrete and provide good release are available.

In case of application of overlays, waterproofing or protective barriers, the surface should be uniform. The size of the defects on the concrete surface which can be tolerated depends on the nature of the barrier system itself. The recommendations of the manufacturer of the system is the best guide. However, for most of the decorative systems all protrusions should be removed and visible holes should be filled. For other types of barriers, protrusion higher than 1.6 mm, the spalls and holes greater than 3.2 mm diameter should be repaired. The size of the holes and the surface conditions determine the material that may be used. Small surface voids are filled with *cement grout* whereas larger voids can be filled with *drypack* or *shotcrete*. A visual inspection may provide an adequate control of these conditions.

In case of any doubt regarding the adequacy of the cleaning process or the presence of undetected contaminants, a *patch test* should be performed. In this test the repair material or barrier should be applied to a typical surface whose adequacy of preparation is to be tested. After curing the adhesion of the material should be checked for its effectiveness.

16.5.2 Repair Techniques

The repair of cracked or damaged structures is discussed under two distinct categories, namely, ordinary or conventional procedures; and special procedures using the latest techniques and newer materials such as polymers, epoxy resins, etc.

1. Ordinary Procedures

Superficial or fine cracks are generally removed by treating the surface with whitewash, soft distempers, silicate cement paints (Snowcem or Aquacem), etc. The methods and materials used for the repair of patches of deteriorated concrete in the structures are described below. The repair is carried out in four steps.

(i) *Preparation of surface* The cracked and deteriorated areas are cut or chipped out to the solid concrete. Application of a sound patch to an unsound surface is meaningless because patch will eventually come out. Any attempt to take short cuts over surface preparation is false economy. The area to be chipped out should be delineated with a saw cut to a depth of about 5 mm in order to provide neat edge. The edges should be cut out as straight as possible and right angled to the surface with corners rounded within the hole. The edges are slightly undercut as illustrated in Fig. 16.4 (c) to provide keys at the edges of the patch. The thickness of edges should not be less than 25 mm to prevent them from breaking under load. The unsound concrete is removed with percussive tools. All the loose material should be cleaned and the surface should preferably be washed off before actual patching work is started. Care should be taken to remove excess water from the cavity. To obtain a good bond it is generally recommended that the surface of concrete be coated with a thin layer of cement grout before placing the patching material. The grout mix and the quantity of mixing water should be same as that of mortar in the replacement material. The patching material should be filled in before the layer of grout dries.

In case of corroded reinforcement the concrete should be removed far enough to ensure that all corroded areas are exposed so that they can be cleaned. Carbonated or chloride contaminated concrete in contact with reinforcement should also be removed and replaced with fresh concrete or an impermeable resin compound. Application of corrosion inhibiting chemicals such as phosphates to exposed reinforcement after cleaning may be recommended. A slurry coating of polymer latex and cement can be used in case of cement-based repair. Resin-based coatings are suitable for the use with both cement-based and resin-based repair materials.

(ii) *Selection of materials* The repair system should be so selected that the mechanical properties of the repair material are similar to those of the structure being repaired. Cement- based repairs can provide fire resistance while resins soften at relatively low temperature. However, the properties of resins may be adjusted within fairly wide margins by suitable formulation so that they can, to some extent, be tailored to fit the job in hand. But the rate of reaction is both exothermic and temperature dependent, the thermal stresses that develop as the materials cool should be kept in mind while planning the repairwork. For conventional repairs, the cement-based materials to be used for patchwork may either be mortar or concrete depending upon the extent of repair.

(iii) *Application of material* The methods generally used for filling the material are (a) drypacking, (b) concrete replacement, (c) mortar replacement, (d) grouting, (e) large volume prepacking of concrete, and (f) shotcreting or guniting.

After the concrete surface has been prepared, a bonding coat should be applied to all the cleaned exposed surfaces. It should be done with minimum delay. The bonding coat may consist of cement slurry or an equal amount of cement and fine sand mixed with water to a fluid paste consistency. Adequate preparation of surface and good workmanship are the ingredients of efficient and economical repairs.

(a) *Drypacking* The method consists of hand-placing of low-water content mortar on the prepared surface followed by tamping or ramming of mortar into place, producing an intimate contact between the mortar and the existing concrete. Because of the low water-cement ratio of the paste, there is little shrinkage, and the mortar provides a durable, strong and watertight patch.

The repair material usually consists of cement and sand mortar in proportions of about 1 : 2.5 or 1 : 3 using medium (passing 1.18 mm IS sieve) or coarse concreating sand. However, for a smooth surface finish, a finer sand for final layer may be used. The first layer of repair material should be applied immediately after the application of bonding coat while latter is still wet. Provision of some mechanical anchorage for the patch by means of dowels drilled and grouted into the surrounding concrete is a wise precaution. The water content for the mix should be carefully chosen because excess water will increase shrinkage which may loosen the replaced material, whereas less water will not make a sound solid pack. The water content should be such as to produce a mortar which will stick together on being moulded into a ball, at the same time not exclude water but leave the hand damp.

To minimize the shrinkage in place, the mortar should stand for 30 minutes after mixing and then remixed prior to use. The repair material should be filled properly in compacted layers of about 10 mm thick and each layer should normally be applied as soon as the preceding one is strong enough to support it. The preceding layer should be scratched before placing the succeeding layer to secure a good bond or key. Each layer is compacted over its entire surface by using a hardwood stick of about 200 to 300 mm length and up to 25 mm diameter with a hammer. The last overflowing layer is struck flush with the surface. There need be no time delays between layers. In case of delay between layers a fresh bonding coat should be applied when work is resumed. The mortar may be finished by laying the flat side of hardwood piece against it and striking it several times with a hammer. Curing is done by covering the patch with absorbent material that is kept damp, preferably covered in turn by polyethene sheets sealed at edges. Shading from sun may be necessary. Sprayed-on curing membranes may be used after complete patch has been applied.

For dormant (non-active) crack repair the portion of crack adjacent to the surface should be widened (routed or chased) to a slot about 25 mm wide and 25 mm deep with a power driven sawtooth bit. The slot should be undercut so that base width is slightly greater than the surface width. After the slot is thoroughly cleaned and dried, the bond coat followed by repair material should be applied. Moist curing can be done by supporting a strip of folded wet burlap along the length of the crack.

(b) *Concrete replacement* In general, this method is used for large and deep patches like those encountered in the repair of old and deteriorated portions of concrete structures where concrete is to be placed to a minimum depth of about 150 mm. The general applications of this method are in the repair of walls, piers, parapets and kerbs, and for refacing walls and relining channels. The method is particularly suitable where the holes extend throughout the concrete section or where the surface area of hole is at least 0.09 m^2 with a depth of 100 mm for plain concrete, or 0.045 m^2 with a depth a little more than reinforcing steel in case of reinforced concrete.

As in case of other types of repairs defective concrete is removed so that the sound surfaces are exposed and reinforcement cleaned. In case the repair material cannot be placed immediately, it may be necessary to apply a protective (corrosion-inhibiting paint) coating to the reinforcement. To ensure good bond to reinforcement, a further coating with a long *open-time* may sometimes be applied.

In plain concrete, the defective area is prepared as explained earlier, but if the repairs are to be made in the reinforced concrete, the reinforcing bars should not be left partially embedded, but at least 25 mm clearance should be provided around each exposed bar. In the case of wall repairs, the top of the hole should be cut to a fairly horizontal line with a 1 to 3 upward slope from back towards face to prevent the formation of air-pockets at the top during vibration. The bottom and the sides of the hole should be cut sharp and approximately square as shown in Fig. 16.5. All interior corners should be rounded to a minimum radius of 25 mm. For repairing the wall for a height more than 500 mm, the back form is built in one piece, the front formwork is constructed in horizontal sections to place concrete conveniently

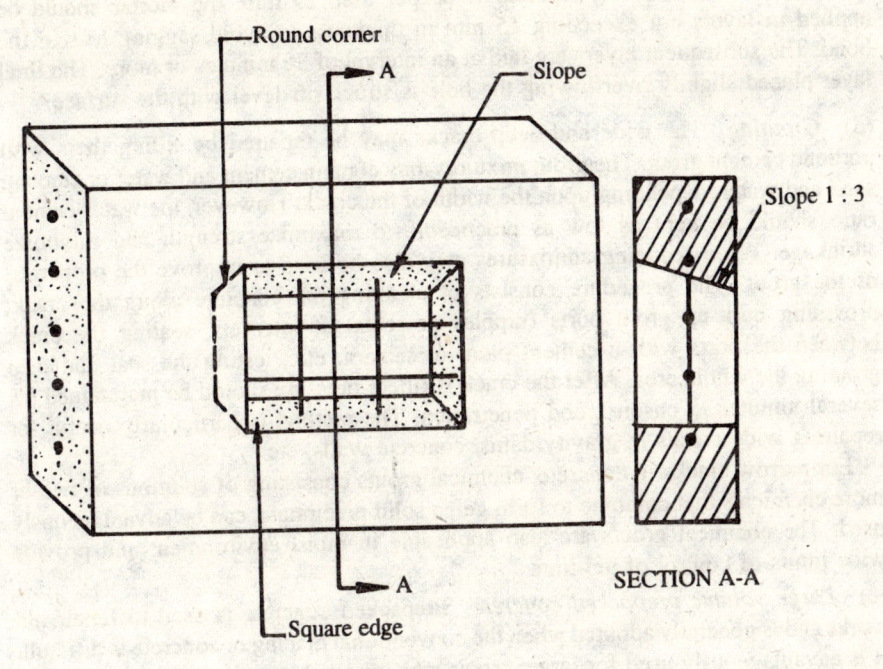

Fig. 16.5 Repair by concrete replacement

in about 300 mm deep lifts. All formwork joints should be mortar tight. Before placing the front sections of formwork for each lift the surface of the old concrete should be coated with cement grout having the same water-cement ratio as that for the mortar in the replacement concrete. To reduce shrinkage, a minimum period of 30 minutes is allowed to elapse between the lifts. The mix proportion and water-cement ratio of the replacement concrete should be the same as that used in the structure. The water content must be as low as possible. Compaction is best achieved by internal vibration if there is access for vibration. If an external vibrator clamped on to the formwork is used, care must be taken not to damage the seals between formwork and existing concrete. The *stripping time* may vary from 20 to 48 hours depending upon the location and extent of repair.

c) *Mortar replacement* The method is suitable for the cavities which are too wide for drypack or too shallow for concrete replacement. Generally it is used for shallow depressions no deeper than that for the side of the reinforcing bars nearest to the surface. For replacement of deteriorated concrete, this method is suitable for minor restorations. The mortar replacement can be done by hand or can be applied pneumatically by using a small pressure gun.

It is preferable to *pre-shrink* the repair mortar by mixing it to a plastic consistency as long in advance of its use as cement permits (the pre-shrinking time ranges from 60 to 120 minutes). For hand placing, the mortar should have the same proportions as the mortar used in the mix of which the structure is made. In the case of a pressure gun, the ratio recommended is 1 part of cement to 4 parts of sand.

In case the hole being repaired is deeper than 25 mm, the mortar should be applied in layers not exceeding 15 mm in thickness to avoid sagging loss in the bond. The subsequent layers are laid at an interval of 30 minutes or more. The final layer placed slightly overflowing the hole is struck off level with the surface.

(d) *Grouting* The wide and deep cracks may be repaired by filling them with portland cement grout. The grout mixtures may contain cement and water or cement, sand and water, depending upon the width of the crack. However, the water-cement ratio should be kept as low as practicable to maximize strength and minimize shrinkage. Water-reducing admixtures may also be used to improve the properties of the grout. The procedure consists of cleaning the concrete along the crack; providing built-up grout ports (nipples or seats) at intervals, sealing the crack between the ports with a cement paint or sealant, etc., testing the seal and then grouting the whole area. After the crack is filled, pressure should be maintained for several minutes to ensure good penetration. The method is particularly useful for repairing wide cracks in gravity dams, concrete walls, etc.

For narrow cracks in concrete, chemical grouts consisting of solutions of two or more chemicals that combine to form gel, a solid precipitate, can be advantageously used. The chemical grouts are also applicable in moist environment and provide wide limits of control of gel-time.

(e) *Large volume prepacked concrete* Prepacked concrete is used to repair old works and is generally adopted when the conventional placing of concrete is difficult. It is advantageously used for larger repair jobs, under-water placement, resurfacing of dams, repair of tunnel linings, piers, retaining wall and spillways. It consists in injecting grout into the voids of compacted mass of clean and well-graded coarse aggregate in the forms. The aggregate is wetted after compacting it and then grout is pumped into the forms. The pumped grout usually contains fine sand, portland cement, a pozzolanic material of low mixing water requirement, a fluidifier designed to increase the fluidity and to inhibit early stiffening of the grout, and mixing water. The forms must constitute a closed system, vented at the top only to avoid trapping of air pockets. The formwork may be of conventional rigid type that either encloses the member to be repaired or is sealed to it at its edges. In grouting aggregate work transparent panels are sometimes provided so that the progress of grouting can be monitored. The aggregate is then placed in the form, and grout lines are attached to inlet or injection plugs fixed to the forms. Pumping of grout should begin at the lowest point as shown in Fig. 16.6 and proceed upwards in order to prevent the formation of air pockets. More than one injection point built into the formwork at different levels may be required if complete filling from bottom requires very high injection pressure. The grout lines should be spaced no more than 1.5 m centre to centre, and the grout level in the mass of aggregate should be brought up uniformly, as determined by the observations of grout levels in the grout pipes. A positive head of at least 1 to 2 m should be maintained in the grout pipes above the level of the outlets. The injection of grout should be a smooth, uninterrupted operation, and a positive head should be maintained in the grout lines after the forms have been filled and until the grout has set.

(f) *Shotcreting or guniting* As explained in Section 14.8 shotcrete or gunite is mortar or concrete conveyed through pressure hose and applied pneumatically at

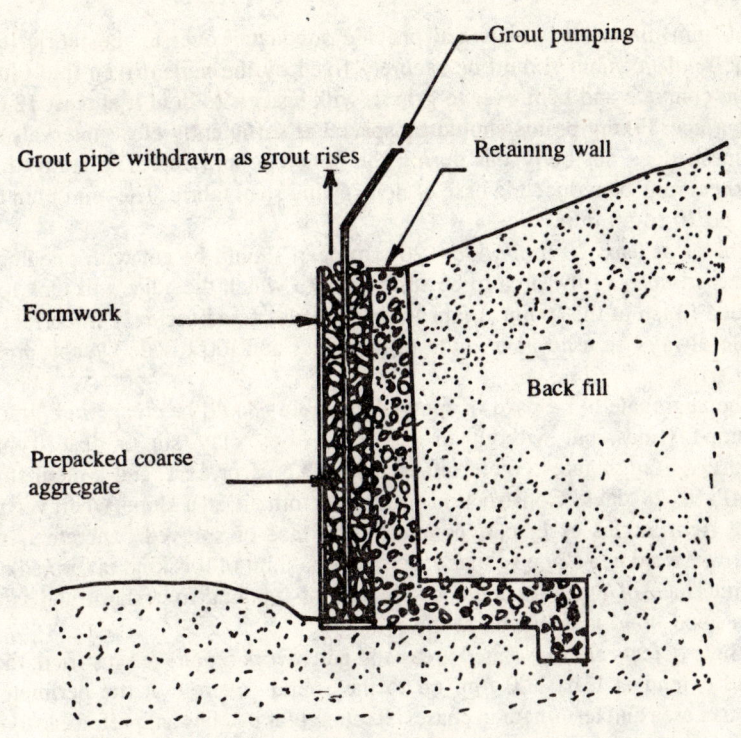

Fig. 16.6 Prepacked concrete repair of a retaining wall

high velocity onto a surface. This material has found wide applications in several major repair works because of ease with which it can be applied on vertical, horizontal or overhead surfaces.

The objective of this type of repair may be to replace concrete that has been lost or removed, and to increase effective cover to the steel reinforcement or to protect against future damage by adding additional concrete. In the preparation of surface to be repaired, all the affected concrete must be cut back to unaffected material.

When a large portion of the structure is defective, it is recommended that repair should encase the whole of the assessible, external surfaces with a specified minimum thickness of sprayed concrete, which shall be brought out to uniform profile, and eyeable lines by additional infilling of the areas from where the damaged concrete has been removed. This can be achieved by applying sprayed concrete over areas of undamaged concrete. The undamaged surface must also be prepared by thoroughly roughening it to remove all original cement laitance, surface deposits and impurities. For the development of good bond, the prepared surface must be sound, rough and homogeneous (i.e. free from shattered or laminated concrete).

Generally, a welded steel farbric is used, or steel fibres are incorporated in case of sprayed concrete without polymer admixtures to minimize the risk of cracks of sufficient size developing (which allows penetration of air and water) by encouraging the development of a large number of very fine, insignificant cracks. A typical fabric for this purpose would be of 3 to 4 mm diameter high tensile steel bars in the form of 75 to 100 mm mesh. With such a fabric the sprayed concrete thickness should be 50 mm minimum which will provide adequate cover to the fabric itself. The fabric reinforcement should be securely fixed by the nails driven into plugs set in parent concrete and bent over to grip it, with spacers to hold it at least 12 mm from the surface. Fixing points should be spaced at sufficiently close intervals such that the mesh does not belly out during the concrete spraying. The sprayed concrete thickness may be reduced in case of non-rusting steel fabric. The minimum thickness may be 30 to 50 mm.

Chases at least 20 mm wide × 20 mm deep should be cut with pneumatic tools at the perimeter of the area to be sprayed, into which the edge will tuck to provide a sound finish at that point. Light timber profiles fixed securely in correct position should always be used as a guide to thickness and to provide eyeable lines on all main arrises.

The aggregate to be used in sprayed concrete should be clean, well graded from 10 mm to fines, but without an excess of fines, clay, silt or dust. Typical mix proportions (aggregate-cement ratios) are be 3.5 : 1 or 4 : 1, and typical strength of 30 MPa at 28 days. Construction joints are formed at a slope wherever possible, not at right angles as in cast concrete. The face of sprayed concrete is normally carried forward at a slope so that a construction joint of this kind is formed naturally. The interface of the joint must be cleaned of rebound, overshoot and laitance so that a good bond is achieved at the joint.

Where defects are in isolated areas and patchwork repairs are justified, the repairs should extend at least 300 mm on to the sound concrete at the perimeter of the defective area and terminate in chases. Rectangular patches are preferable to irregular shapes.

Prior to commencing spraying, the interface must be prepared and wetted so that it does not absorb water from the sprayed concrete, but at the same time is not so wet that there is excess free water on it.

The adequacy of plant and of the properties of the material to be sprayed are confirmed by a *pre-construction test panel*. It consists in placing a test box 600 or 750 mm square and 100 mm thick on the surface simulating the construction conditions of proposed work and then spraying the concrete. Concrete cores of suitable age are taken from the test panels for testing for the specified standards of crushing strength, porosity, water penetration resistance, etc. However, previous experience with similar plant and materials is often accepted in lieu of such pre-construction tests except on very big jobs. Test panels also help to judge the competence of nozzlemen.

Fabric mesh or bar reinforcement should be checked for secure fixing, correct positioning and overlaps, spacers and cleanliness. Profile guides, whether timber formers or stretched wires must be firm, tight and correctly positioned.

The spraying should be carried out with the nozzle held approximately at right angles to the interface, and at such a distance that the concrete compacts effectively. The nozzle should be fanned from side to side, and up and down so that layer is evenly built up over the area in front of the nozzleman. If there is any unevenness in the spray of material, the nozzleman must turn the nozzle away from the work until the spray becomes even again. Where the full thickness has to be built up in more than one layer, it is not advisable to apply more than 50 mm or so at a time on vertical surfaces, or 25 mm or so overhead. The surface of each layer may be lightly trimmed with the edge of a steel float, and must be wetted again once set before applying the next layer.

Curing of the sprayed concrete is even more important than that of conventional cast concrete because thinner section may make water loss easier and more serious. The curing method may include a fine water spray, wetted hessian and curing compounds.

(iv) *Curing of repair work* The curing of patch material requires much more care than that required for a complete structure. There is a tendency of old concrete absorbing moisture from the replacement material. The curing methods have been discussed earlier in detail but the methods commonly used in conventional repairwork are summarized below:

(a) Horizontal repaired surface can be cured by ponding or by placing wet gunny bags.

(b) The vertical or inclined repaired surface may be cured using damp hessian or wet burlap pads.

(c) Where the above two methods are not applicable, membrane curing can be used. Initial curing with water followed by membrane curing is very effective.

(d) Deliquescent salts which hasten curing by keeping the patch moist may also be used.

2. Special Procedures

The polymers have been recently introduced in concrete technology for multipurpose applications in the repair and maintenance of concrete buildings and other structures. The polymers used in concrete repair principally consist of two different types of materials:

(i) Polymers used as modifiers for cementitious systems

(ii) Reactive-thermosetting resins

For the repair of both active as well as dormant cracks, epoxy mortars consisting of an epoxy, hardener and sand, have been used effectively to seal the cracks. The epoxy resins and polymers possess excellent adhesive and sealing properties, although the cost of this repair is quite high. Some materials and techniques for repair of structures are described in the following pages. The physical properties of typical systems used in concrete repair are given in the Table 16.2.

TABLE 16.2 Physical properties of typical products used in concrete repairs

Property	Cementitious grouts, mortars and concretes	Polymer modified cementitious systems	Polyester resin grouts, mortars and concretes	Epoxy resin grouts, mortars and concretes
Compressive strength, MPa	20–70	10–80	55–110	55–110
Compressive modulus E-value, GPa	20–30	1–30	2–10	0.5–20
Flexural strength, MPa	2–5	6–15	25–30	25–50
Tensile strength, MPa	1.5–3.5	2–8	8–17	9–20
Elongation at breaking point, per cent	0	0–5	0–2	0–15
Linear coefficient of thermal expansion per °C	$(7-12) \times 10^{-6}$	$(8-20) \times 10^{-6}$	$(25-35) \times 10^{-6}$	$(25-30) \times 10^{-6}$

Property	Cementitious grouts, mortars and concretes	Polymer modified cementitious systems	Polyester resin grouts, mortars and concretes	Epoxy resin grouts, mortars and concretes
Water absorption, 7 days at 25°C, per cent	5–12	0.1–0.5	0.2–0.5	0–1
Maximum service temperature under load, °C	In excess 300 dependent upon mix design	100–300	50–80	40–80
Rate of development of strength at 20°C	1–4 weeks	1–7 days	2–6 hours	6–48 hours

(i) *Polymer modified cementitious system* The polymers used as admixtures for cementitious systems are normally available as latex (milky white dispersion in water) and are used to gauge the cementitious mortar as a whole or as a partial replacement of mixing water. The polymer latex forms a network of polymer strands interpenetrating the cement matrix and improves the structural properties and reduces permeability of mortar. Such mortars provide the same alkaline passivation protection to steel as do the conventional materials, and can be readily placed in a single application of 12 to 16 mm thickness which gives adequate protective cover. The functions of polymer latex are:

   (a) it acts as a water-reducing plasticizer,
   (b) provides a good bond between repair mortar and concrete surface being repaired,
   (c) improves tensile and flexural strength of the mortar, and
   (d) reduces permeability of repair mortar to water.

There are different types of polymer latexes which have been used as modifiers for cementitious systems. They include polyvinyl acetate (PVAc), styrene butadiene rubber (SBR), polyvinyldiene dichloride (PVDC), acrylics and modified acrylics (generally styrene acrylics). PVDC latexes are not recommended for repair mortars for reinforcement concrete. PVAc latexes are widely used as general-purpose bonding aids and admixtures for building industry for interior applications. SBR, acrylic and modified acrylic latexes are most commonly used as admixtures in concrete repair mortars.

(ii) *Resin-based materials* Resin mortars are extensively used in the locations where the cover is less than 12 mm and areas to be repaired are relatively small. In case of resin mortars the protection of reinforcement depends upon total permeability of envelope. This requires that the reinforcement surface be prepared to a very high standard. Resin repair mortars use reactive-resins filled with carefully graded aggregates.

Epoxy resin mortars are most commonly used in concrete repairs. Polyester and acrylic resin-based mortars are generally used for small area repairs where their very rapid development of strength is required. Since in most repair situations, the polymer-based repair material is bonded directly to concrete or other cementitious material, it is, therefore, important to match the mechanical and physical properties of polymer repair composition and concrete. Polymer bonding aids can assist in achieving a reliable bond between green uncured concrete and cured concrete.

16.5.3 Polymer-based Repairs

As explained earlier in Chapter 14 the polymer concrete includes composites prepared by one of the following methods.

(i) Polymer Inpregnated Concrete (PIC)

This is a portland cement concrete impregnated by a monomer system which is subsequently polymerized by radiation or by heat, and the use of a catalyst.

(ii) Polymer Cement Concrete (PCC)

This is a concrete in which the monomer is added during the mixing of portland cement, water and aggregate, followed by polymerization or curing of the replaced material after its placement.

(iii) Polymer Concrete (PC)

This is a composite material obtained by adding a polymer or its precursor to the aggregate and polymerizing or curing the material after its placement.

The concrete polymer materials provide high strength and improved durability under aggressive conditions as compared to conventional concretes. PIC has proved to be the most successful concrete polymer material for construction. Liquid and gaseous monomers can fully penetrate concrete by external pressure and can be polymerized. To obtain maximum polymer loading in the concretes and hence maximum improvement in the desired properties, it is necessary to dry the concrete to constant mass, remove air, soak in a low viscosity liquid monomer, pressurize with nitrogen and wrap the specimen in polyethylene sheet to reduce evaporation prior to polymerization. The polymer systems commonly used are:

- (a) Methyl-methacrylate (MMA)
- (b) MMA + 10 per cent trimethol propane trimethacrylate (TMPTA)
- (c) Styrene and polyester styrene
- (d) Methanol
- (e) Vinyl monomer

The sulphur impregnated concrete using commercially available (99.9 per cent) pure sulphur may provide a practical and inexpensive substitute of PIC. The polymer cement concretes may be obtained by substituting at least 30 per cent of epoxy resin for cement in ordinary portland cement concrete mixes. The addition of flyash to the epoxy cement mixes is favourable for strength.

The polymer-based crack repair can be affected as described in the following sections.

(a) Polymer Impregnation

The technique consists of flooding the cleaned dried cracked concrete surface with a monomer which is then *polymerized* in place, thus filling and structurally repairing the crack. A monomer system is a liquid that consists of small organic molecules capable of combining to form a solid plastic. The monomer systems used for *impregnation* contain a *catalyst* or an *initiator* and the basic monomer. They also contain a *cross-linking agent*. When heated the monomers join together or polymerize becoming a tough, strong and durable plastic that greatly enhances a number of concrete properties.

Monomers have varying degrees of volatility, toxicity and flammability and do not mix with water. These are low viscosity fluids which may soak into dry concrete, filling the cracks in much the same manner as water does. However, if the cracks contain moisture, the monomer will not soak into the concrete at each crack face, and consequently the repair will be unsatisfactory. If a volatile monomer evaporates before polymerization, the repair will be ineffective.

Polymer impregnation can be used for repairing the fractured elements, by drying the fracture, temporarily encasing it in a watertight sheet metal, soaking the fracture with monomer and polymerizing the monomer. Large voids in compression zones may be filled with fine and coarse aggregates before flooding them with the monomer, thus providing a *polymer concrete repair*.

For treating the concrete surfaces that contain a large number of cracks, vacuum impregnation may be used. The process essentially consists of enclosing the part of structure to be repaired within an air-tight plastic cover and applying vacuum to exhaust air from all cracks within the cover. Resin grout or monomer is then admitted which is forced into cracks and pores in the concrete surface by atmospheric pressure. On completion of impregnation, the cover is removed before the impregnant hardens. The selection of appropriate impregnants and degree of vacuum may be based on experience. The process is extensively used as a means of reducing permeability of weak concrete or masonary.

It can also be used to improve the abrasive resistance of industrial concrete floor slabs. Polymer impregnation has not been used successfully to repair fine cracks.

(b) Drilling and Plugging

The method is only applicable for the cracks running in reasonably straight lines and accessible at one end. It consists in drilling a hole down the length of crack and grouting it to form a key. A hole 50 to 60 mm in diameter should be drilled, centred on and following the crack as shown in Fig. 16.7. The hole must be large enough to intersect the crack along its full length and provide sufficient repair material to structurally take the loads exerted on the key. The drilled hole is then cleaned and filled with grout. For problems concerning watertightness, the drilled hole should be filled with a resilient material of low modulus in lieu of grout. In case where both load transfer and watertightness are desired, two holes are drilled, one filled with resilient material and the second being grouted.

16.5.4 Resin-based Repairs

Cracks in reinforced concrete wider than approximately 0.3 mm may require sealing to prevent the entry of moisture, oxygen and other materials, or for other reasons. The choice of the method and materials will depend upon the cause of cracking and, whether a permanent structural filling of crack is needed to carry out any other required strengthening. For restoring the structure to its original strength, the low-viscosity epoxy resin may be injected. Using pressure injection techniques it is possible to completely fill cracks finer than 5 mm with epoxy resin system. However, the work should be carried out skilfully to avoid blowing of surface seals

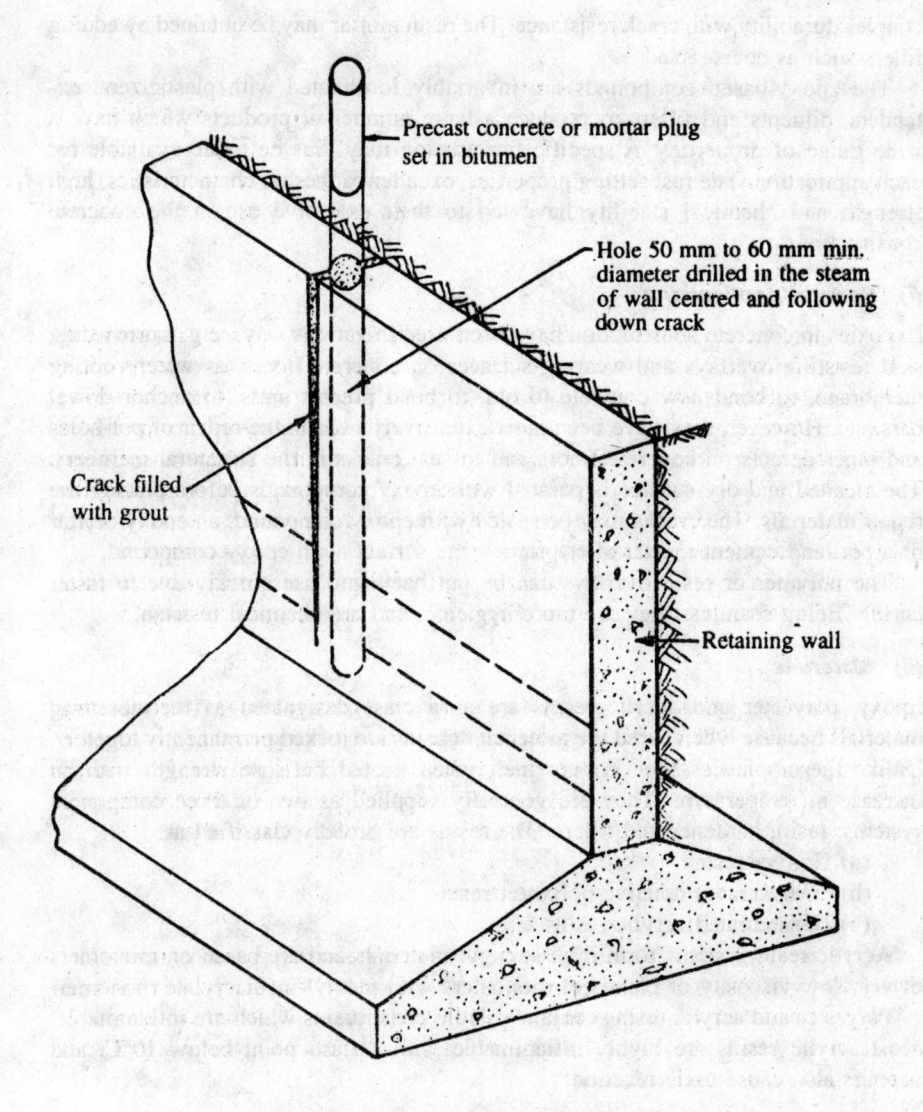

Fig. 16.7 Repair by drilling and plugging

due to back pressure that may develop in case of very fine cracks. Sustained pressure for several minutes may be required to completely fill a fine crack.

The resin and the hardener are generally in a liquid form and each by itself is stable for an indefinite period. When these are mixed a chemical reaction takes place which converts the system from liquid to a tough plastic solid at ambient temperatures. They develop excellent strength and adhesive properties, and are resistant to many chemicals. They have good chemical and physical stability; they harden rapidly and resist water penetration. In all, they provide a toughness that couples durability with crack resistance. The resin mortar may be obtained by adding fillers such as coarse sand.

The epoxy-based compounds are invariably formulated with plasticizers, extenders, diluents and fillers to produce a large number of products which have a wide range of properties. A specific formulation may thus be made available for each application. The fast setting properties, excellent adhesion characteristics, high strength and chemical stability have led to their extensive use in the concrete construction.

(i) General Applications

Expoxies in concrete construction have been used in various ways, e.g. in providing skid resisting overlays and wearing surfaces on concrete floors, as waterproofing membrane, to bond new concrete to old, to bond precast units, to anchor dowel bars, etc. However, they have been most extensively used in the repair of pot-holes and other defects on concrete floors and to seal cracks in the structural members. The cleaned and dry surface is painted with epoxy compounds before placing the repair materials. The cracks may be sealed with epoxy compound, an epoxy mortar or a portland cement mortar after priming the surface with epoxy compound.

The polymer or resin overlays can be put back into use quickly due to faster curing. Being seamless they are more hygienic, and are chemical resistant.

(ii) Materials

Epoxy, polyester and acrylic resins are as a class designated as thermosetting materials because when cured the molecular chains are locked permanently together. Unlike thermoplastics they do not melt when heated but lose strength with an increase in temperature. They are generally supplied as two or three component systems: resin, hardener and fillers. The resins are broadly classified as:
   (a) Epoxy resins
   (b) Unsaturated reactive polyester resins
   (c) Unsaturated acrylic resins

Acrylic resin systems form high strength materials and are based on monomers of very low viscosity or blends of monomers with methyl-methacrylate monomer.

Polyester and acrylic resins contain volatile constituents which are inflammable. Most acrylic resins are highly inflammable with a flash point below 10°C, and vapours also cause toxic reaction.

The properties of commonly used resins are:

Epoxy resins   These have high strength, good bonding characteristics, high impact resistance, high chemical resistance and may be made to provide a non-slip finish.

Polymer resins   These differ from epoxy resins in that these can be laid over wider temperature range and have a better resistance to heat. They are mixed with cement and fine hard aggregate, and laid in thickness up to 15 mm.

Polyvinyl acetate (PVAc)   It is used as a bonding aid when thin mortar overlays are applied to existing concrete. The liquid can be applied straightaway onto a clean, sound surface and allowed to dry. The slight re-emulsification of the film on being re-wetted by application of fresh mortar topping provides a good bond.

Natural rubber latex It is an admixture with excellent adhesive properties and is difficult to mix with ordinary portland cement. It is often used with less alkaline high alumina cement for patching or for underlayments on floors which are to receive vinyl tiles.

Styrene butadiene rubber (SBR) It is an effective alternative to PVAc with high water resistance. Unlike PVAc, the dried film does not develop *grab* on re-wetting so it will act as a *debonding layer* if allowed to dry out. Therefore, mortar mix should be applied while the tack coat of SBR is still wet.

Acrylic resins These admixtures have excellent water resistance and improve bond strength when mixed with mortars. Seamless, non-dusting thin floor overlays can readily be produced with acrylic resins.

Styrene-acrylic resins. A mixture of tough styrene with acrylic resin using 1 : 3 cement-sand mortar can be used to produce hard-wearing floor overlays at a reasonable cost.

(iii) Repair Procedure

(a) *Resin-mortar* The surface preparation requirements are similar to those for cement-based repairs. The constituents of resin-based material must be mixed together thoroughly by use of mechanical mixers or stirrers. Most of the failures of resin-based repairs have been traced to improper proportioning or inadequate mixing. For smaller repair jobs, to obtain correct proportions the constituents are normally available in pre-batched packs. After the preparation of surface a primer or tack coat of unfilled resin is applied to the freshly-exposed surfaces of concrete and reinforcement. In general one coat will be enough, but two may be needed if the substrata is porous. If two coats are used, the second must be applied while the first is still tacky.

The patching material must be applied while the primer is still tacky, and each successive layer of patching material must be applied before the previous one has cured too much. The resin-based materials cure by chemical reaction which starts as soon as the constituents are mixed, so they have a limited *pot-life*. The quantity of materials to be mixed in any one batch is precalculated such that it can be used before it becomes too stiff. The resin-based patches should be well compacted and impermeable.

Normal safety precautions should be observed while using the resins and hardeners, i.e., gloves should be worn, splashes should be washed off the skin but solvents should not be used for this purpose; adequate ventilation should be provided; and smoking, eating or drinking should be prohibited.

(b) *Resin-injection* The injection of polymer under pressure will ensure that the sealant penetrates to the full depth of the crack. The technique in general consists of drilling holes at close intervals along the length of cracks and injecting the epoxy under pressure in each hole in turn until it starts to flow out of the next one. The hole in use is then sealed off and injection is started at the next hole and so on until full length of the crack has been treated. Before injecting the sealant it is necessary to seal the crack at surface between the holes with rapid curing resin. For repair of cracks in massive structures, a series of holes (usually 20 mm in diameter and 20 mm deep spaced at 150 to 300 mm interval) intercepting the crack at a number of location are drilled. Epoxy injection can be used to bond the cracks as narrow as 0.05 mm. It has been successfully used in the repair of cracks in buildings, bridges, dams and other similar structures. However, unless the cause of cracking is removed, cracks will probably recur possibly somewhere else in the structure. Moreover in general this technique is not very effective if the cracks are actively leaking and cannot be dried out.

Epoxy injection is a highly specialized job requiring a high degree of skill for satisfactory execution. The general steps involved are as follows.

(i) *Preparation of the surface*   The contaminated cracks are cleaned by removing all oil, grease, dirt and fine particles of concrete which prevent the epoxy penetration and bonding. The contaminants should preferably be removed by flushing the surface with water or a solvent. The solvent is then blown out using compressed air, or by air drying.

The surface cracks should be sealed to keep the epoxy from leaking out before it has cured or gelled. A surface can be sealed by brushing an epoxy along the surface of cracks and allowing it to harden. If extremely high injection pressures are needed, the crack should be routed to a depth of about 12 mm and width of about 20 mm in a V-shape, filled with an epoxy, and struck off flush with the surface.

(ii) *Installation of entry ports*   The entry port or nipple is an opening to allow the injection of adhesive directly into the crack without leaking. The spacing of injection ports depends upon a number of factors such as depth of crack, width of crack and its variation with depth, viscosity of epoxy, injection pressure, etc., and choice must be based on experience. In case of V-grooving of the cracks, a hole of 20 mm diameter and 12 to 25 mm below the apex of V-grooved section, is drilled into the crack. A tire-valve stern is bonded with an epoxy adhesive in the hole. In case the cracks are not V-grooved, the entry port is provided by bonding a fitting, having a hat-like cross-section with an opening at the top for adhesive to enter, flush with the concrete face over the crack.

(iii) *Mixing of the epoxy*   The mixing can be done either by batch or continuous methods. In batch mixing, the adhesive components are premixed in specified proportions with a mechanical stirrer, in amounts that can be used prior to the commencement of curing of the material. With the curing of material pressure injection becomes more and more difficult. In the continuous mixing system, the two liquid adhesive components pass through metering and driving pumps prior to passing through an automatic mixing head. The continuous mixing system allows the use of fast-setting adhesives that have short working life.

(iv) *Injection of epoxy*   In its simplest form, the injection equipment consists of a small reservoir or funnel attached to a length of flexible tubing, so as to provide a gravity head. For small quantities of repair material small hand-held guns are usually the most economical. They can maintain a steady pressure which reduces chances of damage to the surface seal. For big jobs power-driven pumps are often used for injection. The pressure used for injection must be carefully selected, as the use of excessive pressure can propagate the existing cracks, causing additional damage. The injection pressures are governed by the width and depth of cracks, and the viscosity of resin and seldom exceed 0.10 MPa. It is preferable to inject fine cracks under low pressure in order to allow the material to be drawn into the concrete by capillary action and it is a common practice to increase the injection pressure during the course of work to overcome the increase in resistance against flow as crack is filled with material. For relatively wide cracks gravity head of a few hundred millimeters may be enough.

In the case of vertical and inclined surfaces, the injection process should begin by injecting epoxy into the entry port at the lowest level until the epoxy level reaches the entry port above. The injection tube is then removed and the lower entry port is capped. Using an inert gas, a pressure up to 0.7 MPa is applied for a period of 1 to 10 minutes on the port from which the injection tube has just been removed. This forces the epoxy into hairline cracks. The process is repeated at successively higher ports until the cracks have been completely filled and all ports capped. On the other hand, for horizontal cracks injection should proceed from one end of the crack to the other in the same manner. If the pressure can be maintained, it indicates that the crack is full. If the pressure cannot be maintained, it indicates that the epoxy is still flowing into unfilled portions or leaking out of the crack.

(v) *Removal of surface seal* After the injected epoxy has cured, the surface seal may be removed by grinding or other means as appropriate. Fittings and holes at the entry ports should be painted with an epoxy patching compound.

(vi) *Health and safety precautions* Epoxy materials are toxic and skin-irritant. Contact with skin, inhalation of vapours and ingestion must always be avoided. The following precautions may be helpful.
1. Full-face shield and goggles should be used during all the mixing and blending operations.
2. Protective overalls and polyethylene or rubber gloves should be used.
3. Protective cream for the skin can be used.
4. Adequate fire protection should be provided.

16.6 COMMON TYPES OF REPAIRS

16.6.1 Sealing of Cracks

The crack or joint sealers are very important in concrete structures as every concrete structure has cracks or joints. The crack sealers should ensure the structural integrity and serviceability. In addition they provide protection from the ingress of harmful liquids and gases.

The method consists of enlarging the crack along its length on the exposed surface (called chasing or routing) and sealing it with a suitable joint sealant as illustrated in Fig. 16.8. Omission of routing may effect the permanancy of repair. The routing operation consists of cutting a groove at the surface that is sufficiently large to receive the sealant, using a concrete saw or hand tools. A minimum surface width of routing of 6 mm is desirable, as repairing the narrower grooves is difficult. The surfaces of the routed joints should be cleaned with an air jet and allowed to dry before placing the sealant.

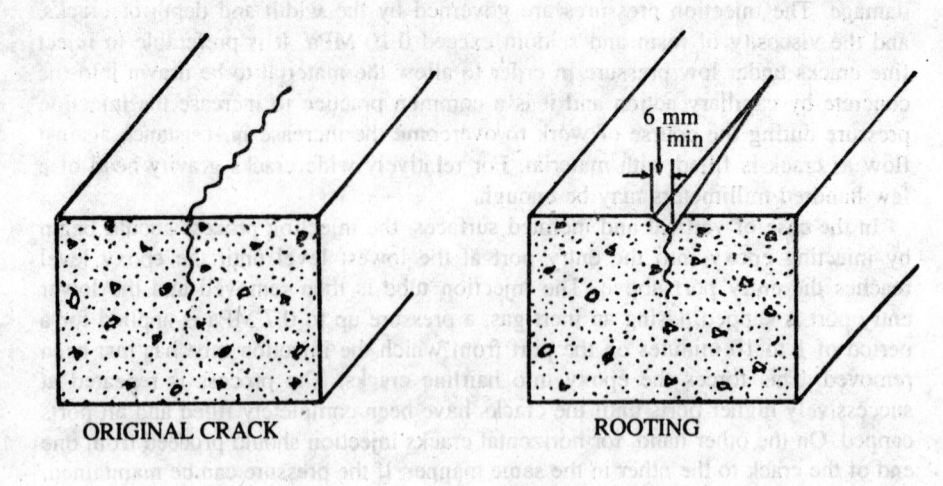

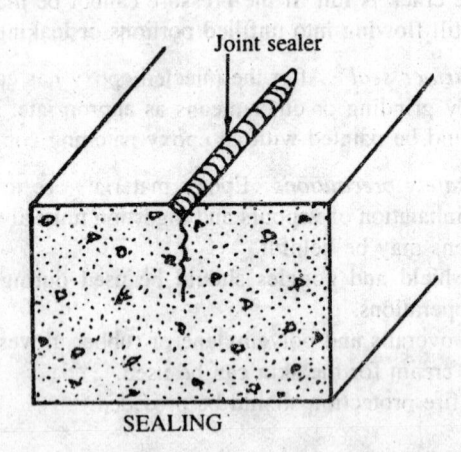

Fig. 16.8 Crack repair by routing and sealing

The function of the sealant is to prevent water from reaching the reinforcement, hydrostatic pressure from developing within the joint, staining the concrete surface, or causing moisture problems on the far side of the member.

The epoxy compounds are often used as sealant material. Hot-poured joint sealants are used when thorough watertightness of joints is not required and the appearance is not important. Urethanes, which remain flexible through large temperature variations, have been used successfully in sealing the cracks up to 20 mm in width and of considerable depth.

16.6.2 Flexible Sealing

For repairing an active crack, it is necessary to provide for its continuing movement. One way to achieve this is to rout or chase the crack along its length. The prepared routed crack is filled with a suitable field moulded flexible sealant with strain capacity being at least as large as the one to be accomodated. A wide crack spreads movement over a greater width so that the resulting strain is compatible with sealant to be used. The sealant must adhere to the sides of the chase but debonded from bottom so that the movement in the crack spreads over the full width of the chase. This can be achieved by providing a *bond breaker* or *debonding strip* of a material such as polyethylene or *pressure sensitive tape* at the bottom of the chase before sealant is applied. This debonding strip does not bond to the sealant before or during cure and allows the sealant to change shape without stress concentration at the bottom. The dimensions of the seal are an integral part of its performance. Fig. 16.9

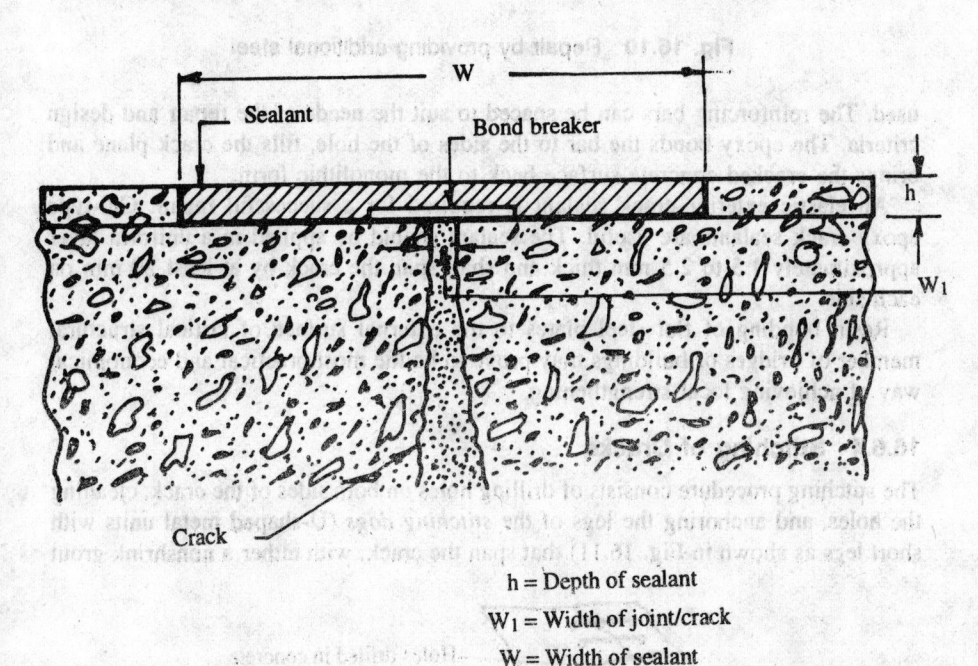

h = Depth of sealant
W_1 = Width of joint/crack
W = Width of sealant

Fig. 16.9 Repair of an active crack by flexible sealing

shows a sectional view of a typical flexible sealing or movement joint. With an increase in chase width, the crack movement which induces shear or tension in sealant will exert considerably reduced stress on the adhesive interface with concrete, and thus enabling the face seal to cope with extensive movement.

16.6.3 Providing Additional Steel

The cracked reinforced elements (usually bridge decks) can be successfully repaired using epoxy injection and reinforcing bars. The technique consists of sealing the crack, drilling holes of 20 mm diameter at 45° to the element surface and crossing the crack plane at approximately 90° as shown in Fig. 16.10. Reinforcing bars are placed into the drilled holes, and the holes and the crack plane is filled with an epoxy pumped under low pressure varying from 0.35 to 0.55 MPa. Typically, 12 or 16 mm diameter bars extending at least 500 mm on each side of the crack are

392 Concrete Technology

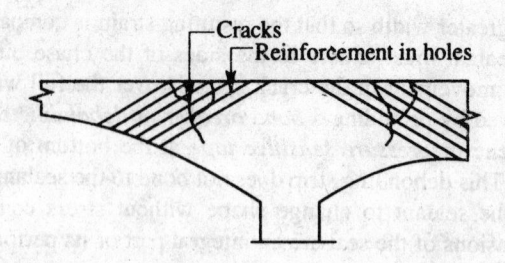

Fig. 16.10 Repair by providing additional steel

used. The reinforcing bars can be spaced to suit the needs of the repair and design criteria. The epoxy bonds the bar to the sides of the hole, fills the crack plane and brings the cracked concrete surface back to the monolithic form.

An elastic exterior crack sealant is required for a successful repair. Gel-type epoxy crack sealants are useful. The sealant should be applied in a uniform layer approximately 1.5 to 2.5 mm thick and shall span the crack by at least 20 mm on each side.

Resin bonding of flat steel plates to the external surface of critical structural member of bridges or buildings may prove to be the most practical and economical way of achieving local strengthening.

16.6.4 Stitching of Cracks

The stitching procedure consists of drilling holes on both sides of the crack, cleaning the holes, and anchoring the legs of the *stitching dogs* (U-shaped metal units with short legs as shown in Fig. 16.11) that span the crack, with either a nonshrink grout

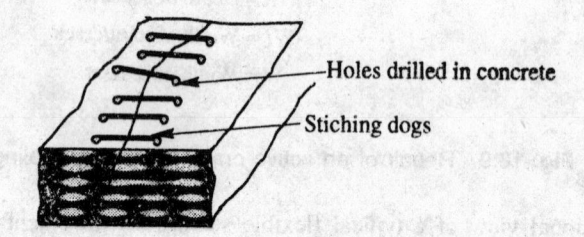

Fig. 16.11 Repair by stitching the cracks

or an epoxy-resin-based bonding system. The stitching dogs should be variable in length and orientation or both, and should be so located that the tension transmitted across the crack is not applied to a single plane within the section but spread over an area. The spacing of stitching dogs should be reduced at the ends of cracks.

Stitching may be used when tensile strength of the member is to be re-established across major cracks. Stitching does not close a crack but can prevent it from propagating further. Stitching tends to stiffen the structure which may accentuate the overall structural restraint, causing the concrete to crack elsewhere. It may, therefore, be necessary to strengthen the adjacent sections using external reinforcement embedded in a suitable overlay.

In the case of bending members, the stitching is done on the tension face where the movement is occurring. If the member is in a state of axial tension, the dogs must be placed symmetrically even if excavation or demolition is required to gain access to opposite sides of the section. The dogs are relatively thin and long and cannot take much compressive force. Accordingly, if there is a tendency for the cracks to close as well as to open, the dogs must be strengthened by encasement in an overlay. In water-retaining structures, the crack must be first made watertight before the stitching begins. The remedial measures for repairing the structural cracking of a slab, and a beam are shown in Figs. 16.12(a) and (b), respectively.

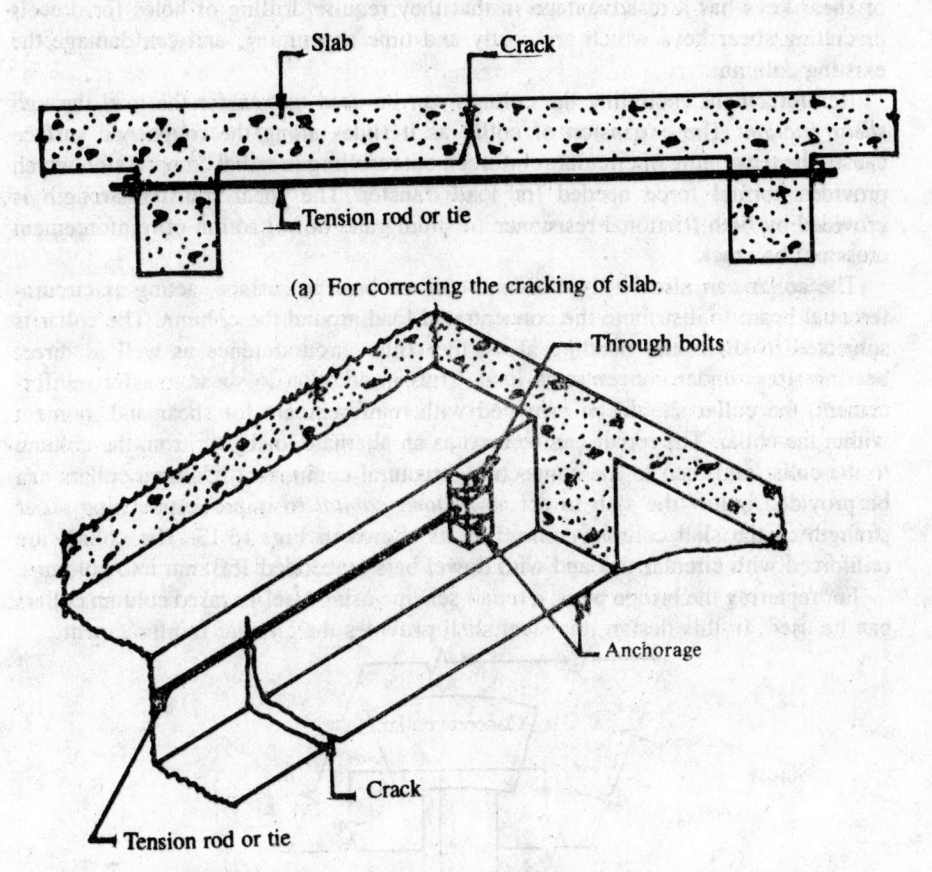

(a) For correcting the cracking of slab.

(b) For correcting the cracking of beam.

Fig. 16.12 (a) and (b) Repair of flexural cracks in slab and beam

16.6.5 Repair by Jacketing

Jacketing is a process of fastening a durable material over the existing concrete and filling the gap with a grout that provides the needed performance characteristics. The jacketing thus restores or increases the section of an existing member by encasement in a new concrete. The technique is applicable for protecting the member against further deterioration as well as for strengthening. In either case, the concrete section may be increased beyond the value required for the design loads to allow for some deterioration in future. The method is particularly useful for the compression members like columns and piles. The *column jacket* can also be used for increasing the punching shear strength of column-slab connections by using it as a *column capital*. When the jacket is provided around the periphery of the column, it is termed a *collar*. In most of the applications, the main function of the collar is to transfer vertical load to the column. Circular reinforcement can be used for load transfer. The practice of transferring load through dowel bars embedded into columns or shear keys has a disadvantage in that they require drilling of holes for dowels or cutting shear keys which are costly and time consuming, and can damage the existing column.

Reinforcement encircling the column can be used to transfer the load through *shear friction*. The expansion of collar as it slides along the roughened surface causes the tensioning of circular reinforcement resulting in radial compression which provides normal force needed for load transfer. The shear transfer strength is provided by both frictional resistance to sliding and dowel action of reinforcement crossing the crack.

The collar can also be used as mid-column bearing surface, acting as circumferential beam to distribute the concentrated load around the column. The collar is subjected to shear and bending along the collar circumference as well as direct bearing stress under concentrated load. Thus in addition to shear transfer reinforcement, the collar should be provided with reinforcement for shear and moment within the collar. The repair can be used as an alternate load path from the column to the collar and then to the connecting structural component. Column collars can be provided below the slab to act as *column capital* to improve punching shear strength of the slab column connection as shown in Fig. 16.13. The collars are reinforced with circular ties and with dowel bars embedded 150 mm into columns.

For repairing the bridge piers, a repair scheme using steel-encased column collars can be used. In this design, the steel shell provides the circular reinforcement.

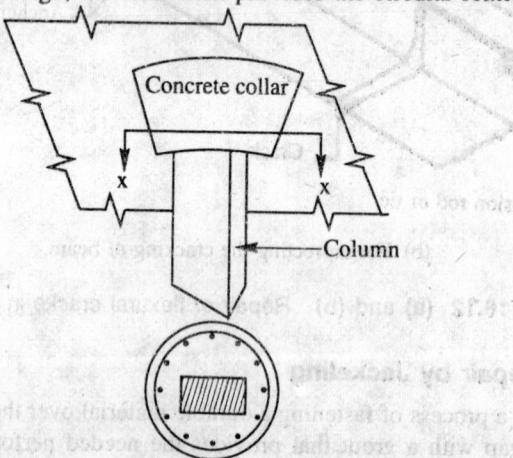

Fig. 16.13 Strengthening of a slab and column connection using a concrete collar

Unless the shear transfer strength is verified by load tests, the collar design and construction should meet the following criteria.

(i) The collar height should be at least 0.8 times of the original column cross-sectional dimension.

(ii) The collar reinforcement is located at or near the outside face. The steel is lapped or welded for full development of strength.

(iii) The concrete strength should be at least 16.5 MPa.

(iv) The column surface should be roughened in the shear transfer zone by bush-hammer. The surface of the column is then cleaned by a wire brush and high-pressure air to remove any loose concrete before placing the concrete in the formwork. Fibre-glass-reinforcement plastics, ferrocement, and other hard materials like polypropylene have been recently used for jacketing.

16.6.6 Autogeneous Heeling

The natural process of crack repair known as *autogenous heeling* has a practical application in closing dormant cracks in a moist environment. Such a case may be found in mass concrete structures. Heeling occurs through the carbonation of calcium hydroxide in cement paste by carbon dioxide, which is present in surrounding air and water. Calcium carbonate and calcium hydroxide crystals precipitate, accumulate, and grow within the cracks. The crystals interlace and twine, producing a mechanical bonding effect which is supplemented by a chemical bonding between adjacent crystals, and between the crystals and the surfaces of the paste and aggregate. As a result, some of the tensile strength of the concrete is restored across the cracked section, and the crack may become sealed.

During the process of heeling it is essential for development of any substantial strength, that the crack and adjacent concrete be saturated with water. The saturation must be continuous for the entire period of heeling. A single cycle of drying and reimmersion will result in a drastic reduction in the heeling strength. Heeling should commence as soon as possible after the crack appears.

16.7 TYPICAL EXAMPLES OF CONCRETE REPAIR

The slabs, beams and columns are the most important components of the concrete structures. Due to large number of cracks or damages observed in these components in the structures built in the past, the repair and rehabilitation of such members assume a greater importance. The materials and techniques described in this chapter can be used to reinstate the structure by repairing these members. Typical repair techniques used for columns, beams and slabs will be discussed in the following sections.

16.7.1 Repair and Strengthening of Columns

The *column jackets* and *collars* discussed in section 16.6.5 can be advantageously used in the repair of deteriorated concrete columns. The material used for the jacket may be metal, plastic or concrete. The arrangement will restore the integrity, and protect the reinforcement from the aggressive environments, and may improve the appearance of the original concrete. The jacketing material is secured to the concrete by means of bolts, screws, dowels, etc.

In case of badly damaged columns additional vertical steel and binding medium is required. Before jacketing the column the concrete surface is suitably prepared, additional reinforcement is fixed and column is then built out to the required profile with gunite. The column surface is roughened and collar reinforcement in the form of rings is placed at the outer face before placing the concrete in the formwork. The *bonding* is ensured by drilling a number of holes about 50 mm deep in the old concrete and placing dowels bonded by epoxy.

16.7.2 Repair of Concrete Floor Slab System

Repair, renovation and upgradation of concrete floor slabs is generally required for increasing their life span, or for change of occupancy. These may require a complete structural upgrading necessitating changes in the surface levels, the repair of visual damage only, and provision of a thin surface topping to existing concrete to produce a wearing surface requiring particularly good resistance to abrasion and wear. Provision of services below the floor surface may also require the replacement of floor surface.

Before repairing the concrete floor slabs, it is necessary as usual to investigate the causes of damage and to draw the specifications for the repair keeping in mind the future use of the floor. The floor should be surveyed for the defects, the existing levels should be recorded with respect to a known datum, the existing services in floors should be plotted and tested to ensure their satisfactory condition. Many situations may require a provision of a bonding coat or topping over the whole floor area, but before this is carried out damaged areas and the joints should be repaired, otherwise these defects may reflect up through the new topping.

The slabs in concrete structures under certain circumstances may deteriorate in selective locations exposing the reinforcement. The *deterioration* in the form of *scaling* may occur as a direct result of an inadequate internal *air-void system* in the concrete. In cold regions in the presence of moisture, *freeze-thaw action* can cause considerable scaling deterioration. The surface delaminations of the concrete in the slab may also be caused by the corrosion of the reinforcement. The delaminations above and around the reinforcement is a condition of concern for long-term structural integrity. The repair schemes to reduce further corrosion of embedded reinforcement by preventing moisture from being absorbed and infiltrating the concrete are recommended.

An evaluation of the live load capacity of the slab should be made to determine the feasibility of adding a thick *overlay system* to the slab. A thick overlay system generally adds 550 to 700 Pa to dead load usually at the expense of live load capacity. The repair schemes consisting of overlays, isolated patches, membrane type systems, sealers, and cathodic protection should be evaluated and estimated cost of each scheme be compared to find a competitive and structurally acceptable solution. For extensively damaged large-slab systems, a repair scheme using a thin methyl-methacrylate polymer concrete overlay may provide definite advantage not offered by the other schemes. In order to assess its application feasibility, sample tests of the polymer concrete of different thicknesses may be carried out for water absorption and chloride penetration characteristics. Usually, polymer concrete does not absorb moisture.

It is seen that the delaminated areas are usually concentrated in the negative bending moment regions of slab and near the reinforcement terminations, and hence well-defined areas may be identified as the repair areas.

Preparation of surface Each repair area should be delineated by saw cut 3 mm wide by 6 mm deep around its perimeter lines at least 100 mm outside the damaged area as shown in Fig. 16.14. The entire surface within this area should be scarified by a scabbling machine to remove the concrete to a level below the wear or damage so as to obtain a sound clean concrete surface suitable for repair. Subsequent to this *scarification* of the repair area, the sounding technique can be used to locate *surface delaminations* by using small chipping guns.

The reinforcement in the delamination area should be exposed and chipping continued until all concrete within 12 mm of the entire exposed portion of the reinforcing bar is removed. The prepared surface should be resounded to ensure that all delaminated unsound concrete has been removed. The exposed reinforcement should be sand-blasted to remove all corrosion byproducts. Finally, the entire repair area is blown off with compressed air to remove any loose corrosion particles, concrete, blasting sand and dust.

In case of cement-mortar repair the prepared reinforcement surface is coated with a cement paste layer which will provide additional protection to the reinforcement. After the coated surface has cured, the repair area is thoroughly kept wet for 24 hours if possible. All surface water must be removed from the area before filling in the cementitious material consisting of 1 : 3 cement-sand mix with sufficient water for firm pressing by hand. When the quantities required for repair are small, the proprietary materials may be used which are carefully batched and their quality controlled. Because of higher workability they require only hand tamping. For other cement-sand mixes vibrating hammers with a square plate on foot are often used, but for a large area a short beam fitted with form vibrator may be used to press the material into the repair area. The repair is finished off with a hand trowel and kept covered with polythene for 7 days.

Dry cementitious mortar materials along with a polymer, such as styrene butadiene rubber latex, can be used for high quality local repairs. The following procedure may be adopted.

The exposed concrete and reinforcement surfaces are coated with a primer compatible with the repair system. The primer can be applied by rolling it on to the surface with a paint roller and allowed to cure. The coat of primer on the reinforcement provides additional protection against corrosion. After the primed surface is adequately cured, the surface becomes impermeable to moisture and could remain unprotected from environmental effects.

An initial bedding layer of polymer concrete is placed in deep areas around the exposed reinforcement. The purpose of this initial bedding layer is to assure that the exposed reinforcement be encapsulated by the polymer concrete. In deeply removed areas beneath the reinforcing bars the polymer concrete is spaded to remove the air pockets. The chipped areas are then back-filled with sand-loaded polymer

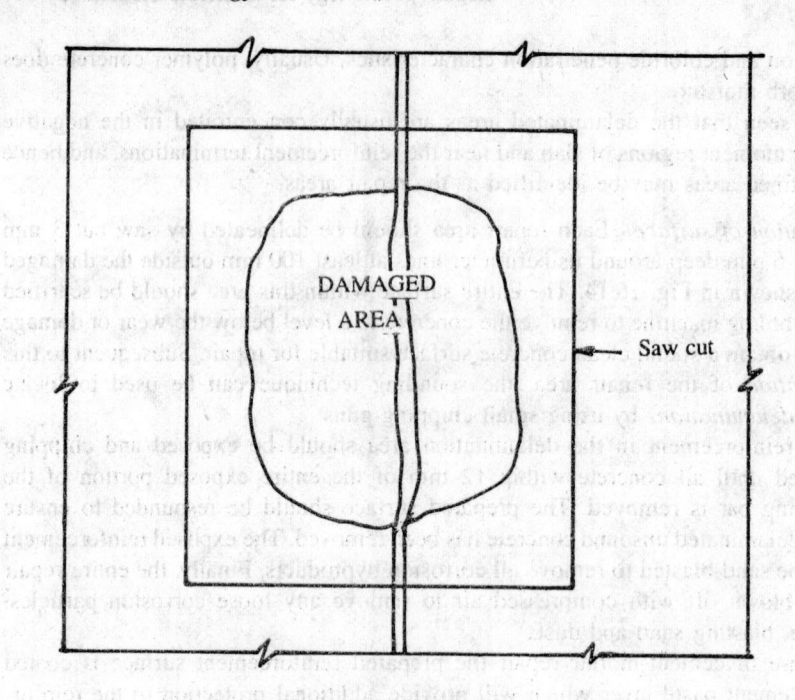

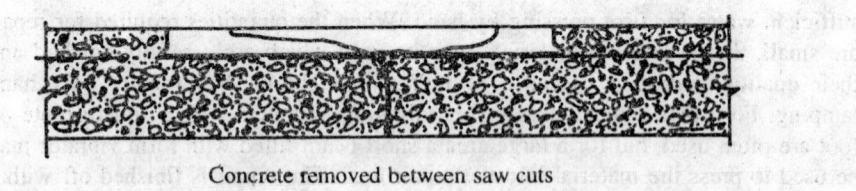

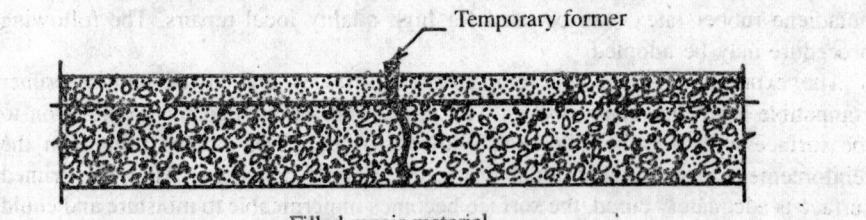

Fig. 16.14 Repair of damaged area on a concrete floor slab

concrete. A skin coat of neat polymer concrete followed by a 6 mm layer is applied over the exposed reinforcement.

The material is allowed to cure before proceeding to the application of the second layer of polymer concrete to build-up the cover thickness in the area over the reinforcement. The final lift of polymer concrete is applied over the entire repair area to provide a minimum thickness of 6 mm. A minimum of 12 mm polymer cover is desired over the exposed reinforcement.

After the final layer of polymer concrete has cured, the joint between the repair material and unrepaired concrete is sealed with a coat of the same primer as placed on original prepared concrete surface.

In deeper patches the material has to be applied in lifts to control shrinkage. At higher temperatures (above 21°C) the rapid evaporation of polymer liquid reduces the working time available before the material gains an initial set. Placing, screeding and trowelling must be done quickly. Working at night or early in the morning for reduced ambient temperature gives a better control of the situation.

16.7.3 Overlays and Surface Treatments

Dormant cracks in both structural and pavement slabs may be repaired using *bonded overlays* or *toppings*. However, most cracks in slab are subjected to movement caused by the variation in loading, temperature, and moisture. These cracks will reflect through any bonded overlays negating the overlay so far as crack repair is concerned, but the one-time occurrence or drying shrinkage cracks can be effectively repaired by the use of overlays.

The slabs and decks containing fine dormant cracks can be repaired by applying an overlay of polymer-modified portland cement concrete or mortar. In highway bridge applications a minimum overlay thickness of 40 mm may be used. Polymers suitable for such applications are latexes of styrene butadiene acrylic, non-reemulsifiable polyvinyl acetate, and certain water-compatible epoxy-resin systems. The minimum resin solids should be 15 per cent by mass of the portland cement.

Before applying an overlay or topping the old floor slab should be thoroughly shot-blasted or scabbled, cleaned and saturated with clean water. The localized damages or depressions are repaired as explained earlier, before bonding the cementitious topping. There are two types of commonly used toppings, namely the bonded toppings and unbonded toppings. The thickness of topping will be governed by the strength and thickness of old floor slab.

Bonded Toppings

They require bonding aids which may include polymers, resins and cementitious grout. A cementitious grout mixed to creamy consistency can be brushed on to the floor slab immediately before placing the topping mix. Usually, a concrete mix having proportions of 1 : 1 : 2 by mass of cement-sand-10 mm aggregate may be adequate for topping. The sand should be medium grade (belonging to Zone II) and coarse aggregate should be hard and clean. Granite aggregates are commonly used; flint or quartzite gravel, ballast and hard limestone can also be used. The amount of water to be added should be minimum to attain full compaction. The topping mix should be laid in 20 to 40 mm thick layers in bays such that the construction joints in old floor must reflect up through the topping. The mix should be compacted on the old floor and trowelled level at the intervals while topping is hardening, and after final trowelling the topping should be cured by covering it with polythene for at least 7 days.

In case of polymer-modified portland cement toppings a bond coat consisting of broomed latex mortar or an epoxy adhesive should be applied immediately before placing the overlay. The polymer-modified overlay should be mixed, placed and finished rapidly (within 15 minutes in warm weather). Such overlays should be cured for 24 hours.

Unbonded Toppings

As the unbonded topping is an additional slab laid over the old floor slab no further surface preparation is required but construction joints in the old floor must be reflected up through the new slab. Properly lapped polythene sheets are laid as damp-proof membrane over the base slab and a M30 grade concrete having a minimum cement content of 350 kg/m^3 should be placed and compacted on it. The concrete mix laid is 100 mm thick in bay sizes of up to 15 m^2. Before this concrete hardens a high strength 10 to 15 mm thick topping mix described above may be placed and compacted on to the surface. The trowelling at intervals during hardening and curing is accomplished as in bonded toppings.

16.7.4 Reconstruction of Slabs

In case of broken slabs, it is preferable to remove the affected slab and reconstruct it. In case of ground floor slabs, the subgrade should be inspected, compacted and brought to correct level by using well-graded crushed rock material or lean concrete 150 mm thick. Before concreting a polythene sheet should be placed over the top of sub-base to act as damp-proof layer. The concrete should be fully compacted with vibrating beam when placed, and finished. It should be cured by covering it with polythene for at least 7 days and should not be loaded for 28 days.

16.7.5 Repairs of Beams

In the case of extensively damaged or deteriorated beams, additional reinforcement at the bottom of beam together with the new stirrups are provided. The stirrups can be anchored by expanding bolts set in the sides of the beam below the slab soffit or may be taken right round the beam through the holes drilled in the slab. The roughened or irregular surface of the prepared concrete usually ensures a good bond between the old and new concretes placed by guniting. However, if required, shear connectors can also be provided by expanding bolts or other means.

16.7.6 Surface Coatings

A layer of material applied to a surface, which forms a continuous membrane is termed *coating*. The coatings generally adhere to the concrete and form films after application. During curing of concrete approximately 25 per cent of water is retained as water of crystallization and 15 per cent as gel. Capillary pores are formed during the evaporation of the remaining 60 per cent and eventually 15 per cent gel water. The *capillary pores* allow carbon dioxide and other gases to diffuse into concrete which may dissolve in pore water to form acidic solutions. The porosity of concrete may also lead to the absorption of water which may carry harmful reagents in solution. Concrete is strongly alkaline, and as such it is susceptible to attack from acidic reagents. Applications of suitable coatings which afford the necessary protection while still allowing free passage of water vapour, is required for the protection of the structure. These preventive measures are described as follows.

(i) Anti-carbonation Coatings

The coatings applied to concrete surface to arrest the carbonation process are known as anticarbonation coatings. These are normally based on chlorinated rubber, polyurethane resin or acrylic emulsions. Anti-carbonation coatings may be effectively used to resist carbonation and general atmospheric deterioration of reinforced concrete. In the situations where corrosion and spalling are more wide-spread, use of anti-carbonation coating will not be satisfactory.

(ii) Coatings to Resist the Effects of Acid Environments

The concrete structures are occasionally subjected to abnormally acidic environments. This may be due to the release of sulphur dioxide from power stations, steel plants and oil refineries into the atmosphere which readily dissolves in rainwater and forms sulphurous acid resulting in direct etching of the concrete. The concrete, in prolonged contact with water due to poor drainage, may disintegrate. The coatings are required to arrest the process of *acid attack* and to provide high degree of chemical resistance to the concrete. Under most circumstances, two part polyurethane coating will be suitable.

(iii) Coatings to Protect Cracked Concrete

The cracks which are very fine and not considered to have structural significance may be protected by applying the coating locally over cracks. These are termed conventional coatings. The coatings should have flexibility as well as ability to bridge cracks. High-build polyurethane and and epoxy-polyurethane formulations have been successful.

16.8 LEAK SEALING

Leakage in the concrete structures causes inevitable damage to the reinforcement. Construction joints, shrinkage and restraint cracks may form leak paths. The amounts of water involved vary from *damp-patches* which tend to evaporate as they are formed, to *running-leaks* which may eventually form pools on undrained surfaces. Damp-patches may also be formed when water passes through the voids along reinforcing bars formed due to *plastic settlement*. The other common routes for larger volume leaks are *honeycombed concrete, movement joints like expansion* and *contraction joints*. In case of water-retaining structures, the extent of leakage may be measured by monitoring loss of liquid from the structure. As per BS : 5337 the structure can be considered watertight if the total drop in surface level does not exceed 10 mm in seven days. For an effective leak sealing it is essential to identify the routes and sources of leakage and due consideration must be given to their likely cause, and their behaviour once the structure is in service.

16.8.1 Leak-sealing Techniques

Leak sealing is expensive, so the operations must be necessary and worthwhile. The leak-sealing methods can be classified as:
  (a) Conventional leak-sealing methods
  (b) Leak-sealing by injection techniques

Conventional Methods

Some sources of minor leakage may dry up by *autogeneous healing* which is an accumulation of calcium salts along the leak path. This will obstruct the passage of water over a period of time and reduce the leakage to negligible proportions.

Once leak spots have been identified, the remedial action may involve the application of local or complete surface seal in the form of a *coating system*. The following format is recommended:
  (i) Surface preparations
  (ii) Filling of surface imperfections with resin-based grouts
  (iii) Application of primer
  (iv) Application of two coats of high-build paint

The procedure may require quite extensive preparatory work including the injection of suspect joints and random shrinkage cracks with a low viscosity resin. Honeycombed concrete if not particularly extensive may be filled out using a resin-based mortar or putty. Laitance and surface contaminants may be removed by sand blasting and power wire brush.

The movement joints can be sealed by filling a resin into the joint which will cure to form *flexible sealant* as explained in Sec. 16.6.2. The concrete joint must be prepared and thoroughly cleaned prior to the application of sealant, and an appropriate bonding coat or primer should be used, if recommended.

Injection Sealing

From liquid flow and pressure considerations the simplest and most cost-effective way is to seal the leakage from the water- retaining side of the structure. However, when the wet side is inaccessible, the leakage must be tackled from the dry side which is considerably more difficult. Successful leak sealing requires injection of sealant (grout) to fill water passages completely, and it is necessary to attain a relatively high flow velocity to achieve this, because of the short *pot-life* or working time of the typical repair material. The first basic step is to restrict or confine the water flow to a tube through which the sealant may be introduced. Once the flow of water has been controlled, the connection between the tube and concrete must be made strong enough to withstand the injection pressure.

However, due to possibility of concrete being stressed during injection, it is preferable to maintain lower pressures. A typical direct method entails the injection of material up the pressure gradient from the down stream side. On the other hand an indirect method involves the introduction of the sealant on the pressure side, so that the pathways are filled under the acting hydrostatic head. The direct methods are very slow due to sealant being pumped slowly through very narrow passages against pressure, and the pressure cannot be maintained for long enough to achieve complete penetration. In many cases water may find another finer pathway leading from the same source. In contrast the indirect methods enable the work to be completed quickly because surface seals are not required and mechanical anchorages can be used.

16.9 UNDERWATER REPAIRS

Many of the method discussed in the preceding sections for above- water (dry) repairs may also be used underwater with only minor modifications. However, the materials specified for use in air are often unsuitable for underwater application. The special features of underwater repair are:

1. Due to high cost and complexity of underwater working, the repair operations need be made as simple as possible. The choice of repair technique is influenced by the available method of access.

2. Adequate preparation of damaged area may require specially adapted techniques.

3. The repair materials must be compatible with underwater application both during placing and curing. Cementitious systems have been found to be better suited for underwater use.

4. Formwork and placement method adopted must minimize mixing between repair material and water.

5. Underwater supervision of repair operations is difficult and costly.

Generally, laboratory trials on both materials and repair methods are used to identify possible problem areas and ensure smooth site operations. Before a repair is undertaken it is necessary to clean the damaged area of marine encrustation (contaminants) to allow detailed inspection to assess the extent of damage. In case of smaller areas, this can be accomplished by using mechanical wire brushes, needle guns or scabbling tools. However, for larger areas a high pressure jet may provide a solution. Once the area has been cleaned, the extent of cracked and spalling concrete may be defined with the help of divers or remote operated vehicles to photograph the area. The steps involved in underwater repair are:

(i) Preparation of Damaged Area

The preparation of surface may require removal of cracked or badly damaged concrete and cutting of distorted reinforcement. If only the damaged concrete is to be removed, high pressure water jetting (with pressure typically between 200 and 1000 atmospheres) directed onto the concrete surface can be used to remove hardened cement paste mortar from the spaces between the aggregate. In this process the reinforcement itself is cleaned but not cut by the water jet, enabling it to be utilized in the repair. If the reinforcement is also to be cut, then an abrasive slurry is injected into the cutting jet.

Alternatively, the damaged concrete can be cut by *splitting techniques*. Wherein hydraulic expanding cylinders are inserted into pre-drilled holes and pressurized until splitting of concrete occurs. The splitting can also be achieved by mixing expansive cement with water to form a paste which is poured into plastic bags. The filled plastic bags are in turn deposited in the pre-drilled holes in the structure. The expansion of the cement over the next 12 to 24 hour period generates stresses approximately 30 MPa which may be adequate to split concrete.

The splitting of concrete may also be achieved by using *soft explosions*. The procedure consists of placing and caulking-in firmly pressurized carbon dioxide cartridges in pre-drilled holes. The pressure is then released by electrically detonating a small initiating charge in each cartridge which produces a comparatively gentle explosion resulting in the controlled splitting of concrete by cracks running between the prepared holes.

The mechanical underwater cutting using hydraulically powered diamond-tipped saws and drills is suitable only for minor works, e.g. core cutting, etc. The reaction force to the tools can be provided by strap or strut arrangement bolted to the structure.

After the removal of damaged concrete, all broken or distorted reinforcement will have to be removed and replaced before reinstating the cover. The commonly used methods for cutting steel underwater are: oxygen-fuel gas cutting, oxy-arc cutting, and mechanical cutting. The first two methods rely upon the burning process which consists of oxidizing the metal (i.e., carbon steel) and removing the oxidized products from working surface.

Once all damaged concrete and distorted steel have been removed, the reinforcing bars are replaced with new lengths joined either by couplers or lapped with existing bars. Immediately prior to reinstating the damaged concrete, the surface must be flushed with clean water to remove any bacterial or microbiological growth which may otherwise significantly reduce the bond between the repair material and structure.

(ii) Application of Materials

(a) Mortar Placement

In case of minor damage or to prevent future deterioration, the defects may be filled with cementitious or resin-based materials. This patchwork is suited to application in small volumes. To reduce washout of cement from conventional cementitious mortars and grouts, adhesive admixtures are used. *High performance mortars* mixed above water can be poured by free fall through water to fill the formwork. The mixes are normally formulated to be self-leveling to ensure good compaction without vibration, and laid in thickness of 20 to 150 mm. To avoid damage due to wave-action on vertical surfaces the repair should be carried out using either formwork and a free flowing grout or a hard epoxy putty.

The normal epoxy or polyester resin-mortars are totally unsuited for underwater applications. However, for underwater use special formulations have been developed. These are normally free-flowing and hence can be poured through water directly into formwork. For vertical work special types of underwater-grade epoxy putty have been developed.

(b) Injection into Cracks

The general principles for injecting grouts in the cracks underwater are the same as for above water injecting. However, owing to the risk of washout of cement, non-conventional epoxy resin injection are normally preferred. Epoxy putty is used to seal the crack between injection points as shown in Fig. 16.15. Epoxy resins

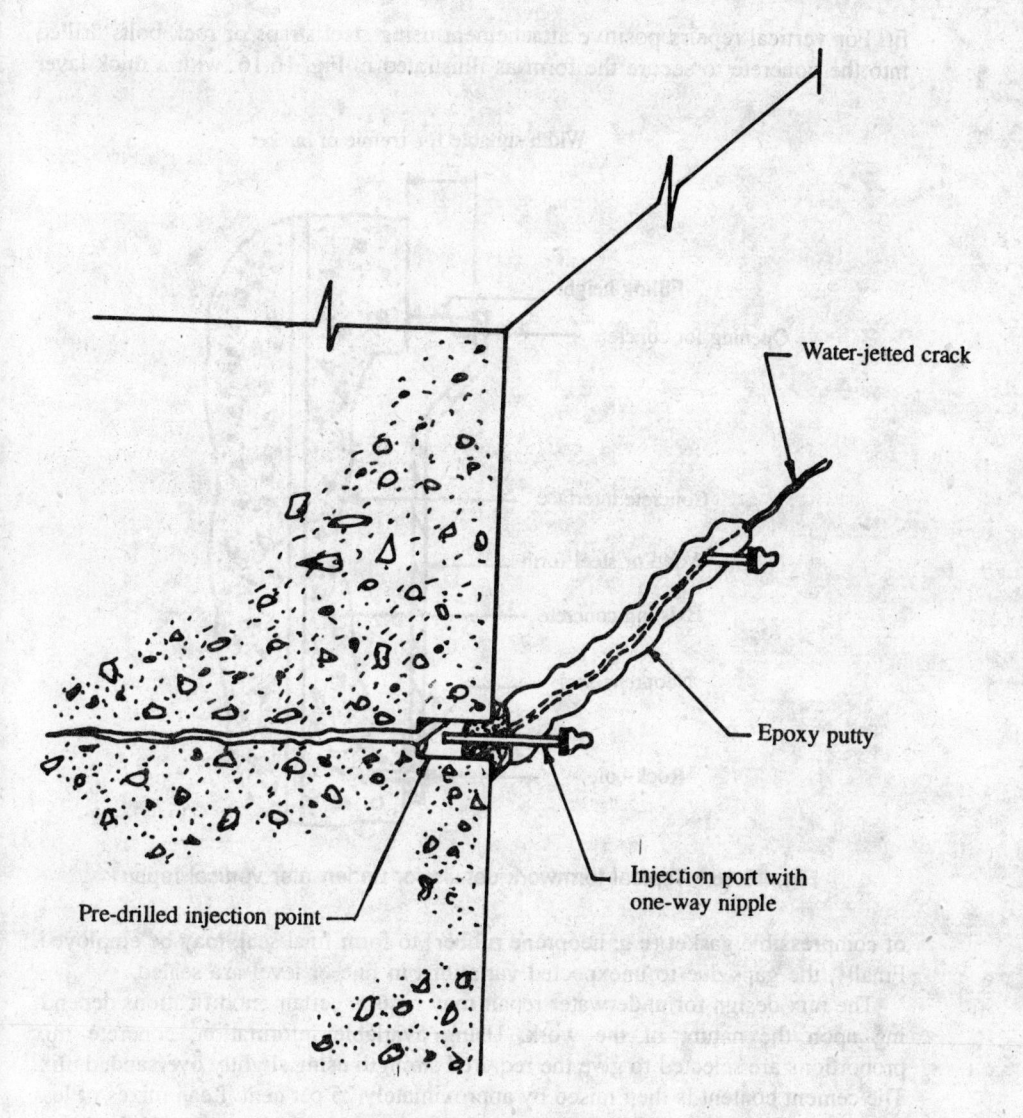

Fig. 16.15 An underwater crack injection arrangement

must be low viscosity solvent free underwater-grades in order that the water in the crack is replaced by a structural material.

For small repairs the use of hand-held cartridge injection guns is a satisfactory procedure.

(c) Large-scale Repairs

For bulk placement of repair material underwater, an easy-to-erect formwork complete with inlet pipes and external vibrators and tolerant of variations in the existing structure, is required. Flexible seals are preferred as they ensure a leak-tight fit. For vertical repairs positive attachement using steel straps or rock bolts drilled into the concrete to secure the form as illustrated in Fig. 16.16, with a thick layer

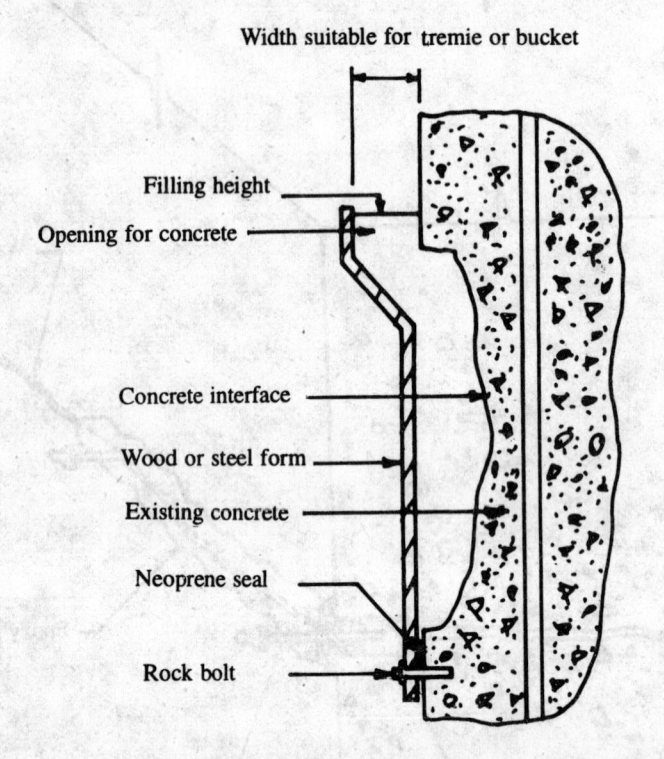

Fig. 16.16 Typical formwork details for underwater vertical repair

of compressible gasket (e.g. neoprene rubber) to form final seal, may be employed. Finally, the gaps due to unexpected variations in line or level are sealed.

The mix design for underwater repair may require certain modifications depending upon the nature of the work. Using available information, concrete mix proportions are selected to give the required strength using slightly oversanded mix. The cement content is then raised by approximately 25 per cent. Lean mixes of less than 350 kg/m^3 are not likely to be suitable due to washout of cement. Well-washed marine dredged aggregates and round river gravels are suitable for the tremie and pump placing methods. For small repairs in heavily reinforced areas, superplasticizing admixtures will be essential to give necessary flow characteristics.

The placement method for underwater repair must be so selected as to minimize the area of contact between concrete and water, and to prevent turbulence. The concrete can be placed by a bottom- opening bucket (skip) or a tremie pipe or by pumping as explained in Section 12.4.1.

The prepacked aggregate concrete method of placement is ideally suited to underwater or tidal works, especially where access conditions are limited or fast flowing water could wash away conventional concrete. This method has been explained in the Sections 12.4.1 and 16.5.2.

16.10 DISTRESS IN FIRE DAMAGED STRUCTURES

A large number of reinforced concrete structures salvaged from destruction in fires by timely fire fighting operations can be put to further service after strengthening and providing some cosmetic repairs since the cost of restoration of such structures is less than that for dismantling and construction of new ones.

The fire may cause different degrees of damage to the structure: the structure may be completely burnt or destroyed; its surface may be slighly damaged or a slight deformation may occur. In the first case, the whole of damaged portion has to replaced during restoration of structure while in the latter, only repair and finishing may be required. The extent of damage caused to the structure during a fire depends on the duration of fire, and the temperature to which the structure was subjected during the fire.

High temperature during a fire reduces the strength of reinforced concrete structures due to change in the strength and deformability of materials, reduction in cross-sectional dimensions, weakening of bond between the reinforcement and concrete, which determines structural action under the load. The maximum temperature reached during a fire is normally estimated indirectly, e.g. from the melting of metallic or other non-combustible articles. A temperature of 1000–1100°C has been observed during fire in residential and administrative buildings. The duration of these fires was mostly between 1 and 2 hours. It has been observed that during fires in theatres and departmental stores, temperature rises up to 1100–1200°C and the fire duration exceeded 2–3 hours in some cases. Still higher temperatures have been observed during fires in industrial buildings and warehouses in which considerable quantities of solid and liquid combustible materials were processed or stored. During a fire in a store of combustible liquid and lubricants which lasted more than 2 hours, a temperature of 1300°C has been reported. Thus duration of a fire and maximum temperature reached can vary over a wide range. Temperatures of 1000–1100°C in fires which lasted for 1–2 hours have been observed more frequently than 1300°C.

An accurate estimate of the *performance characteristics* of structures which have been damaged in a fire helps in developing effective restoration/rehabilitation measures. The performance characteristics take into account the physico-chemical and mechanical properties of the materials burnt, and that of heated concrete. There is an accumulation of irreversible damages of mechanical (cracking, creep, shrinkage and plastic deformations) and *physico-chemical* (corrosion, absorption and degradation) origin. This information improves the reliability of estimation of residual load carrying capacities of structural elements thereby resulting in a considerable saving in the cost of restoration of the structure. This prediction of loss in the carrying capacity of fire-ravaged reinforced concrete elements taking into consideration the physico-chemical characteristics of materials and geometric dimensions of the structure is difficult.

The strength and stiffness of concrete and steel decrease as the temperature of the member increases and dimensional changes occur. The changes in strength, stiffness of concrete are influenced by the type of cement and aggregate, and water content. The stresses due to thermal strain cause the beam, column or slab to crack or spall, reducing concrete area available to resist the applied forces. The following observations will bring out the causes of the damage in fire-ravaged structures and help in obtaining complete performance characteristics of the structure.

Axially loaded columns The column under fire normally fail at mid-height due to brittleness. The failure is accompanied by disintegration of concrete in the whole section and buckling of longitudinal bars. It must be realized that due to considerable drop of temperature (800°C or more) between the periphery and centre of the section, the strength of concrete varies along the cross-section from its initial value at the central portion to zero at the surface. The temperature at which *crushing strength of concrete* is reduced to half its initial value is termed *critical temperature*. The critical temperature depends upon the type of aggregate used in the concrete. This temperature is 550°C for concrete with granite or sandstone aggregates, and 700°C for concrete with limestone.

Due to non-uniformity of temperature in the cross-section, the hottest layers of concrete and main reinforcement bars near the surface of column are relieved due to thermal *creep* and loss of strength, and also due to contraction in the case of concrete. This causes increased stresses in the centre of the section where moderately hot concrete retains its strength and elasticity. Complete failure of column starts when stresses in the central portion of the cross-section becomes equal to the initial prism strength of concrete with deformation approaching its limiting value (0.0025–0.0030).

16.10.1 Restoration of Fire Damaged Elements

The eccentrically loaded columns fail when reinforcement bars in tension heat up. The fire resistance of such elements can be increased by increasing the thickness of protective layer. Heat transmission and temperature of bottom reinforcement are keys to the behaviour of reinforced concrete slab exposed to fire. The reinforcing bars are assumed to retain one half of their original strength. Carrying capacity of slabs can be enhanced by increasing their thickness. For beams, depth and width can be increased. However it should be kept in mind that in case of beams, weakening of bond between transverse reinforcement and concrete on account of heating reduces the residual shear load carrying capacity considerably. The required increase in the dimensions of the beam, longitudinal and transverse reinforcements should be computed by taking into account the change in compressive strength of concrete and modulii of elasticity of concrete and reinforcement.

The carrying capacity of axially loaded column depends upon the cross-section of the column, coefficient of change in strength of concrete under high temperature and corresponding critical temperature. The carrying capacity can be restored by increasing the cross-section with suitable increases in the longitudinal steel.

In retrofitting of the structure a convenient method of anchoring reinforcing bars into existing concrete walls and foundations is to drill a hole in the concrete somewhat larger in diameter than that of the bar and set the bar in an epoxy gel. In slabs epoxy coated bars may be used but epoxy coating thicker than 0.25 mm are not recommended.

APPENDIX
Objective Type Questions

CEMENT

1. Cement is an important ingredient of concrete because
   (a) it is a binding medium for discrete ingredients (b) it is the only scientifically controlled ingredient (c) it is an active ingredient (d) it is a delicate link of the chain (e) any of the above
2. In the manufacture of cement definite proportions of argillacious and calcareous materials are burnt at a temperature of
   (a) 425°C (b) 875°C (c) 1450°C (d) 1650°C
3. During the manufacturing process of portland cement, gypsum or plaster of paris is added
   (a) to increase the strength of cement (b) to modify the colour of cement (c) to adjust setting time of cement (d) to reduce heat of hydration
4. The percentage of gypsum added to the clinker during manufacturing process is
   (a) 0.2 (b) 0.25 to 0.35 (c) 2.5 to 3.5 (d) 5 to 10 (e) 15 to 25
5. The setting and hardening of cement after addition of water is due to
   (a) the presence of gypsum (b) binding action of water (c) hydration of some of the constituent compounds of cement (d) evaporation of water (e) none of the above
6. In terms of oxide composition, the maximum percentage of ingredient in the cement is that of
   (a) lime (b) iron oxide (c) alumina (d) silica (e) magnesium oxide
7. In terms of oxide composition, the minimum percentage of ingredient in the cement is that of
   (a) lime (b) magnesium oxide (c) iron oxide (d) alumina (e) silica
8. In terms of oxide composition, in a cement
   (a) high lime content increases setting time and results in higher strengths (b) high silica content prolongs the setting time and gives more strength (c) presence of iron oxide gives grey colour to the cement (d) presence of unburnt lime and magnesia causes unsoundness (e) all of the above
9. The tricalciumaluminate compound present in cement
   (a) provides weak resistance against sulphate attack (b) is responsible for highest heat of evolution (c) characteristically fast reacting with water (d) all of above (e) none of above
10. The constituents of cement which act as binder are
    (a) tricalcium silicate, dicalcium silicate and sulphur trioxide (b) tricalcium silicate and tetracalcium alumino-ferrite (c) tricalcium silicate, dicalcium silicate and tricalcium aluminate (d) dicalcium silicate, tetracalcium alumino-ferrite, and tricalcium aluminate

11. Which of the following statements in terms of compound composition of cement are incorrect?
    (a) C_3S and C_2S together constitute about 70 to 80 per cent of cement
    (b) Both C_3S and C_2S give the same product on hydration (c) C_2S hydrates slowly and provides much of the ultimate strength (d) C_3S having a faster rate of reaction is accompanied by greater heat evolution (e) C_3S provides more resistance to chemical attacks
12. Following are the statements in terms of compound composition of cement, identify the incorrect one(s)
    (a) Tricalcium aluminate (C_3A) is fast reacting with water and leads to immediate stiffening of the paste (b) C_3A phase is responsible for the highest heat of evolution (c) C_3A provides high resistance against sulphate attack (d) Gypsum is added to prevent flash set of C_3A
13. The constituent compounds of cement in decreasing order of rate of hydration are
    (a) C_2S, C_3S and C_3A (b) C_3S, C_3A and C_2S (c) C_3A, C_3S and C_2S (d) C_3A, C_2S and C_3S
14. The contribution of constituents of cement to the strength of cement is in the decreasing order
    (a) C_3S, C_2S, C_3A and C_4AF (b) C_2S, C_3S, C_3A and C_4AF (c) C_2S, C_4AF, C_3A and C_3S (d) C_3S, C_3A, C_2S and C_4AF
15. Out of the constituents of cement, namely, tricalcium silicate (C_3S), dicalcium silicate (C_2S), tricalcium aluminate (C_3A), the first to set and harden is
    (a) C_3S (b) C_3A (c) C_2S (d) any of the above (e) all set simultaneously
16. The time taken by dicalcium silicate (C_2S) to add to the strength of cement is
    (a) 1–2 days (b) 2–5 days (c) 5–7 days (d) 7–14 days (e) 14–28 days
17. When water is added to the cement
    (a) chemical reaction starts (b) heat is absorbed (c) heat is generated (d) impurities are washed out (e) none of the above
18. Dicalcium silicate (C_2S)
    (a) reacts with water only (b) hydrates rapidly (c) hardens rapidly (d) generates less heat of hydration (e) has no resistance to sulphate attack
19. Snowcem is
    (a) chalk powder (b) powdered lime (c) mixture of chalk powder and lime (d) coloured-cement (e) none of the above
20. In testing the portland cement for the loss on ignition, the sample is heated to
    (a) 100°C (b) 250°C (c) 500°–800°C (d) 900°–1000°C (e) 1250°C
21. In the case of portland cement the loss on ignition should be
    (a) less than 4 per cent (b) less than 10 per cent (c) within 10 to 15 per cent (d) less than 20 per cent (e) more than 20 per cent
22. During the test of OPC for loss on ignition, the loss in weight occurs due to
    (a) decomposition of silicates (b) chemical reaction (c) burning of constituents (d) melting of tricalcium aluminate (e) evaporation of moisture and carbon dioxide
23. The insoluble residue in cement should be
    (a) between 10 to 15 per cent (b) less than 10 per cent (c) between 5 to 10 per cent (d) between 1.5 to 5 per cent (e) less than 0.85 per cent

24. Total heat of hydration of cement is independent of
    (a) ambient temperature (b) composition of cement (c) fineness of cement
    (d) all of the above
25. The length of time for which the concrete mixture remains plastic predominantly depends on the
    (a) setting time of cement (b) amount of mixing water (c) atmospheric temperature (d) equally on all of the above
26. Initial setting time is maximum for
    (a) portland-pozzolana cement (b) portland-slag cement (c) low-heat portland cement (d) high strength portland cement
27. The setting time of cement is influenced by
    (a) percentage of water and its temperature (b) temperature and humidity of air (c) amount of kneading the paste (d) all of the above (e) none of the above
28. For ordinary portland cement
    (a) initial setting time should not be less than 5 minutes and final setting time should not be more than 24 hours (b) initial setting time should not be less than 30 minutes and final setting time should not be more than 600 minutes (c) initial setting time should not be less than 60 minutes and final setting time should not be more than 600 minutes (d) initial setting time should not be less than 5 minutes and final setting time should not be more than 600 minutes (e) none of the above
29. Which if the following statements is incorrrect?
    (a) The microstructure of hydrated cement governs the physical properties of concrete (b) The hydrated crystals form an interlocking random 3-dimensional network called gel (c) The hydrated paste has a porous structure (d) The size of the gel pores is finer than 4×10^{-4} μm
30. In medium-strength concretes the water-cement ratio should not be less than
    (a) 0.25 (b) 0.35 (c) 0.4 (d) 0.45
31. The compressive strength of concrete is basically related to
    (a) water-cement ratio (b) hydrate-space ratio (c) specific surface of cement (d) none of the above
32. The following portland cements have specific surfaces in the descending order
    (a) ordinary, rapid hardening, high strength, low heat (b) rapid hardening, high strength, low heat, ordinary (c) high strength, rapid hardening, low heat, ordinary (d) low heat, ordinary, rapid hardening, high strength
33. The insoluble material in cement is the
    (a) active part of cement and it should be kept to a minimum level (b) active part of cement and it should be kept of the maximum permissible level (c) inactive part of cement but it should be kept to the maximum permissible level (d) inactive part of cement and it should be kept to the minimum level (e) none of the above
34. An excess of free lime in portland cement
    (a) results in an increase in strength (b) increases the initial setting time (c) causes unsoundness in the product (d) improves the quality of the product (e) none of the above

35. In portland cement the quantity of free magnesia should be
    (a) less than 0.5 per cent (b) less than 5 per cent (c) between 5 to 10 per cent (d) between 10 to 15 per cent (e) less than 20 per cent
36. Finer the cement
    (a) higher is the rate of hydration (b) more is the surface area (c) higher is the possibility of prehydration by atmospheric moisture (d) lesser the amount of water required for constant slump (e) all of the above
37. Sieve analysis of portland cement is performed on IS sieve
    (a) No. 1 (b) No. 3 (c) No. 5 (d) No. 7 (e) No. 9
38. In the air permeability method for testing of portland cement for fineness, the apparatus essentially consists of
    (a) permeability cell, sieve, and rotameter (b) permeability cell, manometer, and flowmeter (c) sieve, barometer, and rotameter (d) manometer, sieve, and rotameter (e) sieve, flowmeter, and rotameter
39. In the air permeability test, the specific surface (in mm^2/g) is of the order of
    (a) 1000 (b) 2000–2500 (c) 2500–5000 (d) 225000–350000 (e) 750000–1000000
40. In Vicat's apparatus, the cement paste is said to be of normal consistency, if the rod penetrates by
    (a) 3 mm (b) 5 to 10 mm (c) 23 to 25 mm (d) 33 to 35 mm (e) 43 to 45 mm
41. To ensure that the concrete product does not undergo a large change in volume after setting
    (a) add excess quantity of fine aggregate to the mix (b) add minimum quantity of water to the mix (c) add maximum quantity of water to the mix (d) limit the quantities of free lime and magnesia in the cement (e) use proper curing
42. The absolute minimum water-cement ratio for concrete of medium strength is
    (a) 0.25 to 0.30 (b) 0.31 to 0.35 (c) 0.36 to 0.40 (d) 0.41 to 0.45 (e) 0.46 to 0.50
43. For complete hydration of 100 kg of cement of average composition, the mass of water required, would be
    (a) 15 kg (b) 25 kg (c) 35 kg (d) 40 kg (e) 42 kg
44. The cubes for testing cement in compression are kept at
    (a) $17 \pm 2°C$ and 100 per cent humidity (b) $27 \pm 2°C$ and 90 per cent humidity (c) $37 \pm 2°C$ and 80 per cent humidity (d) 100°C and 70 per cent humidity (e) 0°C
45. The compressive strength of OPC after three days is expected to be more than
    (a) 16 MPa (b) 22 MPa (c) 27.5 MPa (d) 33 MPa
46. For rapid-hardening portland cement
    (a) initial setting time should not be less than 5 minutes and final setting should not be more than 30 minutes (b) initial setting time should not be less than 30 minutes and final setting time should not be less than 600 minutes (c) initial setting should not be less than 60 minutes and final setting time should not be more 240 minutes (d) initial setting time should not be less than 5 minutes and final setting time should not be more than 60 minutes (e) none of the above

47. The heat generated during the setting and hardening of cement is called
    (a) heat of setting (b) heat of evaporation (c) latent heat (d) heat of hydration (e) sensible heat
48. Heat of hydration is determinated by an apparatus called
    (a) hydrometer (b) photometer (c) calorimeter (d) hygrometer (e) none of the above
49. Heat of hydration of cement is expressed in terms of
    (a) colories / cubic centimetre (b) calories (c) farads (d) grams (e) calories/gram
50. A warehouse-set cement is
    (a) cement which is affected by moisture in the warehouse (b) cement which sets due to being stored adjacent to the wall (c) cement which gets compressed due to the load of several bags of cement placed above it (d) cement spoiled in the warehouse (e) there is no such setting
51. An ideal warehouse should have
    (a) water proof masonry walls (b) water proof roof (c) windows limited in number and should not allow seepage of water during the rainy season (d) floor of 150 mm thick concrete slab (e) all of the above
52. The cement from the warehouse is taken out on the basis of
    (a) first in, first out (b) first in, last out (c) last in, first out (d) last in, last out (e) any of the above
53. The field test for the quality of cement consists in putting a small quantity of cement in a bucket containing water. A good quality cement will
    (a) immediately dissolve in the water (b) float on the water surface (c) sink to the bottom of the bucket (d) produce steam (e) produce effervescene
54. In fineness test of rapid hardening portland cement, the residue on IS sieve No. 9 should not be more than
    (a) 1.0% (b) 5% (c) 10% (d) 15% (e) none of the above
55. Identify the incorrect statement (s)
    (a) Expanding cement is used for filling the cracks (b) White cement is mostly used for decorative work (c) Portland pozzolans cement produces less heat of hydration (d) By varying the percentage of four basic compounds of cement, several types of cements can be obtained (e) High strength portland cement is produced from the special materials.
56. The compound constituent of cement abbreviated by C_3A represents
    (a) tetracalcium alumino ferrite (b) tricalcium aluminate (c) tricalcium silicate (d) dicalcium silicate (e) none of the above
57. Argillaceous materials are those
    (a) which have alumina as the main constituent (b) which have lime as the main constituent (c) which evolve heat on the addition of water (d) which easily break when hammered lightly (e) none of the above
58. A sample of cement is said to be *sound* when it does not contain free
    (a) lime (b) silica (c) iron oxide (d) alumina (e) all of the above
59. Initial setting time of rapid-hardening portland cement is nearly
    (a) half a minute (b) 5 minutes (c) 30 minutes (d) 45 minutes (e) 60 minutes

60. Low-heat cement is used for
    (a) repair of roads (b) thin structures (c) thick structures (d) underwater applications (e) all of the above
61. For the repair of roads
    (a) low-heat cement is used (b) rapid-hardening cement is used (c) high-alumina cement is used (d) sulphate-resisting cement is used (e) ordinary portland cement is used
62. Identify the incorrect statement(s)
    (a) White cement is unsuitable for ordinary work (b) Pozzolana cement is white in colour (c) C_3S is tricalcium silicate (d) Strength of cement implies compressive strength (d) Properly stored cement should not be disturbed until it is to be used
63. The number of cement bags in a pile of size $4 \times 3 \times 0.9$ m height in a cement store could be
    (a) 100 (b) 150 (c) 175 (d) 200 (e) 250
64. Total sulphur content of cement is less than
    (a) 0.025% (b) 0.25% (c) 2.5% (d) 5.0% (e) 10%
65. A cement bag stored for two years is likely to result in
    (a) change in the colour of cement (b) increase in the strength (c) loss of strength by 50 per cent (d) formation of lumps (e) swelling by 15 per cent
66. In ordinary portland cement magnesia is restricted to
    (a) 5 per cent (b) 2.5 per cent (c) 1.5 per cent (d) 1.0 per cent (e) 0.5 per cent
67. The cement used in construction of docks and harbours is
    (a) blast-furnace slag cement (b) water proof cement (c) hyrophobic cement (d) sulphate-resisting portland cement (e) high strength portland cement
68. Which cement is used for lining a sulphuric acid plant?
    (a) Super-sulphate cement (b) High-alumina cement (c) Portland-slag cement (d) Rapid-hardening portland cement (e) Portland-pozzolana cement
69. Which cement is used for lining deep tubewells?
    (a) High-alumina cement (b) Blast-furnace slag cement (c) Oil-well cement (d) Sulphate-resisting portland cement (e) Portland-pozzolana cement
70. The cement used for repair of canal banks during the rainy season is
    (a) high-alumina cement (b) rapid-hardening portland cement (c) oil-well cement (d) portland-pozzolana cement (e) portland-slag cement
71. The cement generally used for the construction of road pavements is
    (a) rapid-hardening cement (b) ordinary portland cement (OPC) (c) low-heat cement (d) blast furnace slag cement (e) none of the above
72. For ordinary portland cement the maximum expansion by Le Chatelier's method should not exceed
    (a) 2 mm (b) 5 mm (c) 7.5 mm (d) 10 mm (e) 12 mm
73. Autoclave method is used to determine
    (a) residue (b) expansion (c) heat of hydration (d) sulphur content (e) none of the above

74. Le Chatelier's method can be used to determine
    (a) unsoundness of cement (b) soundness of cement (c) fineness of aggregate (d) sulphur content (e) all of the above
75. The specific surface of OPC is determined by
    (a) Le Chatelier's apparatus (b) air-permeability method (c) autoclave method (d) sieve analysis (e) photo-calorimeter method
76. The specific surface of cement is expressed in
    (a) mm^2 (b) mm^2/g (c) g/mm^2 (d) $mm^3/g\,mm$ (e) any of the above
77. The hydration of concrete ceases at the temperature of
    (a) 0°F (b) 0°C (c) 11°F (d) 11°C (e) none of the above
78. The average specific surface of cement is closer to
    (a) 100000 mm^2/g (b) 200000 mm^2/g (c) 300000 mm^2/g (d) 400000 mm^2/g (e) 500000 mm^2/g
79. Initial setting time of concrete ceases at
    (a) −10°C (b) −4°C (c) 0°C (d) 4°C (e) none of the above
80. Which of the following statement(s) is incorrect?
    (a) Calcium chloride should not be used in pre-stressed concrete (b) Strength of concrete increases below freezing point of water (c) Hardening of concrete takes place rapidly in hot weather (d) The ingredients of concrete should be mixed within three minutes (e) All of the above
81. While _____ is a calcareous material _____ is an argillaceous material.
    (a) limestone, shale (b) clay, limestone (c) shale, limestone (d) slate, laterite (e) marl, chalk
82. The colour of ordinary portland cement is _____ and that of portland-pozzolana cement is _____.
    (a) white, black (b) brown, grey (c) grey, white (d) white, grey (e) grey, black
83. White cement is the _____ cement and low-heat cement is used in _____ structures.
    (a) cheapest, thin (b) costliest, thick (c) costliest, thin (d) cheapest, thick
84. For fineness test of cement IS sieve of _____ is used.
    (a) 90 μm (b) 9 μm (c) 150 μm (d) 300 μm (e) 600 μm

AGGREGATES

85. Aggregate is used in concrete because
    (a) it is a relatively inert material and is cheaper than cement (b) it imparts volume stability and durability to the concrete (c) it provides bulk to the concrete (d) it increases the density of the concrete mix (the aggregate is frequently used in two or more sizes) (e) all of the above
86. The function of fine aggregate is
    (a) to assist in producing workability and uniformity in the mixture (b) to assist the cement paste to hold the coarse aggregate particles in suspension (c) to promote plasticity in the mixture and prevent possible segregation of paste and coarse aggregate (d) all of the above (e) none of the above

416 Concrete Technology

87. An aggregate should
    (a) be of proper shape and size (b) be clean, hard and well graded (c) possess chemical stability (d) exhibit abrasion resistance (e) all of the above
88. An aggregate generally not preferred for use in concrete is one which has the following surface texture
    (a) smooth (b) rough (c) glassy (d) granular (e) honey-combed
89. Aggregate can be classified according to
    (a) geological origin (b) size (c) shape (d) unit weight (e) any of the above
90. The nominal size of particles of graded aggregate is said to be 12.5 mm when most of its passes through a _____ mm IS sieve and is retained in a __ _____ mm IS sieve.
    (a) 16, 4.75 (b) 12.5, 4.75 (c) 12.5, 10 (d) 16, 12.5 (e) 20, 12.5
91. Identify the incorrect statement(s).
    (a) Artificial aggregates namely broken bricks and air-cooled fresh blast-furnace-slag can be used in concretes. (b) Sand is generally considered to have a lower size limit of about 0.07 mm. (c) Aggregates provide about 75 per cent of the body of concrete. (d) Crushed stone sand is produced by crushing of hard stone. (e) None of the above.
92. Classification of aggregate according to size is
    (a) fine aggregate, coarse aggregate, and all-in-aggregate (b) natural sand, crushed stone sand, and crushed gravel sand (c) coarse, medium and fine sands (d) single size aggregate, coarse aggregate, and all-in-aggregate (e) any of the above
93. Spot the odd one(s).
    (a) rounded aggregate (b) irregular or partly rounded aggregate (c) angular flaky aggregate (d) single-size-aggregate (e) elongated aggregate
94. Which of the following statement(s) are incorrect?
    (a) Rounded aggregate requires minimum cement paste to make good concrete. (b) Irregular aggregate requires more cement paste to make a workable concrete. (c) Higher the angularity number, the more angular is the aggregate. (d) The shape and surface texture of the aggregate influence the workability of fresh concrete. (e) An aggregate is termed flaky when its least dimension is less than nine-fifth of its mean dimension.
95. The cyclopan aggregate has a size more than
    (a) 4.75 mm (b) 20 mm (c) 40 mm (d) 60 mm (e) 75 mm
96. If the fineness modulus of sand is 2.5 it is graded as
    (a) very coarse sand (b) coarse sand (c) medium sand (d) fine sand (e) very fine sand
97. Bulking of sand is the
    (a) rodding of the sand so that it occupies minimum volume (b) compacting of the sand (c) increase in the volume of sand due to moisture which keeps sand particles apart (d) segregating sand of particular size (e) none of the above
98. With 4 per cent moisture the bulking of sand may be of the order of
    (a) 2 to 5 per cent (b) 5 to 10 per cent (c) 10 to 15 per cent (d) 15 to 25 per cent (d) 25 to 30 per cent

99. Bulking of coarse aggregate is
    (a) less as compared to sand (b) more than that of sand (c) 15 per cent at 4 per cent moisture content (d) 25 per cent at 4 per cent moisture content (e) negligible

100. Which of the following statement(s) are incorrect?
    (a) Sintered flyash aggregate produces concrete with a density of 12 to 14 kN/m^3. (b) An aggregate with a higher modulus of elasticity generally produces a concrete with a higher modulus of elasticity. (c) The strength of bond between the aggregate and cement paste depends upon the surface texture. (d) The apparent specific gravity of the aggregate is with respect to void free volume. (e) The bulk density is affected by particle shape, size and grading of the aggregate.

101. Which of following statement(s) are correct?
    (a) The surface moisture expressed as a percentage of the weight of saturated surface dry aggregate is termed as moisture content. (b) The empty space between the aggregate particles are termed voids. (c) The thermal properties of the aggregate affect durability of concrete. (d) The alkali-aggregate reaction is a reaction between the active silica constituent of the aggregate and the alkalies in cement. (e) All of the above.

102. Gap grading is one
    (a) in which one or more intermediate fractions are absent (b) in which the particles fall within a narrow limit of size fractions (c) which combines different fractions of fine and coarse aggregates (d) in which all the particles are of uniform size (e) any one of the above

103. Which of the following is/are deleterious material in aggregate?
    (a) Coal (b) Clay lumps (c) Soft fragments (d) Shale (e) All of the above

104. Deleterious substances in aggregate are undesirable because they may
    (a) affect the strength, workability and long term performance of concrete (b) have intrinsic weakness, softness and fineness (c) interfere with the chemical reaction of hydration (d) interfere with the bond between the aggregate and cement paste (e) any one of the above

105. The fineness modulus
    (a) is a numerical index of fineness (b) gives some idea of the mean size of particles present in the entire body of aggregate (c) is a sum of the cumulative percentages retained on the set of specified sieves divided by 100 (d) is regarded as weighted average size of sieve on which material is retained (e) any one of the above

106. Grading of the aggregate
    (a) affects the workability (b) affects the strength of concrete (c) is dependent on the shape and texture of the particles of the aggregate (d) affects the water-cement ratio (e) all of the above is true

WATER

107. Which of the following statements is incorrect?
    (a) Water is the most important and least expensive ingredient of concrete.
    (b) Mixing water is utilized in the hydration of cement and provides

418   Concrete Technology

lubrication between fine and coarse aggregates. (c) Excess water forms a scum or laitance at the surface. (d) Excess water may make concrete honey combed (e) None of the above.

108. For mixing water
(a) suspended particles of clay and silt should be less than 0.02 per cent (b) the quantity of calcium chloride is restricted to 1.5 per cent (c) the pH value should generally be between 6 and 8 (d) free vegetable oil is harmful but mineral oil up to 2 per cent is beneficial (e) all of the above

109. If sea water is used for preparing concrete
(a) it will cause efflorescence (b) it may corrode the reinforcement (c) it will reduce the ultimate strength (d) it may cause dampness (e) all of the above

110. The vegetable oil, if present, in mixing water for concrete
(a) improves strength (b) reduces strength (c) gives more slump (d) gives a smooth surface (e) improves workability due to lubrication

111. The mineral oil, if present, in mixing water for concrete
(a) reduces strength for all concentrations (b) reduces strength for the concentration of oil up to 8 per cent (c) increases strength for the concentration up to 2 per cent (d) increases strength for a concentration beyond 8 per cent (e) does not affect the strength at all

112. The presence of sugar in water for concreting up to _____ per cent has virtually no adverse effect on the strength of concrete.
(a) 0.05 (b) 0.15 (c) 0.20 (d) 0.50 (e) 1.0

113. Presence of 0.20 per cent sugar by weight of cement in the mixing water is likely to
(a) retard the setting of cement (b) reduce the early strength of cement (c) accelerate the setting of cement (d) decrease workability (e) none of the above

114. Which of the following impurities in the mixing water is destructive?
(a) Calcium chloride (b) Lead nitrate (c) Alkalies (d) Algae (e) Sugar

115. With regard to the curing water identify the incorrect statement(s).
(a) Curing water should not produce objectionable stains on the surface. (b) The presence of tannic acid and iron compounds is objectionable. (c) Iron and organic matter are responsible for staining. (d) Water which is suitable for mixing is also suitable for curing. (e) None of the above

ADMIXTURES

116. An admixture
(a) is a basic ingredient of concrete (b) offers improvement not economically attained by adjusting mix proportions (c) is a substitute for good concreting practice (d) in excess quantity may be beneficial to the properties of concrete. (e) none of the above

117. Admixtures could be used to
(a) accelerate initial setting of concrete (b) increase the strength of concrete (c) improve workability (d) reduce heat of evolution (e) any of the above

118. Admixtures could be used to
    (a) retard the initial set (b) increase durability (c) improve pumpability of concrete (d) control the alkali-aggregate reaction (e) any of the above
119. Admixtures could be used to
    (a) improve impermeability (b) inhibit the corrosion of concrete (c) produce cellular concrete (d) produce non-skid surfaces (e) any of the above
120. Adding an accelerator to concrete increases all of the following except
    (a) resistance to alkali-aggregate reaction (b) rate of hydration of cement (c) shrinkage (d) rate of evolution of heat (e) rate of development of strength
121. An accelerator shortens all of the following except
    (a) setting time (b) period of curing (c) period of removal of formwork (d) strength of concrete
122. Following compounds can be used as accelerators except
    (a) $CaCl_2$ (b) $CaSO_4$ (c) $NaCl$ (d) Na_2SO_4 (e) K_2SO_4
123. Addition of a retarder to concrete decreases all of the following except
    (a) rate of hydration (b) water-cement ratio (c) workability and compressive strength (d) rate of development of strength (e) rate of evolution of heat
124. Addition of air-entraining agents to concrete increases all of the following except
    (a) workability (b) strength of concrete (c) durability (d) impermeability (e) uniformity
125. Identify the correct statement which corresponds to accelerator : retarder
    (a) $CaCl_2 : CaSO_4$ (b) $NaCl : CaCl_2$ (c) $NaOH : KOH$ (d) $KOH : NaOH$ (e) $CaSO_4 : CaCl_2$
126. Addition of pozzolana admixtures results in
    (a) improved workability (b) reduction in heat of hydration (c) increased resistance to sulphate attack (d) reduced leaching of calcium hydroxide (e) all of the above
127. Fly ash may be used as
    (a) a part replacement of cement (b) a part replacement of fine aggregate (c) simultaneous replacement of cement and fine aggregate (d) an admixture (e) all of the above
128. Identify the incorrect statement(s).
    (a) Fly ash is a residue from the combustion of pulverized coal collected by electrostatic separators from fuel gases of thermal power plants. (b) Fly ash consists of spherical glassy particles. (c) Specific surface of fly ash is about 350000 to 500000 mm^2/g. (d) Like portland cement, fly ash contains oxides of calcium, aluminium and silicon. (e) All of the above.
129. Superplasticizers or super water-reducers
    (a) result in greatly increased workability (b) result in an increased tendency to segregate (c) facilitate production of flowing concrete (d) contain sulphonated melamine-formaldehyde resin (e) all of the above

130. Air-entraining agents
    (a) are used for entraining air in concrete (b) contain wood resins, fats and lignosulphonates (c) increase durability of concrete to frost-action (d) do not require change in the cement content (e) all of above
131. The following can be used for water-proofing
    (a) potash soaps (b) butylstearate (c) petroleum waxes (d) all of above (e) none of the above

PROPERTIES OF FRESH CONCRETE

132. Fresh concrete should
    (a) be able to produce homogeneous concrete (mixable) (b) not segregate or bleed during transportation and placing (i.e. stable) (c) be cohesive and sufficiently mobile (i.e., flowable) (d) be amenable to thorough compaction and satisfactory surface finishing (i.e., compactable and finishable) (e) all of the above
133. Mixability of a concrete mix governs
    (a) power requirement in the mixing plant (b) reproducibility of concrete batches (c) time of mixing (d) homogeneity of fresh concrete (e) all of the above
134. Workability of fresh concrete is most appropriately defined by
    (a) the composite property satisfying the requirements of mixability, stability, transportability, placeability, mobility, compactability and finishability (b) ease and homogeneity with which it can be mixed, placed, compacted and finished (c) its consistency and plasticity (d) its slump and compaction factor values (e) all of the above
135. The empirical test used for assessing the workability of fresh concrete is
    (a) the slump test (b) the compacting factor test (c) the vee-bee consistency test (d) the flow test (e) all of the above
136. The conventional empirical tests for assessing workability of concrete suffer from the drawback that
    (a) they measure only a particular aspect of workability, i.e. these tests are single-point tests (b) they are operator sensitive (c) none of these tests is capable of dealing with the whole range of workabilities (d) they are not amenable to physical idealized modelling (e) all of the above
137. Identify the incorrect statement(s) with regard to the workability of fresh concrete.
    (a) The change in workability due to a relative change in the water content in concrete is dependent on the mix ratio. (b) An increase in water content may result in a monotonous increase in workability. (c) A high water content may result in segregation and bleeding. (d) Water content is limited to a value given by the water-cement ratio. (e) All of the above
138. Which of the following statement(s) is/are incorrect?
    (a) The use of a larger size and/or rounded aggregate gives higher workability. (b) For the same water content, use of finer sand increases the workability. (c) The grading of fine aggregate is more critical than the grading of coarse aggregate for workability. (d) For high-strength concrete a coarser grading is preferred. (e) fineness of cement has an influence on bleeding.

139. Identify the incorrect statement(s).
    (a) The segregation of coarse particles in a lean dry mix may be corrected by adding a small quantity of water to it.   (b) The tendency to segregate can be minimized by reducing the height of drop of concrete.   (c) The separation of cement paste from the concrete mix is termed segregation.   (d) The aim is to have minimum possible workability consistent with satisfactory placement and compaction of concrete.   (e) All the the above
140. Slump test is the most widely used field test primarily because
    (a) it indicates the behaviour of fresh concrete under action of gravitational forces   (b) of the simplicity of apparatus and test procedure   (c) it measures consistency or wetness of the mix   (d) it ensures uniformity among different batches of similar concrete   (e) all of the above
141. The workability of concrete by slump test is expressed as
    (a) mm^3/h (b) mm^2/h (c) mm/h   (d) mm   (e) hours
142. A concrete having a slump of 70 mm is termed as
    (a) dry   (b) semi-plastic   (c) plastic   (d) flowing   (e) none of the above
143. In case the concrete is to be transported by pumping, the slump should be
    (a) more than 100 mm   (b) between 50 to 70 mm   (c) between 25 mm and 50 mm   (d) more than 25 mm   (e) more than 10 mm
144. If the slump of concrete mix is 75 mm, its workability is considered to be
    (a) very high   (b) high   (c) medium   (d) low   (e) very low
145. The slump test of concrete is used to measure its
    (a) compaction under gravitational force   (b) mobility   (c) consistency   (d) homogeneity   (e) all of the above
146. For a RCC slab the slump of concrete should be
    (a) 0–25 mm   (b) 25–50 mm   (c) 25–100 mm   (d) 50–125 mm   (e) 100–150 mm
147. The slump of concrete to be transported by belt conveyors should be
    (a) 25–50 mm   (b) 50–75 mm   (c) 75–100 mm   (d) 100–125 mm   (e) 0–25 mm
148. Finishing of concrete surface will be difficult when the slump exceeds
    (a) 25 mm   (b) 40 mm   (c) 50 mm   (d) 75 mm   (e) 100 mm
149. Compacting factor test is superior to slump test mainly because it
    (a) gives behaviour of fresh concrete under the action of external forces   (b) measures compactibility of concrete   (c) is more accurate than slump test for concrete mixes of medium and low workabilities   (d) is more sensitive and gives more consistent results   (e) none of the above
150. The compacting factor test for fresh concrete
    (a) is adopted when nominal size of aggregate does not exceed 20 mm   (b) measures the relative effort required to change a mass of concrete from one definite shape to another   (c) measures the compaction obtained by a standard amount of work applied to a standard quantity of concrete   (d) gives an indication of the mobility of fresh concrete   (e) all of the above
151. A compacting factor of 0.88 for a fresh concrete sample indicates a mix of
    (a) high workability   (b) medium workability   (c) low workability   (d) very low workability   (e) none of the above

152. Concrete is considered unsuitable for compaction by vibration if
(a) the compacting factor is more than 0.9 (b) it is of low workability (c) it is very stiff (d) slump is between 25–50 mm (e) none of the above

153. Identify the correct statement(s)

	Degree of workability	Compacting factor	Slump
(a)	high	0.68	125–150 mm
(b)	medium	0.78	25–75 mm
(c)	low	0.84	10–50 mm
(d)	very low	0.90	—
(e)	none of the above		

154. A concrete is said to be workable if
(a) it is of uniform colour (b) it is almost a fluid (c) it can be easily mixed, placed and compacted (d) it has a tendency to segregate and bleed (e) none of the above

155. The Vee-Bee test
(a) is suitable for concrete mixes of low and very low workabilities (b) is a remoulding test (c) is unsuitable for concretes having a slump of 75 mm or above (d) is suitable since the concrete in the test receives a similar treatment as it would in actual practice (e) any of the above

156. Bleeding of concrete is said to occur when
(a) finer particles settle down at the bottom (b) coarser particles get separated (c) cement paste rises to the surface of concrete (d) finer particles collect in isolated pockets (e) none of the above

157. Identify the incorrect statement(s) with regard to the bleeding of concrete.
(a) Bleeding increases the permeability of concrete (b) Bleeding causes laitance at the surface. (c) Bleeding can be corrected by the addition of a small amount of water. (d) Bleeding reduces the durability of concrete. (e) None of the above.

158. Workability of concrete is independent of
(a) mix proportions (b) water content (c) size, shape and texture of aggregate (d) environment conditions (e) none of the above

159. The separation of coarse aggregate from mortar during transportation of concrete is termed
(a) bleeding (b) creeping (c) segregation (d) flow of concrete (e) cohesion

160. Segregation in concrete results in
(a) porous layers (b) honey-combing (c) sand streaks (d) surface scaling (e) all of the above

161. The workability of concrete can be improved by the addition of any of the following except
(a) fly ash (b) copper sulphate (c) calcium chloride (d) plasticizers (e) super plasticizers

162. A retarder plasticizer reduces workability loss
(a) due to slowing down the process of setting (b) due to air-entrainment (c) hydrophobic action (d) through the process of flocculation (e) all of the above

RHEOLOGY OF CONCRETE

163. Rheology of fresh concrete includes the following except
 (a) stability, mobility and compactability of concrete (b) knowledge of water-cement ratio (c) study of forces involved in transmission of stress through concrete mass (d) deformation curve of fresh concrete
164. The conventional workability tests e.g. slump test, compacting factor test, Vee-Bee test and remoulding tests are termed single-point tests because
 (a) they measure one parameter of workability (b) they are conducted at one place (c) each gives complete information about the workability (d) all of the above (e) none of the above
165. Rheological or flow equation of fresh concrete is expressed by
 (a) Newton's model (b) Bingham model (c) Le Chatelier's model (d) Neville model (e) none of the above
166. The following combinations of conventional workability tests help to acheive a better understanding of rheology of concrete
 (a) slump and compacting factor tests (b) Vee-Bee and compacting factor tests (c) slump and Vee-Bee tests (d) compacting factor and remoulding tests
167. Rheological properties of concrete are independent of
 (a) water-content (b) aggregate shape, texture and grading (c) type of mixer (d) temperature (e) type of cement
168. The flow properties of fresh concrete are mainly dependent upon
 (a) the factors affecting resistance to deformation (b) the water-cement ratio (c) the richness of the mixture (d) shape and texture of the aggregate (e) fineness modulii and gradings of the aggregates

PROPERTIES OF HARDENED CONCRETE

169. Concrete may be described as
 (a) an artificial stone obtained by binding together particles of relatively inert fine and coarse materials with cement paste (b) the most widely used man-made construction material tailored to meet the demands of any particular situation (c) an artificial stone in which voids of larger particles are filled by the smaller particles, and the voids of the finer particles are filled with cement paste (d) a material prepared from locally available materials by judicious mix proportioning and proper workmanship to satisfy performance requirements (e) any of the above
170. The main ingredients of concrete are
 (a) cement (b) aggregates (c) water (d) admixtures (e) all of the above
171. After curing normal concrete
 (a) shrinks on drying (b) expands on drying (c) shrinks when still wet (d) may shrink or expand depending upon the proportions of various ingredients (e) neither shrinks nor expands
172. The inert ingredient(s) of a concrete mix is/are
 (a) cement (b) aggregates (c) water (d) entire mix (e) none of the above

173. The best way to specify the concrete is by
    (a) performance-oriented specifications (b) prescriptive specifications
    (c) degree of control (d) unit weight (e) none of the above
174. The most appropriate method to specify the concrete mix is by
    (a) the nominal mix ratio (b) the designed mix ratio (c) the degree of control (d) the grade of concrete (e) none of the above
175. The strength of concrete is influenced by
    (a) size of test specimen (b) moisture conditions (c) type and rate of loading (d) type of testing machine (e) all of the above
176. The strength of concrete depends upon
    (a) type of cement (b) concrete mix proportions (c) degree of compaction (d) type and temperature of curing (e) all of the above
177. The compressive strength of concrete
    (a) decreases with the increase of aggregate-cement ratio (b) increases with the increase in the degree of compaction (c) decreases with entrained air (d) all of the above (e) none of the above
178. The stress and strain curve of concrete in compression is obtained by testing the cylinderical specimen under
    (a) uniform rate of strain (b) uniform rate of stress (c) constant stress condition (d) constant strain condition
179. As compared to the static tests the dynamic tests on concrete give
    (a) higher value of poisson's ratio (b) lower value of poisson's ratio (c) the same value of poisson's ratio (d) all of the above depending upon the test conditions (e) none of the above
180. The shrinkage in concrete is due to
    (a) hydration of cement (b) loss of water by evaporation from the surface (c) withdrawal of water stored in unsaturated air voids of concrete (d) all of the above (e) none of the above
181. Shrinkage increases with
    (a) increase in the water-cement ratio (b) increase in cement content (c) decrease in humidity (d) decrease in the maximum size of aggregate (e) all of the above
182. Permeability of concrete reduces
    (a) with the carbonation of concrete (b) with the strength of cement paste (c) with the decrease in the porosity (d) all of the above
183. The durability of concrete is due to its resistance to
    (a) deterioration from environmental conditions (b) internal desruptive forces (c) chemical attack (d) all of the above
184. The inelastic behaviour of concrete is due to the
    (a) shrinkage in concrete (b) propagation of bond and mortar cracks (c) presence of macro and micro cracks (d) use of aggregates (e) all of the above
185. The thermal conductivity of concrete decreases with the
    (a) light weight concretes (b) increase in the water-cement ratio (c) decrease in the cement content (d) all of the above (e) none of the above

186. For cement concrete the stress–strain curve is linear approximately up to
    (a) 1/4 of ultimate stress (b) 1/3 of ultimate stress (c) 1/2 of ultimate stress (d) 5/8 of ultimate stress (e) 3/4 of ultimate stress
187. The modulus of elasticity of concrete improves with
    (a) age (b) high water-cement ratio (c) shorter curing periods (d) better compaction (e) all of the above
188. Shrinkage of concrete can be reduced by using
    (a) low water-cement ratio (b) water-tight and non-absorbent formwork (c) presaturated aggregates (d) all of the above (e) none of the above
189. The strength of concrete mainly depends upon
    (a) quality of fine aggregate (b) quality of coarse aggregate (c) fineness of cement (d) water-cement ratio (e) none of the above
190. The thermal coefficient of expansion of concrete is approximately
    (a) 3×10^{-8} per °C (b) 3×10^{-6} per °C (c) 3×10^{-5} per °C (d) 3×10^{-4} per °C (e) 3×10^{-3} per °C
191. Creep in concrete is undesirable particularly in
    (a) continuous beams (b) reinforced concrete columns (c) pre-stressed concrete structures (d) all of the above (e) none of the above
192. The knowledge of the fluexural tensile strength is useful in design of
    (a) reinforced concrete members (b) pavement slabs and airfield runways (c) prestressed concrete structures (d) water-retaining structures (e) all of the above
193. Compressive strength of concrete is the most important property because
    (a) it depends upon the water-cement ratio (b) it is related to the structure of hardened cement paste and gives the overall quality of concrete (c) it indicates the extent of voids in the concrete (d) it affects the permeability and durability of concrete (e) none of the above
194. The concrete may attain its 100 per cent compressive strength after
    (a) 7 days (b) 14 days (c) 28 days (d) 1 year (e) 3 years
195. The strength of concrete is decreased by
    (a) vibration (b) impact (c) fatigue (d) all of the above (e) none of the above
196. The permissible stress for concrete subjected to fatigue should be
    (a) 25 per cent (b) 50 per cent (c) 75 per cent (d) 80 per cent (e) 95 per cent
197. According to the Indian Standard specifications, the maximum compressive strength of normal strength concrete can be
    (a) 5 MPa (b) 10 MPa (c) 12.5 MPa (d) 20 MPa (e) 40 MPa
198. The tensile strength of concrete is approximately *what* per cent of compressive strength of concrete?
    (a) 50% (b) 20% (c) 10% (d) 5% (e) 1%
199. The standard size of a concrete cube for compressive strength test is
    (a) 50 mm (b) 100 mm (c) 150 mm (d) 200 mm (e) 250 mm

426 Concrete Technology

200. As per Indian Standard specifications concrete is designated into
 (a) 3 grades (b) 5 grades (c) 7 grades (d) 10 grades (e) 12 grades
201. The porosity of concrete depends largely upon
 (a) cement content (b) grading of aggregate (c) quantity of mixing water
 (d) degree of compaction (e) all of the above
202. The concrete for sea water application should not be leaner than
 (a) $1:2:6$ (b) $1:2:4$ (c) $1:2:3$ (d) $1:1\frac{1}{2}:3$ (e) $1:1:2$
203. In case of plain concrete exposed to sea waves, the grade of concrete should not be lower than
 (a) M15 (b) M20 (c) M25 (d) M30 (e) M40
204. The concrete in contact with alkaline soil or alkaline water should
 (a) use a rich mix (b) have a high water-cement ratio (c) have a high alumina content (d) have a low water-cement ratio (e) have a higher percentage of fine aggregate
205. For high frost resistance the concrete should be
 (a) dense (b) free from cracks (c) steam cured (d) all of the above (e) none of the above
206. For compressive strength determination the minimum number of cubes required in a sample are
 (a) 2 (b) 3 (c) 5 (d) 6 (e) 9
207. The unit weight of plain concrete is generally taken as
 (a) 20 kN/m^3 (b) 22 kN/m^3 (c) 24 kN/m^3 (d) 25 kN/m^3 (e) 16kN/m^3
208. The unit weight of reinforced cement concrete is generally taken as
 (a) 18 kN/m^3 (b) 22 kN/m^3 (c) 24 kN/m^3 (d) 25 kN/m^3 (e) 26 kN/m^3
209. Which of the following has the highest unit weight?
 (a) Common brick work (b) Plain concrete with brick aggregate (c) Plain concrete (d) Reinforced concrete
210. Which one of the following does not react with concrete?
 (a) Sewage water (b) Sulphuric acid (c) Vegetable oil (d) Alcohol (e) none of the above
211. Presence of algae in concrete
 (a) reduces its strength (b) reduces its bond strength (c) causes a large entrainment of air (d) all of the above (e) none of the above
212. The direct methods for calculating tensile strength of concrete suffer due to the
 (a) presence of eccentricity in application of load (b) stress concentration at the jaws (c) difficulty in holding the specimen (d) all of the above (e) reasons other than the above
213. Split tensile strength tests are better than the direct tensile strength tests because
 (a) the test gives more uniform results (b) the results give values closer to the actual tensile strength values (c) same moulds can be used for both compression and tension tests (d) all the above (e) none of the above

QUALITY CONTROL

214. The quality control of concrete is appropriately defined as the
    (a) quality of overall workmanship and supervision at the site (b) assurance that all aspects of materials, equipment and workmanship are well looked after (c) conformity to the specifications, no more no less (d) assurance for the safety and serviceability of structure
215. The quality control can be exercised by
    (a) the field controls i.e. inspection and testing at all the stages (b) adequate compaction and curing (c) by strong motivation to do every thing right the first time (d) by restoring to the acceptance tests
216. Quality control means
    (a) extra cost (b) a rational use of the available resources (c) adequate design to minimize cost (d) all of the above
217. Statistical quality control helps
    (a) in narrowing down the tolerance limits of variability (b) in taking into account the actual variability of concrete (c) to ascertain the range of value that can be expected under existing conditions (d) all of the above
218. The standard deviation is
    (a) the measure of spread of compressive strength test results about the mean strength (b) the measure of variability of test results (c) measure of proportions of all the results falling within or outside certain range (d) all of the above (e) none of the above
219. The target mean strength of concrete mix is given by
    (a) $f_t = kf_{ck} + S$ (b) $f_t = f_{ck} + kS$ (c) $f_t = f_{ck} + S$ (d) $f_t = f_{ck} + k$
220. As per IS : 456–1978 concrete shall be deemed to satisfy the strength requirements if
    (a) every sample has a test strength not less than f_{ck} (b) up to 5 per cent of the samples may have strength less than f_{ck} (c) strength of a sample may be less than $f_{ck} - 1.35S$ (d) strength of sample may be less than $0.80 f_{ck}$
221. Larger the value of standard deviation
    (a) lower will be the variability (b) better will be level of control (c) poorer will be the level of control (d) none of the above

PROPORTIONING OF CONCRETE MIXES

222. The strength of concrete at 28 days as a percentage of strength at one year is
    (a) 98 (b) 90 (c) 80 (d) 75 (e) 60
223. The ratio of tensile strength of concrete to the compressive strength is
    (a) 1/33 (b) 1/25 (c) 1/20 (d) 1/10 (e) 1/5
224. The cube strength of concrete exceeds the cylinder strength by (in per cent)
    (a) 10 to 50 (b) 10 to 15 (c) 15 to 20 (d) 20 to 25 (e) 30 to 40
225. The permissible diagonal tension of M15 concrete is
    (a) 1.5 MPa (b) 1.2 MPa (c) 1.0 MPa (d) 0.75 MPa (e) 0.5 MPa

226. The proportions of materials in a concrete mix may be expressed in the form of

(a) parts (by volume) of cement, fine and coarse aggregates. (b) parts (by weight) of cement, the fine and coarse aggregates. (c) ratio of weight of cement to sum of weights of fine and coarse aggregates, i.e. cement-aggregate ratio (d) cement factor (e) any of the above

227. Identify the incorrect statement(s).

(a) Nominal mix is a mix of fixed proportions which ensure adequate strength. (b) Nominal mixes may result in under or over-rich mixes. (c) Standard mixes are useful as off-the shelf sets of proportions that allow the desired concrete to be produced. (d) Nominal or standard mixes may be used for high performance concrete. (e) Mix design ensures a concrete with the appropriate properties to be produced most economically.

228. The choice of mix proportions of a concrete is independent of

(a) grade designation (b) maximum nominal size of aggregate (c) minimum water-cement ratio (d) batching, mixing, placing and compaction techniques (e) durability and quality control

229. Which of the following statement(s) is correct?

(a) The aim of mix design is to produce concrete that satisfies the job requirements, namely of compressive strength, workability and durability as economically as possible. (b) Compressive strength is governed by the water-cement ratio. (c) For the given aggregates, the workability of concrete is governed by its water content. (d) Mix design is a basis for making an initial guess about the optimum combination of ingredients. (e) Final mix proportion is obtained on the basis of further trial mixes. (f) All of the above

230. The maximum nominal size of the coarse aggregate is determined by sieve analysis and is designated by the sieve size higher than the largest size on which the material retained is more than

(a) 5 per cent (b) 15 per cent (c) 25 per cent (d) 50 per cent (e) 75 per cent (f) none of the above

231. The larger maximum size aggregate

(a) is beneficial for high strength concrete (b) requires a smaller quantity of cement for a particular water-cement ratio (c) results in reduced workability (d) reduces stress concentration in mortar aggregate interfaces (e) all of the above

232. Identify the correct statement(s).

(a) For air-entrained concretes, the compressive strengths are approximately 80 per cent that of non air-entrained concrete. (b) Grade designation gives the characteristic compressive strength requirements of concrete. (c) Depending upon the degree of control, mix is designed for a target mean compressive strength of concrete. (d) Durability of concrete means its resistance to the deteriorating influences of environment. (e) Permeability of cement paste increases exponentially with an increase in water-cement ratio above 0.45. (f) All of the above

233. Water-cement ratio in concrete is the ratio of
    (a) volume of water to volume of cement   (b) volume of water to the weight of cement   (c) weight of water to the weight of cement   (d) weight of water to the volume of cement   (e) weight of water required for chemical reaction to the weight of water required to wet the cement

234. Lower water-cement ratio in concrete
    (a) increases the compressive strength   (b) improves the frost-resistance of concrete   (c) reduces the permeability of concrete   (d) reduces the shrinkage and creep   (e) all of the above

235. Water in excess of that required for chemical reaction in concrete results in
    (a) bleeding   (b) segregation   (c) cracks   (d) voids on drying   (e) honey combing

236. Most of the methods of concrete mix design are based on
    (a) the water-cement law as a criterion of strength   (b) the assumption that workability is solely dependent on the water content   (c) the assumption that durability is independent of the cement content   (d) principle that there is no air-entrainment in the mix   (e) all of the above

237. The common mix design method for medium strength concrete is
    (a) the trial and adjustment method   (b) the DoE (British) mix design method   (c) the ACI mix design method   (d) the mix design according to Indian Standard recommended guidelines   (e) any of the above

238. The trial and adjustment method
    (a) aims at producing a concrete mix which has minimum voids and hence maximum density   (b) requires sufficient quantity of cement paste to fill the voids in the mixed aggregate   (c) indicates that the optimum percentage of sand is lower for lower water-cement ratios   (d) all of the above   (e) none of the above

239. The DoE mix design method
    (a) determines aggregate-cement ratio   (b) uses free water content determined by the size and type of aggregate, and the level of workability   (c) uses free water-cement ratio based on target mean compressive strength   (d) all of the above   (e) none of the above

240. The ACI method of mix proportioning
    (a) uses bulk volume of coarse aggregate estimated for maximum nominal size of aggregate and fineness modulus of sand   (b) takes into account the air-content of concrete   (c) is suitable for normal and heavy weight concretes in the workability range of 25–100 mm slump   (d) can be used for the concrete having a target mean compressive strengths of up to 75 MPa   (e) all of the above   (f) none of the above

241. Mix design by Indian Standard recommended guidelines
    (a) is suitable for medium and high strength concretes   (b) requires data on the characteristic strength, degree of workability, limitations on the water-cement ratio, and maximum nominal size of aggregate   (c) calculations are based on an absolute volume basis   (d) all of the above   (d) none of the above

430 Concrete Technology

242. For a constant aggregate-cement ratio, if the coarse aggregate is increased at the expense of sand, the total surface of the aggregage
(a) remains constant   (b) is reduced   (c) is increased   (d) depends on other factors   (e) none of the above

243. In a trial concrete mix, if the desired slump is not obtained, the adjustment in the water content for each 10 mm difference in slump (in per cent) is
(a) 0.5   (b) 1.0   (c) 2.0   (d) 5.0   (e) 10.0

244. In a concrete mix design, while making adjustments for the air-entrainment of amount e, the quantity of water is reduced by v, then the reduction in the solid volume of sand is given by
(a) $e-v$ (b) $1/2(e-v)$ (c) $e-v/2$ (d) $1/2(e+v)$ (e) $e/2-v$

245. If the trial mix gives a higher 28-days compressive strength value than the design value, then for the next trial
(a) cement content is reduced   (b) water content is increased   (c) water-cement ratio is increased   (d) proportion of sand is increased   (e) curing period is decreased

246. When water is added in an increasing amount to a fixed mass of dry mortar mix, the volume of mortar
(a) initially increases then decreases to a minimum value   (b) does not change as the water simply fill the voids   (c) decreases   (d) increases   (e) increases proportionately more than the volume of water added

247. The volume of water which corresponds to a minimum volume of mortar is termed
(a) saturation water content   (b) basic water content   (c) lowest water content   (d) highest water content   (e) hygroscopic water content

248. The amount of water mixed in mortar should be always
(a) more than the basic water content   (b) equal to the basic water content   (c) less than the basic water content   (d) 50 per cent of the basic water content   (e) none of the above

249. A water content of 1.25 for a mortar mix means
(a) 1.25 litre of water has been added per litre of cement   (b) 1.25 litre of water has been added per litre of mortar   (c) 25 per cent more water has been added than the basic water content requirements   (d) 1 litre of water has been added in 1.25 litre of mortar   (e) none of the above

250. The nominal mix corresponding to M20 grade concrete is
(a) $1:1:2$ (b) $1::1\frac{1}{2}:3$ (c) $1:2:3$ (d) $1:2:4$ (e) $1:3:6$

251. The grade of concrete corresponding to nominal mix proportions of $1:3:6$ is
(a) M35   (b) M25   (c) M15   (d) M10   (e) M7.5

252. The total number of grades of ordinary concrete stipulated in IS : 456–1978 are:
(a) 10   (b) 8   (c) 7   (d) 6   (e) 5

253. The volume of sand per cubic metre of $1:2:4$ (by volume) concrete would be approximately
(a) 0.2 to 0.4   (b) 0.4 to 0.6   (c) 0.6 to 0.9   (d) 0.8 to 1.0   (e) none of the above

254. The number of bags of cement required per cubic metre of 1 : 2 : 4 concrete would be approximately
 (a) 5 to 6 (b) 4 to 5 (c) 3 to 4 (d) 2 to 3 (e) 1 to 2
255. For slabs and beams, the concrete of nominal mix generally used is
 (a) 1 : 1 : 2 (b) 1 : $1\frac{1}{2}$: 3 (c) 1 : 2 : 4 (d) 1 : 3 : 6 (e) 1 : 2 : 4
256. For water retaining structures the nominal mix generally used is
 (a) 1 : 1 : 2 (b) 1 : $1\frac{1}{2}$: 3 (c) 1 : $1\frac{1}{2}$: 4 (d) 1 : 2 : 4 (e) 1 : 2 : 6
257. To take into account the variation of individual samples, the laboratory design strength can be obtained by increasing the target mean strength by (per cent)
 (a) 5 (b) 5 to 10 (c) 10 to 15 (d) 15 to 20 (e) 20 to 25
258. To ensure proper quality control, the number of cube specimens to be made for 5m³ of concrete are
 (a) 3 (b) 6 (c) 9 (d) 12 (e) 15
259. After moulding, the test specimens of trial mix are placed at a temperature of
 (a) 10 ± 2° (b) 15 ± 2° (c) 23 ± 2° (d) 27 ± 2° (e) 100°C
260. As per Indian Standard specifications concrete should be cured under a humidity of (per cent)
 (a) 10 (b) 25 (c) 50 (d) 75 (e) 90
261. It is often difficult to place the whole of the concrete in one operation and hence joints are provided. To have proper joints
 (a) the joint should be provided along the line of minimum shear (b) at the joint the old surface should be treated with a rich cement mortar paste before the new concrete is laid (c) the reinforcement of old concrete should extend into the new one (d) all of the above (e) none of the above
262. Which of the following statements is correct?
 (a) Bulking of sand always decreases with an increase in the quanlity of water. (b) While batching by weight, the effect of bulking of sand is not considered. (c) For mass concrete in the foundation the mix proportions are 1 : 2 : 4. (d) The water content in ordinary concrete is 5 per cent by weight of cement and 30 per cent by weight of aggregate. (e) All of the above
263. For a water-cement ratio of 0.6 the water content per bag of cement is
 (a) 10 kg (b) 20 kg (c) 30 kg (d) 40 kg (e) 50 kg

For questions 264 to 268 refer to the data given below:
A concrete mixer of 400 litres capacity is used to prepare concrete of mix proportions 1 : 2 : 4 by volume with a water-cement ratio of 0.6. The surface moisture of coarse and fine aggregates are 6 and 2 per cent (by volume), respectively. The bulking of sand is 15 per cent.

264. Volume of sand required for a batch will be (in litre)
 (a) 100 (b) 200 (c) 300 (d) 400
265. Volume of cement required for a batch would be
 (a) 100 (b) 200 (c) 300 (d) 400
266. Volume of wet sand required will be (in litre)
 (a) 170 (b) 195 (c) 215 (d) 230
267. Volume of wet aggregate required will be (in litre)
 (a) 224 (b) 336 (c) 400 (d) 505

268. Water required to be added is (in litre)
    (a) 44   (b) 52   (c) 58   (d) 65
269. A concrete has mix proportions of 1 : 2 : 4 by dry volume with a water-cement ratio of 0.6. The bulk densities of cement, sand and coarse aggregate are 1500, 1725, 1615 kg/m^3, respectively. The mix proportions by weight are
    (a) 1 : 1.74 : 3.78   (b) 1 : 1.9 : 3.90   (c) 1 : 2 : 4   (d) 1 : 2.3 : 4.3
270. Erntroy and Shacklock's empirical method
    (a) is used for proportioning high strength concrete mixes   (b) relates compressive strength to an arbitrary reference number   (c) is limited to the aggregate containing up to 30 per cent, materials which can pass through a IS : 4.75 mm sieve   (d) uses a water-cement ratio corresponding to the reference number for the given workability   (e) all of the above

PRODUCTION OF CONCRETE

271. Which of the following statement(s) are incorrect?
    (a) The design of a satisfactory mix proportions ensures quality concrete work.   (b) The production of concrete consists of mixing ingredients to obtain some kind of plastic mass.   (c) Batching, mixing, transportation, placing, compaction, finishing and curing are independent operations for the production of concrete.   (d) All of the above   (e) None of the above
272. Identify the incorrect statement(s).
    (a) Batching, mixing, transportation, placing, compaction, finishing and curing are complimentary operations for the production of quality concrete.   (b) Good quality concrete is a homogeneous mixture of ingredients obtained by a scientific process based on well established principles.   (c) The aim of quality control is to ensure the production and continuous supply of concrete of uniform strength.   (d) A proper and accurate measurement of ingredients is essential to ensure uniformity of proportions and aggregate grading in successive batches.   (e) None of the above
273. In the batching of materials, the ingredients should be measured to a tolerance (as a percentage of batch quantity) of
    (a) ±1.0   (b) ±2.0   (c) ±3.0   (d) ±5.0   (e) 0.0
274. In weight batching the weight of surface water carried by the wet aggregate
    (a) can be ignored   (b) must be taken into account   (c) may or may not be taken into account depending upon the type of job   (d) is taken care off by drying the aggregate
275. The choice of a proper batching system depends upon
    (a) size of job   (b) required production rate   (c) required standard of batching performance   (d) availability of resources   (e) all of the above
276. A mobile mixing plant is particularly useful
    (a) as it can be kept close to the site where concreting is required   (b) where concrete is required to be laid over a very large area   (c) as it can also be used to carry the materials   (d) all of the above   (e) none of the above

277. Identify the incorrect statement(s).

(a) In volume batching it is generally advisable to set the volumes in terms of whole bags of cement. (b) In volume batching, allowance has to be made for the moisture present in the sand. (c) While filling the measuring boxes, no compaction is to be allowed. (d) Volume batching is adopted for small jobs. (e) All of the above (f) None of the above

278. Mixers are normally classified on the basis of

(a) the technique of discharging the mixed concrete (b) capacity of batch handled (c) the number of drums (d) any of the above

279. The capacity of a concrete mixer is expressed in terms of

(a) total volume of concrete produced per day (b) total volume of concrete produced in 8 hours (c) total volume of concrete produced per hour (d) volume of concrete mix handled per batch (e) weight of aggregate per batch

280. A mixer designated 400 NT indicates that

(a) it is a non-tilting type mixer (b) its nominal mix batch capacity is 400 litres (c) both of the above (d) it is a non-tilting type mixer requiring 400 revolutions for proper mixing

281. The objectives of mixing concrete materials are the following except

(a) coat the surface of all aggregate particles with cement paste (b) blend all the ingredients into a uniform mass (c) obtain concrete of uniform colour and grading (d) obtain concrete of desired workability

282. Identify the incorrect statement(s).

(a) The size of the mixer is designated by a number representing its nominal mix batch capacity in litres. (b) Most of the mixers can handle a 15 per cent overload satisfactorily. (c) In the tilting type mixer the chamber is tilted for discharging. (d) The efficiency of the mixing operation depends upon the shape and design of vanes fixed inside the drums. (e) A non-tilting type mixer rotates about a horizontal axis and cannot be tilted.

283. The pan mixer consisting of a circular pan rotating about a vertical axis is suitable

(a) as a mobile mixer (b) as a central mixing plant (c) for ready mixed concrete (d) any of the above (e) none of the above

284. The mixing time

(a) is the time required to produce uniform concrete (b) is reckoned from the time when all the solid materials have been put in the mixer (c) is independent of the number of revolutions (d) may be ignored in favour of number of revolutions (e) all of the above

285. In machine mixing, the recommended minimum mixing time for mixers up to 750 litre capacity reckoned from the time when all the materials have been added is (in minutes)

(a) 1.0 (b) 1.5 (c) 2.0 (d) 2.25 (e) 5.0

286. Identify the incorrect statement(s).

(a) Delays in laying the concrete after the initial set has taken place are not injurious provided the concrete retains adequate workability for compaction.

(b) The specifications permit a maximum of two hours between introduction of mixing water to the dry mix and the discharge if concrete is transported in a truck mixer or agitator.   (c) During transportation of concrete segregation should be prevented and the concrete should remain uniform.   (d) All of the above   (e) None of the above

287. The freshly mixed concrete can be transported by
(a) barrows   (b) trippers and lorries   (c) truck mixers or agitator lorries   (d) dump buckets   (e) any of the above

288. Pumpable concrete
(a) is tranported through completely filled delivery pipelines   (b) should be very cohesive and fatty having a slump of 50 to 100 mm   (c) should have mix proportions with total fines passing 200 μm sieve not less than 350 kg/m^3 (d) is high slump, flowing concrete obtained by using superplasticizers   (e) all of the above

289. While pumping concrete
(a) care should be taken to reduce the number of bends in the delivery pipe   (b) the pipe should be cleaned immediately after use   (c) initially a 1 : 3 cement-sand mortar should be pumped to lubricate the pipeline   (d) all of the above   (e) none of the above

290. (A) When concrete is pumped by a pump of 60 h.p., the maximum horizontal distance that can be covered would be
(a) 150 m   (b) 200 m   (c) 300 m   (d) 350 m   (e) 400 m
(B) A 90° bend in the pipeline reduces the effective pumping distance by approximately
(a) 10 m   (b) 5 m   (c) 3 m   (d) 2 m   (e) 1.5 m

291. Ready mixed concrete (RMC)
(a) is weigh batched and mixed in a centrally located plant, transported in a track mixer or agitator and delivered in a condition ready to used   (b) is produced under site conditions   (c) does not require control of all operations of manufacture and transportation of fresh concrete   (d) all of the above   (e) none of the above

292. Sometime when the concrete is partially mixed at the central plant and mixing is completed enroute, the concrete is known as
(a) transit-mixed concrete   (b) ready-mixed concrete   (c) shrink-mixed concrete   (d) any of the above   (e) none of the above

293. Identify the incorrect statement(s) with respect to placing of concrete.
(a) Concreting should begin at the ends or corners of forms and continue towards the centre.   (b) In large openings concreting should end around the perimeter.   (c) On a slope concreting should begin at the lower end of slope.   (d) The concrete in columns should be allowed to stand for at least two hours before concrete is placed in slab or beams.   (e) None of the above

294. The effect of delay in placing of concrete
(a) is the gain in compressive strength provided concrete can still be adequately compacted   (b) varies with the richness of the mix and the initial slump   (c) is automatically taken into account by controlling the uniformity of concrete as delivered for placement   (d) all of the above   (e) none of the above

295. Identify the incorrect statement(s).
(a) The process of removal of entrapped air and uniform placement of concrete to form a homogeneous dense mass is termed as compaction.
(b) Compaction is accomplished by doing external work. (c) Presence of even 5 per cent voids in hardened concrete due to incomplete compaction may reduce compressive strength by about 40 per cent. (d) All of the above (e) None of the above

296. Compaction by mechanical vibrations is suitable for
(a) all the grades of concrete (b) all the structural elements (c) all the mixes except very plastic mixes (d) all of the above (e) none of the above

297. The acceleration imposed on the particles during compaction of concrete by high frequency vibrations is of the order
(a) up to g (b) g to 2g (c) 4g to 7g (d) 7g to 9g (e) none of the above

298. Which type of vibrator is generally used for compaction of concrete?
(a) form vibrator (b) needle vibrator (c) surface vibrator (d) screen vibrator (e) none of the above

299. For compacting thin reinforced concrete slabs following vibrator is recommended
(a) immersion vibrator (b) surface vibrator (c) vibrating table (d) any of the above (e) none of the above

300. Surface vibrator is effective only when the thickness of concrete member does not exceed
(a) 100 mm (b) 125 mm (c) 150 mm (d) 200 mm (e) 500 mm

301. A surface vibrator for compaction of concrete is preferred for all of the following except
(a) raft footings (b) columns (c) R.C.C. slab (d) road pavements

302. While using vibrators for compacting concrete mixes
(a) vibrations are used for spreading concrete in the form (b) vibrations reduce entrained air (c) vibrations cause smaller and lighter constituents to rise to the surface and give better finish (d) prolonged vibrations reduces chances of segregation (e) all of the above

303. Rotational instability occurring during compaction of concrete using vibration technique is due to
(a) mix being under-sanded (b) mix being oversanded (c) entrapped air (d) all of the above (e) none of the above

304. Curing of concrete
(a) governs the resultant microstructure of the hydrated cement (b) provides adequate moisture within concrete to ensure sufficient water for continuing hydration process (c) provides warm temperature to help chemical action (d) all of the above (d) none of the above

305. Maturity of concrete is the
(a) 28-day strength of concrete (b) 365-day strength of concrete (c) product of period of curing and temperature of curing (d) percentage of strength of concrete cured at 18°C for 28 days (e) none of the above

306. Membrane curing of the concrete is the
(a) process of providing plastic sheeting as a protective cover for curing concrete (b) process of applying a membrane forming compound on the

concrete surface (c) process of spraying the sodium silicate on the concrete surface (d) any of the above (e) none of the above

307. In steam curing of concrete
 (a) mixes of high water-cement ratio respond more favourably than mixes with low water-cement ratio (b) the heating of concrete products is caused by steam at low pressure or at high pressure (c) the steam curing is followed by water curing for a period of at least 21 days (d) all of the above (e) none of the above

308. The following methods may be used for the curing of concrete except
 (a) membrane curing (b) electrical curing (c) mechanical curing (d) infrared radiation curing (e) chemical curing

309. The following sealing compounds can be used for the membrane curing of concrete except
 (a) rubber latex emulsions (b) asphaltic emulsion or cutbacks (c) sodium silicate solution (d) emulsions of paraffin (e) varnishes

310. The standard moist curing of concrete for the first 7 to 14 days may result in a compressive strength of _____ per cent of 28-day moist curing.
 (a) 60 to 70 (b) 70 to 80 (c) 80 to 90 (d) 90 to 95 (e) none of the above

311. The following conditions of concrete placement are termed as extreme environmental conditions.
 (a) When concreting operations are carried out at temperature beyond 40°C. (b) When concreting operation are done at temperature below 5°C. (c) Underwater concreting. (d) Any of the above. (e) None of the above.

312. Concreting in hot weather
 (a) reduces handling time of fresh concrete and strength of hardened concrete (b) increases tendency to cracking (c) make it difficult to control air-content (d) all of the above (e) none of the above

313. In hot weather concreting it is recommended to
 (a) use cold mixing water (b) have minimum cement content consistent with other functional requirements (c) use cements with lower heat of hydration and use water reducing admixtures (d) reduce period between mixing and placement to a absolute minimum (e) all of the above

314. Concreting in cold weather
 (a) reduces rate of development of strength (b) delays removal of formwork (c) temperature differential within the concrete mass may promote cracking (d) freezing and thawing during the prehardening period may reduce strength by 50 per cent (e) all of the above

315. In cold weather concreting it is recommended to
 (a) heat the water for mixing (b) use insulating formwork and delay removal time (c) use additional quantity of cement (d) use air-entraining agents (e) all of the above

316. In cold weather curing of concrete should be continued for
 (a) 7 days (b) 14 days (c) 21 days (d) 28 days (e) 45 days

317. For placing the concrete underwater the principal technique(s) used are
 (a) tremie method (b) bucket placing (c) placing in bags (d) prepacked concrete (e) any of the above

318. The timber formwork for concrete should be made of
    (a) teak wood (b) Shisham wood (c) soft wood planks (d) green timber
    (e) hard wood
319. For a concrete slab for a 3.75×4.75 m room the stripping time of form should be
    (a) 3 days (b) 7 days (c) 14 days (d) 21 days (e) 28 days
320. To take care of any sag in the beams, the forms are given a camber of
    (a) 1 : 200 (b) 1 : 300 (c) 1 : 500 (d) 1 : 650 (e) 1 : 750
321. For a medium income group big housing project which type of formwork you would recommend?
    (a) timber formwork (b) plywood formwork (c) steel formwork (d) other type

INSPECTION AND TESTING

322. Identify the incorrect statement(s).
    (a) The testing of representative concrete does not give the quality of actual in-place concrete. (b) Quality control can be exercised by testing three concrete cubes at 28 days. (c) The quality control is carried out much before any cubes become available for testing. (d) Cube tests relate to concrete specimens specially prepared for testing. (e) The influence of workmanship in placing, compaction and curing can be judged by testing the concrete in the structure.
323. The concept of performance oriented specifications suffers due to difficulty in
    (a) defining what constitutes satisfactory performance (b) setting appropriate performance limits (c) the absence of tests to monitor the performance (d) all of the above (e) none of the above
324. Which of the following statement(s) are incorrect?
    (a) Uniform workability ensures uniform strength. (b) The ball-penetration test can be performed on concrete as placed in the forms. (c) Slump test is more accurate than compacting factor test and the results can be reproduced. (d) Vee-Bee test is suitable for low and very low workabilities.
    (e) None of the above.
325. The permissible variation in the compacting factor measurement is
    (a) ± 0.02 for C.F. values of 0.90 or more (b) ± 0.04 for C.F. values between 0.90 and 0.80 (c) ± 0.06 for C.F. values of 0.80 or less (d) ± 0.07 for C.F. values below 0.70 (e) none of the above
326. The allowable variation in the slump measurement is
    (a) ± 25mm (b) one-third of the required value (c) lesser of the above values (d) greater of the above values (e) none of the above
327. The cement content in a sample of fresh concrete can be determined by
    (a) rapid analysis machine (b) EDTA titration method (c) HCl heat of solution method (d) accelerated strength method (e) any of the above

328. Identify the incorrect statement(s).

(a) The accelerated strength test results are of doubtful nature as far as potential 28-day strength of concrete is concerned.   (b) The maturity test gives valid results provided concretes have initial temperature between 15°C to 26°C.   (c) The ultrasonic pulse velocity method can assess the quality of in-place fresh concrete.   (d) At the time of initial setting the fresh concrete has ultrasonic pulse velocity of the order of 2000 m/s.   (e) All of the above.

329. The quality and strength of concrete in a structure can be assessed by

(a) the concrete core test   (b) the pull out test   (c) the ultrasonic method   (d) the schmidt test hammer method   (e) any of the above

330. In ultrasonic test for hardened concrete good quality of concrete is indicated if the pulse velocity is

(a) below 3.0 km/s   (b) between 3.0 to 3.5 km/s   (c) above 3.5 km/s   (d) above 4.5 km/s   (e) none of the above

331. The ultrasonic pulse velocity test is based on the assumption that

(a) the time taken by a pulse in passing through a concrete mass is proportional to the modulus of elasticity of the concrete   (b) the frequency of pulse is proportional to the compressive strength of concrete   (c) the amplitude of the pulse is proportional the compressive strength of the concrete   (d) due to internal flaws the pulse velocity is reduced   (e) none of the above

332. The resonant frequency method is based on the assumption that

(a) pulse velocity depends primarily upon the materials and mix proportions of the concrete   (b) the modulus of elasticity of concrete improves with the quality of concrete   (c) resonant frequency is directly proportional to the square of strength of concrete   (d) strength of concrete increases with the age   (e) all of the above

SPECIAL CONCRETE AND CONCRETING TECHNIQUES

333. Special concretes are obtained by improving the properties of concrete by

(a) modification in the microstructure of the cement paste   (b) reduction in the overall porosity   (c) improvements in the strength of aggregate-matrix interface   (d) control of extent and propagation of cracks   (e) all of the above

334. The microstructure of the concrete can be improved by

(a) application of high pressure during moulding   (b) moulding at a temperature up to 150°C   (c) application of high pressure during moulding at high temperature   (d) prolonged steam curing   (e) none of the above

335. Sulphur-impregnated concrete is obtained by

(a) emptying the pores in the conventional concrete under vacuum and sucking the liquid sulphur in the pores   (b) mixing the sulphur powder as an ingredient of normal concrete and heating the cured concrete at high temperature   (c) applying a coating of molten sulphur on the surface of the concrete   (d) any of the above   (d) none of the above

336. The extent and propagation of cracks in concrete can be controlled by
(a) providing reinforcement bars   (b) incorporating fibres in the concrete   (c) polymer impregnation   (d) all of the above   (e) none of the above

337. Light-weight concrete is used
(a) for reducing the dead weight of structures   (b) for improving thermal insulation   (c) in filler wall panels in multistorey buildings   (d) for non-load bearing and partition walls   (e) any of the above

338. The light-weight concrete may be produced by
(a) incorporating air in its composition   (b) using light aggregates   (c) omitting the finer sizes from the aggregate grading that is using no fines concrete   (d) formation of air voids in the cement slurry by the addition of substances causing sponge-like cellular forms   (e) any of the above

339. Aerated concrete is produced by addition of
(a) copper sulphate   (b) aluminium powder   (c) sodium silicate   (d) zinc sulphate   (e) none of the above

340. Light-weight concrete has all the following beneficial characteristics except
(a) high strength to mass ratio   (b) high thermal insulation   (c) high sound insulation   (d) excellent fire resistance   (e) reduced drying shrinkage

341. Light-weight aggregate are produced by
(a) bloating clays with or without additives   (b) sintering fly ash   (c) using blast furnace slag   (d) any of the above   (e) none of the above

342. In mass concrete
(a) a large size aggregate and low slump is adopted   (b) mix being harsh and dry requires immersion type power vibrators   (c) heat of hydration may lead to a considerable rise of temperature   (d) there is early high strength but lower later age strength   (e) all of the above

343. Vacuum concrete
(a) is obtained by vacuum treatment of fresh concrete involving the removal of excess water and air by suction   (b) is the normally cured hardened concrete involving removal of air from the voids of the concrete by suction   (c) is no-fine-concrete where finer sizes are omitted from the aggregate grading producing uniformly distributed voids in the concrete mass   (d) has a low wear and abrasion resistance   (e) none of the above

344. The ferrocement is a composite material obtained by
(a) random dispersal of short, dis continuous fibres in the conventional concrete   (b) reinforcing the cement mortar with steel fibres in the form of wire mesh   (c) blending ferrous compounds in the ordinary portland cement   (d) any of the above procedures   (e) none of the above

345. The cement-sand ratio in the ferrocement matrix should not be leaner than
(a) 1 : 1.5   (b) 1 : 2.0   (c) 1 : 3.0   (d) 1 : 4.0   (e) 1 : 6.0

346. The volume of reinforcement in ferrocement (per cent) normally varies between
(a) 1–2   (b) 2–5   (c) 5–8   (d) 8–10   (e) none of the above

347. The aggregate i.e. the sand recommended for ferrocement mixes is
(a) with maximum sizes of 2.36 mm and 1.18 mm with optimum grading zones II and III   (b) with maximum size 4.75 mm and grading zone I   (c) with maximum size of 600-micron and grading zone IV   (d) any of the above   (e) none of the above

348. The water-cement ratio for ferrocement mix should be
    (a) less than 0.35   (b) between 0.35 to 0.40   (c) between 0.40 and 0.50
    (d) between 0.50 and 0.60   (e) greater than 0.60

349. For ferrocement structures exposed to corrosive environements
    (a) apply asphaltic and bituminous coatings on the exposed surface
    (b) apply rust proof paint on the wire mesh   (c) apply vinyl and epoxy coatings on the wire, mesh   (d) any of the above   (e) none of the above

350. Identify false statement(s).
    (a) The cracking resistance, ductility, flexibility, impact and fatigue resistances of ferrocement are higher than those of concrete.   (b) The impermeability of ferrocement products is far more superior than ordinary R.C.C. product.   (c) Ferrocement has high resistance to cyclic loading.   (d) Ferrocement is suitable for manufacturing the precast units.   (e) none of the above

351. Fibrous ferrocement
    (a) is obtained by adding short fibres to plain mortar matrix of ferrocement
    (b) has improved toughness and impact resistance over conventional ferrocement   (c) panels can withstand very high stresses compared to those in conventional ferrocement   (d) all of the above   (e) none of the above

352. Identify the true statement(s)
    (a) Fresh fibre concrete has reduced workability.   (b) Fibre reinforced concrete is more cohesive and less prone to segregation.   (c) High modulus fibres improve both flexural and impact resistance simultaneously.   (d) Low modulus fibre improve only the impact resistance of the concrete.   (e) All of the above

353. Fibre reinforced concrete
    (a) is used for precast products, airport runways, blast and impact resistant structures, tunnel lining and hydraulic structures   (b) has superior crack resistance, improved ductility, high impact resistance and toughness   (c) uses indented, crimped or bent fibres for improved bond   (d) is more vulnerable to corrosion damage than conventional steel reinforcement   (e) all of the above

354. Which of the following statement(s) are incorrect?
    (a) The ratio of diameter of fibre to its length is called aspect ratio.   (b) The improvement in structural performance of fibre reinforced concrete depends on strength characteristics of fibres, volume of fibres, shape and aspect ratio of fibres.   (c) The ultimate strain of steel fibre reinforced concrete is 20 to 50 times that of plain concrete.   (d) With 4 per cent of steel fibres, the flexural strength increases by 2.5 times the strength of unreinforced composite.   (e) None of the above

355. Identify the incorrect statement(s).
    (a) Asbestos fibres cause health hazards.   (b) With vegetable fibres long term durability is doubtful.   (c) Organic fibres may decay.   (d) Carbon fibres are very expensive and are more vulnerable to damage than glass fibres, hence are treated with resin coating.   (e) Glass fibres are chemically stable in cement paste matrix.

356. Which of the following fibres give highest improvement in the impact strength of fibre reinforced concrete?
    (a) Polypropylene, nylon and other organic fibres (b) Glass fibres
    (c) Carbon fibres (d) Asbestos fibres (e) Vegetable fibres
357. The fundamental requirement of fibre reinforced concrete is
    (a) uniform distribution of fibres throughout the mix (b) mix should have sufficient paste to coat the fibres and aggregate (c) mix should have optimum content of fibres for workability (d) all of the above (e) none of the above
358. The performance of hardened fibre reinforced concrete basically depends upon the
    (a) specific fibre surface (b) critical concentration of fibres (c) aspect ratio (d) all of the above (e) none of the above
359. The following range of parameters is associated with commonly encountered steel fibre reinforced concrete mixes except
    (a) water-cement ratios: 0.45 to 0.60 (b) cement content: 300–500 kg/m^3
    (c) fibre aspect ratio: 10 to 100 (d) fibre content: 1.0 to 2.5 per cent
    (e) fine/total aggregate ratio: 0.5 to 1.0
360. Polymer-impregnated concrete is obtained by
    (a) impregnating low viscosity prepolymers or monomers into the pore systems of hardened concrete and polymerizing it by heating (b) replacing the cement-water matrix in cement concrete by pre-polymer and polymerizing it (c) incorporating a polymeric material into concrete during the mixing state (d) any of the above processes (e) none of the above
361. The partial or surface polymer impregnation of bridge deck concrete mainly improves
    (a) its structural properties (b) durability and chemical resistance (c) the riding of the surface (d) all of the above (e) none of the above
362. Factors affecting the monomer loading are the
    (a) extent of moisture in concrete and air in the voids (b) degree of vacuum applied and its duration (c) viscosity of monomer and external pressure (d) all of the above (e) none of the above
363. The polymerization by heating the catalyzed monomer to the required level can be done by
    (a) heating under water (b) low pressure steam injection (c) infra-red heaters (d) heating in an air-oven (e) any of the above
364. Identify the incorrect statement(s).
    (a) Thermoplastic monomers lose their effectiveness at high temperatures.
    (b) Thermosetting resins are more viscous and difficult to impregnate into concrete, but can withstand higher temperatures without softening. (c) The heating decomposes the polymer before polymerization by cross-linking.
    (d) All of the above (e) None of the above
365. The repair of damaged concrete structures by polymer impregnation is
    (a) cost effective for normal structures (b) suitable for restoration and preservation of stone monuments (c) suitable for strengthening piles in sea water (d) all of the above (e) none of the above

366. In polymer concrete (resin concrete) the polymerization can be achieved by
(a) thermal-catalytic reaction (b) catalyst-promotor reaction (c) radiation (d) any of the above (e) none of the above

367. Polymer concrete can be used
(a) for overlays (b) rapid repair of damaged airfield pavements and industrial structures (c) treating sluiceway and stilling basin of the dam (d) for the manufacture of electrical transmission poles, etc (e) all of the above

368. Polymer modified concrete
(a) is obtained by incorporating a polymeric material into concrete during the mixing stage (b) constitutes an interpenetrating matrix that binds the aggregate (c) is least expensive and the processing is simplest (d) all of the above (e) none of the above

369. Shotcrete
(a) is mortar or very fine concrete deposited by jetting it with high velocity on to the prepared surface (b) is frequently more economical than conventional concrete (c) is very useful for restoration and repair of fire damaged concrete structures (d) is used for stabilization of rock slopes, etc (e) all of the above

370. Shotcrete differs from conventional concrete with regard to
(a) materials, proportions and void system (b) consolidation or compaction (c) application procedure (d) nature of failure (e) all of the above

371. Identify the incorrect statement(s).
(a) If the water content is more then the concrete tends to slump when jetted on to the vertical surface. (b) If the water is deficient, the material which will rebound from the surface will be excessive. (c) The water-cement ratio should be between 0.45 to 0.60. (d) In the wet mix process all the ingredients are mixed before entering the chamber of delivery equipment. (e) Dry mix process is preferred in case of light weight concrete.

372. Shotcretes suffer due to
(a) environmental hazards because of dust problem (b) necessity of cleaning and hauling of rebound material to the approved waste area (c) high cost of shotcrete and wastage due to rebound (d) spalling of shotcrete due to corrosion of reinforcement and peeling of sound shotcrete because of bond failure due to lack of surface preparation (e) all of the above

373. Special shotcretes can be obtained
(a) by adding up to 2 per cent of steel fibres (by volume) (b) by using calcium aluminate cement (hydraulic cement) as binding agent (c) by using air-entraining cements (d) any of the above (e) none of the above

374. While repairing deteriorated concrete by shotcreting it is essential that
(a) all unsound material be removed (b) at the perimeter of the cavity square shoulders should be provided (c) shotcrete is applied on the moist surface (d) the fresh layer of shotcrete is applied before the receiving layer takes it initial set (e) all of the above

375. Guniting
(a) is the technique of depositing very thin layers of mortar in each pass of nozzle than that available with the shotcrete (b) mix is 1 : 3 to 1 : 4.5 with a water-cement ratio of about 0.30 (c) requires careful and skilful handling of nozzle for high quality finish work (d) reduces permeability and enhances resistance to weathering and chemical attack (e) all of the above

DETERIORATION OF CONCRETE AND ITS PREVENTION

376. Identify the incorrect statement(s).
(a) The durability of concrete is its resistance to the deteriorating influences of both external and internal agencies. (b) the gradual deterioration of concrete by chemically aggressive agents is called corrosion of concrete. (c) corrosion of steel reinforcement results in internal stresses which promote destruction of concrete. (d) microcracks present at the aggregate-cement interface do not affect durability as long as they are small in size and discontinuous. (e) none of the above.

377. Following are the principal forms of destruction of concrete except
(a) decomposition of concrete due to leaching action (b) chemical reaction between hardened cement constituent of concrete and chemicals of a solution (c) grinding action of sand storms (d) crystallization resulting in the increase of volume of solid phase within the pore structure

378. Deterioration of concrete can be prevented by
(a) making a denser concrete (b) ensuring sufficient alkalinity to provide passive environment against corrosion of reinforcement (c) limiting water-cement ratio (d) impregnating the pores of concrete with a suitable polymer (e) all of the above

379. The corrosion of reinforcement can be prevented by the following except
(a) the use of corrosion inhibitors, coating on steel or concrete (b) proper design of concrete for the intended environmental exposure (c) use of stainless steel (d) impregnating the pores of concrete by suitable polymer (e) increase thickness of concrete cover over reinforcement

380. The corrosion of steel reinforcement embedded in concrete is rapid when the member is immersed in
(a) acidic solution (b) alkaline solution (c) water with dissoved oxygen (d) sea-water (e) water with chloride ion concentration

381. To reduce the corrosion of reinforcement the chloride ions should be limited to its threshold or critical value of (per cent by mass of cement):
(a) 0.05 (b) 0.10 (c) 0.15 (d) 0.20 (e) 0.25

382. Identify the incorrect statement(s).
(a) In the regions having relative humidity of 50 per cent or less corrosive action may be neligible. (b) Structures permanently immersed in water exhibit little corrosion of the reinforcement. (c) The distress due to corrosive action may be in the form of deep pitting and a severe loss of cross-section of reinforcement. (d) For prestressed concrete the total amount of chloride ions in concrete should be limited to 0.10 per cent by mass of cement. (e) For reinforced concrete members totally immersed in sea-water additional cover of 40 mm should be provided.

383. The following compounds may be used as inhibitors of iron corrosion

(a) potassium dichromate, zinc and lead chromates   (b) calcium and sodium chlorides   (c) calcium sulphate   (d) fly ash   (e) molasses

REPAIR TECHNOLOGY FOR CONCRETE STRUCTURES

384. The distress in concrete structures may be due to

(a) poor construction practices   (b) errors in design and detailing   (c) environmental loads exceeding design stipulations   (d) shrinkage, thermal stresses and corrosion of reinforcement   (e) all of the above

385. Blow-holes are caused by

(a) improper design of form work   (b) lack of compaction   (c) inadequate workability   (d) excess of water-content   (e) none of the above

386. The repair of bulges, projections, bolt-holes and blow-holes should be done within _____ of stripping off the forms.

(a) 24 hours   (b) 3 days   (c) 7 days   (d) 21 days   (e) 28 days

387. The most common symptom of distress in a concrete structure is

(a) spalling of concrete   (b) cracking of concrete   (c) scaling of concrete   (d) surface crazing   (e) none of the above

388. Identify the incorrect statement(s).

(a) Cracks may represent the total extent of damage or they may point to problems of greater magnitude.   (b) The cracking in concrete structure is not necessarily a cause for blaming the designer, builder or supplier.   (c) A crack where movement is observed to continue is termed dormant.   (d) Cracks in concrete members usually occur due to incompatible dimensional (volume) changes.   (e) None of the above.

389. The cracking in plastic concrete is caused by

(a) relative volume changes between surface and interior concrete by very rapid loss of moisture due to low humidity, high wind or high temperature   (b) local restraint to the continuing consolidation or settlement of the concrete   (c) leaking or highly flexible forms   (d) any of the above   (e) none of the above

390. The cracking of plastic concrete can be controlled by

(a) using plastic sheeting to cover the surface between final finishing operations   (b) using lowest possible slump   (c) adequate vibration and proper form design   (d) providing sufficient time interval between placement of concrete in various elements   (e) all of the above

391. The cracking due to weathering in structural concrete can be controlled by the following measures except

(a) use lowest practical water-cement ratio   (b) use lowest practical water content   (c) use non air-entrained concrete   (d) reduce the temperature differences within a concrete structure   (e) none of the above

392. The cracking due to corrosion of reinforcement is characterized by

(a) exposed reinforcement   (b) splitting and spalling of concrete in definite patterns   (c) longitudinal cracks parallel to the bar or spalling of concrete   (d) rust staining   (e) all of the above

393. Before proceeding with the repair of the cracks, diagnosis is made to determine the

    (a) location and extent of cracking (b) causes of cracking (c) likely extent of further deterioration (d) suitability of various remedial measures (e) all of the above

394. The weak spots in a concrete member can be identified by

    (a) tapping the surface and observing the sound for hollow areas (b) opening up (by chipping) the suspected weak concrete (c) observing spalled areas, exposed reinforcement, surface deterioration and rust staining (d) by non-destructive tests (e) all of the above

395. Microstructure cracks in concrete have a size, i.e. width/depth ranging form

    (a) 0.01 to 0.05 mm (b) 0.05 to 0.1 mm (c) 0.1 to 0.3 mm (c) 0.3 to 0.5 mm (e) 0.5 to 1.0 mm

396. The selection of repair technique is based on following objective(s):

    (a) to restore load carrying capacity (b) to improve functional requirements (c) to improve durability (d) to prevent access of corrosive agents to the reinforcement (e) one or more of the above

397. The preparation of surface for repair consists of

    (a) complete removal of unsound material (b) undercutting with formation of smooth edges (c) removing crack from the surface (d) providing rough but uniform surface (e) all of the above

398. The laitance at the surface of the concrete can be best removed by

    (a) grinding, scarifying and sand blasting (b) sand blasting only (c) acid etching (d) cleaning with detergents or caustic soda solution (e) any of the above

399. For repair of large and deep patches of deteriorated concrete in a structure, the material filling is done by

    (a) dry packing (b) concrete replacement method (c) mortar replacement method (d) grouting (e) prepacked concrete

400. Wide and deep cracks in concrete members may be repaired by

    (a) grouting (b) shotcreting or guniting (c) mortar replacement (d) epoxy injection (e) any of the above

401. The technique of epoxy injection is used for

    (a) sealing actively leaking cracks (b) sealing narrow cracks in structural members (c) repairing water tanks/hydraulic structures (d) any of the above (e) none of the above

402. For sealing the cracks in concrete structures by using epoxy, the minimum width of routing required is

    (a) 3 mm (b) 6 mm (c) 9 mm (d) 15 mm (e) 20 mm

403. The stitching of cracks is done to accomplish all of the following except

    (a) to re-establish tensile strength across major cracks (b) to prevent the crack from further propagation (c) to stiffen the structure (d) to strengthen the cracks which have a tendency to close as well as to open

446 Concrete Technology

404. The bonded overlays of polymer-modified portland cement concrete or mortar
(a) can be used for repairing slab and decks containing fine dormant cracks
(b) have minimum overlay thickness of 75 mm   (c) should be mixed, placed and finished within 30 minutes   (d) should be cured for 72 hours   (e) none of the above

405. Polymer impregnation of concrete is not effective in the following cases except
(a) when cracks are dry   (b) when cracks contain moisture   (c) when volatile monomer evaporates   (d) when cracks to be repaired are fine

406. Sulphur impregnated concrete can be used
(a) as a practical and inexpensive substitute of polymer impregnated concrete (PIC)   (b) for repairing fractured elements   (c) for repairing slab and decks containing fine dormant cracks   (d) as a practical substitute of polymer-modified cement concrete overlay   (e) none of the above

407. The polymer system(s) commonly used for polymer concrete is
(a) latexes of styrene butadiene acrylic   (b) methyl-methacrylate   (c) 99.9 per cent pure sulphur   (d) all of the above   (e) none of the above

408. Jacketing
(a) is a process of fastening a durable material over concrete and filling the gap with grout   (b) increases the section of an existing member by encasement in a new concrete   (c) is used for compression members like columns and piles   (d) along with collars can be advantageously used for repairing deteriorated concrete columns   (e) all of the above

409. Autogenous healing
(a) is a natural process of crack repair occurring in the presence of moisture
(b) cannot take place in continuous saturation   (c) is particularly effective in cycles of drying and re-immersion   (d) all of the above   (e) none of the above

ANSWERS TO THE OBJECTIVE TYPE QUESTIONS

1. (e)	2. (c)	3. (c)	4. (c)	5. (c)	6. (a)	7. (b)	8. (e)	
9. (d)	10. (c)	11. (e)	12. (c)	13. (c)	14. (a)	15. (b)	16. (e)	
17. (c)	18. (d)	19. (d)	20. (d)	21. (a)	22. (e)	23. (e)	24. (c)	
25. (b)	26. (c)	27. (d)	28. (b)	29. (d)	30. (c)	31. (b)	32. (c)	
33. (d)	34. (c)	35. (a)	36. (e)	37. (e)	38. (b)	39. (d)	40. (d)	
41. (d)	42. (c)	43. (b)	44. (a)	45. (a)	46. (a)	47. (d)	48. (c)	
49. (e)	50. (c)	51. (e)	52. (a)	53. (b)	54. (b)	55. (e)	56. (b)	
57. (a)	58. (a)	59. (b)	60. (c)	61. (b)	62. (a)	63. (d)	64. (c)	
65. (c)	66. (a)	67. (d)	68. (b)	69. (c)	70. (b)	71. (a)	72. (d)	
73. (b)	74. (b)	75. (b)	76. (b)	77. (c)	78. (c)	79. (c)	80. (b)	
81. (a)	82. (c)	83. (b)	84. (a)	85. (e)	86. (d)	87. (e)	88. (c)	
89. (e)	90. (c)	91. (e)	92. (a)	93. (d)	94. (e)	95. (e)	96. (d)	
97. (c)	98. (d)	99. (e)	100. (d)	101. (e)	102. (a)	103. (e)	104. (e)	
105. (e)	106. (a)	107. (e)	108. (e)	109. (e)	110. (b)	111. (c)	112. (a)	
113. (c)	114. (b)	115. (e)	116. (b)	117. (e)	118. (e)	119. (e)	120. (a)	

Appendix

121. (d)	122. (b)	123. (c)	124. (b)	125. (a)	126. (e)	127. (e)	128. (e)	
129. (e)	130. (e)	131. (d)	132. (e)	133. (e)	134. (a)	135. (e)	136. (e)	
137. (a)	138. (b)	139. (c)	140. (b)	141. (d)	142. (c)	143. (b)	144. (c)	
145. (c)	146. (d)	147. (b)	148. (c)	149. (d)	150. (c)	151. (b)	152. (a)	
153. (c)	154. (c)	155. (e)	156. (c)	157. (c)	158. (a)	159. (c)	160. (e)	
161. (b)	162. (a)	163. (b)	164. (a)	165. (b)	166. (b)	167. (c)	168. (a)	
169. (e)	170. (e)	171. (a)	172. (b)	173. (a)	174. (d)	175. (e)	176. (e)	
177. (d)	178. (a)	179. (a)	180. (d)	181. (e)	182. (d)	183. (d)	184. (b)	
185. (d)	186. (c)	187. (a)	188. (d)	189. (d)	190. (d)	191. (c)	192. (b)	
193. (b)	194. (e)	195. (d)	196. (b)	197. (e)	198. (c)	199. (c)	200. (c)	
201. (e)	202. (b)	203. (c)	204. (c)	205. (d)	206. (b)	207. (c)	208. (d)	
209. (d)	210. (d)	211. (d)	212. (d)	213. (d)	214. (b)	215. (a)	216. (b)	
217. (b)	218. (d)	219. (b)	220. (a)	221. (c)	222. (c)	223. (d)	224. (d)	
225. (e)	226. (e)	227. (d)	228. (d)	229. (f)	230. (b)	231. (b)	232. (f)	
233. (c)	234. (e)	235. (d)	236. (b)	237. (e)	238. (d)	239. (d)	240. (e)	
241. (d)	242. (b)	243. (b)	244. (a)	245. (c)	246. (a)	247. (b)	248. (a)	
249. (c)	250. (b)	251. (d)	252. (c)	253. (a)	254. (a)	255. (c)	256. (b)	
257. (c)	258. (a)	259. (d)	260. (e)	261. (d)	262. (b)	263. (c)	264. (b)	
265. (a)	266. (d)	267. (c)	268. (d)	269. (d)	270. (e)	271. (d)	272. (e)	
273. (c)	274. (b)	275. (e)	276. (b)	277. (f)	278. (a)	279. (d)	280. (c)	
281. (d)	282. (b)	283. (b)	284. (e)	285. (a)	286. (e)	287. (e)	288. (e)	
289. (d)	290a.(e)	290b.(a)	291. (a)	292. (c)	293. (b)	294. (d)	295. (d)	
296. (d)	297. (c)	298. (b)	299. (b)	300. (d)	301. (b)	302. (b)	303. (b)	
304. (d)	305. (c)	306. (b)	307. (b)	308. (c)	309. (c)	310. (b)	311. (d)	
312. (d)	313. (e)	314. (e)	315. (e)	316. (d)	317. (e)	318. (c)	319. (c)	
320. (c)	321. (c)	322. (b)	323. (d)	324. (c)	325. (b)	326. (c)	327. (e)	
328. (a)	329. (e)	330. (c)	331. (a)	332. (b)	333. (e)	334. (c)	335. (a)	
336. (b)	337. (e)	338. (e)	339. (b)	340. (c)	341. (d)	342. (e)	343. (a)	
344. (b)	345. (c)	346. (c)	347. (a)	348. (b)	349. (a)	350. (c)	351. (d)	
352. (e)	353. (e)	354. (a)	355. (e)	356. (b)	357. (d)	358. (a)	359. (c)	
360. (a)	361. (b)	362. (d)	363. (e)	364. (c)	365. (b)	366. (d)	367. (e)	
368. (d)	369. (e)	370. (e)	371. (c)	372. (e)	373. (d)	374. (a)	375. (e)	
376. (e)	377. (c)	378. (e)	379. (c)	380. (a)	381. (c)	382. (d)	383. (a)	
384. (e)	385. (a)	386. (a)	387. (b)	388. (c)	389. (d)	390. (e)	391. (c)	
392. (e)	393. (e)	394. (e)	395. (c)	396. (e)	397. (e)	398. (c)	399. (b)	
400. (a)	401. (b)	402. (b)	403. (c)	404. (a)	405. (a)	406. (a)	407. (b)	
408. (e)	409. (a)							

Bibliography

1. ACI Committee 116, "Cement and Concrete Terminology", SP-19, *American Concrete Institute*, 1967, p. 146.
2. ACI Committee 201, "Guide for Making a Condition Survey of Concrete in Service", *Journal of American Concrete Institute*, Vol. 65, No. 11, November 1968, pp. 905–918.
3. ACI Committee 201, "Guide to Durable Concrete", *Journal of American Concrete Institute*, Vol. 74, No. 12, December, 1977, pp. 573–609.
4. ACI Committee 211, "Recommended Practice for Selecting Proportions for Normal and Heavy Weight Concrete", ACI 211.1-77, *Journal of American Concrete Institute*, Vol. 66, No. 8, 1969, pp. 612–629; Vol. 70, No. 4, 1973, pp. 253–255; Vol. 71, No. 11, 1974, pp. 577–578; Vol. 74, No. 2, 1977, pp. 59–60.
5. ACI Committee 212, "Guide For Use of Admixtures in Concrete", *Journal of American Concrete Institute*, Vol. 68, No. 9, September, 1971, pp. 646–676.
6. ACI Committee 214, "Recommended Practice for Evaluation of Strength Test Results of Concrete", ACI 214–77, *Journal of American Concrete Institute*, Vol. 73, No. 5, 1976, pp. 265–278.
7. ACI Committee 224, "Control of Cracking in Concrete Sturctures", *Journal of American Concrete Institute*, Vo. 69, No. 12, December, 1972, pp. 713–753.
8. ACI Committee 304, "Recommended Practice for Measuring, Transporting and Placing Concrete", ACI 304–73 (Reaffirmed 1978), *Journal of American Concrete Institute*, Vol. 69, July, 1972, pp. 374–414; Vol. 70, January, 1973, p. 55; Vol. 70, May, 1973, p. 322.
9. ACI Committee 305, "Hot Weather Concreting", *Journal of American Concrete Institute*, Vol. 74, August, 1977, pp. 317–332.
10. ACI Committee 313, "Guide for Structural Lightweight Aggregate Concrete", ACI 213 R-79, *Concrete International*, Vol. 1, No. 2, February, 79, pp. 33–62.
11. ACI Committee 403, "Guide for Use of Epoxy Compounds with Concrete", *Journal of American Concrete Institute*, Vol. 59, No. 9, September, 1962, pp. 1121–1142.
12. ACI Committee 506, "Specifications for Materials, Proportioning and Application of Shotcrete", ACI-506, 2–77, *Journal of American Concrete Institute*, Vol. 73, December, 1976, pp. 679–685; Vol. 74, November, 1977, p. 563.
13. ACI Committee 544, "State-of-the-Art Report on Fibre Reinforced Concrete", *Journal of American Concrete Institute*, Vol. 70, November, 1973, pp. 729–744.
14. ACI Committee 544, "Measurement of Properties of Fibre Reinforced Concrete", *Journal of American Concrete Institute*, Vol. 75, 1978, pp. 283–289.
15. ACI Committee 548, "Polymers in Concrete", *Journal of American Concrete Institute*, 1971.
16. ACI Committee 612, "Curing of Concrete", *Journal of American Concrete Institute*, Vol. 30, No. 2, August, 1958, pp. 161–172.
17. Allen, R.T.L. and S.C. Edward (ed.), *Repair of Concrete Structure*, Blackie, London, 1987.
18. American Concrete Institute, "Epoxies with Concrete", *Special Publication No. 21* (SP-21), 1968, p. 145.

19. American Concrete Institute, "Polymers in Concrete", *Special Publication No. 40* (SP-40), Detroit, 1973, p. 368.
20. American Concrete Institute, "Polymers in Concrete: International Symposium", *Special Publication No. 58* (SP-58), Detroit, 1978, pp. 420.
21. American Concrete Institute, "Ferrocement: Material and Applications", *Special Publication No. 61* (SP-61), Detroit, 1979.
22. American Concrete Institute, "Recommended Practice for Cold Weather Concreting", ACI 306–1966, *ACI Manual of Concrete Practice*, Part I, 1979, pp. 306–319.
23. American Concrete Institute, "Testing of Hardened Concrete: Non-Destructive Methods", *ACI-Monograph No. 9*, 1974, p. 153.
24. Angles, J.G., "Measuring Workability", *Concrete*, Vol. 8, No. 2, 1975.
25. Anon, "Air-Permeability of Concrete", *Concrete and Constructional Engineering*, London, May, 1965, p. 166.
26. Bansal, T.K. and V.V. Sastry, "Effect of Corrosive Environment on Concrete Structures", *Symposium on Corrosion and Its Prevention*, Institute of Technology, BHU, Varanasi, India, October, 1984.
27. Beeby, A.W., "Cracking, Cover and Corrosion of Reinforcement", *Concrete International: Design and Construction*, Vol. 5, No. 2, February, 1983, pp. 35.
28. Berry, E.E. and V.M. Malhotra, "Fly Ash for Use in Concrete: A Critical Review", *Journal of American Concrete Institute*, Vol. 77, No. 8, 1980, pp. 59–73.
29. Blick, R.L., C.F. Petersen and M.E. Winter, "Proportioning and Controlling High Strength Concrete", *Proportioning Concrete Mixes, ACI Special Publication SP 46*, American Concrete Institute, 1974.
30. Bloem, D.L., "Effect of Maximum Size of Aggregate on Strength of Concrete", *National Sand and Gravel Association*, Washington, Circular No. 74, February, 1959.
31. Bloem, D.L. and Gaynor, R.D., "Effect of Aggregate Properties on Strength of Concrete", *Journal of American Concrete Institute*, Vol. 60, October, 1963, pp. 1429–1455.
32. Bogue, R.H. *Chemistry of Portland Cement*, Reinhold, New York, 1955.
33. Brook, J.J., P.J. Wainwright and A.M. Neville, "Time Dependent Properties of Concrete Containing a Superplasticizing Admixture", *Superplasticizers in Concrete*, American Concrete Institute, Special Publication No. 62, 1979, pp. 293–314.
34. BS : 1881–Part 201, *Guide to use of Non-destructive Methods of Test for Hardened Concrete*, British Standards Institution, London.
35. BS : 5168–1975, *Glossary of Rheological Terms*, British Standards Institution, 1975.
36. BS : 6089, *Guide to Assessment of Concrete Strength in Existing Structures*, British Standards Institution, London.
37. Buckingham, E., "Model Experiments and the Form of Empirical Equations", *Transactions of American Society for Mechanical Engineers*, Vol. 37, 1915, pp. 263–296.
38. Bungey, J.H., "Concrete Strength Variation and In-place Testing," *Proceedings of 2nd Australian Conference on Engineering Materials*, University of New South Wales, Sydney, 1981, pp. 85–96.
39. Bungey, J.H., *Testing of Concrete in Structures*, 2nd edn, Surrey University Press, London, 1989.
40. Chanda, J.N. and P.C. Sharma, "Quality Cement Mortar For Ferrocement Structures", *Proceedings, Asia-Pacifc Symposium on Ferrocement for Rural Development*, Roorkee, April, 1984.
41. Concrete Society (London), "Admixtures for Concrete", *Technical Report TRCS1*, December, 1967, p. 12.
42. Concrete Society (London), "Concrete Core Testing for Strength", *Technical Report No. 11*, May, 1976, p. 44 and 87.
43. Concrete Society (London), *Polymers in Concrete*, Construction Press Limited, 1976.

44. Cooke, A.M., "A Guide to the Design of Concrete Mixes", *Technical Report No. TR 36*, Cement and Concrete Association of Australia, 1974.
45. CP : 110 (Part I)-1972, *Code of Practice for the Structural Use of Concrete, Part I—Design, Materials and Workmanship*, British Standards Institution, London, November, 1972.
46. CRI, "Development and Application of Polymer Concrete Composites in India", Cement Research Institute of India, New Delhi, Special Publication SP-12, 1978.
47. Cusens, A.R., "The Measurement of the Workability of Dry Concrete Mixes", *Magazine of Concrete Research*, Vol. 8, No. 22, March, 1956, pp. 23–30.
48. Dahl, G., "Vacuum Concrete", Swedish Cement and Concrete Research Institute, *CBI Report 7 : 75*, Part I, 1975, p. 10.
49. Danke, P.S., "Role Played by Quality of Cement in Concrete Making", *Indian Concrete Journal*, Vol. 52, No. 9, 1978, pp. 236–40.
50. Dass, A., "Dampness in Building: A General Appraisal of Its Causes and Remedies", *Journal of Institution of Engineers* (India), Vol. 65, Part C 15, March 1985, pp. 181–184.
51. Davey, N., "Concrete Mixes for Various Building Purposes", *Proceedings of Symposium on Mix Design and Quality Control of Concrete*, Cement and Concrete Association, London, 1954, pp. 28–41.
52. Dewar, J.D., "Some Effects of Prolonged Agitation of Concrete", *Technical Report 24.367*, Cement and Concrete Association, London, 1962.
53. Dewar, J.D., "Relation Between Various Workability Control Tests for Ready Mixed Concrete", Cement and Concrete Association, U.K., *Technical Report TRA-375*, 1964.
54. Dibean, J.T., "Development and Use of Polymer Concrete and Polymer-Impregnated Concrete", *Progress in Concrete Technology*, CANMET, Energy, Mines and Resources, Canada 1980, pp. 539–583.
55. Double, D.D., "Studies of Hydration of Portland Cement", *Proceedings of Admixtures Congress, C180* Construction Press, Lancaster, 1980, pp. 32–48.
56. Edmeades, R.M. and P.C. Hewlett, "Superplasticized Concrete: High Workability Retention", *Proceedings of Admixtures Congress*, Construction Press, Lancaster, C180, 1980, 49–72.
57. Erntroy, H.C. and B.W. Shacklock, "Design of High-strength Concrete Mixes", *Proceedings of Symposium on Mix Design and Quality Control of Concrete*, Cement and Concrete Association, London, May 1954, pp. 55–73.
58. Erntroy, H.C., "Variation of Works Test Cubes", *Research Report No. 10*, Cement and Concrete Association, London, November, 1960.
59. "Fibre Reinforced Cement Composites", Concrete Society, London, *Publication No. 51.067*, 1973.
60. Forbes, W.S., *Ready Mixed Concrete*, Contractors Record Limited, London, 1958.
61. Fulton, A.A. and W.T. Marshall, "The Use of Fly Ash and Similar Materials in Concrete", *Proceedings of Institution of Civil Engineers*, Part I, Vol. 5, No. 6, November, 1956, pp. 714–730.
62. Fulton, P.S., *Concrete Technology*, The Portland Cement Institute, Johannesburg, 1977.
63. Gambhir, M.L., *Concrete Manual: Laboratory Testing For Quality Control of Concrete*, 4th Edn. Dhanpat Rai and Sons, Delhi, 1992.
64. Gaynor, R.D., "Ready Mixed Concrete", *ASTM Special Publication No. 169B*, 1978, pp. 471–502.
65. Ghosh, R.K., M.R. Chatterjee and R. Lal, "Flexural Strength of Concrete: Its Variations, Relationships with Compressive Strength and Use in Concrete Mix Design", *Road Research Bulletin No. 16*, Indian Road Congress, 1972.
66. Ghosh, R.K., M.R. Chatterjee and R. Lal, "Accelerated Strength Tests for Quality Control of Paving Concrete", *ACI Publication SP-56:* Accelerated Strength Testing, 1978, pp. 169–182.

67. Gonnerman, H.F., "Effect of Size and Shape of Test Specimen on Compressive Strength of Concrete", *Proceedings of ASTM*, Vol. 25, Part II, 1925, pp. 237–250.
68. Hannant, D.J., *Fibre Cements and Fibre Concretes*, John Wiley and Sons Ltd., Chichester, 1978.
69. Hansen, T.C., "Cracking and Fracture of Concrete and Cement Paste", *Symposium on Causes, Mechanism and Control of Cracking in Concrete*, American Concrete Institute, *Special Publication No. 20*, 1968, pp. 5–28.
70. Hewlett, P.C. and R. Rixom, "Superplasticized Concrete", *Concrete*, Vol. 10, No. 9, September, 1976, pp. 33–42.
71. Hewlett, P.C. (Ed), "Superpasticizing Admixtures in Concrete", Cement and Concrete Association, Slough, *Publication No. 45.030*, 1976
72. Higginson, E.C., "The Effect of Cement Fineness on Concrete", *ASTM Special Publication 473*, 1970.
73. Highway Research Board, "The Alkali-Aggregate Reaction in Concrete", *Research Report 18C*, 1958.
74. Houston, J.T. et al., "Corrosion of Reinforcing Steel Imbedded in Structural Concrete", Centre of Highway Research, Texas, 1972.
75. Hughes, B.P. and B. Bahramian, "Workability of Concrete: A Comparison of Existing Tests", *Journal of Materials*, Vol. 2, No. 3, 1967, pp. 519–536.
76. Hughes, B.P., "The Rational Design of High Quality Concrete Mixes", *Concrete*, Vol. 2, No. 5, 1968, pp. 212–222.
77. Hughes, B.P. and N.F., Fattuhi, "Improving the Toughness of High Strength Cement Paste with Fibre Reinforcement", *Composites*, Vol. 7, July 1976, pp. 185.
78. I.C.E. and I. Struct. E., Joint Committee, "The Vibration of Concrete", London, 1956.
79. IRC: 44-1976, "Tentative Guidelines for Cement Concrete Mix Design for Road Pavements", The Indian Road Congress, New Delhi, 1976.
80. ISO: 3893-1977, "Concrete Classification by Compressive Strength", International Organization for Standardization, 1977.
81. IS: 269-1976 (Third Revision), Specifications for Ordinary and Low Heat Portland Cements, Bureau of Indian Standards, New Delhi.
82. IS: 383-1970 (Second Revision), Specifications for Coarse and Fine Aggregates from Natural Sources for Concrete.
83. IS: 455-1976 (Third Revision), Specifications for Portland Slag Cement.
84. IS: 456-1978 (Third Revision), Code of Practice for Plain and Reinforced Concrete.
85. IS: 460 (Part I)-1978 (Second Revision), Test Sieves: Part I—Wire Cloth Test Sieves.
86. IS: 460 (Part II)-1978 (Second Revision), Test Sieves: Part II—Perforated Plate Test Sieves.
87. IS: 516-1959, Methods of Test for Strength of Concrete.
88. IS: 650-1966 (Reaffirmed 1980), Standard Sand for Testing of Cement.
89. IS: 1199-1959, Methods of Sampling and Analysis of Concrete.
90. IS: 1489-1976 (Second Revision), Specifications for Portland Pozzolana Cement.
91. IS: 2386 (Part I)-1963, Methods of Test for Aggregate for Concrete: Part I—Particle Size and Shape.
92. IS: 2386 (Part IV)-1963, Methods of Test for Aggregates for Concrete: Part IV—Mechanical Properties.
93. IS: 2386 (Part V)-1963, Methods of Test for Aggregates for Concrete: Part V—Soundness.
94. IS: 2386 (Part VII)-1963, Methods of Test for Aggregates for Concrete: Part VII—Alkali Aggregate Reactivity.
95. IS: 2430-1969, Sampling of Aggregates.
96. IS: 2770 (Part I)-1967, Method of Testing Bond in Reinforced Concrete: Part I—Pull Out Test.

97. IS : 3812 (Part II)-1981, Specification for Fly Ash: Part II—For use as Additive.
98. IS : 4634-1968, Method for Testing Performance of Batch Type Concrete Mixer.
99. IS : 4845-1968 (Reaffirmed 1980), Definitions and Terminology relating to Hydraulic Cement.
100. IS : 5640-1970, Method of Test for Determining Aggregate Impact Value of Soft Coarse Aggregate.
101. IS : 5816-1970, Method of Test for Splitting Tensile Strength of Concrete Cylinders.
102. IS : 6441 (Part II)-1972, Methods of Test for Autoclaved Cellular Concrete Products: Part II—Determination of Drying Shrinkage.
103. IS : 6441 (Part IV)-1972, Methods of Test for Autoclaved Cellular Concrete Products: Part IV—Determination of Compressive Strength.
104. IS : 6452-1972, Specifications for High Alumina Cement for Structural Use.
105. IS : 6461 (Part I)-1972, (Reaffirmed 1980), Glossary of Terms Relating to Cement Concrete: Part I—Concrete Aggregates.
106. IS : 6461 (Part IV)-1972, Glossary of Terms Relating to Cement Concrete: Part IV—Types of Cements.
107. IS : 6461 (Part VII)-1973, Glossary of Terms Relating to Cement Concrete: Part VII—Mixing, Laying, Compaction, Curing and other Construction Aspects.
108. IS : 6461 (Part VIII)-1972, Glossary of Terms Relating to Cement Concrete: Part VIII—Properties of Concrete.
109. IS : 6909-1973, Specifications for Super Sulphated Cement.
110. IS : 7861 (Part I)-1975, Code of Practice for Extreme Weather Concreting: Part I—Recommended Practice for Hot Weather Concreting.
111. IS : 7861 (Part II)-1981, Code of Practice for Extreme Weather Concreting: Part II—Recommended Practice for Cold Weather Concreting.
112. IS : 8041-1978 (First Revision), Specifications for Rapid Hardening Portland Cement.
113. IS : 8112-1976, Specifications for High Strength Ordinary Portland Cement.
114. IS : 8142-1976, Method of Test for Determining Setting Time of Concrete by Penetration Resistance.
115. IS : 9013-1978, Method of Making, Curing and Determining Compressive Strength of Accelerated Cured Concrete Test Specimen.
116. IS : 9103-1979, Specifications for Admixtures for Concrete.
117. IS : 10262-1982, Recommended Guidelines for Concrete Mix Design.
118. Johansson, A. and K. Tuutti, "Pumped Concrete and Pumping of Concrete", *CBI Research 10*, Swedish Cement and Concrete Research Institute, 1976.
119. Johnson, S.M., *Deterioration, Maintenance and Repair of Structures*, McGraw-Hill Book Company Ltd., New York.
120. Jones, R., and E.N. Gatfield, "Testing of Concrete by an Ultrasonic Pulse Technique", *Road Research Technical Paper No. 34*, H.M.S.O., London, 1955.
121. Jones, T.E.R., G. Brindley, and B.C. Patel, "A Study of Rheological Properties", *Conference Proceedings, University of Sheffield*, Cement and Concrete Association, Slough, April, 1976, pp. 135-149.
122. Kameswara Rao, C.V.S., "Effectiveness of Random Fibres in Composites", *Cement and Concrete Research*, Vol. 9, November, 1979, p. 685.
123. Kaplan, M.F., "The Effects of the Properties of Coarse Aggregate on the Workability of Concrete", *Magazine of Concrete Research*, Vol. 10, No. 29, August, 1958, pp. 63-74.
124. Kar, J.N. and A.K. Pal, "Strength of Fibre Reinforced Concrete", *Journal of Structural Division, American Society for Civil Engineers*, Vol. 98, May 1972, pp. 1053.
125. Kelly, J.W., "Cracks in Concrete: The Causes and Cure", *Concrete Construction*, Vol. 9, April, 1964, pp. 89-93.

126. Kempster, E., "Pumpable Concrete", *Current Paper 26/69, Building Research Station*, Garston, 1968.
127. Kolek, J., "The External Vibration of Concrete", *Civil Engineering*, London, Vol. 54, No. 633, March 1959, pp. 321–325.
128. Kong, F.K., R.H. Evans, E. Cohen and F. Roll (Ed.), *Handbook of Structural Concrete*, Pitman Publishing Inc., 1983.
129. Kreijger, P.C., "Plasticizing and Dispersive Admixtures", *Proceedings of Admixtures Congress, C 180*, Construction Press, Lancaster, 1980, pp. 1–16.
130. Krishna Raju, N., *Pre-stressed Concrete*, Tata McGraw-Hill Publishing Company, New Delhi, 1981.
131. Krishnaswamy, K.T., "Strength and Microcracking of Plain Concrete Under Triaxial Compression", *Journal of American Concrete Institute*, Vol. 65, October 1968, pp. 856–863.
132. Kukerja, C.B., *Structural Characteristics of Steel Fibre Reinforced Concrete*, Ph.D. Thesis, University of Roorkee, India, 1981.
133. Lamond, J.F., "Accelerated Strength Testing by the Warm Water Method", *Journal of American Concrete Institute*, Vol. 76, No. 4, 1979, pp. 499–512.
134. Lane, R.C. and J.F. Best, "Properties and Use of Fly Ash in Portland Cement Concrete", *Concrete International*, July, 1982, pp. 81–92.
135. Lea, F.M. and N. Davey, "The Deterioration of Concrete in Structure", *Journal of Institution of Civil Engineers*, No. 7, May, 1949, pp. 248–95.
136. Lerch, W., "Plastic Shrinkage", *Journal of American Concrete Institute*, Vol. 53, February, 1957, pp. 797–802.
137. Lydon, F.D., *Concrete Mix Design*, Applied Science Publication, London, 1972.
138. Maiti, S.C., P.G. Devadas and A.K. Mullick, "Correlation of Strength of Accelerated Cured and Normally Cured Concrete", *Research Bulletin RB-12*, 79, Cement Research Institute of India, New Delhi, November, 1979.
139. Malhotra, V.M., "Effect of Specimen Size on Tensile Strength of Concrete", *Journal of American Concretes Institute*, Vol. 67, June, 1970, pp. 467–469.
140. Malhotra, V.M., "Maturity Concept and the Estimation of Concrete Strength: A Review", *Indian Concrete Journal*, Vol. 48, No. 4, April, 1974, pp. 122–126; Vol. 48, No. 5, May, 1974, pp. 155–159.
141. Malhotra, V.M., "Development of Sulphur-infiltrated High Strength Concrete," *Proceedings of the Journal of American Concrete Institute*, Vol. 72, No. 9, September, 1975.
142. Malhotra, V.M., "Testing of Hardened Concrete: Non-destructive Methods", *American Concrete Institute Monograph No. 9*, 1976, pp. 188.
143. Malhotra, V.M., "Superplasticizers in Concrete", *CANMET, Report MRP/MSL-77–213*, Canada Centre for Mineral and Energy Technology, Ottawa, August, 1977.
144. Malhotra, V.M., and G. Garette, "Comparison of Pull-out Strength of Concrete with Compressive Strength of Cylinders and Cores; Pulse Velocity and Rebound Number", *Journal of American Concrete Institute*, Vol. 77, No. 3, 1980, pp. 161–170.
145. Mather, B., "The Partial Replacement of Portland Cement in Concrete", *Special Technical Publication No. 205*, American Society for Testing and Materials, 1958, pp. 37–73.
146. McCoy, W.J., "Mixing and Curing Water for Concrete", *ASTM Special Technical Publication No. 169-B*, 1978, pp. 665–673.
147. McIntosh, J.D., *Concrete Mix Design*, 2nd Ed., Cement and Concrete Association, London, 1966.
148. McIntosh, J.D., "The Selection of Natural Aggregate for Various Types of Concrete Works", *Reinforced Concrete Review*, Vol. 4, No. 5, March, 57, pp. 281–305.
149. McIntosh, J.D., *Concrete and Statistics*, Contractors Record Limited, London, 1963.

150. Meyer, U.T., "Measurement of Workability of Concrete", *Proceedings of Journal of American Concrete Institute*, Vol. 59, No. 8, 1962, p. 1071–1080.
151. Morinaga, S., "Pumpability of Concrete and Pumping Pressure in Pipelines in Fresh Concrete: Important Properties and their Measurements", *Proceedings of RILEM Seminar*, Leeds University, March, 1973, pp. 7.3.1–7.3.39.
152. Murata, J. and H. Kikukawa, "Studies on Rheological Analysis of Fresh Concrete: Important Properties and Their Measurement", *Proceedings of RILEM Seminar*, March, 73, pp. 1.2.1.–1.2.23.
153. Murdock, L.J., "The Workability of Concrete", *Magazine of Concrete Research*, Vol. 12, No. 36, November, 1960, pp. 135–144.
154. Narang, C.N. and A.K. Sinha, "Use of Polymer Resins in Restoration of Concrete Structures in Distress", *Journal of Institution of Engineer (India)*, Vol. 55, Part CI 4, March, 1975, pp. 141–144.
155. Nasser, K.W. and H.M. Marzouk, "Properties of Mass Concrete Containing Fly Ash at High Temperatures", *Journal of American Concrete Institute*, Vol. 76, No. 4, April, 1979, pp. 537–550.
156. National Ready-Mixed Concrete Association, "Control of Quality of Ready Mixed Concrete", Publication No. 44, June, 1957.
157. Neville, A.M., "The Failure of Concrete Compression Test Specimens", *Civil Engineering*, Vol. 52, July, 1957, pp. 773–774.
158. Neville, A.M., "The Relation Between Standard Deviation and Mean Strength of Concrete Test Cubes", *Magazine of Concrete Research*, Vol. 11, No. 32, July, 1959, pp. 75–84.
159. Neville, A.M., "Shrinkage and Creep in Concrete", *Structural Concrete*, Vol. 1, No. 2, March, 1962, pp. 49–85.
160. Neville, A.M., *Properties of Concrete* 3rd Edn., Pitman Publishing Company, 1981.
161. Newman, K., "Properties of Concrete", *Structural Concrete*, Vol. 2, No. 11, 1965, pp. 451–482.
162. Nilsson, S., "The Tensile Strength of Concrete Determined by Splitting Tests on Cubes", *RILEM Bulletin No. 11*, June, 1961, pp. 63–67.
163. Orchad, D.F., *Concrete Technology*, Vol. I, 3rd Edn., Applied Science Publishers Limited, London, 1973.
164. Owens, P.L., "Basic Mix Method: Selections of Proportions for Medium Strength Concrete", *Cement and Concrete Association*, London, 1973.
165. Owens, P.L., "Flyash and Its Usage in Concrete", Concrete, Vol. 13, No. 7, July, 1979, pp. 21–26.
166. Paul, B.K., and B.P. Pama, *Ferrocement*, International Ferrocement Information Centre, Asian Institute of Technology, Bangkok, 1978.
167. Plowman, J.M., "Maturity and the Strength of Concrete", *Magazine of Concrete Research*, Vol. 8, No. 22, March, 1956, pp. 13–22.
168. Powers, T.C., "The Physical Structure and Engineering Properties of Concrete", *Bulletin No. 90*, Portland Cement Association Research Department, Chicago July, 1958, p. 39.
169. Powers, T.C., *The Properties of Fresh Concrete*, John Wiley and Sons Inc., 1968.
170. Price, W.H., "Factors Influencing Concrete Strength", *Journal of American Concrete Institute*, Vol. 47, February, 1951, pp. 417–432.
171. Ravina, D. and R. Shalon, "Plastic Shrinkage Cracking", *Journal of American Concrete Institute*, Vol. 65, April, 1968, pp. 282–292.
172. Rehsi, S.S., S.K. Garg and P.D. Kalra, "Accelerated Testing for 28-day Strength of Concrete", *ISI Bulletin*, Vol. 23, No. 6, 1971, pp. 273–76.
173. Richards, O., "Pull-Out Strength of Concrete", *ASTM Special Publication No. 626*, 1977. pp. 32–40.
174. Rixom, M.R., "Chemical Admixtures for Concrete", *Span*, London, 1978.

175. Rixom, M.R. (Ed.), *Water Reducing Admixtures for Concrete-Concrete Admixtures: Use and Applications*, The Construction Press Limited, Lancaster, England, 1977.
176. Road Research Laboratory, "Design of Concrete Mixes", DSIR *Road Note No. 4*, HMSO, London, 1950.
177. Shacklock, B.W., *Concrete Constitutents and Mix Proportions*, Cement and Concrete Association, London, 1954.
178. Shacklock, B.W., "Comparison of Gap-and Continuously-Graded Concrete Mixes", *Cement and Concrete Association London, Technical Report No. TRA/240*, September, 1959.
179. Shah, S.P. and B.V. Rangan, "Fibre Reinforced Concrete Properties", *Journal of American Concrete Institute*, Vol. 68, No. 2, February, 1971, pp. 126–135.
180. Sharma, S.P., and P.C. Sharma, "Ferrocement Applications Developed at SERC Roorkee", *Asia-Pacific Symposium on Ferrocement Applications for Rural Development*, Roorkee, Sarita Prakashan, April 23–25, 1984, pp. 101–110.
181. Special Issue on Ferrocement, *Journal of Structural Engineering*, Vol. 2, No. 4, January, 1975.
182. SP : 23–1982, *Handbook on Concrete Mixes*, Bureau of Indian Standards, New Delhi, 1982.
183. Steinour, H.H., "Concrete Mix Water: How Impure Can it be?," *PCA Research Department Bulletin 119*, Portland Cement Association, Research and Development Laboratories, September, 1960, pp. 32–50.
184. Subrahmanyam, B.V. and E.A. Karim, "Ferrocement Technology: A Critical Evaluation", *The International Journal of Cement Composites*, Vol. 1, No. 3, 1979, pp. 125–140.
185. Swamy, R.N., "Fibre Reinforced Concrete: Mechanics, Properties and Applications", *Indian Concrete Journal*, Vol. 48, No. 1, 1974, pp. 7–16.
186. Swamy, R.N., "The Technology of Steel Fibre Reinforced Concrete for Practical Application," *Proceedings of Institute of Civil Engineers*, Vol. 56, May, 1974, pp. 143–159.
187. Swamy, R.N. (ed), *Concrete Technology and Design-Vol. 1: New Concrete Materials*, Surrey University Press, 1983.
188. Swamy, R.N. (ed), *Concrete Technology and Design-Vol. 2 : New Reinforced Concretes*, Surrey University Press, 1984.
189. Tattersall, G.H., "The Workability of Concrete", *View Point Publication No. 11.008*, Cement and Concrete Association, Slough, 1976.
190. Tattersall, G.H. and P.F.G. Banfill, *The Rheology of Fresh Concrete*, Pitman Advanced Publishing Program, 1983.
191. Taylor, W.H., *Concrete Technology and Practice*, McGraw-Hill, Sydney, 1977.
192. Teychenne, D.C., R.E. Franklin and H. Erntroy, *Design of Normal Concrete Mixes*, Department of Environment, HMSO, London, 1975, p. 31.
193. Thaulow, S., "Tensile Splitting Test and High Strength Concrete Test Cylinder", *Journal of American Concrete Institute*, Vol. 53, January, 1957, pp. 699–706.
194. Tomsett, H.N., "Ultrasonic Pulse Velocity Measurements in the Assessment of Concrete Quality", *Magazine of Concrete Research*, Vol. 32, No. 110, March 1980, pp. 7–16.
195. Troxell, G.E., H.E. Davis and J.W. Kelly, *Composition and Properties of Concrete*, 2nd Edn., McGraw-Hill, 1968.
196. Tuthill, L.H., "Performance Failure of Concrete Materials and of Concrete as a Material", *Concrete International*, Vol. 2, No. 1, January, 1980, pp. 33–39.
197. Tylor, I.L., "Uniformity, Segregation and Bleeding", *ASTM Special Technical Publication No. 169*, 1956, pp. 37–41.
198. Underwater Concrete, *Heron*, Delft, Vol. 19, No. 3, 1973, p. 52.

199. U.S. Bureau of Reclamation, *Concrete Manual*, 8th Edn, Department of the Interior, Bureau of Reclamation, U.S.A., Denver, Colorado, 1975.
200. Valore, R.C., "Pumpability Aids for Concrete", *ASTM Special Publication No. 169B*, 1978, pp. 860–872.
201. Verbeck, G.J., "Hardened Concrete-Pore Structure", *ASTM Special Technical Publication No. 169*, 1955, pp. 136–142.
202. Verbeck, G.J., "Carbonation of Hydrated Portland Cement", *ASTM Special Publication No. 205*, 1958, pp. 17–36.
203. Visvesvaraya, H.C. and A.K. Mullick, "Relation Between Water Content in Concrete Mixes and Compressive Strength", Second International CIB/RILEM *Symposium on Moisture Problems in Buildings*, Netherlands, 1974.
204. Visvesvaraya, H.C., and A.K. Mullick, "Compressive Strength of Concrete in Winter: A Probabilistic Simulation", *Second International Symposium on Winter Concreting, RILEM*, Moscow, Vol. 2, October, 1975, pp. 83–103.
205. Vollick, C.A., "Effects of Revibrating Concrete", *Journal of American Concrete Institute*, Vol. 54, March, 1958, pp. 721–732.
206. Whitehurst, E.A., "Soniscope Tests for Concrete Structures", *Journal of American Concrete Institute*, Vol. 47, February, 1951, pp. 433–444.
207. Whorlow, R.W., *Rheological Technique*, John Wiley and Sons Inc., Chichester, 1980.
208. Wigmore, V.S., "Ready Mixed Concrete", *Reinforced Concrete Review*, Vol. 5, No. 12, December, 1961, pp. 793–816.
209. Wills, M.H., "Accelerated Strength Tests", *ASTM Special Technical Publication No. 169B*, 1978, pp. 162–179.
210. Wright, P.J.F., "Entrained-Air in Concrete", *Proceedings of Institution of Civil Engineers*, London, Parts 1, 2, No. 3, May 1953, pp. 337–358.
211. Wright, P.J.F., "The Design of Concrete Mixes on the Basis of Flexural Strength", *Proceedings of Symposium on Mix Design and Quality Control of Concrete*, Cement and Concrete Association, London, May, 1954, pp. 74–76.
212. Wright, P.J.F., "Comments on an Indirect Tensile Test on Concrete Cylinders", *Magazine of Concrete Research*, Vol. 7, No. 20, July, 1955, pp. 87–96.

Index

Abram's law 83
Absolute
  density of aggregate 31
  relation of specific gravity 31
  specific gravity of aggregate 31
  volume calculations 154
  volume of mix ingredients 196
Absorbed water 32, 165
Absorption
  of aggregate 28, 30, 32
Accelerated
  curing test 157, 251
  boiling water method 157–159, 251
  warm water method 158, 159, 251
  strength 108, 157, 159, 251
  relation to 28-day strength 157, 158
Accelerating admixture 41–51
Acceleration of hardening 239
Accelerators 61, 82
Acceptance of concrete
  test 108, 252
  criteria 113, 197
ACI method of mix design 149–151, 163
Acid
  attack 100, 346
  cleaning 373
  resistance test 347
Acrylic fibres 311
Admixtures 11, 48–62, 80, 81, 122
  accelerating 49, 60
  air-detraining 49, 54, 60
  air-entraining 49, 52, 55
  bonding 49, 55
  classification 49
  colouring 49, 59
  corrosion inhibiting 49, 55
  fungicidal 49, 55
  gas forming 49, 54
  grouting 49, 52

  hardening 49, 59
  permeability reducing 49, 55
  pozzolana 49, 56, 57
  retarding 49, 51, 60
  water proofing 49, 55
  water reducing 51, 60
Aerated concrete 52
Aggregate 1, 23–43, 67, 107
  abrasion 29, 30, 32
  absorption 28, 30, 32
  angular 26, 30, 80
  artificial 24, 27, 29, 288
  bond with mortar phase 30, 273
  bulk density 31
  bulking 33
  brick 24, 289
  characteristics 29–34
  classification 23–29
  coarse, see coarse aggregate
  compressibility 30
  content 144, 165
  continuously graded 40, 125
  crushed 23–26
  crushing value 29
  elongated 26, 40
  fine, see fine aggregate
  flaky 26, 30, 40
  gap-graded 40
  glassy 30
  grading 31, 38–42, 80
  heavyweight 27
  impact value 29
  impurities in 34, 35
  influence on durability of concrete 30
  irregular 26, 30
  lightweight 27–29
  maximum size, see maximum aggregate size
  mechanical properties 29–34

Aggregate (contd.)
  modulus of elasticity 30
  moisture content 31–33
  natural 23–24
  normal weight 27
  porosity 30, 32
  reaction with alkali 36
  red mud 290
  roughness 27, 29, 30
  rounded 26, 30
  saturated 32
  saturated and surface dry 32, 33
  shape 26, 27, 30, 31, 80
    influence
      on strength 26, 27
      on workability 26, 27, 67, 69
  size 24–26, 31, 69
  soundness 35
  specific
    gravity 31–33
    surface 35, 38–41
  strength 29
  surface texture 30, 31, 76, 80
  toughness 29
  void ratio 32
  voids 32, 38
  volume in concrete 143, 150, 154
    for wearing surfaces 29, 30
Aggregate-cement ratio 40, 91, 122, 125, 135, 136, 138–140, 143, 198
  relation to workability 138, 139
  and water-cement ratio 138, 139
Air
  bubbles size 52, 81
  content 108, 149, 239
    influence on strength 92
  detrainment 54
  entrained concrete 52, 53, 130, 133
  entraining agent 52, 61, 81, 239, 293
  entrainment 29, 48, 79, 82, 294
    influence
      on permeability 55
      on strength 52, 92
      on water content 52
      and workability 52
  voids 52, 55, 396
Algae in water 46
Alkali
  aggregate reaction 36, 56, 362, 366
  carbonation 362
  chloride 47
  reactive aggregate 36

  silica reaction products 362, 369
  silicate gel 36
Alkalinity 348, 349
Alkalis
  in aggregate 36
  in water 46
All-in-aggregate 25, 37, 41, 42
  grading 42
Aluminium powder 54
Aluminimum stearate 19
Amplitude of vibration 107, 218, 219
Analysis of fresh concrete 250
Angularity
  of aggregate 26
  index 26
Apparent specific gravity of aggregate 31
Argellaceous material 6, 7
Artificial aggregate 24, 27, 29
Asbestos cement 323
Asbestos fibres 323
Aspect ratio 309, 311, 313
Attrition 29
Autoclave products 232
Autogenous
  curing 395
  healing 395
Average strength 113

Barrows 211
Batching 205–207
  tolerance 205
  volume 206
Batching plant 206, 207
Bauxite 19
Bibliography 448
Bingham model 77, 78
Black concrete 288
Blast-furnace
  cement 16
  slag 7, 16, 24, 276
Bleeding 3, 39, 55, 70, 72, 82
Blended cement 124, 156
Bloated clay aggregate 28, 275
Blockage in pumping concrete 213
Block outs 54
Blow-crete 291
Blow-holes 356, 357
Bond
  agent 55, 56
  of aggregate 26, 36
  cracks 104

Index

Brick aggregate 24, 289
Bulk density of aggregate 31
Bulking
  allowance for batching 34
  factor 34
  of sand 33, 34
Buttering of mixer 208

Calcareous aggregate 7
Calcium
  carbonate efflorescence 101, 232
  chloride 46, 49, 50
    for cold weather concrete 50, 242
  hydroxide 8, 10, 11, 56, 232
  silicate hydrates 8, 10, 57
  stearate 19
  sulphate 8, 347
Capillaries, see Capillary pores
Capillary
  action 44
  pores 10, 32, 52, 99, 223, 400
    effect of entrained air 52
    influence on strength 10, 52
Carbon fibres 323
Carbonation 98, 362
Cellular concrete 52
Cement 1, 6–22
  -aggregate ratio, see aggregate-cement ratio
  coloured 19, 59
  composition 8
  compounds 8, 16
  concrete 6
  content 91, 96, 131, 132, 142, 280
    for durability 131, 132
    in mass concrete 291
    factor 122
  fineness, see fineness of cement
  jet 338
  manufacture 7
  mortar 6, 299
  oxide composition 7, 16
  paste 1, 38
  physical requirements 12, 20
  strength development 9–13
  types 16–19
Cement-water ratio, see water-cement ratio
Centrifugation 217, 222
Chalk 7, 19
Characteristic strength 112, 114, 124, 133, 134, 223
  and standard deviation 110–114, 153, 197

Chemical
  composition of cement 7, 8, 16
  curing 230
  interaction 346
  properties of cement 15
  testing of hardened concrete 269–271
Clay in aggregate 35
Clinker 7, 16–18
Coal in aggregate 35
Coarse aggregate 1, 6, 23, 25, 37–39, 69, 70, 106, 107, 135, 136
  grading 125, 126, 128
Coefficient
  of permeability 99
  of thermal expansion of concrete 36, 103
  of variation 111, 133
Cohesion 74
Colcrete 247
Cold weather concreting 241
Collar 394
Collodial membrane 11
Coloured portland cement 19, 59
Column
  jackets 394, 395
  collars 394, 395
Combining fine and coarse aggregate 125–129
Compactability of concrete 64, 65, 74, 76, 77
Compacting factor 65, 68, 70, 73, 76
  effect of time 71
  and slump 65, 68
  and Vee-Bee 68
  and workability 68
Compacting energy 218
Compaction
  of concrete 107, 216–223
  influence on strength 76, 90
  methods 216
    centrifugation 217, 222
    jolting 222
    rolling 223
    spinning 222
    vibropressing 222
Compliance to specification 248, 252, 254, 260, 269
Composition
  of cement 7, 16
  of concrete 1, 2, 6
Compound composition of cement 8, 16
Compression tests 14

Compressive
  strength   3, 13, 14, 17, 18, 20, 30, 49, 51,
    83, 109, 123, 129, 141, 149, 150, 186,
    226, 228, 251, 262, 264
    polymer concrete   331
    target   124, 130
      and flexural strength   85
    relation to tensile strength   85
    test   14, 83
    water-cement ratio   129, 130
Concrete
  black   88
  brocken-brick   289
  controlled   3
  core test   253, 263
  glare-free   288
  grades   3
  light-weight   52, 274–284, 293
  matrix   310
  ordinary   3
  placing   213
  production   205
  reference   157
Concreting
  extreme weather   238–242
  underwater   243–247
Conductivity of concrete   101
Consistency   64–66
  of concrete   79
  for pumping   212
Construction joints   215
Content of cement   131, 132
Control
  influence of coefficient of variation   114
    degree   48, 114
  ratio   113, 114
Cooling of aggregate   240
Cooling period   231, 232
Cone penetration test   313, 314
Core
  strength relation to cube strength   266
  test   263, 269
Corrosion
  of concrete   343–349
  of steel   35, 131, 349–355, 363
    effect
      of carbonation   352
      of chlorides   352
      of humidity   353
      of inhibitors   353
      of moisture   352
      of oxygen   352
      of pH   352
    with calcium chloride   352, 354
    mechanism   350
Cover to reinforcement   353, 354
Crack   358
  analysis and evaluation   365
  arrestor   308
  bond   104, 309
  detection by pulse velocity   366, 369
  first   309
  initiation   104, 361
  multiple   309
  patterns   68, 367
  propagation   104, 274
  resistance   308
  types
    active   359
    dormant   359, 376
    static   359
    structural   359
    superficial   359
    thermal   39, 361
Cracking   51, 98, 359–65
  causes and control   357
  due to alkali aggregate reaction   362
  due to applied loads   365
  due to chemical reaction   362
  due to construction overloads   364
  due to corrosion   349, 353, 363
  due to passive protection   363
  due to poor construction practice   363
  due to settlement   360
  due to sulphate attack   362
  due to temperature   39, 362
  due to weathering   362
  corrosion-interaction   350
  of plastic concrete   360
  plastic shrinkage   239, 360
Crazing   51, 361
  effect of carbonation   353
Crack repair, see repair of concrete structures
Creep   30, 93, 98
  recovery   93, 98
Critical value of chloride ions   352, 354
Crystallization   347
Cumulative percentage   37
Curing   223–233, 239, 240
  accelerated   157, 227
    boiling water method   159
    warm water method   159
  chemical   230
  conditions   224, 228

Curing (contd.)
  electrical 233
  in hot weather 239
  influence
    of delayed curing 233
    on strength 226–228
  infrared radiation 233
  membrane 229
  methods 227–233
  periods 225, 227
  steam 230–232

Dampproofers 55
Deflocculation 81
Degree of control 3, 114, 133, 153
Delay of placing of concrete 215
Delayed setting 241
Deleterious substances
  in aggregate 34, 35
  in water 44–47
Design
  of mixes 123–204
  as a system 202
  strength 112
Deterioration of concrete 10, 35, 343–349
  prevention 348
  types 344
Diatomite 27, 276
Dicalcium silicate 8–11
Direct tension test 87
Disruption of concrete 100, 295
Distress
  in concrete 356
  symptoms 356–358, 407
Distribution curve 109
DoE mix design method 141–149, 162
Dormant period 10
Drilling and plugging 384
Dry
  mix process
    of cement manufacture 7
    of shotcreting 293
  packing 375
Drying shrinkage 39, 57, 79, 95, 295, 296, 360
Ductility 295, 306, 308, 316
Dumpers 211
Durability
  of concrete 1, 30, 34, 37, 57, 76, 83, 99, 105, 107, 117, 131, 133, 134, 186, 335, 343, 344
  of shotcrete 294

effect
  of pore size 32, 99, 223
Effects of creep 98
Efflorescence 35, 45, 46, 101, 232, 337
Elastic
  deformation, *see* elastic strain modulus 94
  strain 93
Elastomeric polymers 332
Emulsion 56
Entrained air 52, 122, 153
  due to admixture 52, 53
  content 53, 135, 153, 154
  influence
    on durability 52
    on mix proportion 153
    on strength 52, 92
Epoxy
  coated steel bars 355
  grouts 381
  injection 387, 388
  mortar 381, 387
  resins 381, 386
Emtroy and Shacklock's method 187–195
Evaporation 223
  influence on plastic shrinkage 233
  relation to cracking 239
Expanded shale 28, 275
Expanded-polystyrene 284
Expansion 36, 98
Expansion producing admixtures 54
Extreme weather concreting
  in cold weather 241
  in hot weather 238

False set 13
Fatigue strength 216, 298, 316, 317
Ferrocement 298
  applications 299, 306
  construction 302–304
  fibrous 307
  properties 299, 304
Fibre
  aspect ratio 309, 311–313
  concentration 313
  content 310, 312, 314
  -reinforced concrete 307–325
    applications 318
    design illustration 320
    factors affecting workability 313
    mechanism 308
    mix design procedure 319

Fibre (contd.)
  practical mix proportion 320
  production operation 324
  properties 312, 316–318
 specific surface 314
 types 311
Fibrous ferrocement 307
Fine 6
 aggregate 1, 6, 23, 24, 33, 35, 37–39, 41, 42
 bulking 33–34
 grading 38–42, 125–128
 influence on voids 32
 grading zones 41, 42
Fineness
 of aggregate 33
 of cement 12, 13, 15, 16, 19, 20
  influence
   on rate of heat development 12
 modulus 37, 42
 of sand 33, 300
Finishability 64
Fire resistance 101, 283, 284
 damaged structure 407
 restoration 408
Flakiness
 index 27, 30
 influence on workability 30
Flaky particles 26, 30
Flash coat 294
Flash set 8, 241
 with calcium chloride 50
Flexible sealing 390, 391
Flexural strength and
 compressive strength 50, 84, 85
 tensile strength 89
Flocculated cement 11
Flocculation 11, 81
Flow
 table 67
 test 66
Flowability 64, 74, 75
Flowing concrete 81, 213
Fly ash 28, 57, 276, 290
Foamed slag 27, 276
Formwork 233–237
 design loads 236
 planning 234
 requirements 234
 stripping 236
  time 237
 types 235

Freezing and thawing 30, 32, 133, 232, 336, 396
Frequency of vibration 107, 218
Fresh concrete 63
Fungicidal agents 55

Gap-graded concrete 40, 325, 341
Gaussian distribution 109
Gel
 pores 10, 99, 223
 -space ratio 12, 83
Glass fibres 322
Grade of concrete 3, 114, 122, 153
Graded standard sand 14, 17
Grading 38
 of coarse aggregate 38–41, 125–129, 136
 of concrete 258
 curves 38–40, 125–129, 137–140
 ideal 40
 of fine aggregate 38–42, 125, 129, 152
 limits 40–42, 136
 in mix design 125, 136–139
 zones 42
Granulated
 blast-furnace slag 24, 58
 iron 54
Gravel 24, 25, 138, 139
Grout 378
 epoxy 381
Grouting admixtures 49, 52
Gunite 291, 297, 298, 378
Gypsum 8, 9, 19

Hand
 mixing 210
 rodding 217
Hardened
 concrete 3, 83
 cracking 361
Hardening rate 8, 9
Harshness 39
Healing of concrete 395
Heat of hydration 8, 15, 18, 20, 38, 57, 242, 361
Heating of concrete ingredients 241
Heavy weight aggregate 27
Height/diameter ratio of specimens 84, 265
High-alumina cement 18
 heat of hydration 18
 setting 18
High
 density concrete 27

High (*contd.*)
  durability 232
  early strength 225, 232
  pressure steam curing 232
  strength cement 18, 20, 21
Honey-combed concrete 44, 99
Honey-combing 34, 369
Hot pressing technique 272
Humidity 96, 238
  relative 97, 239
Hydrate/space ratio 12
Hydrated
  calcium silicate 8, 287
  cement 9, 10, 223
Hydration 8-11, 15, 16, 44, 57, 223, 224, 287
  effect
    of admixtures 11
    of humidity 223
    of temperature 226, 239
    of vapour pressure 223
  mechanism 11
Hydrophilic compounds 81
Hydrophobic
  cement 19
  compound 81
Hydroxylated carboxylic acid 51, 61

Ideal grading 40
Igneous rocks 24
Ignition loss of cement 15, 21
Ignition test 15, 21
Impact
  strength 29, 305, 307
  test on ferrocement 307
Impermeability 107, 298, 300, 306
Impregnation
  mortar 303
  polymer 325, 326, 383
Impurities
  in aggregate 34
  in water 45-47
Induction period 10, 11
Inert matrix 6
Indirect method 87
Industrial waste
  concrete 289
  water 46
Infrared radiation curing 233
Initial
  elastic strain 93
  set 210

shrinkage 232
tangent modulus 94
  and secant modulus 94
Insoluble residue 16, 21
Instantaneous
  recovery 98
  strain 98
Inspection testing 248-271
Iron oxide 10, 19, 21

Jacketing
  repair 393
Jet
  cement 338
  -crete 291
Joints
  contraction or expansion 98

Kiln 7

Laitance 44, 72, 215 221, 243, 293, 373
  removal by acid etching 373
Latex-modified concrete 332
Leaching 57, 232, 337, 346
Leak sealing 401
Light-weight aggregate 27-29, 274-284
Light-weight concrete 52, 274-284, 293
Lignosulphonic acid 51, 61
Lime
  free 14
  saturation factor 21
Lime stone aggregate 7, 19, 27
Loading uniaxial 104
Load tests 267
Loss on ignition 15, 21
Low
  heat cement 16, 20, 21
  heat portland blast-furnace cement 17
  temperature concreting 49, 241

Macrocracks 103
Magnesia 10, 21
Magnesium
  sulphate 46
  sulphate attack 36, 100, 348
Magnetite aggregate 27
Marl 7
Mass concrete 124, 291, 361
Matrix 125, 273, 299, 303, 308
Maturity 225, 226, 251
  meter 251

464 Index

Maximum
  aggregate size 37, 38, 68, 80, 96, 125, 132, 136, 137, 140, 152, 153
    in mix design 124, 125
    influence
      on strength of concrete 37, 125
      on water cement ratio 37, 125
      on workability 38, 69, 125
Mean
  size 37
  strength 110, 112
    relation to minimum and standard deviation 110, 112
  target strength 124, 130, 134, 152, 156
Mechanical vibrations 217
Mechanism
  of fibre-matrix interaction 308
Membrane curing 229, 296
  sealing compound 229
Microcracking 103
  influence on stress-strain relation 104
Microcracks 103, 344
Microfillers 330
Microstructure of gel 223, 238, 272
Mix
  design
    ACI method 149–151, 163
    computer-aided design 166–185
    design example 140, 144, 150, 155, 160, 162
    DoE (British) method 141–149
    Emtroy and Shacklock's method 187–195
    Indian Standard
      recommended guidelines 152–156, 164
      Rapid method 156–160
    Road Note No. 4 136–141, 161
    trial and adjustment method 135, 136
  proportions 122, 123
    for ferrocement 300
    nominal 22
    standard 123
  harsh 218
  plastic 218
Mixer 207
  capacity 208
  choice 210
  types 207
    dual drum 209
    non-tilting 207, 208
    pan 207, 208
    tilting 207, 208

Mixability 63, 64
Mixing 205, 207
  by hand 210
  time 209, 210
  water 44, 241
Mobility of mix 30, 64, 74–77
  effect of air-entrainment 53, 64
Modulus
  dynamic 94
  of elasticity 94, 102
  of polymer concrete 331
  of rupture 84
  static 94
Moist curing 223, 233
Moisture 240
  content of aggregate 31–33, 135
  influence
    on strength 223
  movement 95, 232
Monomer 273, 328, 329
Mortar
  aggregate interface 38, 125, 273
  cracks 103
  impregnation 303
  replacement 377

Natural aggregate 24
No-fines concrete 274, 342
Nominal 152, 153
  maximum size 125, 132, 136–139, 143, 150
  mix 3, 34, 122
Non-destructive tests 251, 253, 365, 366
  pulse echo method 369
  rebound hammer test 254, 366
  soniscope 368
  ultrasonic pulse velocity 251, 257, 258, 366
Normal
  distribution 109, 110
  of strength 109
  weight aggregate 27
Nozzle 292

Objective type questions 409
Oil contamination 47
Oleic acid 19
One hour cement 338
Optimum
  frequency 218
  percentage 135
  period of vibration 219

Organic fibres 323
Ordinary portland cement 7, 8, 16, 20, 21
Oscillations 217
Overlays
   systems 396, 399
Over vibration 220
Oxide composition of cement 7

Pan-type mixer 208
Partial replacement of cement by pozzolana 57
Particle
   roughness 30
   shape of aggregate 26, 30, 39
   size distribution 38
   texture of aggregate 30, 39
Pachometer 368
Pattern cracking 56, 362
Permanent set 93
Permeability
   of aggregate 32
   and capillary pores 72, 99
   coefficient 99
   of concrete 55, 72, 76, 83, 99, 131, 273, 349, 363
   effect of carbonation 98
   reducer 55
   test 18
p.f.a., *see* fly ash 57
pH value 46
Pigments 19, 59
Pitting and scaling 35
Pceability 64
Placing concrete 39, 64, 205, 213
   under water 243-247
Plastic
   concrete 3, 64
   shrinkage 95, 360
   viscosity 77
Plasticity 64
Plasticizers 62, 70, 81, 117
   super 62, 81
Poisson's ratio 94
Polymer 326, 381
   cement concrete 273, 325, 328-331, 383
   impregnated concrete 273, 325, 327, 329, 334, 383
Polymerizing 229, 326, 330, 383
Polymer modified concrete 331, 383
Polypropylene fibres 308, 321, 323
Polyester fibres 311
Polythelene 311

Polyester-styrene system 331
Pop-outs 35, 366
Pore pressure 224
Pores
   in aggregate 30, 32
   capillary 99, 223
   in cement paste 325
   gel 99, 223
Porosity
   of aggregate 30, 32, 36
   of concrete 273, 348
   of gel 12
Portland
   blast-furnace cement 16, 20, 21, 233
      resistance to chemical attack 100
   cement 6, 7, 16, 20, 21, 100
      storage 19
   pozzolana-cement 17, 20, 21, 100, 130
Potash as admixture in cold water 51
Potassium oxide, *see* alkali
Power barrows 211
Pozzolana 7, 21, 56, 287
   and admixture 56
Pozzolanic
   actiivity 56, 57
   cement 17, 57, 100, 349
Pre-cooling 240
Pre-heating 231, 232, 241
Pre-packed concrete 246, 378
Prepolymer concrete 333
Prepolymer modified concrete 332
Prescribed mixes 2, 122
Presteaming period 231
Prestressed concrete 2
Prevention of deterioration 248
Probability
   factor 111, 112, 114
   based specifications 106, 108
Procedure of mix design 134
Prolonged vibration 219
Promoter 330
Proportioning
   of mixes 1, 122-204
Pull-out
   test 260
   strength 261, 262
Pull-out resistance of fibres 309
Pulse echo method 253, 369
Pulse velocity 256, 258, 369
Pulverized-fuel ash, *see* fly ash
Pumice 27, 276
Pumped concrete 212

Pumping of concrete   212
Quality
  assurance   119
  audit   118, 120
  of concrete   3, 38, 105, 258
  control   105–117
  management   117
  statistical   108
  of mixing water   44
  variation   4, 106
Quantity of aggregate   31, 196
Quick set   51

Random samples   109, 113
Rapid anaylsis machine   250
Rapid hardening cement   16, 20, 21, 70, 80, 141, 237
Rapid mix design method   156
Rate
  of gain of strength   12, 49, 225
  of hydration   8, 11, 49
Reactive
  aggregate   36, 361
  silica   36, 287
Ready-mixed concrete   1, 209, 213, 292
Rebound
  hammer test   254, 366
  number   254, 255
  in shotcreting   292, 293, 295, 296
  in guniting   292
Recovery of creep   98
Recycled concrete   288
Reference mix   157
Refractory shotcrete   296
Reinforced concrete   2, 46
Reinforcement   297
  corrosion   349–355
  effect of
    carbonation   352
    chloride ions   352
    cover   352
    humidity   353
    inhibitor   353
    moisture   352
    oxygen   352
  electrochemical   350
  for ferrocement   300
Rejection of cement   22
Relative humidity *see* humidity   97
Remixing   13
Remoulding test   66
  and compacting factor   66

Repair   371
  of concrete structures   356
  by additional steel   391
  by concrete replacement   376
  by drilling and plugging   384
  by drypacking   375
  by epoxy grouts and mortars   381, 386–388
  by grouting   378
  by jacketing   393
  by mortar replacement   377
  by prepacking   378
  by routing and sealing   389, 390
  by sealing of cracks   389, 401
  by shotcrete and guniting   291, 293, 297, 378
  by special procedures   381
  by stitching of cracks   392
  of crack   371
  technique selection   370, 374
  preparation of surface   371, 374, 388, 397, 403
  with polymer materials   381, 383, 384
    polymer impregnation   383
    polymer concrete   331, 383
Resin concrete   328
Revibration   219, 220
Rheology of concrete   74–82
Rheological models   77
Rheological techniques   67
Ring specimen   89
Road Note No. 4   136–141, 161
Routing and sealing   389, 390
Rust   349–351

Safety in repairwork   389
Salt crystallization in pores   344, 347
Sampling   109, 197
Sand 24, 25, *see also* fine aggregate
  bulking   33
  blasting of concrete   373
  grading   38, 40
  zones   25, 41, 42, 152, 154
  standard   14
Saturated
  and surface dry aggregate   32
Scarification   373, 397
Schmidt hammer   254
Scum   44
Sea water   46, 100
Sealing of cracks
  flexible sealing   390, 391

Index

Secant modulus 94
  and tangent modulus 94
Segregation
  of concrete 3, 23, 27, 34, 39, 41, 63, 67, 208, 211–214,
  and bleeding 3, 63, 69, 70, 72, 74
  effect of aggregate size 72
Separators 57
Serviceability 105
Set controlling 51
Setting 239, 251
  time 13, 18, 20, 48, 49, 51
    final 3, 18, 20, 252
    initial 3, 18, 20, 252
Settlement cracking 360
Shape of aggregate 26, 27
Shear strain 76
Shotcrete 291, 322, 374, 378
  disadvantages 296
  failure 295
  types 292
  properties 294
  procedure 293
  special 295
Shrinkage 94, 98
  carbonation 98, 362
  compensating concrete 54, 361
  cracking 38, 79, 98, 360, 361
  deflection 98
  drying 20, 39, 54, 57, 58, 79, 95
  effect
    of admixture 54, 97
    of aggregate 38, 96, 97
    of cement content 96, 97
    of curing 98
    of grading 97
    of humidity 96, 97
    of shape of specimen 97
    of time 95
    of water-cement ratio 96
    of workability 96
Shrinkless concrete 54, 361
Shrink mixed concrete 213
Shrinkage strain 95
Sieve analysis 37
Sieves 37
Silica-alkali reaction 36
Silica-fume concrete 290
Single point test 76, 77
Single-size-aggregate 25, 41
Sintered flyash 27, 28, 276
Skip hoists 212

Slag *see* blast-furnace slag
Slump 65, 66, 68, 73, 131, 135, 150
Sodium
  chloride 46, 51
  silicate 55
  sulphate 36, 51
Sound insulation 283
Soniscope 368
Soundness
  of aggregate 35
  of cement 14, 20
Special concretes 272, 342
Specific
  gravity 143
    of aggregate 31, 33, 154
    of cement 15, 154
  heat 37, 103
  surface
    of aggregate 35, 39, 69
    of cement 12, 17, 18, 20
    of fly ash 57, 59
    of reinforcement 304
Specifications
  performance oriented 2, 105
  prescriptive 2
  probability based 106, 108
Splitting 253
  in compression test 86–88
  tensile strength 86–88
    relation to compressive strength 88
    relation to flexural strength 89
  test 89
    for cubes 87
    for cylinders 86
Spinning of concrete 222
Stability 63, 64, 74
Staining 47, 253
Standard
  concrete mix 157
  deviation 110–112, 114, 153, 197
  mix 123
  sand 14, 17
Steam
  curing 230
    high pressure 232
    low pressure 231
  cycle 231, 232
Steel
  fibres 295, 311–320
  reinforced concrete 311–321
Skeleton 301
Stitching of cracks 392

Storage conditions 19, 22
Strain
  elastic 93
Strength
  by accelerated curing 157, 230
  of aggregate 29
  of concrete 1, 34, 37, 83, 89, 230, 231
  development 10, 11
  effect
    of admixtures 11, 89
    of age 89, 226, 227
    of aggregate 89
    of aggregate-cement ratio 91
    of air content 92
    of air voids 92, 93
    of cement 89, 90
    of cement content 91
    of cement-water ratio 90
    of compaction 89–90
    of curing 90, 224–228
    of degree of hydration 10
    of delay in placing 215
    of density 90
    of height/diameter ratio 84
    of loading rate 89
    of maturity 225, 226
    of revibration 220
    of temperature 101, 225, 226
  mean 110
  target mean 112, 124, 130, 152
  variability 4, 108, 111
Strengthening
  of beam 400
  of column 395
  of slab 296, 400
Stress concentration 38, 86, 125, 364
Stress-strain
  curve 92, 94, 104
  relation 92, 93, 104
Stripping off 2, 356
  period 237, 377
  strength 232
Sugar
  retarding action 46
Sulphate
  attack 100
  resistance 100, 297
Sulphur concrete 333, 335
Sulphur impregnated concrete 273, 333–336, 383
Super plasticizers 81, 82

Supersulphated cement 18, 349
Surface
  area 125
    of aggregate 35, 38, 39
    of cement 12, 20
  coating 34, 400
  cracking 360, 361
  crazing 361
  delamination 397
  scarification 397
  staining 35, 349
  treatment 399
Swelling 29

Tangent modulus 94
Target
  design strength 69
  mean strength 112, 124, 130, 152, 156, 197
Tensile
  cracking strain 308
  strength 84, 86, 88
    of polymer concrete 330, 331
Test
  cores 262–267, 369, 370
  cubes 83
  cylinder 83
  prisms 83, 84
  results variation 108–111
  sieves 37
Tetra calcium aluminoferrite 8
Texture of aggregate 30, 31
Thermal
  coefficient of expansion 36, 101, 102, 103, 336
  conductivity 37, 101, 102, 274
  cycle 336
  diffusivity 103
  insulation 283, 284
Thermoplastics 326
Thermosetting 326
  resins 326, 381
  polymers 326
Tippers 211
Toppings
  bonded 399
  unbonded 400
Toughness of aggregate 29
  index 295
Transient mixed concrete 213
Transportability 64
Transporting concrete 210

Tremie 243
Trial mixes 123, 124, 135, 195
Tricalcium
  aluminate 8–11
  hydrate 8
  silicate 8–11

Ultimate
  load testing 269
  strain 309
  strength 317
Ultra light-weight concrete 284
Ultra-rapid hardening cement concrete 338–341
Ultrasonic pulse velocity 256–258
  test 256–259, 366
Under-sanded 41
Under water
  concreting 243–247
  repairs 403
Unit weight 108
Unsound particles 35
Unsoundness 10, 14

Vacuum
  concrete 285
  treatment 285
Variability
  of concrete 110
  limits 108
  mean 110
  measurement 110
  range 111
  of test specimens 115, 116
Variance 111
Variation
  of strength 110
  coefficient 111
Vee-Bee
  test 65, 66, 76
  time 66, 68, 73, 310, 312
Vegetable fibres 324
Vermiculite 28, 276
Vibrating table 220
Vibration 217
  acceleration 218
  amplitude 107, 218
  frequency 107, 218–220
  optimum frequency 218
  period 218
  prolonged 220
  re- 219, 220

Vibropressing 72, 222
Vibrators 107, 218
  choice 218
  external 219
  form 219
  immersion or internal 218
  pan 219
  screed 219
  shutter 219
  surface 219
Viscosity 75, 77, 81
Viscous forces 76
Void 133
  content 67, 76
  ratio 32
Voids 32, 38, 80, 83, 135, 325, 337
Volcanic ash 7, 27

Waste material concrete 286–291
Water 44–47
  content 69, 80, 123, 133, 143, 150, 153
  for curing 44, 47
  emulsions 56
  for hydration 44
  for mixing 44
  permeability 99
  quality 44
Water-reducing admixtures 51, 60, 61
Watertightness 99
Water-cement ratio 12, 16, 18, 30, 37–40, 44, 51, 69, 70, 74, 76, 79, 80, 83, 90, 96, 99, 100, 107, 108, 113, 117, 121–123, 125, 129–132, 134, 135, 145–147, 149, 156, 198, 228, 230, 279, 292, 294, 303, 343
Waterproofed cement 19
Waterproofing agent 55, 62
Weighted average size 37
Wet
  curing 223
    of shotcrete 292
  mix process of cement manufacture 7
Wetness of mix 65
White death of concrete 346
White portland cement 19
Winter concreting 241
Wire mesh 301, 302
Workability 11, 27, 30, 33, 37, 39–41, 48, 51, 58, 59, 64–73, 77, 80, 82, 96, 108, 124, 125, 131, 134–136, 143, 186, 210, 239, 249, 310, 312, 313, 314

Workability (contd.)
  cone penetration test 313
  control of mix proportions 67
  control of water content 70.
  effect
    of aggregate 69, 72
    of aggregate-cement ratio 65
    of fines 69
    of flakiness of aggregate 26
    of grading 69, 80
    of mix proportions 67
    of placing time 70, 71
    of shape 30, 69, 76, 80
    of surface area 30, 69
    of temperature 70
    of time 70, 71
  factors influencing 67, 69, 70
  loss with time 70, 71
  measurement 64
  relation with water content 69
  and segregation 72
  and slump 65
  single point test 67, 76, 77
  test 64
    comparison 68
    errors 73
  tolerance 249
  two point test 77

Yield of concrete 34
Yield stress 77, 78
Young's modulus 316

Zinc for gas concrete 54
Zones
  for sand grading 25, 42, 152, 154